3rd National Colloquium on Wind and Earthquake Engineering and International Seminar on Sustainable Construction Engineering 2019

IOP Conference Series: Materials Science and Engineering
Volume 712

Kuala Lumpur, Malaysia
12 - 13 July 2019

ISBN: 978-1-7138-2571-5
ISSN: 1757-8981

Printed from e-media with permission by:

Curran Associates, Inc.
57 Morehouse Lane
Red Hook, NY 12571

Some format issues inherent in the e-media version may also appear in this print version.

Institute of Physics
Dirac House, Temple Back
Bristol BS1 6BE UK

Phone: 44 1 17 929 7481
Fax: 44 1 17 920 0979

techtracking@iop.org

Additional copies of this publication are available from:

Curran Associates, Inc.
57 Morehouse Lane
Red Hook, NY 12571 USA
Phone: 845-758-0400
Fax: 845-758-2633
Email: curran@proceedings.com
Web: www.proceedings.com

TABLE OF CONTENTS

IOP Conf. Series: Materials Science and Engineering **712** (2020) 011001 doi:10.1088/1757-899X/712/1/011001

PREFACE NCWE & ISSCE 2019

After the success of the 1st and the 2nd NCWE in 2017 and 2018 respectively, once again we have organized the 3rd NCWE from 12th to 13th July 2019. But time we have paired up with seminar entitled ISCCE (International Seminar of Sustainable Construction Engineering) that is co-organised with our international partner (*Beijing Jiaotong University*, China).

Just like the previous conferences we have co-organized with the School of Civil Engineering of *Universiti Sains Malaysia* (USM) and also proud to team up for the first time with another university from Indonesia which is *Universitas Syiah Kuala* (UNSYIAH). Co-organizing with not one but two universities would certainly add more values and symbionts like mutual benefits.

All the four speakers are well known for their tremendous contributions in terms of knowledge and research outputs. Prof Yukio Tamura and Prof Shuyang Cao will be sharing on their expertise in wind engineering, Dr Ade Faisal in earthquake disaster meanwhile Associate Prof Guoqing Jing on his innovation related with railway engineering.

In line with the theme of the conference, ***'Built to last'***, we have upgraded and emerged from colloquium to a conference status. This time around we have received quite a number of papers hitting the record from around the globe. This number has given the right momentum for this conference and the organizing team to give the best to all the participants. I believe from the selection of highly prominent key note speakers to the first class conference venue in Kuala Lumpur will certainly leave each participants with full satisfaction.

Last but not least I would like to thank all the support from the presenters and participants. Special gratitude goes to our International Advisory committee members and reviewers whom have spent their valuable time and effort in making this conference a success. Not to forget our sponsors main sponsors Doshin Rubber Engineering and SOH Wind Engineering too.

Chairman, NCWE & ISSCE 2019

Dr. Mohamad Idris Ali

IOP Conf. Series: Materials Science and Engineering **712** (2020) 011002 doi:10.1088/1757-899X/712/1/011002

Peer review statement

All papers published in this volume of *IOP Conference Series: Materials Science and Engineering* have been peer reviewed through processes administered by the proceedings Editors. Reviews were conducted by expert referees to the professional and scientific standards expected of a proceedings journal published by IOP Publishing.

IOP Conf. Series: Materials Science and Engineering **712** (2020) 012001 doi:10.1088/1757-899X/712/1/012001

Approximation of Shear Wave Velocity Using MASW Method for Seismic Hazard Mitigation: A Preliminary Study of Banda Aceh City Indonesia

K Jamaluddin[1,2], B Chairullah[1], I Irwandi[2] and T Majid[3]

[1]Department of Civil Engineering, Universitas Syiah Kuala, 23111 Banda Aceh, Indonesia
[2]Department of Geophysical Engineering, Universitas Syiah Kuala, 23111 Banda Aceh, Indonesia
[3]School of Civil Engineering, Universiti Sains Malaysia, 14300, Nibong Tebal Penang, Malaysia

Abstract. Earthquake 9.1 SR happened in the Aceh Province Indonesia that followed tsunami on December 2004. After this event, the development of infrastructures in Banda Aceh city has shown rapidly progress. Therefore, learning of the experience of this natural event that has caused many casualties and infrastructures, the Banda Aceh city has been being taken initiative as a city that is friendly of disasters. The research objective analysed of influence sedimentation behaviour at buildings. The shear wave velocity parameters using MASW method were used for sediment characteristic assessment. The sediment characteristics are very soft, soft, dense and very dense that it's have level different of shaking, respectively. The detail of 1D profiles of shear wave velocity have obtained. In this case, we analysed VS30 parameters of eleven locations of all measurements at the Ulee Kareng Sub District, Banda Aceh city. The ground shaking level of sediment layer depend of sedimentation characteristics. The sediment characteristic of Ulee Kareng Sub District, Banda Aceh has shown medium dense and soft soil.

1. Introduction

The geomorphology of Sungai Krueng Aceh basin was formed by the Sumatra fault. In geomorphological dynamic, the basin has filled by sediment material with different age of geology. The surface of the sedimentation layers have formed by geomorphic fluvial process within halocen period. The sedimentation layers are heterogenous and anisotropic. Knowledge of the sediment material characteristic and its spatial distribution within the basin area are a thing interesting to study in earthquake hazard mitigation and urban infrastructures planning. In seismic active zone such as the Sungai Krueng Aceh basin, detail estimation of shear wave velocity parameters (Vs) of sedimentation layer are important part of the earthquake hazard assessment of infrastructure safety. Recently, there are two main approximation of determination of the near surface geological structures: one of them is invasive method such as bore hole, standard penetration test (SPT), cone penetration test (CPT) while the other method is non-invasive. The non-invasive method is multichannel analysis of surface wave (MASW). The borehole, SPT and CPT methods were used and accepted by geotechnical engineers in order to estimate of the characterization and stiffness of the soil columns in the site. Recently, the development of seismic exploration technology such as MASW is a promising solution for near surface investigation. The method is relative cheap, fast and can be implemented in urban

IOP Conf. Series: Materials Science and Engineering **712** (2020) 012001 doi:10.1088/1757-899X/712/1/012001

environment. In seismic wave theories, there are two types of wave velocity: compressive wave velocity (Vp) and shear wave velocity (Vs). However, in the geotechnical and geohazard assesment, Vs parameter from MASW method is rarely using. The Active and passive surface waves methods have used to characterize sediment layers at Himalayan mountains in the context of earthquake hazard microzonation analysis [1]. For other applications, the surface wave's method for geotechnical characterization of dyke in the central Europe [2] has used. Gun et al. [3] have used surface wave's method to measure the dynamic stiffness characteristics of railway embankment. For bridge design purpose, Jamaluddin et al. [4] also have used surface method. The shear wave velocity parameters for earthquake hazard analysis in Takengon basin Aceh Province was analysed by [5]. The research objective is to obtain sediment characteristics using Vs detail for earthquake hazard assessment. In sediment materials, earthquake waves can be occur attenuation and wave amplification. The one dimensional of shear wave velocity was obtained for the interpretation of heterogeneous material behavior based from Uniform Building Code (UBC) and Eurocode 8 (EC8) standards (Table 1).

Table 1: Vs30 based on UBC and EC8 [11]

Ground Profile (soil) type (UBC) of Subsoil Class (E8)	Ground Description (UBC)	Description Stratigraphic Profile (EC8)	Shear Wave Velocity Vs30 (m/s)
SA (UBC)	Hard Rock		>1500 (UBC)
SB (UBC) or A (EC8)	Rock	Rock or other rock-like geological formation, including at most 5 m of weaker material at the surface	760-1500 (UBC) or > 800 (EC8)
SC (UBC) or B (EC8)	Very dense soil and soft rock	Deposit of very dense sand, gravel or very stiff clay, at least several tens of m in thickness, characterized by a gradual increase of mechanical properties with depth	360-760 (UBC) nor 360-800 (EC8)
SD (UBC) or C (EC8)	Stiff soil	Deep deposits of dense or medium-dense sand, gravel or stiff clay with thickness from several tens to many hundreds of m	180-360 (UBC and EC8)
SE (UBC) or D (EC8)	Soft soil	Deposits of loose-to medium cohesionless soil (with or without some soft cohesive layers), or of predominantly soft-to cohesive soil	<180 (UBC and EC8)

1.1. Field description and local geology

This research has been carried out in the Ulee Kareng Subdistrict, Banda Aceh city. The location of Ulee Kareng subdistrict is in northern of the Sungai Krueng Aceh river. The sungai Krueng Aceh river is main river in Banda Aceh basin. This region is urban area which the types of infrastructure are roads, housing and educational buildings. The Ulee Kareng Subdistrict is part of the Banda Aceh basin which had limited by two faults: the Seulimeum fault and the Aceh fault. Both Seulimeum and Aceh faults are the secment of Sumatra fault. Therefore, Banda Aceh city is the active seismic region. The sedimentary layer of the Banda Aceh area is fluvial landscape. Generally, sediment material from the fluvial process such as clay, silt, sand and gravel. These fluvial deposits are the quarternary unit. The informations about geology of Banda Aceh were studied by [6-7].

IOP Conf. Series: Materials Science and Engineering **712** (2020) 012001 doi:10.1088/1757-899X/712/1/012001

2. Methodology

Data acquisition of MASW active field seismic testing was done similar to seismic reflection testing layout. Record signal used 24 channel PASI 16S-24P seismograph. The frequency of vertical geophone was used 10 Hz which was plotted linearly with a space of 1 m. This space had considered the effect of possible signal aliasing between geophone. Surveys that had been carried out along the linear array used the fixed along mode method with a sampling time of 250 micro seconds with a total record length of 1024 milli second for each signal record. Source offset from the first geophone 5 m. Sledgehammer 7 kg was used with a height of approximately 1.5 m falling with a metal plate size 25 cm x 25 cm and 2.5 cm thick. To produce a 2-D profiles, a sledgehammer beat was done at offset and along between the geophone. Figure 1 show locations about MASW testing. The study of 2D profile data in detail has been studied by [8]. Configuration of MASW testing is showed at Figure 2.

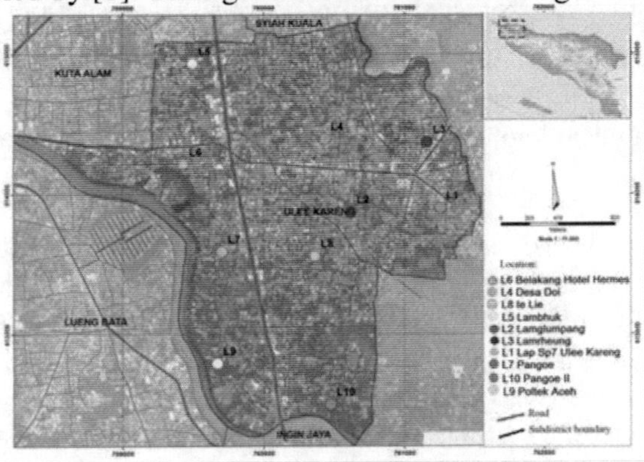

Figure 1: The locations of measurement of shear wave velocity (Ulee Kareng-Banda Aceh)

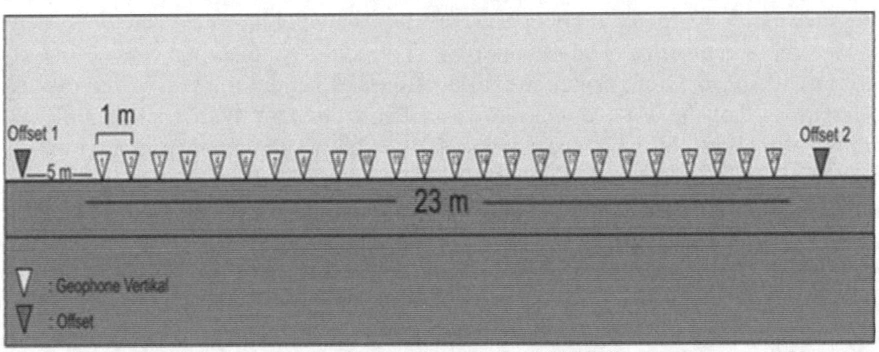

Figure 2: Geophone configuration of the MASW measurement [5].

The surface wave dispersion curve was obtained from MASW data through processing carried out with the SeisImager program. Raw seismic data was processed in the data format with a combination of all shots from a single file. The processing of seismic record in the form of overtone images where represented amplitudo is a function of the frequency with the phase velocity. From image, the dispersion curve was identified by reference to the frequency and maximum phase velocity. In surface wave analysis, a 2-3 Hz signal frequency could be recorded with 4.5 Hz geophone. [9] say investigations using a 10 Hz geophone can detect 5 Hz frequency of wave. It was observed that low frequency wave detection was not limited by the natural frequency of the geophone. The use of a 10 Hz geophone gave results that were almost identical to the 4.5 Hz geophone [10]. Figure 3 shown Rayleigh wave data at offset position that was aquisited at Gampong Doi village.

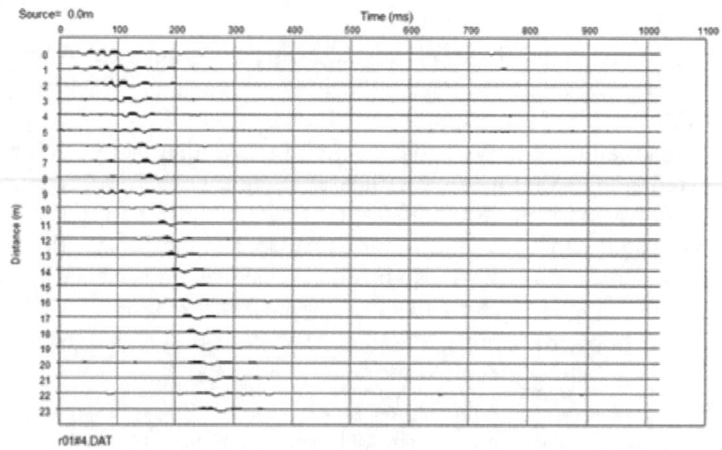

Figure 3: An example of raw Rayleigh wave data at Gampong Doi village at offset position

3. Results and Discussion

The aquisition of seismic data, in the eleven locations at Ulee Kareng Sub district as preliminary study for Banda Aceh city, had been adjusted due to the field condition. The locations of aquisition data were chosen at the open field and avoid from noise that can reduce the accuration of the seismic data. The aquisitions were not due to the direction of north but randomly carried out according to the field situation. The signal processing was carried out through waveform dispersion and followed by inversion of the dispersion curve. The results of the inversion process are a one-dimensional (1-D) Vs curve and its information is shear wave velocity and depth. In the framework of this research, it only was taken up to depth of 30 meters from the earth surface. The shear wave velocity (Vs30) is average of Vs from surface until 30 meters. The surface wave data processing used SeisImager program. Figure 4 is shown 1-D Vs profiles. The results of 1-D Vs data were shown three locations only. The data are Lambhuk village location, Poltek Aceh location and Lamglumpang village location. The Vs data of three locations that the Vs average show smaller from 150 m/s. It show soft soils layer and another data of Vs also show soft soil. The results of Vs30 have calculated from average of shear wave velocity from each layers (Equation 1). The *di* parameter is depth of each layer and *n* is total of layer. However, the Vs30 was obtained from the eleven locations and they was analysed further using the Arcgis program. The Vs30 data distributed or mapping that was calculated by linear interpolation is shown in the Figure 5. However, the research is a preliminary study of Banda Aceh city that is eleven laocations data only. The paper can not conclude overall of Banda Aceh city area but the information can be early information for natural hazard reduction of Banda Aceh city.

$$Vs30 = \frac{\sum_{i=1}^{n} di}{\sum_{i=1}^{n} \frac{di}{vsi}} \qquad (1)$$

IOP Conf. Series: Materials Science and Engineering **712** (2020) 012001 doi:10.1088/1757-899X/712/1/012001

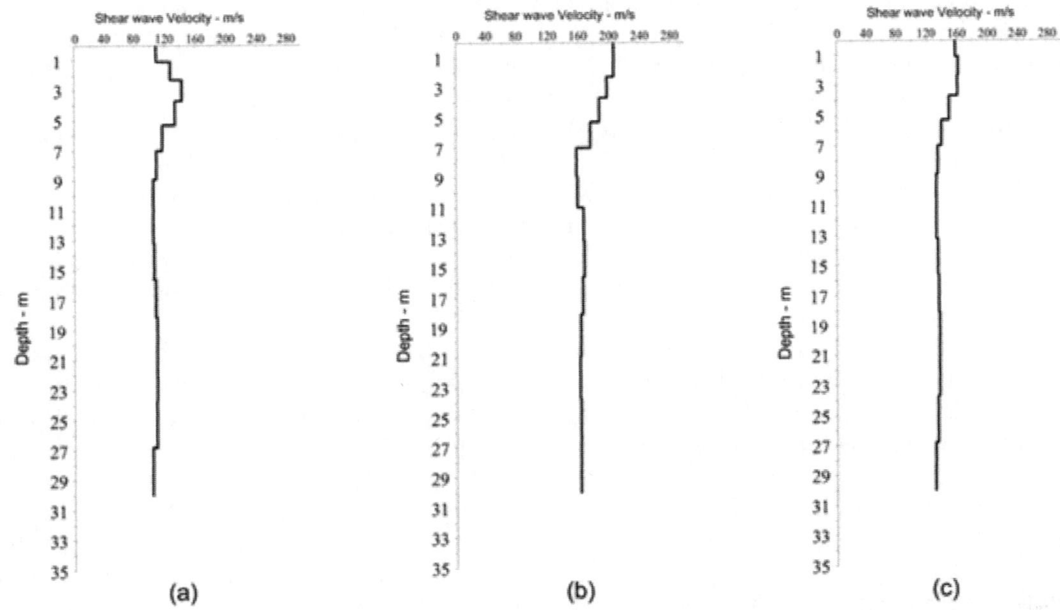

Figure 4: 1-D Vs data of three locations: a). Lambhuk location, b). Lamglumpang location, c). Poltek Aceh location.

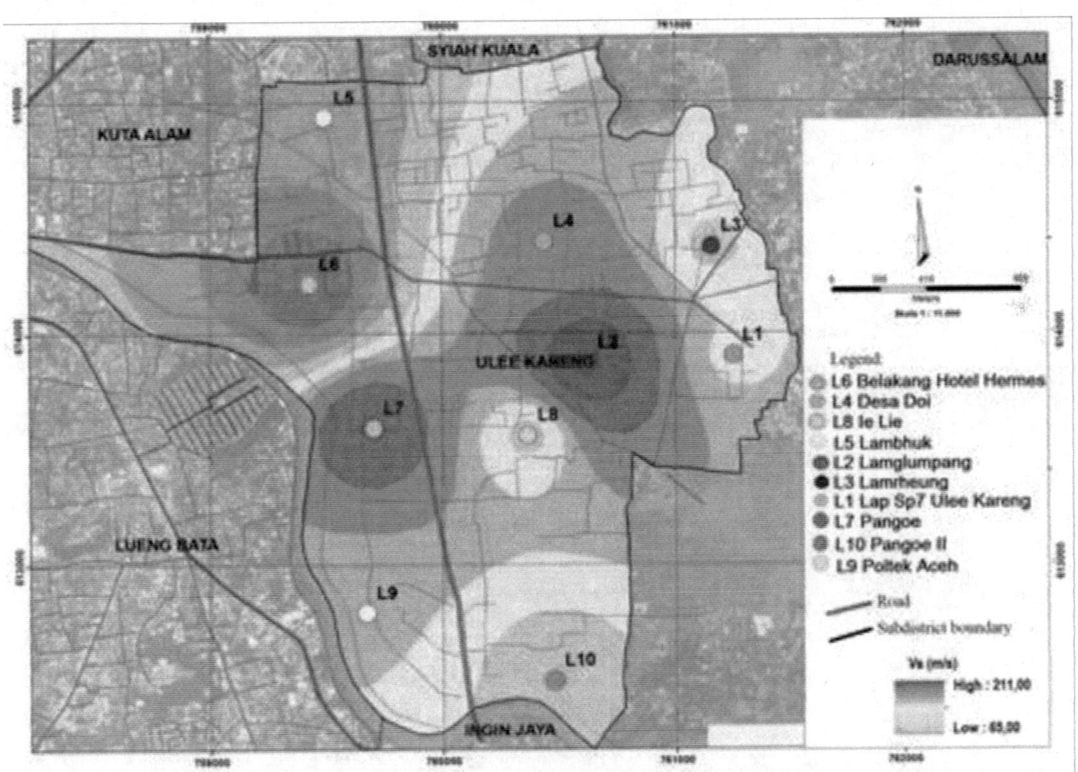

Figure 5: Vs30 value distribution was obtained of Ulee Kareng Subdistrict, Banda Aceh for eleven locations measurement.

IOP Conf. Series: Materials Science and Engineering **712** (2020) 012001 doi:10.1088/1757-899X/712/1/012001

At Figure 5, it can be seen that the distribution of parameters Vs30 consists of three colours, namely redness, yellowish and greenish. The style of colours was resulted by linear interpolation of Vs30 values at each location using MASW method. According to Figure 4 for redness, the range Vs30 are 160 m/s until 210 m/s. For yellowish the range Vs30 are 140 m/s until 159 m/s and greenish range are 65 m/s until 140 m/s. Based on UBC and EC8, the theirs Vs30 range were categorized into two soil characteristics for amplification assessment. The reddish was inserted into the medium dense and the green and yellowish were inserted into soft soil where these soil type have an impact on amplification. The amplification level is relative small in the soils which have high until very high hardness level. In the geotechnic, the value of SPT of high to very high hard soils is 40 until 60. These soil would be less risk of amplification. The building construction on hard soil will also be safe against failure due to amplification behaviour.

Figure 6 shows the results of the seismic vulnerability of Vs data that classified based UBC and EC8 standarts. The results show that there are two seismic susceptibility zones which are dense medium for zone I and soft soil for zone II. The study of material behaviour against natural hazard of seismic waves is paramount important in the civil engineering problem. A occurence of strong earthquake can be many victims due to the collapse of buildings that were not built due to seismic building code. A structure engineer designs the building with very strong according to the strength of concrete material but they do not understand that there are other factors that can be the building failure when the earthquake occure. Those factors are geology, seismic waves and the locally material conditions where the building is constructed [11]. Therefore, the assessment of the seismic hazard of building is things that is very important to reduce casualties of human when catastrophic earthquake occure in a region. However, knowledge of geomorphology and geology also very important of assessment of earthquake behaviour of sedimentation layers. The old layers are a good stiffness and turn the layers will be resistant of earthquake wave amplification.

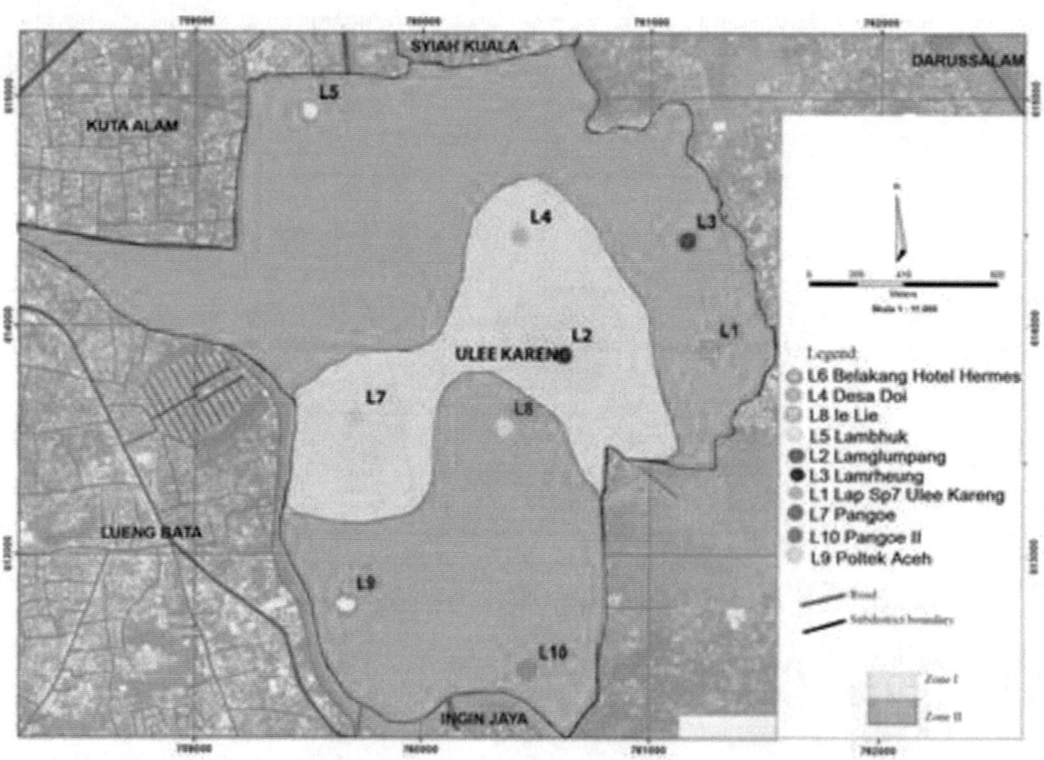

Figure 6: The seismic vulnerability zonation map of Ulee Kareng Subdistrict Banda Aceh that is classified under UBC and EC8.

4. Conclusions

The research was carried out at Ulee Kareng Subdistrict, Banda Aceh city using the MASW method to obtain shear wave velocity and Vs30 parameters. The results were obtained Vs30 map and soil vulnerability zonation map at Ulee Kareng subdistrict based on UBC and EC8 standards. The classification of the soil layers is two zones, namely medium dense (zone I) and soft density soil (zone II). The soft soil was more amplified by earthquake seismic waves than medium dense soil, hard and very hard soils. The medium dense soil was distributed at around of the central of the Ulee Kareng Subdistrict, Banda Aceh city.

5. References

[1] Mahajan A K, Galiana-Merino J J, Lindholm C, Arora B R, Mundepi A K, Rai N and Chauhan N 2011 Characterization of the sedimentary cover at the Himalayan foothills using active and passive seismic techniques *J. Appl. Geophys.* **73** 196-206

[2] Karl L, Fechner T, Schevenels M, Francois S and Degrande G 2011 Geotechnical characterization of a river dyke by surface waves *J. Near Surf. Geophys.* **9** 515-527

[3] Gun D A, Nelder L M, Chambers J E, Raines M G, Reeves H J., Boon D P, Pearson S G, Haslam E P, Carney J N, Stirling A, Ghataora G S, Burrow M P N, Tinsley R D and Tilden-Smith R 2006 Assesment of railway embankment stiffness using continuous surface waves *Proc., Inter., Conf., 1st on Railway Foundations Birmingham UK*, Expanded Abtracts 94-106

[4] Jamaluddin K, Nawawi M, Rusydy I, Irwandi I, Chairullah B 2019 Geotechnical investigation using surface waves method: A Case Study of Sabet Bridge, Lamno, Aceh Province, Indonesia *IOP Conf. Series: Earth and Environm.* **318**(1) 012004

[5] Rusydy I, Jamaluddin K, Fatimah E, Syafrizal, Andika F 2016 Studi Awal: Analisa Kecepatan Gelombang Geser (Vs) Pada Cekungan Takengon Dalam Upaya Mitigasi Gempa Bumi *Jurnal Teknik Sipil Universitas Syiah Kuala* 1-12

[6] Culshaw M G, Duncan S V, Sutarto N R 1979 Engineering Geological Mapping of The Banda Aceh Alluvial Basin, Northern Sumatra, Indonesia *Bull. Inter. Assoc. Engineering Geology*

[7] Bennet J D, Bridge D M, Djunuddin A, Ghazali S A, Jeffrery D H, Kartawa W, . . . Whandoyo R 1982 Geologic 1:250,000 Map of Banda Aceh Quadragle, Sumatra. Bandung: *Geological Research and Develop. Centre (GRDC)*

[8] Steinel H, Hausmann J, Welban U, Dietrich P 2014 Reability of MASW profilling in near surface aplication *J. Near Surf. Geophys.* **12** 731-737

[9] Xia J, Miller R D and Park C B 2000 Estimation of near surface shear wave velocity by inversion of Rayleigh waves *J. Geophys.* **64** 691-700

[10] Park C B, Miller R D and Xia J 1999 Multichannel analysis of surface waves *J. Geophys.* **64** 800-808

[11] Kanli A I, Tildy P, Pronay Z, Pinar A, Hermann L 2006 Vs30 Mapping and Soil Classification for Seismic Site Effect Evaluation in Dinar Region, SW Turkey *J. Geophys. International* **165** 223-235

Acknowledgements

The paper is part of our project in 2019 (investigation of sediment in the earthquake hazard mitigation and liquifaction using MASW method of Banda Aceh location). The authors would like to thanks to the Universitas Syiah Kuala, Banda Aceh Indonesia for funding support this research (166/UN11.2/PP/PNBP/SP3/2019). Our thanks also to Head of Department of Geophysical Engineering has been permitted of seismic equipments for field data aquisitions. Especially Mr. Muhammad Saputra and other our undergraduate students who helped of field data aquisition.

IOP Conf. Series: Materials Science and Engineering **712** (2020) 012002 doi:10.1088/1757-899X/712/1/012002

Effect of Ground Granulated Blast Furnace Slag as Partial Replacement in Fly Ash-Based Geopolymer Concrete

M F A Abdul Sani[1], R Muhamad[1] and K H Mo[2]

[1]Razak Faculty of Technology and Informatics, Universiti Teknologi Malaysia, Jalan Sultan Yahya Petra 54100 Kuala Lumpur, Malaysia
[2]Department of Civil Engineering, Faculty of Engineering, Universiti of Malaya, 50603 Kuala Lumpur, Malaysia

Abstract. Replacement of Ordinary Portland cement (OPC) with geopolymer concrete (GPC) using fly ash is currently utilized in the construction industry to reduce the excessive carbon footprint. However, fly ash has lower strength development compared to OPC. Due to that reason, ground granulated blast furnace slag (GGBFS) is chosen to be blended with fly ash to produce GPC with improved mechanical strength. The analysis of the mechanical properties includes workability, compressive strength and splitting tensile strength of hardened GPC specimens have been carried out. The specimens were prepared with different percentages of GGBFS, from 0 % up to 20 %, partially replacing the fly ash-based GPC. Experimental results showed that the slump, compressive strength and splitting tensile strength of the GPC increase as the GGBFS percentage increases and found to be suitable for structural application.

1. Introduction

Fly ash is an industrial by-product produced from coal combustion, whereas ground granulated blast furnace slag (GGBFS) produced from the burning of melting iron ore [1]. Those by-products have the potential to be used as cementitious materials in reducing the excessive carbon footprint from Ordinary Portland cement (OPC) manufacturing. Recently, GGBFS is introduced in geopolymer concrete (GPC) to partially replace fly ash-based geopolymer concrete as a substitution for OPC in construction industries. It is one of the ways to reduce OPC manufacturing more effectively by reducing carbon footprint up to 9% [2]. Nowadays, there are millions of tons of those by-products generated and are unutilized or underutilized, leading to environmental issues. It may be caused by storage problem and pollution to the surrounding field [3-4]. Since the environmental problems and health issues have arisen, the concern among researchers to investigate the utilization of such by-products as potential construction materials has been drastically increased [2,4-5]. Although the application of GGBFS is still limited compared to fly ash, the durability and mechanical performances of both by-products are promising in replacing OPC as construction materials. Thus, this paper will assess the effect of GGBFS utilization as a partial replacement in fly ash-based GPC as a sustainable alternative for construction materials.

2. Experimental Methods

2.1. Raw Materials and Process of Design Mix

The base materials for manufacturing GPC were fly ash and GGBFS. They were obtained from Jimah Power Plant, Port Dickson in Negeri Sembilan and YTL Cement Malaysia, Kapar in Klang, respectively. GGBFS was utilized as a partial replacement in fly ash-based GPC. The chemical composition of both materials was analysed through X-ray fluorescence (XRF) test at Analytical Laboratory in Universiti Teknologi Malaysia, Kuala Lumpur. Four main chemical oxides could affect the performances of both fly ash and GGBFS in GPC, known as calcium oxide (CaO), silica oxide (SiO_2), aluminium oxide (Al_2O_3) and iron (III) oxide (Fe_2O_3). From previous studies, the variations of the compositions were considerably contributed to the performance of raw materials in GPC manufacturing [3,6-7]. Figure 1 and Figure 2 show the physical characteristics of both low-calcium fly ash and GGBFS samples obtained from the manufacturers. Details on XRF results governing the performance of fly ash-based GPC will be explained further in the next section.

Figure 1. Class F fly ash sample [4]

Figure 2. Ground granulated blast furnace slag (GGBFS) sample [8]

As preparation for alkaline activator in GPC, sodium hydroxide (NaOH) solution is produced by diluting a portion of NaOH pellets with distilled water in a volumetric flask. For instance, NaOH solution with a concentration of 14 M to be achieved consists of 14 x 40 = 560 g, where 40 is the molar mass of the NaOH. In other words, 560 g of NaOH pellets are mixed with 1 litre of distilled water to produce 14 M concentration of NaOH solution. The solution is then left to cool down to laboratory temperature so that the dissolution of the solution took place thoroughly before mixed with sodium silicate (Na_2SiO_3). The Na_2SiO_3/NaOH ratio is fixed to 1.5, where every 150 g of Na_2SiO_3 is mixed with 100 g of NaOH [9]. During the process, an exothermic reaction took place when the solution is stirred gradually, resulting in an extreme amount of heat released. Then, the activator is left to ambient temperature for 24 hours [10-11]. As a safety precaution, safety goggles, gloves and mask are recommended to be worn during the preparation of alkaline activator. For the GPC samples preparation, fly ash, GGBFS and aggregates are dry mixed in the laboratory pan mixer for about 3 to 4 minutes. Then, the alkaline activators prepared, together with some extra water, are added to the dry mixture. The mixing continues for another four minutes. After placed in greased moulds, all the samples are subjected to oven curing at 60°C for 24 hours as optimum temperature [9]. Table 1 shows the details of the mix proportions of fly ash and slag-based GPC. The symbol C annotates concrete, whereas the following acronym, TF, refers to Class F fly ash [12]. The following numbers 14 annotates 14 M concentration of the alkaline activator utilized whereas the last acronym, S0, S10 and S20 annotate the percentage replacement of the total binder with GGBFS consisting of 0 %, 10 % and 20 %, respectively.

Table 1. Design mix of fly ash and slag-based GPC (kg/m^3)

Name	Fly ash	GGBFS	Fine aggregates	Coarse aggregates	Alkaline activator	Additional water
C-TF-14-S0	500	-	707	1060	150	74
C-TF-14-S10	450	50	707	1060	150	74
C-TF-14-S20	400	100	707	1060	150	74

2.2. Specimens and Testing

Concrete cubes of 100 mm x 100 mm x 100 mm (27 numbers of samples) are used to evaluate the compressive strength of fly ash and slag-based GPC whereas concrete cylinders of 150 mm x 300 mm (27 samples) are used to assess the splitting tensile strength for the GPC. The compressive strength test is conducted based on BS EN 12390-6: 2009 using three cubes for each percentage of 0 % to 20 % and curing age of 7, 14 and 28 days, while splitting tensile strength are carried out in accordance to BS EN 12390-4: 2000 [13-14]. Three cylindrical specimens are tested for each age and percentage as well [10]. Figure 3 and Figure 4 show those specimens under compressive and splitting tensile test, respectively. Fly ash-based GPC without GGBFS addition (C-TF-14-S0) is considered as the control concrete specimens since there is no replacement of fly ash in the concrete.

Figure 3: Compressive strength test

Figure 4: Splitting tensile strength test

3. Results and Discussions

3.1. X-ray Fluorescence (XRF) Analysis

Fly ash is generally pozzolanic since SiO_2 content reacts with calcium hydroxide from the cement hydration process to produce calcium silicate hydrate (CSH) gel thus produced cementitious compounds suitable for GPC manufacturing. Also, fly ash with higher CaO content can increase the compressive strength of the concrete in the early ages since the calcium content present plays a significant role in developing the compressive strength [3]. The presence of calcium ions produced a quick reactivity; thus, the geopolymer will yield rapid hardening in shorter curing time under heat curing. Although the strength for the fly ash-based GPC achieved is comparable to the standard OPC strength, improvement in strength development is still yet to be achieved [15]. The addition of GGBFS in the fly ash matrix is then applied to increase CaO content in the fly ash geopolymeric gel. This will result in the enhancement of compressive strength and overall performances of the concrete. The GGBFS is contributing in compressive strength due to its compactness of microstructure and thus enhance the hardening of fly ash and GGBFS blended geopolymer through C-S-H and C-A-S-H formation. The hardening is then followed by the formation of C-S-H, N-A-S-H and C-A-S-H [6]. Table 2 shows the results of XRF from current and some of the previous studies for both fly ash and GGBFS. The differences in chemical composition show varieties in terms of the mechanical performances that will be discussed in the next subsection.

Table 2. Chemical composition of fly ash and GGBFS (%)

Oxides	Fly ash (Current study)	GGBFS (Current study)	Fly ash [5]	GGBFS [5]	Fly ash [16]	GGBFS [16]	Fly ash [3]	Fly ash [18]
SiO_2	46.8	28.7	65.6	30.6	53.6	35.2	38.8	63.4
Al_2O_3	18.4	12.3	28.0	16.2	33.0	21.4	14.7	30.5
Fe_2O_3	6.08	0.5	3.0	0.6	5.5	1.8	19.5	3.0
CaO	3.32	46.6	1.0	34.5	1.8	31.2	18.1	1.0

3.2. Slump Test

When the addition of GGBFS of 10 % and 20 % are applied, there are increments of 3.96 % and 4.90 % of the slump, respectively. Figure 5 shows the workability values are slightly increased as the replacement of GGBFS increased. Higher GGBFS content will tend to produce a higher GPC slump with a higher strength [16,17]. It was suggested that the slump value is between 65 mm and 100 mm for fresh GPC [3]. Studies performed by Abdullah et al. and Gupta and Chandrakar using fly ash-based GPC are compared with slump value in the current study. Based on Figure 5, the slump values of fly ash-based GPC observed are in line with the previous studies performed, with 3 % and 33.1 % higher in slump values for Abdullah et al. and Gupta and Chandrakar, respectively [3,18]. The variations of slump values may be attributed to the higher total of silica, alumina and calcium content in fly ash samples [19]. Table 2 shows Abdullah has 71.6 %, whereas Gupta and Chandrakar have 94.9 % in the total content, which is higher than the current study with only 68.52 %. This also proves that different chemical composition of fly ash will produce a variation of the mechanical properties of GPC, including slump values. Addition of GGBFS is then applied and found to be effective in increasing the workability of the GPC, as shown in Figure 5. However, for 10 % and 20 % GGBFS replacement, slump result data are not shown since none of them are displayed in previous studies.

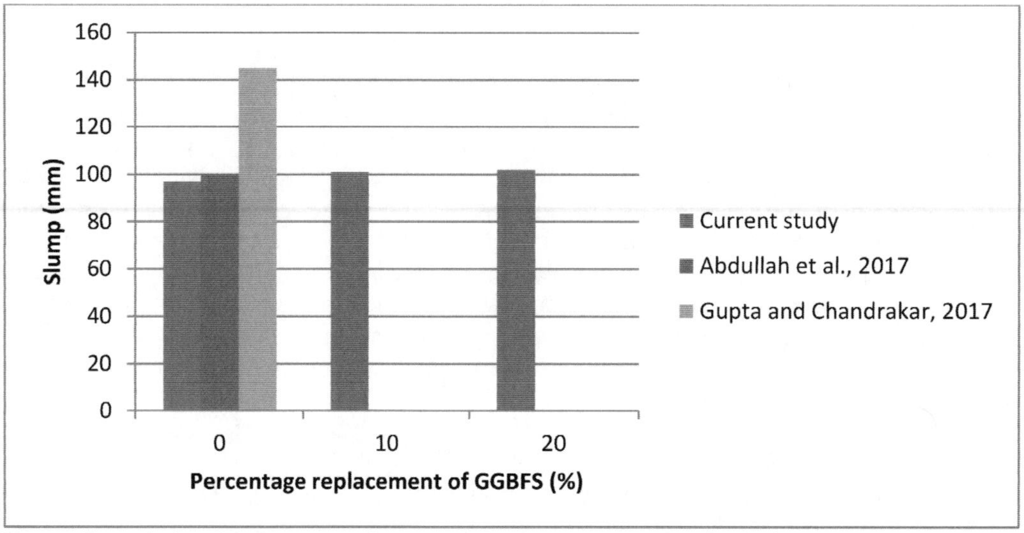

Figure 5: Relationship between slump and GPC containing a different percentage of GGBFS

3.3. Compressive Test

Table 3 and Figure 6 summarized the values of compressive strength for all mixes and the relationship between compressive strength and GGBFS percentage replacement. The database of S50, S75 and S100 annotate 50 %, 75 % and 100 % GGBFS replacement, respectively. They are collected from previous studies for comparison purpose [5]. It can be observed that the graph trend shows the strength increased significantly along with the percentage replacement of GGBFS.

Table 3. Compressive strength for the GPC mix

Mechanical property	Age (days)	Mix type					
		C-TF-14-S0 (Current study)	C-TF-14-S10 (Current study)	C-TF-14-S20 (Current study)	S50 [5]	S75 [5]	S100 [5]
Compressive strength, f'_c (MPa)	7	28.4	44.1	47.7	40	44.4	52.4
	14	30.4	48.6	52.9	46.5	48.2	56.2
	28	32.8	48.9	53.2	53.5	55.5	58.6

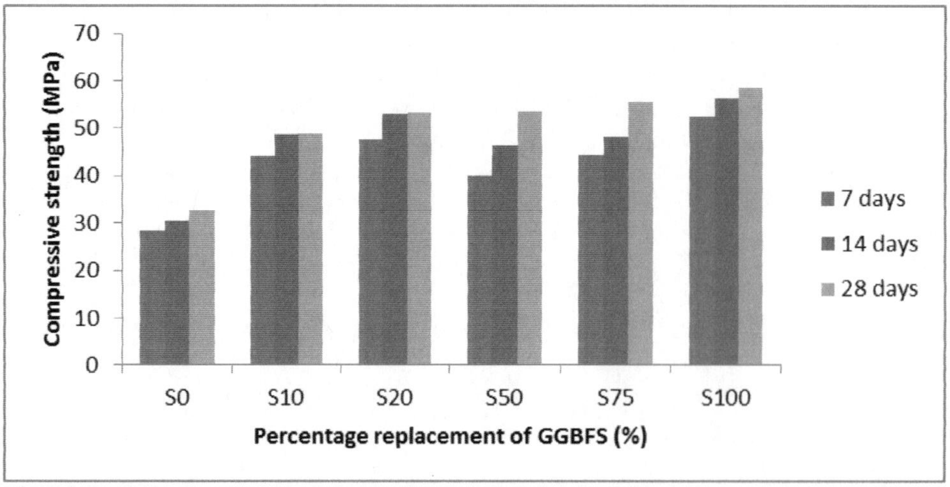

Figure 6: Compressive strength vs. GPC containing a different percentage of GGBFS

For fly ash-based GPC, there is an improvement in the compressive strength along the curing days, from 6.58 % to 13.41 %. After GGBFS replacement of 10 % to 20 % are applied on the fly ash-based GPC, the strength enhancement could be observed. Figure 6 shows 32.90 % to 37.45 % increment when 10 % GGBFS is added whereas 38.35 % to 42.53 % increment after 20 % GGBFS is added to the GPC along the curing days. The increment of GGBFS replacement in GPC for current study achieved a significant compressive strength compared to the control concrete even at seven curing days. Table 3 and Figure 6 show that the GPC for both current and previous studies achieved a range of 86.59 % to 90.18 % of the strength at 28 days, which is suitable for structural applications [5]. Besides that, setting time is an essential aspect for the fresh concrete before setting for proper compaction. Replacement of GGBFS in the fresh concrete will reduce the setting time since the calcium content increases along with the replacement percentage of GGBFS. Higher GGBFS will induce a shorter setting time which is not preferable during casting and compacting. Thus, 20% is considered as an optimum replacement percentage for the GPC casting and thus induce a higher mechanical performance for structural applications [1].

3.4. Splitting Tensile Test

Splitting tensile strength results in Table 4 indicates that the performance of GPC is comparable with conventional concrete, where they are strong in compression but not in tension [2]. Table 4 and Figure 7 display the details and graph of the relationship of splitting tensile strength with GGBFS for all ages, respectively. For the current study, the increment of splitting tensile strength is very slight compared to the studies performed by Jawahar and Mounika. This may be attributed to the higher GGBFS replacement in the study conducted by them compared to the current study and also mishandling during casting that leads to such behavior. Besides, the database for S50, S75 and S100 for seven days and 14 days are not provided, making it difficult for further analysis [5].

Table 4. Splitting tensile strength for the GPC mix

Mechanical property	Age (days)	Mix type					
		C-TF-14-S0 (Current study)	C-TF-14-S10 (Current study)	C-TF-14-S20 (Current study)	S50 [5]	S75 [5]	S100 [5]
Splitting tensile strength, f_{ct} (MPa)	7	2.11	2.31	2.54	-	-	-
	14	2.09	2.56	2.68	-	-	-
	28	2.08	2.65	2.77	3.25	3.39	3.54

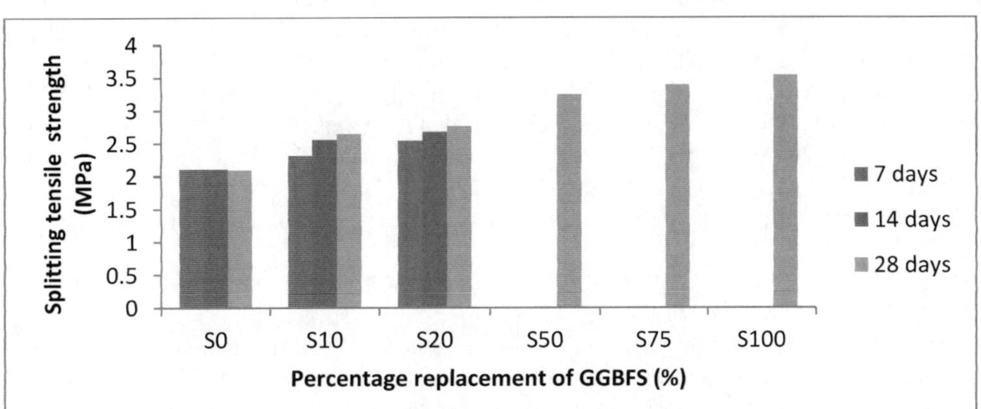

Figure 7: Effect of GGBFS percentage on splitting tensile strength

Based on Table 4, the application of 10 % and 20 % GGBFS in concrete for current study achieved 21.51 % and 24.91 % increment of splitting tensile strength compared to the control concrete, respectively. However, the strength of control concrete is decreasing along the ages for about 1.42 %, may be attributed to lesser compaction and mishandling during casting that leads to such behaviour. The strength of 10 %, and 20 % GGBFS replacement in fresh concrete, however, cannot surpass 3.0 MPa for splitting tensile strength like previous studies. There is 21.75 % higher in maximum splitting tensile strength by Jawahar and Mounika for 100 % replacement of GGBFS compared to the maximum strength achieved in the current study for 20 %. Nonetheless, the replacement percentage is capable of achieving splitting tensile strength as high as that conventional concrete for structural application [5]. The increase in GGBFS replacement level improves the microstructure of GPC, thus leads to enhancement of splitting tensile strength of GPC. From the results it is observed that GGBFS and fly ash blended GPC concretes attained enhanced mechanical properties after cured as in the case of only fly ash based GPC concrete but with more significant results [20,21] attributed to the strong bonding of geopolymer paste and aggregates that tends to increase the properties of GPC [5]. Still, the optimum percentage of GGBFS in concrete to be used as fly ash replacement is 20 % considering the workability as mentioned in the previous section.

4. Conclusion

Based on the investigations, GGBFS has the potential to be utilized in the construction industry. It is one of the ways in mitigating environmental issues, including carbon footprint and cost-saving as well. GGBFS also found to be suitable as partial replacement of cement in concrete production, especially in fly ash-based GPC. Most of the researchers suggested up to 50 % replacement of either GGBFS as cement or fly ash-based GPC replacement in the concrete production [5]. However, to achieve a comparable strength with OPC concrete for bond assessment of GPC, GGBFS replacement up to 20 % is adequate for such application. Some researchers suggested that replacement up to 50 % is recommended but the setting time should be carefully controlled [17]. It is also suggested to use water retarders to prolong the setting time before concrete casting, but still, there will be some defect on the mechanical properties [11]. Although the application of geopolymer in practical construction has already begun in some parts of the world, it will take time to understand the material and its technology and make it accepted worldwide [7,16]. For future works in sustainability enhancement, carbon footprint produced from oven curing could be eliminated as the addition of GGBFS can accelerate the hardening time of the fresh concrete of GPC. Besides, the bond assessment recommended being performed as it is crucial in determining bond behaviour and material properties of the GPC to be utilized in structural applications.

5. References

[1] Barnard R 2014 Mechanical properties of fly ash/slag based geopolymer concrete with the addition of macro fibres *Masters Thesis* Stellenbosch University.

[2] Turner L K and Collins F G 2013 Carbon dioxide equivalent (CO2-e) emissions: A comparison between geopolymer and OPC cement concrete *Const. Build. Mater.* **43** 125–130.

[3] Abdullah M M, Tahir M F, Tajudin M A, Ekaputri J J, Bayuaji R, Khatim N A 2017 Study on The Geopolymer Concrete Properties Reinforced with Hooked Steel Fiber *IOP Conf. Series: Mater. Sci. Eng.* **267** 012014.

[4] Islam M M U, Alengaram U J, Jumaat M Z and Bashar I I 2014 The development of compressive strength of ground granulated blast furnace slag-palm oil fuel ash-fly ash based geopolymer mortar. Mater. Des. **56** 833–841.

[5] Jawahar J G and Mounika G 2016 Strength properties of fly ash and GGBS based geopolymer concrete *Asian J. Civ. Eng. (BHRC)* **17** 127–135.

[6] Kumar S, Kumar R and Mehrotra S P 2010 Influence of granulated blast furnace slag on the reaction, structure and properties of fly ash based geopolymer J. Mater. Sci. **45** 607–615.

[7] Palomo A, Krivenko P, Garcia-Lodeiro I, Kavalerova E, Maltseva O and Fernández-Jiménez A. 2014 A review on alkaline activation: new analytical perspectives. Mater. Const. **64** 022.

[8] Krypton Marketing (2014). Sustainable Concrete Building - Supplementary Cementitious Materials. blog.kryton.com/2014/09/supplementary-cementitious-materials/.

[9] Chang E H, Sarker P, Lloyd N, Rangan B V 2009 Bond behaviour of reinforced fly ash-based geopolymer concrete beams *Proceedings of the 24th Biennial Conference of the Concrete Institute Australia* Concrete Institute of Australia.

[10] Mo K H, Yeap K W, Alengaram U J, Jumaat M Z and Bashar I I 2018 Bond strength evaluation of palm oil fuel ash-based geopolymer normal weight and lightweight concretes with steel reinforcement *J. Adh. Sci. Tech.* **32** 19–35.

[11] Siddique R and Kaur D 2012 Properties of concrete containing ground granulated blast furnace slag (GGBFS) at elevated temperatures *J. Adv. Res.* **3** 45–51.

[12] ASTM C 2012 *C618-12a* Standard specification for coal fly ash and raw or calcined natural pozzolan for use in concrete.

[13] BS EN 12390-4 2000 British standard for testing hardened concrete–Part 4: Compression strength-Specification of test machines.

[14] BS EN 12390-6 2002. Testing Hardened Concrete. Tensile Splitting Strength of Test Specimens. British Standard Institution, London.

[15] Sofi M, Van Deventer J S J, Mendis P A and Lukey G C 2007 Bond performance of reinforcing bars in inorganic polymer concrete (IPC) *J. Mater. Sci.* **42** 3107–3116.

[16] Shah A 2017 Optimum Utilization of GGBS in Fly Ash Based Geopolymer Concrete *Kalpa Pub. Civ. Eng.* **1** 431–440.

[17] Xie J, Wang J, Rao R, Wang C and Fang C 2019 Effects of combined usage of GGBS and fly ash on workability and mechanical properties of alkali activated geopolymer concrete with recycled aggregate. *Comp. B: Eng.* **164** 179–190.

[18] Gupta S and Chandrakar G 2017 Experimental Studies on Fly Ash Based Geopolymer Concrete without Portland Cement-An Eco Friendly Construction *Int. J. Eng. Sc. Comp.* **7** 11514–11520.

[19] Al-Azzawi M, Yu T, Hadi M N 2018 Factors affecting the bond strength between the fly ash-based geopolymer concrete and steel reinforcement *Structures* **14** 262–272.

[20] Siddiqui K S 2007 Strength and Durability of Low-Calcium Fly Ash-based Geopolymer Concrete, *Final Year Honours Dissertation*, The University of Western Australia, Perth.

[21] Sreenivasulu C, Ramakrishnaiah A and Guru Jawahar J 2015 Mechanical Properties of Geopolymer Concrete Using Granite Slurry as Sand Replacement *Int. J. Adv. Eng.Tech.* 83–91.

Acknowledgement
This work was financially supported by the Universiti Teknologi Malaysia GUP Tier 1 [grant no: Q.K130000.2540.20H40].

An Investigation on Urban Circle Rail Transit: A Case Study on the Alignment of Mass Rapid Transit 3 (Circle Line)

A A Mohamed[1], E J Wei[1], P J Ramadhansyah[2], M A Farooqi[3], M K I M Satar[4] and N Mashros[4]

[1]Faculty of Engineering, Department of Civil Engineering, the University of Nottingham, 43500 Selangor, Malaysia
[2]Faculty of Civil Engineering Technology, Universiti Malaysia Pahang, 26300 Gambang, Pahang, Malaysia
[3]Department of Civil Engineering, Zakir Hussain College of Engineering and Technology, Aligarh Muslim University (AMU), Aligarh-202002, India
[4]Faculty of Engineering, School of Civil Engineering, Universiti Teknologi Malaysia, 81310 Johor Bahru, Malaysia

Abstract. Mass Rapid Transit (MRT) is one of the important components of the Greater Kuala Lumpur (KL) development plan. It consists of three line-systems and would be finally operated by Klang valley public transport system. These three lines are aligned on the concept of wheel and spoke pattern that encompasses two north-south radial lines as well as a circular line (also known as ring line) circulating around KL and enclosing the central business district. However, fixing the locations of the stations has become an issue. The aim of the paper is to find out the optimal alignment for the ring line with respect to the total passenger benefit. A numerical-analytical model has been used to determine the optimum radius for the proposed ring line in a city. Passenger route choice for many-to-many, origin-destination (OD) pairs is analysed using radial-only rail network and rail network with ring lines of varying radii. The route choice takes into consideration three separate options including a) direct access to destination, b) taking only radial lines, and c) taking a combination of radial and ring lines. Several important parameters such as access cost, ride cost, and transfer penalty, are the main aspects to evaluate the optimality of the ring line. Microsoft Excel has been used to run the cost-benefit optimization model with the aim of obtaining an optimal radius of ring line. Results show that the net benefit of presenting a circle line in the KL by manipulating the passenger's cost for each OD pair. Finally, the optimum alignment of MRT circle line in KL has been proposed.

1. Introduction

Kuala Lumpur (KL), the capital city of Malaysia, is one of the fastest-growing cities in Asia. The inherent problem faced by such rapidly growing cities is increased level of traffic volume and traffic delays. Historically, KL has been a large exporter of tin minerals to United Kingdom (UK). To minimize the transportation time to the nearest port, the very first rail route from KL to Bukit Kuda (near Klang) was built by British engineers and operations commenced on September 15th 1886. The transportation facilities in KL have since been constantly improving [1]. The rail transit system in Klang valley dramatically evolved in the late 20th century, when the routes were electrified, and currently it is operating 8 different lines spanning over 375.5km. The evolution timeline of Klang Valley Integrated Transit System (KVITS) are shown in Figure 1. In tune with this evolution, three more lines are being proposed to be built by 2025, which would be an ambitious enhancement and would definitely complement the efforts to ease off the traffic congestion problem of KL. The rail transportation development is not keeping pace with the land property development in KL, hence, making many places unreachable by train. This leads to increased use of private cars, clogging roads and overriding public transportation [2]. Kuala Lumpur has been a frontline profit earner from the rapid economic growth. Increasing incomes are another key force to the escalating car ownership Likewise, as stated by Ministry of Transport Malaysia Official Portal [3], the upsurge in population as well as LRT commuters, there is a drastic spike in the demand to enhance the public transportation capacity, particularly rail transportation, in KL. Thus, instead of roads, more rail lines should be constructed keeping in view the convenience and benefit of the passengers

1.1. Mass Rapid Transit in Malaysia

Mass Rapid Transit (MRT) is a major segment of development plan in Greater Kuala Lumpur [4]. It is proposed to have three lines, arranged in the wheel and spoke pattern, encompassing two north-south radial lines as well as one circle line (also known as ring line) circulating around KL. After completion, MRT system would be operated as part of Klang Valley public transport system. This project would excessively enhance the capacity of the existing Klang Valley Integrated Transit System (KVITS) by integrating with the current rail network, which would alleviate the traffic congestion problem in Greater KL.

1.2. Klang Valley Transportation Connectivity

Although, significant improvement has been done in rail transportation, however, when it comes to KVITS, Klang Valley is trailing behind. Problem with traffic congestion in KL has kept on rising probably due to inadequate rail connectivity and increase in population and car ownership. In case of road system, a purely radial road network would cause people to travel along the radial roads that pass through central business district (CBD) to reach a destination away from it. This would clearly lead to unwanted vehicle load in passages, extra travel time, and high transfer of vehicle load at CBD. To address this problem, government has built several ring roads that brought suburban areas in KL together as well as to lower down the vehicles transfer loads in CBD. However, problem with traffic congestion still remains unsolved due to the increased car ownerships in KL. Building ring roads does not actually seem to provide an effective solution. On account of these reasons, it may be concluded that beside building new roads, efforts would have to be made on the development of rail transport to actually mitigate the long-unsolved traffic congestion issue. All the existing and proposed rail lines in Klang Valley are considered to be radial lines. Therefore, it will again cause the same problem as is inherent in the purely radial roads network [5].

1.3. Circle Rail Line

Referring to the solutions that other countries have adopted, circle line seems to be an ideal solution to the problems that mentioned earlier. A circle line can help to improve the connectivity along with diminishing the network vulnerability, given that alternative routes are provided to all destinations. Moreover, the network connectivity, operational efficiency and directness can be benefitted from the circle line and these have been confirmed by several studies [6,7,8]. The attempts to come out with a solution to the problem of rail transit, planning may be traced back to circa [9]. This planning of the transit network has been practised in different type of modes to generate a configuration of optimal transit network. Numerous studies have been undertaken on the planning of rail transit [10,11,12], are indirectly applicable to circle line.

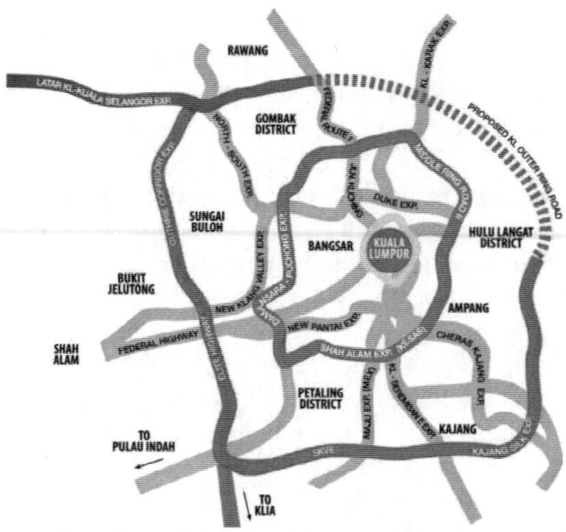

Figure 1. Existing and proposed ring roads around Kuala Lumpur City Centre [4]

2. Methodology

2.1. Passenger Route Choice

The minimum passenger cost of each origin-destination (OD) pair plays an important role in determining the total passenger benefit. It is determined by considering routes of different passengers of each OD pair that gives the lowest passenger cost to access the destination from origin. KVITS is simplified to be made of two types of lines, which are radial line and circle line. For radial-line-only network, passengers can either head to a destination either taking the same line without having to transfer, or possibly have to change to another radial line at CBD. There are some assumptions made in this paper.

List of different alternative paths of any OD pairs in the circle-radial network model are as below:

a) Access directly from origin to destination without using rail network.
b) Take only radial line;
c) Take circle line (radial line may also be used);

A deterministic route choice was adopted to evaluate the passenger cost for various paths for each OD pair. Cost of each alternative passenger route choice for each OD pair has to be calculated, and the one with the minimum passenger cost was used in determining total passenger cost. Passenger cost for the alternative routes has been classified into:

$D\ (Direct\ alternative)$ (1)

$\gamma_a D_o + \gamma_R D_R + \xi_{RL} k H_{RL} \gamma_w + (\xi_{RL} - 1)\,\gamma_T + \gamma_a D_d$ (Radial Line) (2)

$\gamma_a D_o + \gamma_R D_R + \xi_{RL} k H_{RL} \gamma_w + \xi_{RN} k H_{RN} \gamma_w + (\xi_{RL} + \xi_{RN} - 1)\,\gamma_T + \gamma_a D_d$ (Circle Line) (3)

2.2. Total Passenger Benefit of Circle Rail Line

Total passenger cost for radial line only network and the total passenger cost with circle line network (R) are computed depending on the minimum passenger cost and the travel demand of each OD pair. The equations for total passenger cost are:

$$C_{RL} = \Sigma_{i,j}^{N} (PC_{\min i,j} * OD_{i.j}) \quad (4)$$

$$C_{CL} = \Sigma_{i,j}^{N} (PC(R)_{\min i,j} * OD_{i.j}) \quad (5)$$

Where C_{RL} is total passenger cost for radial line network only; C_{CL} is total passenger cost for circle line network only; $PC_{\min i,j}$ is the minimum cost of passenger from zone i to j for network with radial line only; $PC(R)_{\min i,j}$ is the cost of train rides per kilometre per passenger (RM/km/passenger) and $OD_{i.j}$ the cost of train rides per kilometre per passenger (RM/km/passenger).

$$\text{Total Passenger Benefit } (R) = C_{RL} - C_{CL} \quad (6)$$

From equation 6, it is noticeable that total passenger benefit is a function of radius, implying the total passenger benefit changes as the radius of circle line changes. All the unit cost values that were used in the cost benefit analysis are shown in Table 1.

Table 1. Summary table of all unit cost values for KL.

Parameters	Unit Cost (Malaysian Ringgit (MR))
Access cost, γ_a	RM2.17/km/passenger
Ride cost, γ_R	RM1.00/km/passenger
Wait cost, γ_w	RM0.58/min/passenger
Transfer disutility, γ_T	RM1.91/transfer/passenger
Train headway, H_{RL} & H_{RN}	KTM – 22.5 mins STAR and PUTRA LRT – 5.3 mins Monorail – 6 mins MRT – 7 mins
Wait time factor, k	0.5

3. Result and Discussion

3.1. Total passenger benefit

The total passenger benefit from the circle line in terms of cost with respect to the radius of the circle line centred at CBD is as shown in Figure 2. The higher the cost, the more cost-beneficial it is for the passengers to have a circle line at such radius. It is interesting to see that as the radius goes further away from CBD to a certain distance, more passengers will get benefit from the circle line. It is also found out that the passengers benefitted from small radius of circle line are significantly lesser despite higher population density near CBD. This is simply due to the circle line at such small values of radius would only give better coverage to a small bunch of people inside and outside of the small circumferential line. As KL is experiencing an economic boom, it is believed that the growth in population and the development of land property will lead to a more feasible circle line with a higher total passenger benefit. Values in the analysis have shown that the total passenger cost for radial line only network is always higher as compared to the rail network with single ring line. This indicates that having a circle line in a rail network will deliver more advantages to the passengers in overall. The core reason behind this was attributable to the absent of circle rail line connecting suburban areas which causes unnecessary distance travelled, rail transferred at the CBD. However, after conducting the analysis, it is found that at a radial distance of 6km from CBD, the circle line appeared to be most favourable for passengers.

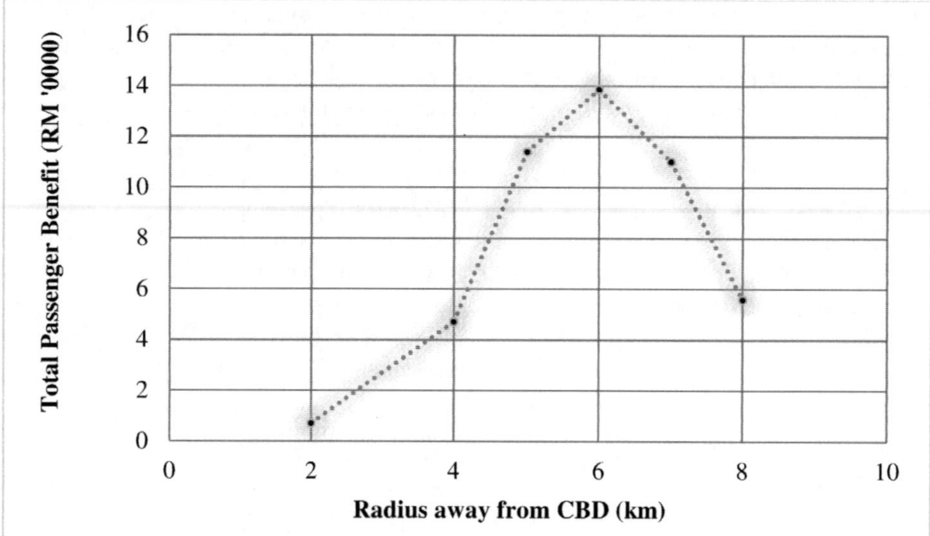

Figure 2. Total passenger benefit against radius away from CBD

3.2. Effect of Access Cost

A sensitivity analysis has also been done to study the effect of access cost on the optimality of the circle line in KL. By fixing all other unit cost values, different unit values of access cost have been input into the same model. As predicted, the higher the access cost, the larger the total passenger benefit for a circle line to be constructed, as shown in Figure 3. When the access cost is high compared to the ride cost, passenger will tend to travel to their destination by train where possible. Circle line in the case is used as a substitution for the passengers who previously chose to access directly to radial line or to the destination. Inversely, when the access cost is equal to ride cost (GammaA = 1), passenger will unquestionably choose to access directly to destination instead of taking trains. This explains the relatively flat line found in the graph when GammaA is equal to 1, which literally indicate building a circle line does not bring much benefit to the passengers. In spite of the changes in access cost, the optimum alignment for circle line still remains at 6km from CBD.

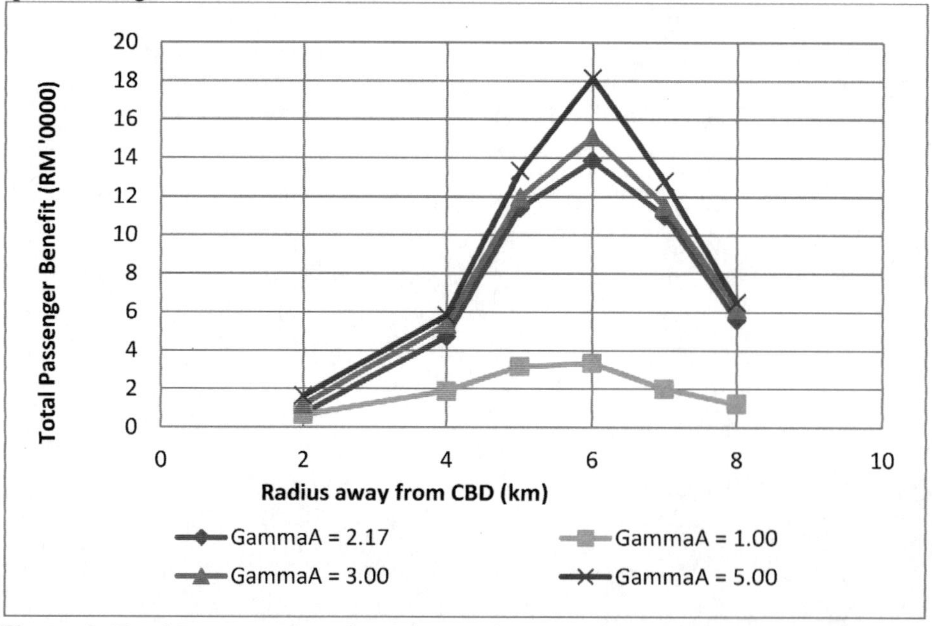

Figure 3. Total Passenger benefit against radius away from CBD (access cost factor)

3.3. Effect of Ride Cost

Another sensitivity analysis has been done by assuming all other unit costs to be constant, then finding the changes of the alignment optimality for circle line by manipulating the ride cost. The unit access cost, in this case, is fixed to be at Ringgit Malaysia 2.17/km/passenger; subsequently it is observed that the nearer the ride cost to the access cost, the lesser the benefit of introducing a circle line in general, as shown in Figure 4. In this scenario where the ride cost increases till the same value as the access cost, passengers will obviously choose to access directly to the destination and save the cost incurred in the wait time and transfer disutility. Accordingly, ride cost also plays a deterministic factor in finding the optimality radius of a circle line. If there is reduction in ride cost, access cost to their nearest train station would be the only concern for a passenger. The figure has shown that the reduction in ride cost will make circle line more feasible and efficient. Nevertheless, due to the comparatively lower number of zones used in the analysis, the total passenger cost is therefore not showing a substantial rise when the access cost is being varied.

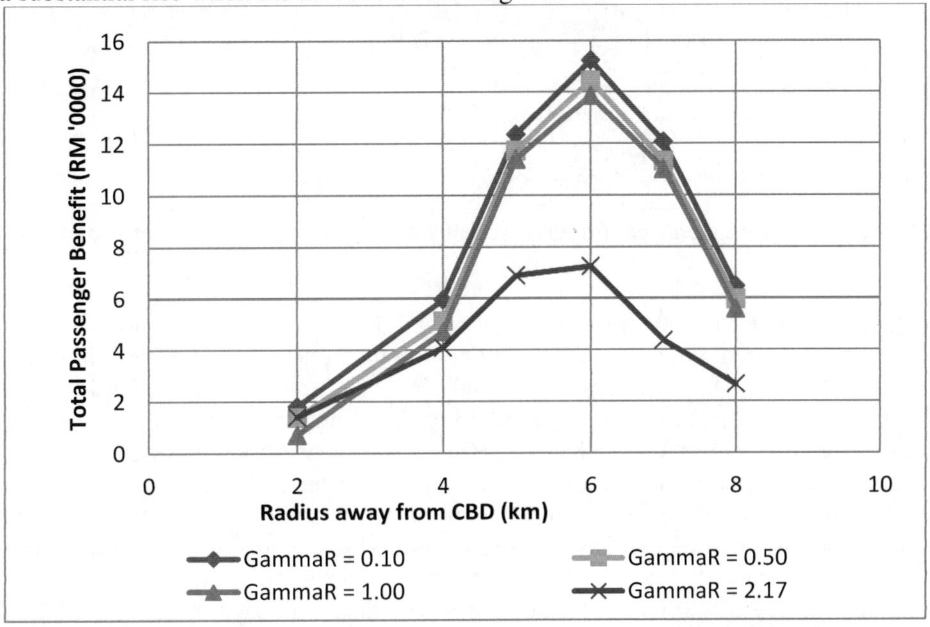

Figure 4. Total passenger benefit against radius away from CBD (ride cost factor)

3.4. Effect of Value of Time

Figure 5. presents the changes in total passenger benefit due to different value of time (VOT). It can be clearly seen that the higher the VOT, the more efficient the implementation of a circle line. Assuming all other variables to be fixed, the VOT will be proportional to the mean per capita income, and the higher mean per capita income implies a higher VOT. According to Department of Statistics Malaysia [13], mean monthly household income of KL has a compound annual growth rate of 10.7%, which means that the VOT in KL is predicted to be growing annually. This has again proven that building a circle line will bring more advantages in the long run. Although the change of VOT does not shift the optimum radius of circle line away from its original optimum radius, a decrease in VOT is evidently making the circle line undesirable and of no benefit to the passengers.

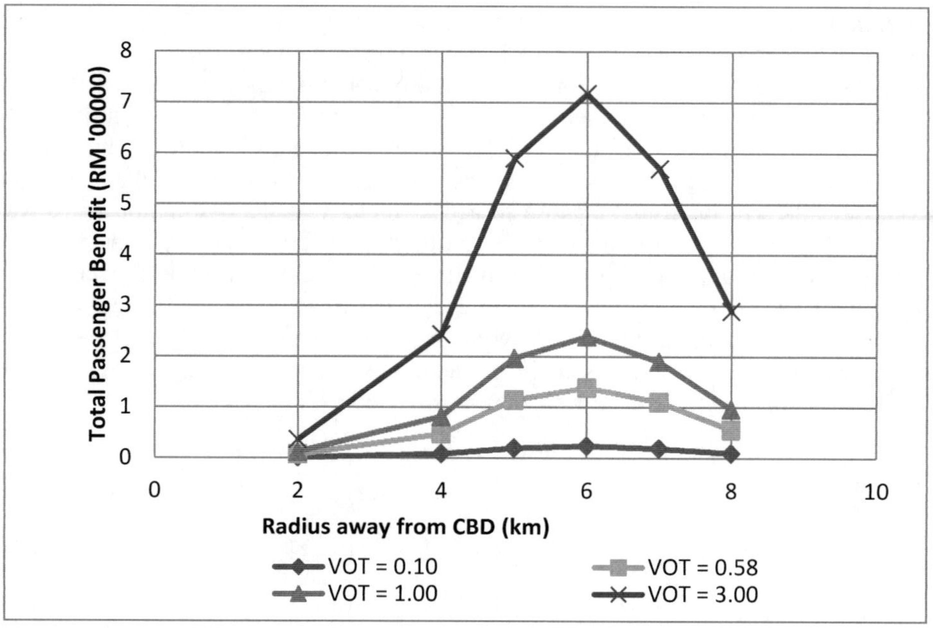

Figure 5. Total passenger benefit against radius away from CBD (value of time factor)

4. Conclusion

In this paper, an approximate mathematics analytical model for circle-radial rail network has been applied in finding the optimal alignment of circle line in Kuala Lumpur, Malaysia. This model is said to be comparable and easily applicable to any radial-circle rail network, especially in a monocentric city. In this study, only a single circle line, based on the passenger route choice that gives the lowest minimum passenger cost. Parameters in KL that affect the cost of a journey of a passenger for each OD pair, such as access cost and ride cost have been utilized to determine the optimum alignment of a circle line. Despite a uniform travel demand distribution is used in this study. Nonetheless, a prospective net benefit of a circle line has also been acquired by evaluating the projected reduction in total passenger cost. Thus, the passenger route choice plays an extremely important role in manipulating the outcome of the result. In the case where most passengers travel to CBD, the circle line is only acting as a feeder service to the nearby radial line rather than providing cross-suburban movements. Certainly, a decrement in the ride cost, plus the increase in the access cost and value of time will lead to a greater attractiveness of introducing a circle line.

5. References

[1] Phang S N, Kuppusamy S and Norris M W 1996. *Metropolitan Management of Kuala Lumpur*. Ruland Jurgen (Ed.), The Dynamics of Metropolitan Management in Southeast Asia, Institute of Southeast Asian Studies, Singapore, pp. 133-167.

[2] Mohamad J and Kiggundu A 2007 *Int. Assoc. Traffic. Safety. Sci.* **31**(1) pp. 69-77.

[3] Ministry of Transport Malaysia Official Portal 2015 *Transport Statistic Malaysia* 2015, s.l.: Ministry of Transport Malaysia.

[4] MRT Corp 2018 MRT Corp [Online] Available at: http://www.mymrt.com.my.

[5] Saeid S, Yuxiong J, Cheng C, Jinping G, Shengchuan J, Lina K, Yuchuan D and Wirasinghe S C 2016 *Transport. Res. Rec.* **2540**(1) pp. 1-20.

[6] Kennedy C and Derrible S J 2009 *Transport. Res. Rec.* **3** pp. 17-25.

[7] Vuchic V 2014 *Planning, Design and Operation of Rail Transit Networks*. Metro Report International, pp. 48-53.

[8] Laporte G, Mesa J A and Ortega F A 1997 *Top* **5**(1) pp. 95-104.

[9] Baaj M H and Mahmassani H S 1995 *Transportation research. Part C, Emerging technologies*. **3**(1) pp. 31-50.

[10] Wirasinghe S C and Vandebona U 1999 *Proc. 14th Int. Symposium on Transp. Traffic Theory*. pp. 759-777.
[11] Lee Y J and Vuchic V R 2005 *J. Transp. Eng.* **131**(1) pp. 1-10.
[12] Chen H, Gu W, Cassidy M J and Daganzo C F 2015 *Transport Res B-Meth.* **81** pp. 755-774.
[13] Department Of Statistics Malaysia 2014 *Report of Household Income and Basic Amenities Survey*. Kuala lumpur, Malaysia.

Acknowledgments
The support provided by Malaysian Ministry of Higher Education, University of Nottingham and Universiti Malaysia Pahang in the form of a research grant (RDU/UMP) vote number RDU1803160 for this study are highly appreciated.

IOP Conf. Series: Materials Science and Engineering **712** (2020) 012004 doi:10.1088/1757-899X/712/1/012004

Assessment of safety performance level on simple urban residential building: Case study at Bogor city Indonesia

M Lutfi[1], H Arien[2], M Hendrawati[3], P J Ramadhansyah[4] and N I Ramli[4]

[1]Civil Engineering Study Program, Faculty of Engineering & Science, Ibn Khaldun University, 16162 Bogor, Indonesia
[2]Institute for Research and Community Services, Ibn Khaldun University, 16162 Bogor, Indonesia
[3]Student of Civil Engineering Study Program, Faculty of Engineering & Science, Ibn Khaldun University, 16162 Bogor, Indonesia
[4]Faculty of Civil Engineering Technology, Universiti Malaysia Pahang, 26300 Gambang, Pahang, Malaysia

Abstract. Simple urban residential building must meet the technical requirements set forth in the building and structure law number 28 of 2002. In the city of Bogor, especially in the densely populated village of Kayumanis, there are areas that have the potential of landslides. In that area, there are some houses that have been damaged by the movement of land and landslides that can threaten the safety of its inhabitants. Therefore they need to stay aware about potential collapse of several parts of house structures, and then understand how to evacuate to safer area, outside of the house after the incident was stopped. Based on these conditions, it is necessary to evaluate buildings in accordance with FEMA on the safety level of buildings. The structure and modeling analysis using SAP 2000-V10 and Indonesia Standard. Specification of simple urban residential building using the reinforced concrete material with the quality of concrete (fc') 17 MPa, 15x15 cm column dimension, 15x18 cm beam dimension and 4D12 BJTP main reinforcement, and shear reinforcement D8-20. The results of the analysis of urban residential building structures shown in 3D form, obtained values for column structure elements almost close to failure are marked with orange color, while for the element of the beam structure is declared safe marked with green color. From result of modeling analysis referring to FEMA 356 obtained value $C_0 = 1.0$; $C_1 = 1.378$; $C_2 = 1.0$; $C_3 = 1.0$; Sa = 1.011; and Te = 0.2206, which states the value of the displacement target is 0.02 m which is categorized to be at 'The Life Safety' (LS) performance level. This means that if there is a movement of land, the structure of the house is declared stable and has adequate service capacity and damage non-structural part is still controlled.

1. Background

Losses due to earthquakes/disaster in Indonesia are mainly caused by lack of public awareness, knowledge, preparedness, and low quality of home construction [1,2]. Simple urban residential buildings should meet the technical requirements set forth in the Building and Building Law number 28 of 2002 [3,4]. One of the residents residing in landslide area of RT02 RW02 Kayumanis Sub-District Tanah Sareal Sub-district of Bogor City has been damaged caused by the movement of land that can threaten the safety of its occupants [5,6]. Therefore they need to stay aware about potential collapse of several parts of house structures, and then understand how to evacuate to safer area, outside of the house after the incident was stopped. Based on these conditions, it is necessary to evaluate the building to know the safety performance of the building, and then plan the evacuation route. Evaluation of residential buildings refers to FEMA 356 standards regarding building safety performance. Initiation of evaluation begins with the analysis of building structures using SAP2000 V10 applications and based on SNI 2847-2013, where the results of structural analysis of the building can be said to be safe or not. This study was conducted with limitations of problems such as, reviewing only one of the houses that were damaged by the movement of the land, the loads observed due to dead load, live load, and earthquake load on beams and columns.

2. Research Methods

The research place conducted at Kayumanis Road RT02 RW02 Kayumanis Sub-District Tanah Sareal Sub-Province Bogor. Analysis structure and modeling using application SAP2000 V10 and SNI 2847-2013 which then evaluated in accordance with FEMA 356 on building safety performance level. Stages of research starting from the literature study are the assessment or first step in doing research, the usual reference in the use of books, theses, journals, thesis and research related to the title of research. The next stage is to collect data in the form of existing data of simple residential buildings. The data obtained are 15x15 cm column dimension, 15x18 cm beam dimension, 4D12 BJTP main reinforcement, and shear reinforcement D8-20 using 17 MPa concrete qualities. Next, illustrate the existing construction with the help of AutoCAD Aplication 2007 and the SAP2000 V10 application for structural modeling. Finally, evaluate the modeling results referring to the FEMA 356 building safety criteria level. The Federal Emergency Management Agency (FEMA) method is a calculation performed to determine the safety level of a building by modifying the linear elastic response of Single Degree Of Freedom (SDOF) system equivalent to coefficient factor C_0, C_1, C_2, and C_3 so that maximum global displacement can be obtained (elastic and inelastic) called the displacement target (δT). The resulting displacement target will result in the building deformation limit shown in Table 1 that will be categorized at the level of building safety performance. The target value of the transfer is determined by the following formula [7]:

$$\delta_T = C_0.C_1.C_2.C_3.S_a.\left(\frac{Te}{2\pi}\right)^2.g \qquad (1)$$

Where:
C_0 = modification factor to relate spectral displacement of an equivalent SDOF system, C_1 = modification factor to relate expected maximum inelastic displacements to displacements calculated for linear elastic response, C_2 = modification factor to represent the effect of pinched hysteretic shape, stiffness degradation and strength deterioration on maximum displacement response, C_3 = modification factor to represent increased displacements due to dynamic P-Δ effects, Sa = response spectrum acceleration, Te = fffective fundamental period of the building in the direction under consideration, and g = acceleration of gravity.

Table 1. Limit of building deformation [8]

	Performance Level			
Interstory Drift Limit	Immediate Occupancy	Damage Control	Life Safety	Structural Stability
Maximum Total Drift	0.01	0.01 – 0.02	0.02	0.33 Vi/Pi
Maximum Inelastic Drift	0.005	0.005 – 0.0015	No Limit	No Limit

Based on FEMA, there are 4 levels of building safety performance as shown in Table 2.

IOP Conf. Series: Materials Science and Engineering **712** (2020) 012004 doi:10.1088/1757-899X/712/1/012004

Table 2. Building safety performance level [9]

No.	Performance Level	Description
1	Operational Level	Utility equipment still works, there is little damage
2	Immediate Occupancy Level	The building receives a green sign (safe to use), the results of the examination need a little improvement
3	Life Safety Level	The structure remains stable and has adequate service capacity, nonstructural parts damage is still controlled
4	Collapse Prevention Level	The building stood, almost collapsed. Other damages are still permitted

3. Result and Discussion

The result of modeling and structural analysis according to SNI 2847-2013 in a simple residential building is obtained for column structure element almost near the failure which is marked with orange color, while the beam structure element is declared safe marked with green color. Furthermore, the result of evaluation using SAP2000 V10 application refers to the FEMA 356 obtained value of displacement target of 0.02 m, which is categorized at the level of a safety performance of the building is at Life Safety (LS) level.

3.1 Structural modeling and analysis

The specifications of the existing residential buildings are 520 cm X-direction and 1190 cm Y-direction, 300 cm height Z-direction using 15x15 cm column dimension, 15x18 cm dimension beam, 17 MPa concrete (FC), 4D12 BJTP main reinforcement, and shear reinforcement D8-20, is in quake zone 4 with the medium type of soil (Figure 1).

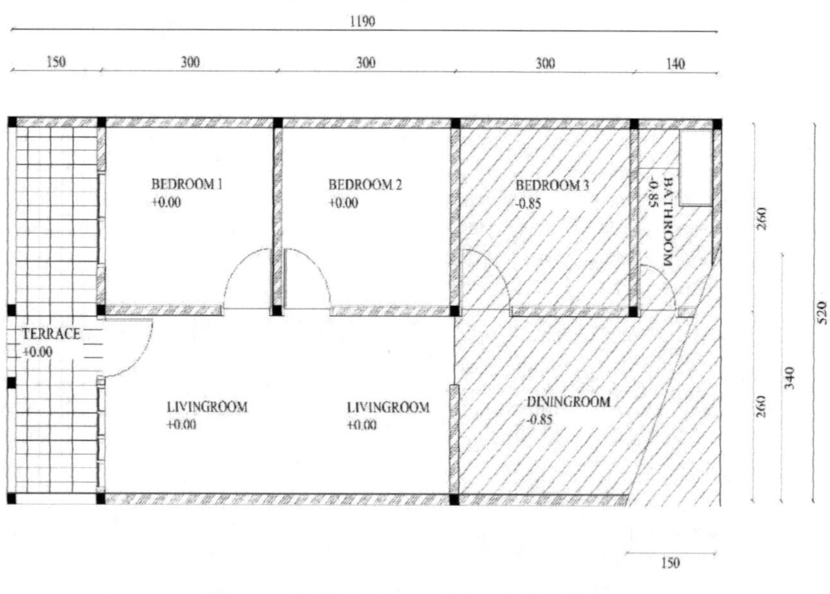

Figure 1. Existing residential building

The chart is then modeled in 3D using the SAP2000 V10 application, shown in Figure 2.

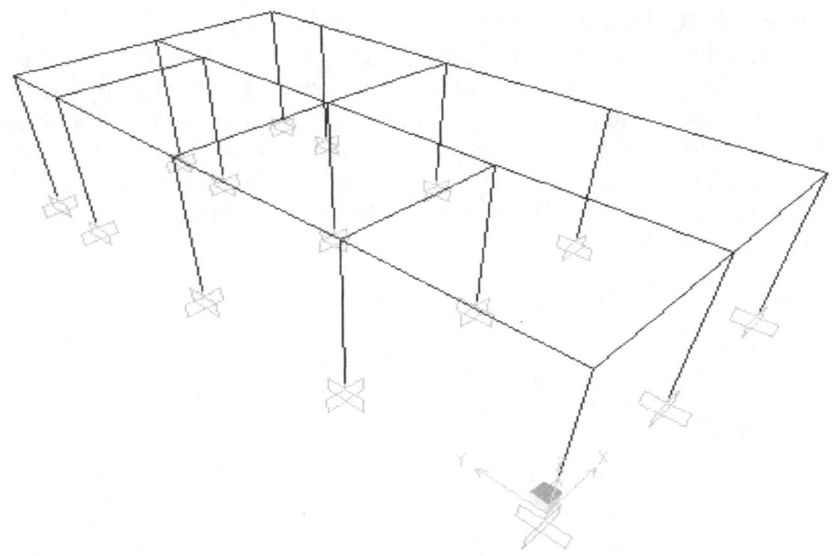

Figure 2. Structured modelling

From the modeling, we performed a structural analysis with a load combination of dead load covering roof 3 m x 10.4 m x 50 kg/m^2=150 kg/m^2, ampig wall (1.54 m x 250 kg/m^2) / 2 = 192.5 kg/m^2, hanging = 11 kg/m^2. Live load on roof = 100 kg/m^2. Earthquake loads using spectrum response data based on the Public Works Research and Development Center [10,11,12]. The results of the load combinations analysis obtained for the column structure elements are almost close to failure which is indicated by orange color, while for the beam structure elements expressed which are indicated by the green color shown in Figure 3.

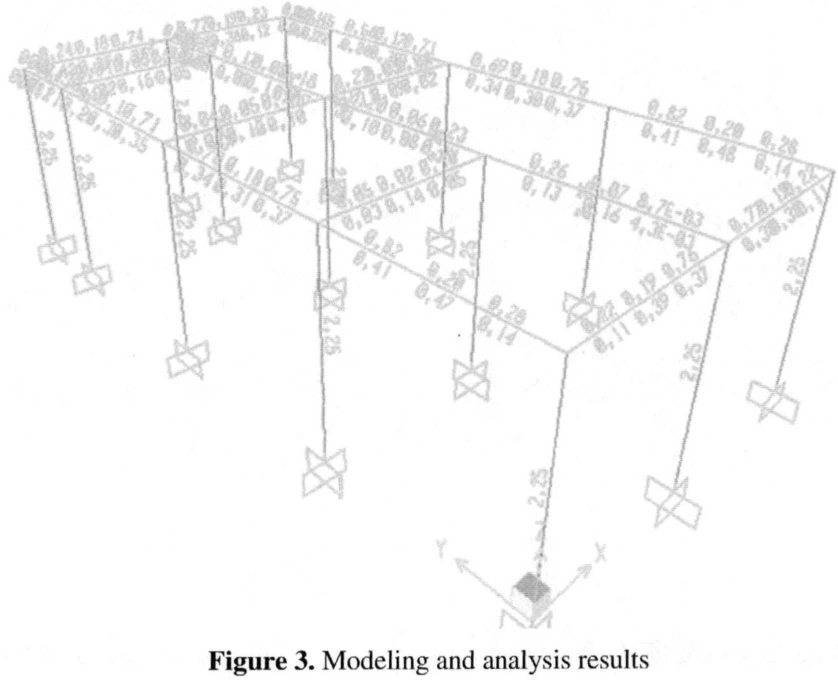

Figure 3. Modeling and analysis results

IOP Conf. Series: Materials Science and Engineering **712** (2020) 012004 doi:10.1088/1757-899X/712/1/012004

3.2 Modeling referring to FEMA 356

Furthermore, modeling with FEMA 356 was obtained from the pushover method found in SAP2000 V10 to produce plastic joints in the beam and column structure elements. In steps 1 up to step 6 with colorful nodes shown in Figure 4 and Figure 5. Description of pushover color nodes in plastic joints is based on Table 2 which is divided into 7 levels described in Table 3. According Figure 5, the front part of the house is not able to withstand the shear and shattering forces, while the back part is still function. Therefore some reinforcement is necessary to keep the evacuation route through the house front.

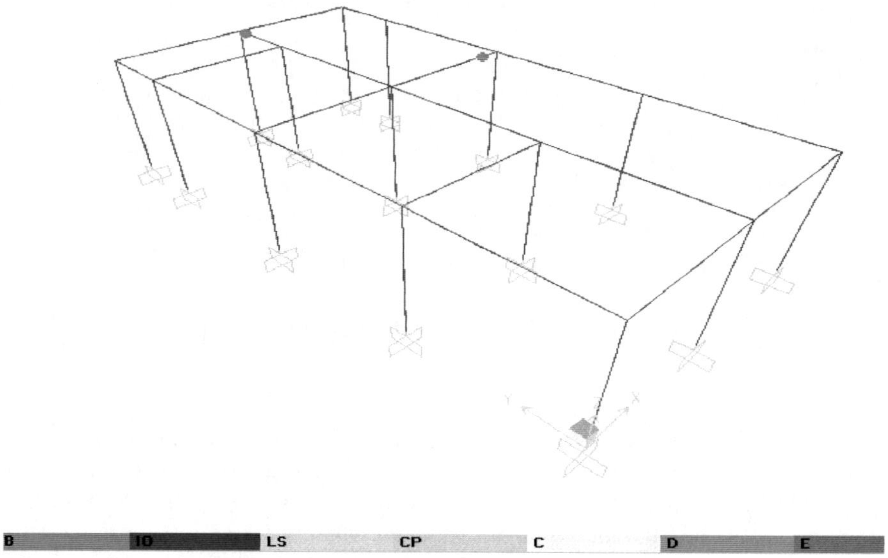

Figure 4. Plastic joints that occurs in step 1

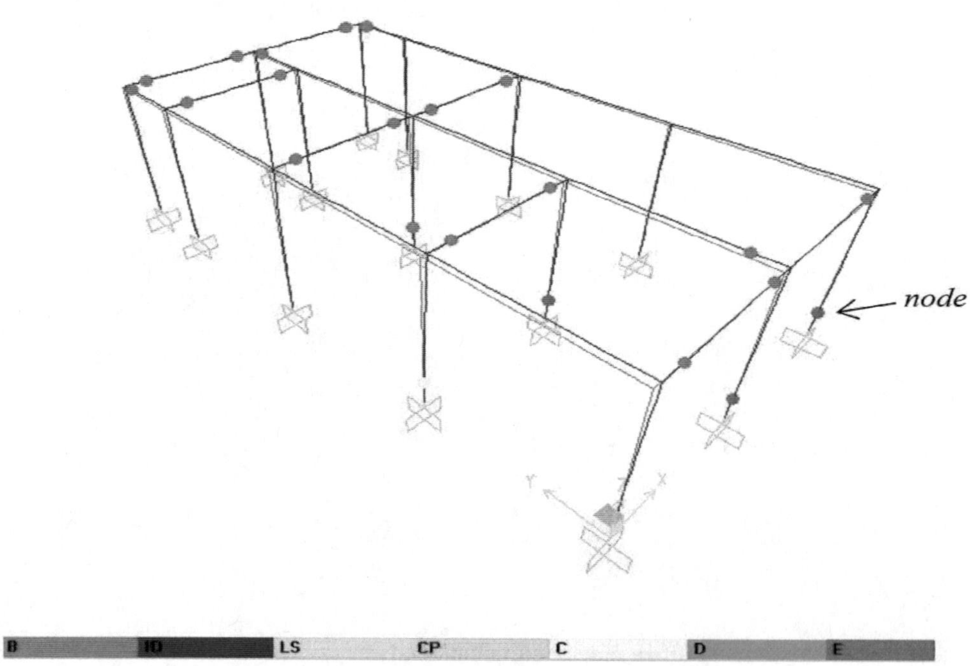

Figure 5. Plastic joints that occurs in step 6

Table 3. Node color pushover

Level of Performance	Symbol	Explanation/description
B	●	There is no significant structural and non-structural damage, the building can still function
IO	●	No structural damage occurs, non-structural components are still in place and buildings can still function without undoing repair problems
LS	●	Structural damage occurs but no collapse occurs, non- structural components are not working but buildings can still be used after repair
CP	●	Damage occurs in structural and non-structural components, buildings are almost collapsed, and accidents due to the collapse of building materials are very likely
C	●	The maximum limit of shear forces that can still be retained
D	●	The degradation of large structural trajectories, resulting in unstable structural conditions and almost collapse
E	●	The structure is not able to withstand the shear and shattering forces

Pushover also produces curves referring to FEMA 356 shown in Figure 6 and generates the values required for the displacement target (δT). The values obtained from the FEMA 356 curve are $C_0 = 1.0$; $C_1 = 1.378$; $C_2 = 1.0$; $C_3 = 1.0$; Sa = 1.011; and Te = 0.221.

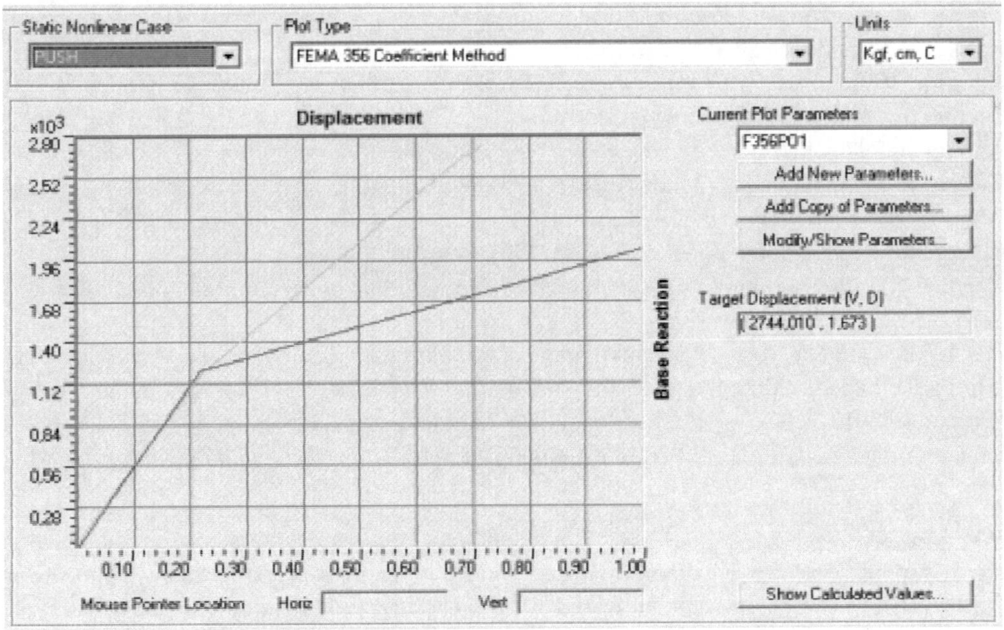

Figure 6. Pushover curve which refers to FEMA 356

Furthermore, the values are calculated using the equation (1) so that the result of the displacement target value is 0.02 m categorized under table 1 at the Life Safety (LS) performance level. This means that if there is a movement of land, the structure of the house is stable and has adequate service capacity and damage non-structural part is still controlled.

4. Conclusion

Based on the analysis and discussion, by modeling with SAP2000 V10 application in simple residential building obtained the following conclusion. The result of evaluation and modeling of the simple residential building structure in RT02 RW02 of Kayumanis Sub-District of Tanah Sareal Sub-Province of Bogor, it is found that the column structure element almost closer to the failure is characterized by orange color, for the element of a beam structure is declared safe with green color. The modeling results referring to FEMA 356 resulted in a building displacement target value of 0.02 m located at the Life Safety (LS) building performance level.

Reference

[1] Badan Standarisasi Nasional 2013 *Persyaratan Beton Struktural Untuk Bangunan Gedung,* SNI 03-2847-2013, Jakarta.

[2] Badan Standarisasi Nasional 2012 *Tata Cara Perencaaan Ketahanan Gempa Untuk Struktur Bangunan Gedung dan Non Gedung,* SNI 03-1726-2012, Jakarta.

[3] Departemen Pekerjaan Umum 1987 *Pedoman Perencaaan Pembebanan Untuk Rumah Dan Gedung tahun 1987,* Jakarta.

[4] Gerry F Waworuntu, M.D.J. Sumajouw, R.S. Windah., 2014. *Evaluasi Kemampuan Struktur Rumah Tinggal Sederhana Akibat Gempa,* Jurnal Sipil Statik, Vol. 2 No. 4, April 2014, Universitas Sam Ratulangi Manado, Sulawesi Utara.

[5] Handana M A P, Karolina and Steve 2018 *Performance evaluation of existing building structure with pushover analysis IOP Conf. Ser.: Mater. Sci. Eng.* **309** 012039.

[6] Sudarman, Hierico M, Reky S. W, and Servie O. D 2014. *Analisis Pushover Pada Struktur Gedung Bertingkat Tipe Podium*, Jurnal Sipil Statik. Vol. 2 No. 4, April 2014, Universitas Sam Ratulangi Manado, Sulawesi Utara.

[7] Anonim 1996 *Seismic Evaluation and Retrofit of Concrete Building Volume 1,* Applied Technology Council (ATC-40), California.

[8] Anonim 1997 *NEHRP Guidelines for the Seismic Rehabilitation Of Buildings,* Federal Emergency Management Agency (FEMA 273), Washington, USA.

[9] Anonim 2000 *Prestandard Commentary for the Seismic Rehabilitation Of Buildings,* Federal Emergency Management Agency (FEMA 356), Washington, USA.

[10] Boen, T 2016 *Belajar dari Kerusakan akibat Gempa Bumi: Bangunan Tembokan Nir-Rekayasa di Indonesia*, Yogyakarta.

[11] Pusat Penelitian dan Pengembangan Permukiman Kementerian Pekerjaan Umum, http://puskim.pu.go.id/Aplikasi/desain_spektra_indonesia_2011. Diakses pada tanggal 01 September 2017.

[12] Undang-undang RI Nomor 28 2002 *Undang-Undang Tentang Bangunan Gedung,* Jakarta.

Acknowledgments

The support provided by Ibn Khaldun University and Universiti Malaysia Pahang in the form of a research grant (RDU/UMP) vote number RDU1803160 for this study are highly appreciated.

IOP Conf. Series: Materials Science and Engineering **712** (2020) 012005 doi:10.1088/1757-899X/712/1/012005

Flash Flood Susceptibility Modelling: A Review

A Saleh[1], A Yuzir[1] and I Abustan[2]
[1]Disaster Preparedness & Prevention Centre (DPPC), Malaysia-Japan International Institute of Technology (MJIIT), Universiti Teknologi Malaysia (UTM), 54100 Kuala Lumpur, Malaysia
[2]School of Civil Engineering, Engineering Campus, Universiti Sains Malaysia, 14300 Nibong Tebal, Penang, Malaysia

Abstract. Flash flood is common problem in the developed or developing country especially in urban areas. This is due to the changes of land use, poor planning of drainage system, deforestation and unplanned land use and etc. Flash flood susceptibility modelling (FFSM) for urban area is important to avoid the loss and etc. FFSM can be vital component and essential tools for planning and management of natural disaster and environmental. The objective of this paper is to review selected paper on flash flood susceptibility modelling using Geographic Information System (GIS) coupling empirical models. A lot of methods have been coupling with GIS to developed flood susceptibility modelling, for example, Weight of Evidence, Random Forest, etc. However, there is still lack of susceptibility model for flash flood. The GIS coupling method can improve the accuracy of flood susceptibility modelling. Thus, this paper will review the ability of GIS coupling with empirical models for the flash flood susceptibility modelling.

1. Introduction

The occurrences of flash flood cannot be predicted and it is worst hydro-meteorological disaster [1]. Flash flood disasters occur frequently due to global climate change [2]. Flood events commonly cause destruction to agricultural crops and property, and may even result in the loss of human lives [3]. According to Elkhrachy [5], flash flood can be defined the flood which is begin in a short period of time and normally show high peak discharges. Elkhrachy [5] added usually the geomorphic low-lying zones will hit by flash flood when heavy rainfall.

By develop the flood susceptibility mapping and modelling, it can help local authority in flood management to identify the most sensitive zones for civil protective actions, assess damages, and make valid urban planning [5]. By taking the definition of landslide susceptibility mapping, Santangelo et al. [6] defined the susceptibility mapping as the probability that a risk occurrence happens in a particular area and in a not determined date. Susceptibility mapping was developed based on the relationship of the conditioning factors with the distribution of previous events [6]. That means susceptibility modelling is actually just an estimate of "where" disasters such as landslide or flood are likely to occur. Susceptibility map also known as natural hazard potential map for some groups.

According to Reichenbach et al. [7], there are confusion between "susceptibility" and "hazard" modelling. Even though these two terms usually used as same meaning but it these two words stating different models [7]. Reichenbach et al. [7] added that susceptibility model does not consider the size of natural hazard e.g., the length, width, depth, area or volumes. Meanwhile, hazard model would predicting "where" a flood or landslide will take place, "when" or "how frequently" it will happen, and "how large" it will be [8]. Flood susceptibility map will manage any future flood problems [9] and important step to predict and manage the future flood event [10].

IOP Conf. Series: Materials Science and Engineering **712** (2020) 012005 doi:10.1088/1757-899X/712/1/012005

2. Modelling Methods

There are many methods of flash flood susceptibility mapping; traditional bases hydrological methods, statistical, and machine learning based methods. Rule-based and automated modelling methods have outperform outdated traditional flood models due to their more suitable for hazard analyses [11]. In recent times, researchers have coupling many methods with GIS to increase the accuracy prediction of flooded areas (flood susceptibility). These methods are including qualitative method such as analytical hierarchy process (AHP), quantitative techniques such as weight of evidence (WoE) and frequency ratio (FR), and machine learning method such as artificial neural networks (ANN). However, there are still lack of studies have been carried out for flash flood susceptibility modelling. This paper aims to review selected study of flash flood susceptibility modelling.

3. Flash Flood Susceptibility Modelling Study

Bui et al. [12] is using feature selection method (FSM) and tree based ensemble methods. The study area located at the Bao Yen district and the Bac Ha district of Lao Cai Province, Vietnam. Bui et al. [12] were used 654 floods and 12 conditioning factors (stream density, toposhade, slope, curvature, stream power index (SPI), elevation, topographic wetness index (TWI), rainfall, aspect, normalize difference vegetation index (NDVI), soil type cover and lithology). This method used a fuzzy rule based algorithm (FURIA) which is as attribute evaluator, while Genetic Algorithm (GA) as the search method. This is to get best set of conditioning factors used in modelling assessments. Then, the FURIA-GA method was combined with LogitBoost, Bagging and AdaBoost ensemble algorithms for prediction model. Based on the result, FURIA-GA-Bagging (93.37%) outperformed the other ensemble algorithm, FURIA-GA-LogitBoost (92.35%) and FURIA-GA-AdaBoost (89.03%). The flash flood susceptibility mapping developed by using FURIA-GA-Bagging method is shown in Figure 1.

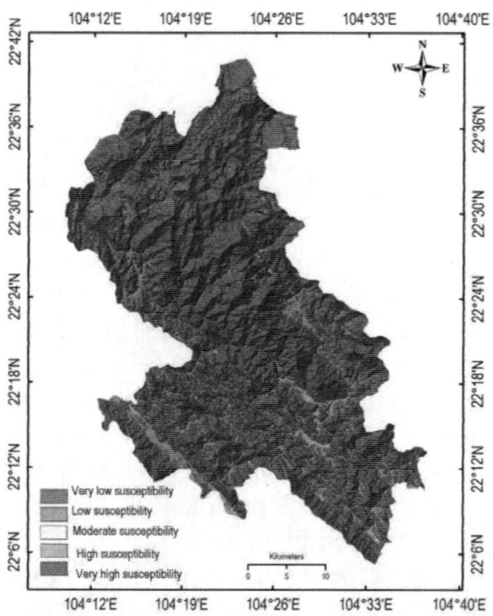

Figure 1. Flash flood susceptibility mapping developed by using FURIA-GA-Bagging method [12]

Youssef and Hegab [13] were applied Analytic hierarchy process (AHP) which is can categorize in qualitative method. The main objective of this study is to examine the effectiveness and reliability of AHP. The location of study area is Ras Gharib, Egypt. This study also using high-resolution images obtained after previous flood events in 2016. These high-resolution images are used to validate the susceptibility model by using a slicing technique and historical flood data. Youssef and Hegab [13] only used 7 flood factors - distance from streams, slope, curvature, lithological units, angle, elevation, and topographic wetness index (TWI). However, results of this study shown that AHP give good result for flash flood susceptibility which is 83.3%. The flash flood susceptibility mapping developed by using AHP method is shown in the Figure 2.

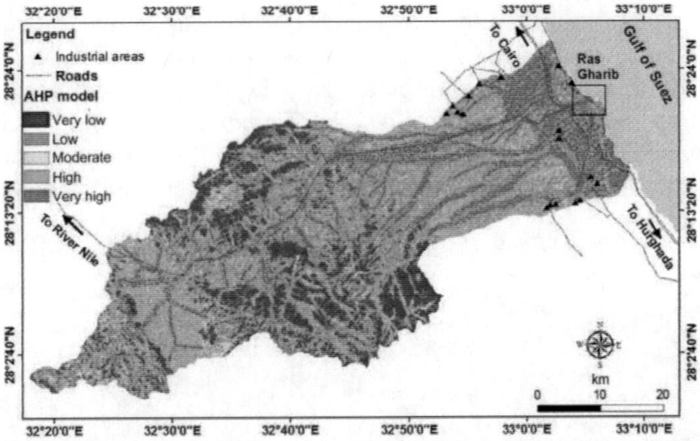

Figure 2. Flash flood susceptibility mapping developed by using AHP method [13]

Ngo et al. [14] analysed flash flood in two districts Bac Ha and Bao Yen, Vietnam by using machine learning techniques which are Firefly algorithm (FA), Levenberg–Marquardt (LM) Backpropagation, and an artificial neural network (FA-LM-ANN), LM-ANN, FA-ANN, support vector machine (SVM) and classification tree (CT). For this study, Ngo et al. [14] used 12 flood factors which are aspect, elevation, stream power index (SPI), slope, topographic wetness index (TWI), stream density, rainfall, curvature, normalized difference vegetation index (NDVI), lithology, toposhade, and soil type and 654 flash flood locations were identified to evaluate the flash flood model. According to the results, the integrated FA-LM-ANN gives the good results which is 97.0% following by LM-ANN - 92.6%, FA-ANN – 91.9%, SVM – 92.9% and CT – 90.8%. Figure 3 shows the flash flood susceptibility mapping developed by using FA-LM-ANN method.

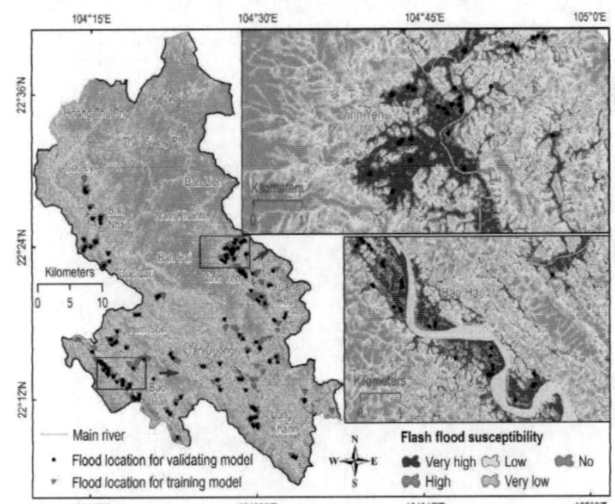

Figure 3. Flash flood susceptibility mapping developed by using FA-LM-ANN method [14]

Khosravi et al. [15] has done the comparisons between Logistic Model Trees (LMT), Reduced Error Pruning Trees (REPT), Naïve Bayes Trees (NBT), and Alternating Decision Trees (ADT) which is these four methods is decision trees algorithms. The study area is located in Haraz watershed, northern Iran. The 201 flood points were identified and 11 flood factors were selected which are altitude, ground slope, curvature, river density, normalized difference vegetation index (NDVI), topographic wetness index (TWI), distance from river, land use, stream power index (SPI), rainfall, and lithology. The results show that the ADT is the highest AUC value – 97.6 %, followed by NBT – 97.4 %, LMT –

97.1% and REPT - 81.1%, respectively. Before that, Khosravi et al. [16] compare the results between statistical index, Shannon's entropy, and weighting factor models for flash flood susceptibility model. For this study, Khosravi et al. [16] was used 211 flood locations were identified and 10 flash flood conditioning factors was used in this study area which are plan curvature, stream power index (SPI), land use, geology, distance from river, topographic wetness index (TWI), rainfall, slope angle, altitude, and normalized difference vegetation index (NDVI). The result show that the statistical index model with the prediction of 98.72% is the highest rate of prediction followed by weighting factor and Shannon's entropy models with the 97.6% and 92.42% for the prediction rate. The flash flood susceptibility mapping produced by using Alternating Decision Trees (ADT) and statistical index (SI) are presented in the Figure 4.

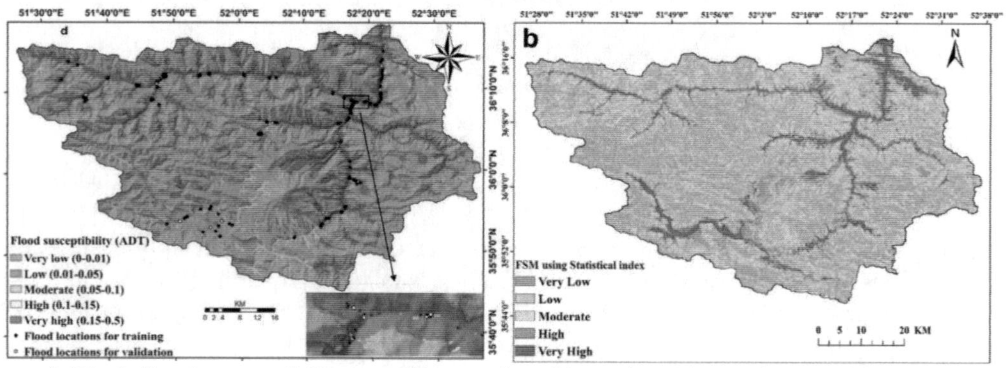

Figure 4. Flash flood susceptibility mapping produced by using Alternating Decision Trees (ADT) [15] and statistical index (SI) [16]

Youssef et al. [17] were used bivariate (frequency ratio) and multivariate statistical models (ensemble frequency ratio and logistic regression) to develop flash flood susceptibility modelling in Jeddah, Saudi Arabia. In the study, 127 flood locations were identified and 7 flood factors were used to develop flash flood model which are slope, elevation, curvature, geological units, land use, soil drain, and distance from streams. The results showed that the prediction accuracy achieved using ensemble technique between Frequency Ratio (FR) and Logistic Regression (LR) has the best (91.3%) followed by the traditional Frequency Ratio method - 89.6 %. This results show that bivariate statistical analysis (BSA) method can improve by integrated with Multivariate statistical analysis (MSA) method. Table 1 summarizes the methods used in flash flood susceptibility model. The flash flood susceptibility mapping developed by using multivariate statistical models (ensemble frequency ratio and logistic regression) is shown in Figure 5.

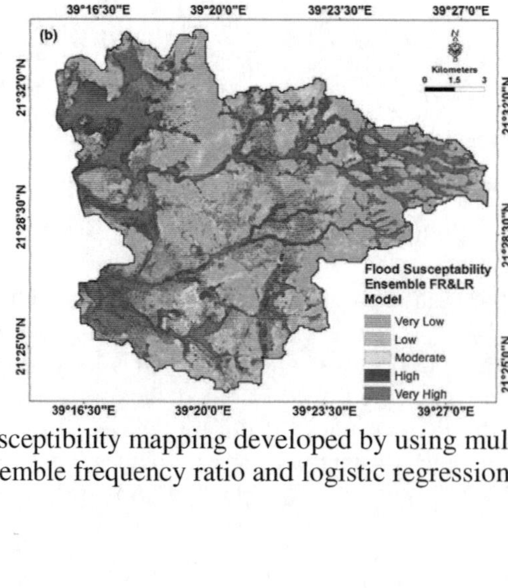

Figure 5. Flash flood susceptibility mapping developed by using multivariate statistical models (ensemble frequency ratio and logistic regression) [17]

Table 1. The summarizes of previous flash flood susceptibility model studies.

Authors	Method	Data/Conditioning Factors	Finding/Result
Bui et al. [12]	The feature selection method (FSM), used a fuzzy rule based algorithm Fuzzy Unordered Rules Induction Algorithm (FURIA), as attribute evaluator, whereas Genetic Algorithms (GA) were used as the search method, in order to obtain optimal set of variables used in flood susceptibility modeling assessments. The novel FURIA-GA was combined with LogitBoost, Bagging and AdaBoost ensemble algorithms.	654 flood locations were identified 12 flood factors - Elevation, slope, aspect, curvature, TWI, SPI, toposhade, stream density, rainfall, NDVI, soil type cover and lithology cover	FURIA-GA-Bagging - 93.37% FURIA-GA-LogitBoost - 92.35% FURIA-GA-AdaBoost -89.03%
Youssef and Hegab [13]	Analytical hierarchy process (AHP)	232 flood locations were identified 7 flood factors - Slope, angle, elevation, distance from streams, lithological units, TWI, and curvature.	AHP - 83.3%
Ngo et al. [14]	Firefly algorithm (FA), Levenberg–Marquardt (LM) Backpropagation, and an artificial neural network (FA-LM-ANN), LM-ANN, FA-ANN, support vector machine (SVM) and classification tree (CT)	654 flash flood locations were identified 12 flood factors - Elevation, slope, aspect, curvature, TWI, SPI, toposhade, stream density, rainfall, NDVI, soil type, and lithology	FA-LM-ANN -97.0% LM-ANN - 92.6% FA-ANN – 91.9% SVM – 92.9% CT – 90.8%
Khosravi et al. [15]	Decision trees algorithms Logistic Model Trees (LMT), Reduced Error Pruning Trees (REPT), Naïve Bayes Trees (NBT), and Alternating Decision Trees (ADT)	201 present and past flood locations were identified 11 flood factors - Ground slope, altitude, curvature, SPI, TWI, land use, rainfall, river density, distance from river, lithology, and NDVI.	ADT – 97.6 % NBT – 97.4 % LMT – 97.1% REPT - 81.1%
Khosravi et al. [16]	Shannon's entropy, statistical index, and weighting factor models	211 flood locations were identified 10 flood factors - Slope angle, plan curvature, altitude, TWI, SPI, distance from river, rainfall, geology, land use, and NDVI	SI - 98.72 % WF - 98.1% SE - 92.53%
Youssef et al. [17]	bivariate and multivariate statistical models (frequency ratio and ensemble frequency ratio and logistic regression)	127 flood locations were identified 7 flood factors - Slope, elevation, curvature, geological units, landuse, soil drain, and distance from streams	FR - 89.6 % FR+LR - 91.3%

4. Conclusion

In summary, the purpose of this paper was to review selected study on flash flood susceptibility modelling (FFSM). Based on the selected studies, the ability of integration between GIS and analytical model shows the good result in developing flash flood susceptibility model. The ensemble and hybrid model of machine learning can increase the accuracy of result. The methods of FFSM are also play important role. The studies carried out by using machine learning method shows that improvement of result when they applied the machine learning technique at same study area. It also found that the result of machine learning method has a higher accuracy than qualitative and quantitative method (WoE, FR, and AHP) even the number of flooded points was decrease compare with previous study. It is proved that the ability of integration between GIS and analytical model is able to generate the flash flood susceptibility modelling which are very useful tools to plan and manage the flood.

5. References

[1] Mahmood S and Rahman A 2019 Flash flood susceptibility modeling using geo-morphometric and hydrological approaches in Panjkora Basin, Eastern Hindu Kush, Pakistan *Environ. Earth Sci.* **78** 0

[2] Li H, Zhang S, Li Q, Zhang X and Guo L 2019 Research on flash flood disaster warning index: Case study of Luoning County *IOP Conf. Ser. Earth Environ. Sci.* **218** 012075

[3] Dahri N and Abida H 2017 Monte Carlo simulation-aided analytical hierarchy process (AHP) for flood susceptibility mapping in Gabes Basin (southeastern Tunisia) *Environ. Earth Sci.* **76** 1–14

[4] Elkhrachy I 2015 Flash Flood Hazard Mapping Using Satellite Images and GIS Tools: A case study of Najran City, Kingdom of Saudi Arabia (KSA) *Egypt. J. Remote Sens. Sp. Sci.* **18** 261–78

[5] Sadek M and Li X 2019 Low-Cost Solution for Assessment of Urban Flash Flood Impacts Using Sentinel-2 Satellite Images and Fuzzy Analytic Hierarchy Process : A Case Study of Ras Ghareb City , Egypt **2019**

[6] Santangelo N, Santo A, Di Crescenzo G, Foscari G, Liuzza V, Sciarrotta S and Scorpio V 2011 Flood susceptibility assessment in a highly urbanized alluvial fan: The case study of Sala Consilina (southern Italy) *Nat. Hazards Earth Syst. Sci.* **11** 2765–80

[7] Reichenbach P, Rossi M, Malamud B D, Mihir M and Guzzetti F 2018 A review of statistically-based landslide susceptibility models *Earth-Science Rev.* **180** 60–91

[8] Basith A 2011 Landslide Susceptibility Modelling Under Environmental Changes: A Case Study of Cameron Highlands, Malaysia 1–290

[9] Kourgialas N N, Karatzas G P, Kourgialas N N and Karatzas G P 2011 Flood management and a GIS modelling method to assess flood-hazard areas — a case study Flood management and a GIS modelling method to assess flood-hazard areas — a case study **6667**

[10] Kourgialas N N and Karatzas G P 2017 A national scale flood hazard mapping methodology: The case of Greece – Protection and adaptation policy approaches *Sci. Total Environ.* **601–602** 441–52

[11] Tehrany M S, Kumar L, Neamah Jebur M and Shabani F 2019 Evaluating the application of the statistical index method in flood susceptibility mapping and its comparison with frequency ratio and logistic regression methods *Geomatics, Nat. Hazards Risk* **10** 79–101

[12] Bui D T, Tsangaratos P, Ngo P T, Pham T D and Pham B T 2019 Flash flood susceptibility modeling using an optimized fuzzy rule based feature selection technique and tree based ensemble methods *Sci. Total Environ.* **668** 1038–54

[13] Youssef A M and Hegab M A 2019 Flood-Hazard Assessment Modeling Using Multicriteria Analysis and GIS: A Case Study—Ras Gharib Area, Egypt *Spatial Modeling in GIS and R for Earth and Environmental Sciences* pp 229–57

[14] Ngo P T, Hoang N and Pradhan B 2018 A Novel Hybrid Swarm Optimized Multilayer Neural Tropical Areas Using Sentinel-1 SAR Imagery and

[15] Khosravi K, Pham B T, Chapi K, Shirzadi A, Shahabi H, Revhaug I, Prakash I, Tien Bui D, Shahabi H, Chapi K, Shirzadi A, Pham B T, Khosravi K and Revhaug I 2018 A comparative assessment of decision trees algorithms for flash flood susceptibility modeling at Haraz watershed, northern Iran *Sci. Total Environ.* **627** 744–55

[16] Khosravi K, Pourghasemi H R, Chapi K and Bahri M 2016 Flash flood susceptibility analysis and its mapping using different bivariate models in Iran: a comparison between Shannon's entropy, statistical index, and weighting factor models *Environ. Monit. Assess.* **188**

[17] Youssef A M, Pradhan B and Sefry S A 2016 Flash flood susceptibility assessment in Jeddah city (Kingdom of Saudi Arabia) using bivariate and multivariate statistical models *Environ. Earth Sci.* **75** 1–16

Acknowledgments
The first author would like to acknowledge Malaysia-Japan International Institute of Technology (MJIIT), UTM for providing Incentive Scheme to undertake his PhD program.

IOP Conf. Series: Materials Science and Engineering **712** (2020) 012006 doi:10.1088/1757-899X/712/1/012006

Exploration on Pavement Surface Conditions Attributed To Mineral Freight and Logistics Operations on Kuantan Road Network

A Ismail[1,2], M R Intan Suhana[1], K A Masri[1] and N H Rapar[1]

[1]Faculty of Civil Engineering Technology, Universiti Malaysia Pahang, 26300 Gambang, Pahang, Malaysia
[2]Earth Resources & Sustainability Centre, Universiti Malaysia Pahang, 26300 Kuantan, Pahang, Malaysia

Abstract. The most active mineral mining operations in Malaysia are located in Kuantan, Pahang. The rapid increases of the minerals operations are of concern due to its significant impacts on the environment and human health. While effects to the environment and human health are a great of concern, the effect to the transportation system also should not be neglected. With the continuous growth of these industries, there will be more heavy vehicles generated on road transporting minerals related products leading to several issues especially on road pavement damages. Pavement damages are always associated with its optimum function and may contribute to road crashes. There are several factors that influenced the pavement damages such as weather changes, quality of material use, improper maintenance and traffic operations. This study was conducted to explore the pavement damages occurrence based on traffic operation which focusing on heavy vehicles composition as centre for critical analysis. To achieve this study, the identification of regular routing used for minerals logistic operation from the mining source to the port was chosen as study location. This study was conducted by using visual observation survey and traffic volume survey. A set of historical data on route inspection supplied by Road Care also been used for analysis. Then, data obtained were analyzed using road disability level and road rating assessment. The result shows that the road rating leading to Kuantan Port was less satisfactory (rating C) with the marks of 7.27 and the highest type of damage found was cracking with 36% followed by rutting (33%) and potholes (31%). The outcomes from this study were hoped to bring some understanding on the impact of heavy vehicles towards transportation system within a local road network in Kuantan.

1. Introduction

Malaysia has a long past of mining, exclusively tin, but until very recent times it scarcely listed on global markets as a source of mineral. That changed suddenly in 2013 when bauxite mining in Indonesia ceased their operation. Kuantan, Pahang, a district in eastern Peninsular Malaysia, has become a hot spot for bauxite mineral mines in Malaysia to fill the supply gap after Indonesia barred exports of mineral ores in 2014. Since then, this industry continues to evolve over time from the early days to present times, is one of Malaysia's most important industries that has been debated for their pros and cons as widely reported by mainstream and social media.

Almost all minerals activities contribute to intense problems to the surrounding area. As in Kuantan, the boomed of this mineral mining industry has led to several issues between year 2013 to 2016 before moratorium of bauxite was implemented somewhere in June 2016. Several studies had been done especially on the impact of bauxite towards environmental quality degradation and health condition among the public. [1] and [2] in their studies have investigating the impact of bauxite activities towards psychosocial and health conditions among communities that being affected.

While effects to the environment and human health are a great of concern, the effect to the traffic and transportation system also should not be neglected. The route of minerals freight distribution and their traffic affect the environment through which they pass. Consequently, it will also affect traffic performance on the roads accordingly. The complexity of the minerals hauling has indeed brought several issues to the surrounding areas especially environmental factors in terms of traffic and land use interactions that has been generated by the heavy trucks operations that transporting the minerals [3]. These issues include travel behaviour, environmental and health impacts, damage of pavement and other infrastructure and also road accident.

As in Kuantan, the development of the mineral industry has led to the growth of new industries that generate state and population revenues. Along with these developments, the increase in the number of vehicles especially for transporting the minerals has also increased. The significance growth of vehicular road traffic especially the heavy trucks will indirectly contributed to an increase of road safety issues. Moreover, if with uncontrolled mining activities will just lead to worsen the condition. Therefore, this study was conducted to explore on road condition and environment that focusing on pavement surface conditions attributed to heavy vehicles operation which perceived to have an association on pavement damage.

2. Literature Review

The transport of minerals from the field to the port is an important step in the overall operation of the minerals freight distribution. As for example, the typical bauxite freight distribution processes involved the loading onto trucks and hauling to a crushing facility. In this process trucks ranging in size from 30 to 180 tonnes are used depending on the size mines [4]. The mine site in Kuantan were located at Balok, Bukit Goh, Bukit Sagu and Sungai Karang where the most progressively mineral mining area occurred in the vicinity of Bukit Goh in about 18000 ha. Meanwhile, the minerals product storage areas were found to be located around Bukit Goh, Bukit Kuantan, Bukit Sagu, Kampung Bukit Kubang Badak, Jeram, and Beserah before transported to Kuantan Port before delivery to export or to local refineries. These mineral's product were transported to its stockpile areas mostly via Kuantan Bypass road (Federal Road 3) or through local state road penetrating Felda Bukit Goh. Due to the location of mining activities in Kuantan that totally different scenario comparing to the other common area in mining in other countries has led to intense suffering to the local even though on the other hands, it offers some exciting economic opportunities for various parties. In other country, remote locations and well defined zoning area are common for bauxite mining activities. While in Kuantan the location of mining activities is scattered and occurring near to and within community settlements without a clearly defined zone [3]. With the rapid developments of this industry especially in the freight and logistics operation are of concern due to its significant effects on the road condition and environment. Considering the scale of the minerals mining activities, it is not surprising that they also have a wide range of impacts on the road condition and environment such as air pollution, noise pollution, pavement damages, crashes and so on. This is especially true for the uncontrolled mining activities. As mention by [3] the extensive and uncontrolled mining activities have great potentials to cause adverse impacts on the environment, health and quality of life of the people living in the affected areas. Local newspaper had reporting the news regarding traffic issues caused by minerals transportation by highlighting fatal accident cases and congestion that linked to the heavy trucks generated on the road to transporting the minerals. The changes of road surface such as potholes also occur along with these developments where it has changed the condition of the road surface. As stated by [5] damage of pavement are considered among the main contributing factors to fatal accident. This is also in line with statement pointed out by Peter Larsson from The Swedish Transport Agency during his presentation to Ministry of Work (MOW) Malaysia, which more than 29% of road crashes factors are caused by road condition and environment. Road condition and environment itself covers many aspects of the physical elements of the road system such as roadside conditions, road surface conditions traffic volumes, operational speed and not to be missed out the driving ambience itself.

An extensive research done by [1] reported on the environmental problems arises due to bauxite mining activities as shown in Table 1 that involved community in Felda Bukit Goh Kuantan which is within 5 km radius from the boundary of the mining site. The finding shows that road damages were among the top five serious issues that have been reported by the respondents (69.8%). This was believed associated with the heavy trucks since they were heavily used the route for transportation of minerals.

Table 1. The environmental problems reported by the respondents

Environmental Problems	Frequency (n)	Percentage (%
Increased air pollution	142	87.7
Dustier house	142	87.7
House cleaning problem	129	79.6
Road damage	113	69.8
Agricultural problem	86	53.1
Loss of calmness	75	46.3
Land erosion/landslide	60	37.0
Water resource problem	50	30.9
Food resource problem	12	7.4

N=162

As mentioned by [6], damages on the roads are primarily caused by the heavier axle loads associated with large commercial vehicles. One of the defects caused by heavy trucks on the road is the deformation of the pavement surface due to overloading that is more than the design load. As stated by [7], the deterioration of pavement arises from deformation generally associated with cracking under heavy trucks. The increased traffic loading will then cause failures such as cracks and depression on the pavement. Other than that, the issue of overloading solely also causes roads in Malaysia to have their mending period earlier than the expected lifespan. Vehicle overloading has been identified as one of the major contributors to road pavement damage in Malaysia [8]. Overloaded vehicles produce higher kinetic energy, resulting in greater impact forces and damages to other vehicles or to the infrastructure.

3. Methodology

Analysis of traffic data often begin with some type of statistical analysis and often involves some type of graphical display. These data should be transformed in appropriate way in order to provide information. In this report, the data collected provide an overview of traffic volumes and pavement surface conditions.

3.1 Field Data

To accomplish research objectives, two routes have been chosen to serve as a study area namely Station 1 and Station 2. Station 1 was on Kuantan Bypass Federal Route 3 and Station 2 was on Route Jalan Kuantan heading to Kuantan Port. Station 1 was an urban arterial of dual two-lane carriageway similar to the East Coast Expressway in the number of lanes but a rank lower in the hierarchy of roads in Malaysia. Meanwhile for Station 2 was a main distributor road that connects Kuantan Port with the outside areas. These two routes were selected as a study area by considering the function of both routes as a main route that provide essential link to and fro Kuantan Port where high volume of heavy trucks could be expected travelling on these routes. Having said that, it clearly shows that the traffic flow catered along both routes was relatively heavy and becoming much heavier especially due to the traffic generated from the growth of the mineral industry in Kuantan in combination with the rapid development along the road. The data collection processed has been performed during weekdays in September and October 2018.

Before proceed with the data collection, a historical data that contain information on pavement damage and traffic volume data from 2014 to 2016 has been collected to provide a basis for further analysis. The pavement damages data was supplied by Road Care Sdn. Bhd. and the peak hour traffic volume data was obtained from Road Traffic Volume Malaysia.

3.2 Traffic Volume Data

Traffic volume is defined as the number of vehicles that pass a given section of road during a given time under prevailing roadway and weather conditions. The purposes of traffic volume study are to evaluate the present traffic flow condition for document planning activities in highway usage and also for design and operation analysis. Traffic volume counts by vehicle class have been conducted for three days on Monday, Wednesday and Friday for AM peak from 7.00 am to 10.00 am at Station 1 and Station 2. An electronic device called Metro Count was used to assist the data collection process. As to achieve the objective of this study, the traffic volume was segregated into vehicle classification as to look in detail the pattern compositions of heavy vehicles on both stations 1 and 2.

3.3 Visual Observation and Physical Measurement

This research employed a quantitative approach to achieve the study's aim. A survey to identify the pavement surface conditions was done by visual observation and physical measurements was made at both location stated above. Only the common type of road damage was considered in this study which were cracking, rutting and potholes. Observation was made within 3 km around the identified stations. Meanwhile the traffic volume based on vehicles class also was taken as to explore the heavy trucks operation which perceived to have an association on pavement damage.

3.4 Road Rating Assessment

Data obtained from the site was then analyzed to identify the frequency of each damage type and road rating assessment was used to determine the seriousness of the road damaged based on road disability level and rating as shown in Table 2 and 3 below.

Table 2. Road Disability Level

Type of road defect	Mark for disability level		
	Low	Moderate	High
Cracking (m)	< 50	50 -150	> 150
Rutting (m)	< 0.012	0.012 – 0.025	> 0.025
Potholes	0	1	> 1
MARKS	1	2	3

Table 3. Road Rating

Marks	Rating	Remarks
< = 3.00	Very Satisfying	A
3.00 – 6.00	Satisfying	B
> 6.00	Less Satisfying	C

4. Results And Discussion

4.1. Traffic Volume Characteristic

As shown in Table 4 below was a volume variation characteristic of site location on weekday basis of Monday, Wednesday and Friday for AM peak from 7.00am to 10.00 am. Basically from the data collected it is found that there was a little different in traffic flow on each weekday for both station with maximum volumes observed was on Monday. However, the volume pattern for both stations was distinctly different with the maximum average volume was at Station 1. This was resulting from the function characteristics of route mention as preceded that station 1 was operated as trunk road that

IOP Conf. Series: Materials Science and Engineering **712** (2020) 012006 doi:10.1088/1757-899X/712/1/012006

connecting Kuala Lumpur and Johor to East Coast region meanwhile Station 2 was a main distributor road form Kuantan Port to the outside area.

Table 4. Traffic Volume Counts

Days	Station 1	Station 2
Monday	6421	816
Wednesday	5159	783
Friday	5819	558
Average	5800	719

In order to achieve the objective of this study, the traffic volume was then segregated into vehicle classification as to have a detail overview on the pattern compositions of heavy vehicles on both stations during those 3 days period. In Table 5 indicated the average traffic volume that was separated according to its vehicles class. C1 class is referred to cars/van/pick up, C2 for medium lorry, C3 for heavy lorry, C4 for bus while C5 is for motorcycle. Based on the tables 5, a total of 10% commercial vehicles of C2, C3 and C4 were observed in Station 1 meanwhile in Station 2 was about 50%. This was especially true because Station 2 was somehow the main distributor route for logistics activities as this route connecting outside areas to Kuantan Port where heavy vehicles were expected using this road more regular compare to the other vehicles class.

Table 5 Average Percentage of Vehicles Composition

Types of Vehicle	Station 1	Station 2
C1	78%	25%
C2	7%	36%
C3	2%	14%
C4	1%	0%
C5	12%	25%

4.2 Pavement Disability Level and Road Rating

The distress on the pavement surface at Station 1 and Station 2 has been conducted by visual observation and physical measurement. From the observation, the frequency and percentage of pavement damages within 3 km of identified station was recorded as shown in Table 6 and 7. The highest frequency of damages was seen on Station 2 (42) compare to Station 1(16). This is corresponding to several literature stated above where heavy vehicles have a significant impact on pavement fatigue life and distress. As in Station 2, where 50% of the volume of vehicles was significantly commercial vehicles with C2 36%, C3 14%, meanwhile in Station 1 the commercial vehicles only comprises of 10% with the C2 7%, C3 2% and C3 1%. The pattern of damage of both stations was quite similar where cracks were found to be the main type of damage with 75% in Station 1 and 36% in station 2. Cracks are one of the common types of road damages especially on federal roads [9]. There was no rutting found in station 1 but potholes were observed about 25%. Meanwhile in station 2 rutting was about 33% followed by potholes 31%. Based on the observation that has been done, rutting usually occurred near to the patching area and at the edge of the road.

Table 6. Pavement Damage at Station 1

Types of Damages	Frequency	Percentages %
Cracks	12	75
Potholes	4	25
Rutting	0	0
Total	16	100

Table 7. Pavement Damage at Station 2

Types of Damages	Frequency	Percentages %
Cracks	15	36
Potholes	13	31
Rutting	14	33
Total	42	100

The pattern of pavement of damages in this study was further assess to determine the seriousness of the road damaged based on road disability level and rating as shown in Table 1 and 2 above. As mentioned earlier the observation was made within 3 km at each identified station. This 3 km road section was then segregated into each 200 m that lead into 15 segmentation of road section. It is to make sure a uniformity of understanding on the damages types was under control. Figure 1 shows the illustration on the pavement damages within 3 km length at station 1 and 2. Marks was then given based on the level of seriousness of the damage as 1 (low), 2 (moderate) and 3 (high). Hence the road rating could be classifying as very satisfying condition to less satisfying condition.

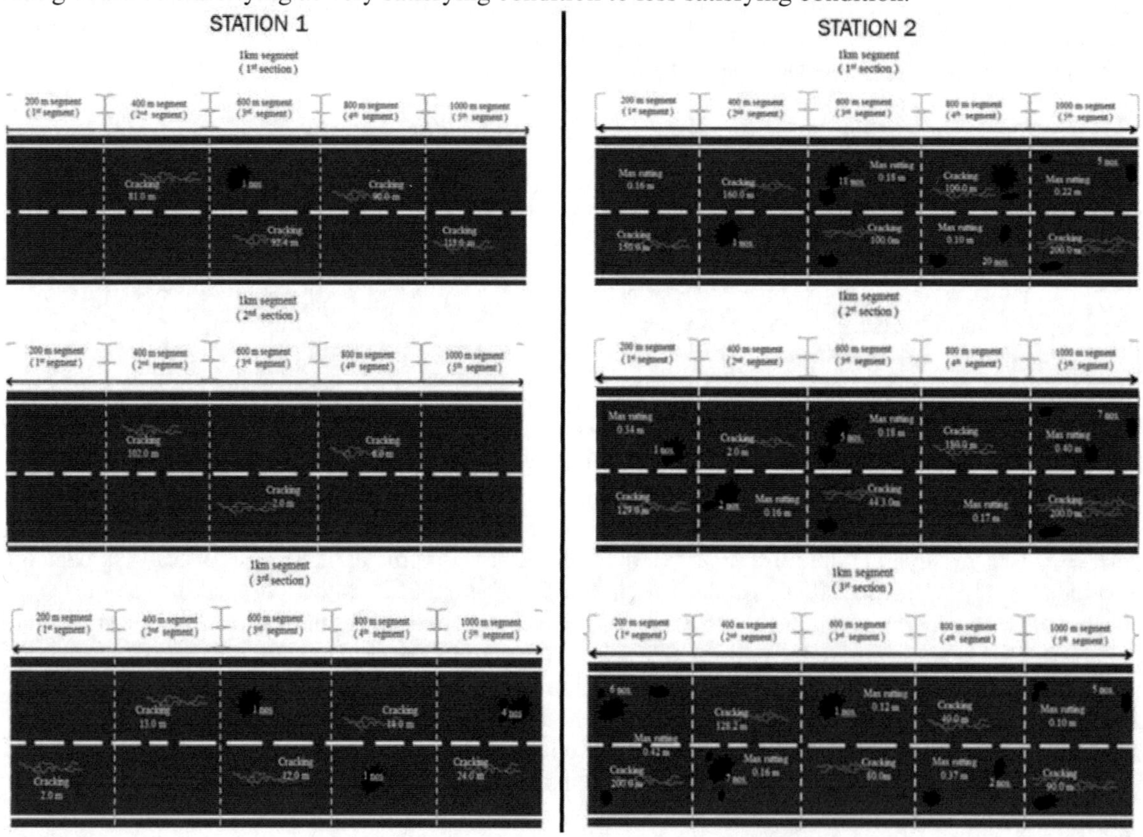

Figure 1: Illustration on Pavement Damages at Station 1 and Station 2

From the analysis it indicates that road rating for Station 1 as 1.73 which was in rating A indicates very satisfying condition. Meanwhile for Station 2 the road rating was 7.27 (rating C) which means less satisfying condition. It was shown that most cracking and rutting occurred at Station 2 was severe with maximum physical measurement of cracking was 200 m and rutting was 0.42 m in the 200 m road segmentation. In other words, along 200 m length of road at Station 2 the pavement surface condition was badly in damage conditions. This is in line with the study done by [10] mentioned that physical road surface condition of the main road leading to Port Klang was not satisfactory as a results of load bearing movement of logistics operations. Moreover, [11] and [12] also mentioned that increasing in vehicular traffic especially heavier vehicles and also traffic loading was among the factors influencing pavement damage. As shown in Table 4 above, though the analysis stated that Station 1 has the highest volumes of vehicles of 5800 compared to Station 2 of 719 however the significant different in patterns of vehicles composition on both station do plays some role in the road condition and environment as in this study was focusing on pavement surface condition.Vehicles compositions in Station 1 were dominantly conquer by class 1 by 78% followed by 10% commercial vehicles of C2, C3 and C4 and remaining 12% was C5. This is in contrast with Station 2 where the existence of light and heavy vehicles was at the same par of 50%. At Station 2 the light vehicles compositions of C1 and C5 was 50% meanwhile the remaining 50% was governing by commercial vehicles of C2 and C3. Consequently, high volumes of heavy vehicles encountered at Station 2 compare to Station 1.

This finding provides some overview on the operation of heavy vehicles which perceived to have an association on pavement damage. As stated by [13], pavement damage is the process by which distress develop in pavement under the combined effect of traffic loading. Deterioration of pavement greatly affects serviceability, safety and riding quality of the road. The pavement deterioration over time is caused by a combination of factors however; traffic loads play a key role in consumption of pavement life. As trucking has become the most popular mode of freight transportation because of its efficiency and convenience it has also contributed to most of pavement deterioration [14]. Therefore the possibility of heavy vehicles damaging the road is high.

According to [15] highlighted that the using of trucks with single and tandem axles have high potential of fatigue damage compared with the use of tridem and multiples axles of trucks. This is because the load carry by the trucks will distribute evenly among the axles hence the stress imposed on the pavement is reduced. In the case of mineral logistic operations, most trucks used were single and tandem axles. If the trucks is exceed the legislated maximum axle load limit or overloading, it will worsen the pavement condition. Figure 2 below shows a historical data on the association of peak hour traffic volume, heavy vehicles composition and number of damages occurred at Station 1 from 2014 to 2016. As discussed preceded, the minerals mining industry in Kuantan was well ascertained during those years. Operated as main road that provides essential link to Kuantan Port, the expectation on heavy vehicles generated on the road will supposed increasing tremendously. However the heavy vehicles composition was observed in decreasing pattern at which the lowest compositions was in 2016 with 8.5% of heavy vehicles. Meanwhile the number of pavement damage occurred was found the highest during that year. Based on this finding, it shows that heavy vehicles compositions attributed to minerals industry in Kuantan alone not really a significant factor contributes to the pavement damages.

Figure 2 below shows a historical data on the association of peak hour traffic volume, heavy vehicles composition and number of damages occurred at Station 1 from 2014 to 2016. As discussed preceded, the minerals mining industry in Kuantan was well ascertained during those years. Operated as main road that provides essential link to Kuantan Port, the expectation on heavy vehicles generated on the road will supposed increasing tremendously. However the heavy vehicles composition was observed in decreasing pattern at which the lowest compositions was in 2016 with 8.5% of heavy vehicles. Meanwhile the number of pavement damage occurred was found the highest during that year. Based on this finding, it shows that heavy vehicles compositions attributed to minerals industry in Kuantan alone not really a significant factor contributes to the pavement damages.

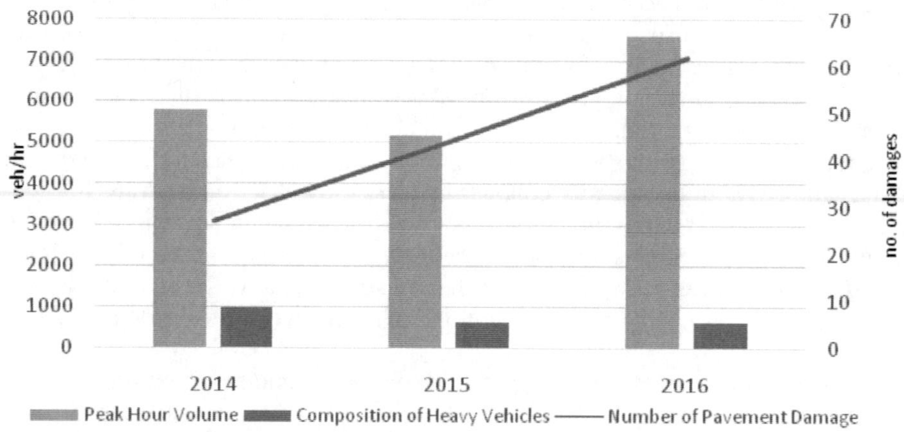

Figure 2. Association of Traffic Volume, Heavy Vehicles Composition and Pavement Damage

Nevertheless, based on the finding obtained from the field data by comparing two routes (Station 1 and Station 2) of significant different of heavy vehicles compositions with number of damage do provide insight to performance of the pavement. Looking at these scenarios, it is recommended to explore closely other factors that contribute to pavement damages.

5. Conclusion

From the analysis, it can be concluded that existence of heavy vehicles attributed to the minerals industry in Kuantan did not clearly the significance factors affect the performance of the pavement. However the compositions of heavy vehicles on the road do play a significant impact on the pavement surface condition. Based on the finding in this study, Station 1 comprises 10 % commercial vehicles with the C2 of 7%, C3 of 2% and C4 of 1%. Meanwhile for Station 2, 50 % of the vehicles class was significantly commercial vehicles with C2 of 36% and C3 of 14%. The road rating of pavement condition was very satisfying with rank A at Station 1. On the other hand, in Station 2 the road rating was less satisfying with rank C. This study gives insight on how compositions of commercial vehicles or heavy vehicles subjectively associated towards pavement condition. In order to reinforce the results, it is recommended to further analyze the pavement condition by looking at other factors such as quality of materials use, the details in design and construction process including weather and surrounding condition. With an addition of traffic generated by minerals industry in Kuantan, this study may provide an overview to the road authorities to enhance the quality of road environment and to improve Kuantan road network system thus creating sustainable land transport thereby helping to boost up the economic activities and quality of life.

6. References

[1] Hussain NH, Hashim Z, Hashim JH, Ismail N and Zakaria J 2016 *Int. Journal of Public Health and Clinical Sciences* **3**(5):174-89

[2] Lee KY, Ho LY, Tan KH, Tham YY, Ling SP, Qureshi AM, Ponnudurai T and Nordin R 2017 *Int. Med. Journal Malaysia* **16**(2)

[3] Abdullah NH, Mohamed N, Sulaiman LH, Zakaria TA and Rahim DA 2016 *The Malaysian Journal of Medical Sciences* (3):1

[4] Donoghue AM, Frisch N and Olney D 2014 *Journal Of Occ. and Env. Medicine* **56**(5 Suppl):S12

[5] Fares H, Shahata K, Elwakil E, Eweda A, Zayed T, Abdelrahman M and Basha I. 2012 *Str. and Inf. Engineering* **8**(11):**1067**-79

[6] Croney D and Croney P 1991 *The design and performance of road pavements* McGraw-Hill, Incorporated

[7] Adlinge SS and Gupta AK 2013 *Int. Journal of Innovative Research and Development.* **2**(4):437-50
[8] Karim MR, Ibrahim NI, Saifizul AA and Yamanaka H 2014 IATSS *research* **37**(2014):124-9
[9] Kordi NE, Endut IR, Wahab MY and Baharom B 2012 *IEEE Business, Engineering & Industrial Applications Colloquium* (pp. 430-434)
[10] Anor N and Ahmad Z 2010 *Malaysian Universities Transportation Research Forum and Conference.*
[11] Abdullah AS, Karim MR and Yamanaka H 2010 *17th ITS World Congress* ITS Japan ITS America ERTICO
[12] Judin AK 2011 *Road infrastructure development in Malaysia* Public Work Department Malaysia Kuala Lumpur
[13] Zumrawi MM 2016 *Journal of Civil Engineering and Technology* **7**(2), PP 203 – 214
[14] Bai Y, Schrock SD, Mulinazzi TE, Hou W, Liu C and Firman U 2010 *Estimating highway pavement damage costs attributed to truck traffic* Mid America Transportation Centre
[15] Chatti K, Salama H and El Mohtar C 2004 *Proceedings 8th International Symposium on Heavy Vehicle Weights and Dimensions* Document Transformation Technologies

Acknowledgments
The authors would like to express their gratitude to Universiti Malaysia Pahang (UMP) for providing financial assistance to this research through grant number RDU172205.

Risk Factors Influencing the Construction Projects in Yemen from Expert's Perspective

R A Bahamid[1], S I Doh[1], M A Al-Sharafi[2] and A R Rahimi[1]

[1]Faculty of Civil Engineering Technology, Universiti Malaysia Pahang, 26300 Gambang, Kuantan, Pahang, Malaysia
[2]Faculty of Computer Systems and Software Engineering, Universiti Malaysia Pahang, 26300 Gambang, Kuantan, Pahang, Malaysia Malaysia

Abstract. This purpose of this study is to validate the important risk factors influencing the construction projects in a developing country of Yemen by using experts' evaluations and literature review. These factors influence the project's quality, time and cost. The semi-structured interviews were employed through online means using thirteen experts to achieve the set objective. Depending on the expert's overviews, out of 56 considered factors, 42 factors were assigned as the most related risk factors influencing the construction projects in Yemen. The obtained outcomes from this study can serve as the risk factors for building blocks in establishing a conceptual model to manage risks of the construction projects in Yemen.

1. Introduction

Risk management in a construction project is very essential to achieve the objectives of the projects without any cost overrun or delay in order to attain achievement of high quality based on the design and planning of the project. The risk management system assists project managers in prioritizing resource allocation and helps them obtain reliable decision-making, which contributes to achieving the objectives and success in the project [1]. Implementing a risk management not only result in an enormous degree of awareness of the corollaries of risk but pivot on more structured strategy, improved conduct of risk data between parties and more efficient centralized control [2]. Furthermore, the construction industries play an essential responsibility in the economic state of the nation [3]. No construction project can be without a risk; this risk can either be managed and controlled, reallocation and shared, accepted or transferred, minimized; however, it cannot be overlooked [4]. Construction projects are exposed to external and internal risks. All these risks can be affected by quality, schedule or cost of the project in gloomy ways [5]. Just like other developing countries, the construction projects in Yemen are dramatically experiencing a high risk prone due to their dynamic and complex environments. Hence, such nature is creating an atmosphere of high risks and uncertainties [6]. The construction project in Yemen is the fourth substantially employer of personnel in the nation; this equivalent to 9-10% of the working population, the average annual growth rate of the sector is about 5.4% which contribute to the nation's economic growth. The service demand in this sector is bounded to improve evenly. Nonetheless, the actual output in terms of quality, progress and fun utilization is reduced primarily because of the lack of proper project management, efficient monitoring and supervision and greatly depend on the contractors [6]. Therefore, it is very infrequent that the construction project will be finished using the allocated cost and time [7]. Several construction projects in Yemen are suffering from a delay in completion or mostly suffer abandonment or suspension. Continuing in this type of situation can cause a widened practices and construction costs between the developed and developing countries; resulting in enormously difficult for construction workers in Yemen to implement, adopt and afford the implementation of modern sustainable construction technology [8]. Thus, the risk management needs to be well identified and managed as an integrated function of the project management at project lifecycle. Here lies the importance of this study in identifying the risks facing the construction projects in Yemen to benefit the construction industry and the Yemeni economy at large.

2. Methods

The first step is the identification of risk factors affecting the construction projects in developing nations; this was carried out through the literature review [9]. The aim is to recognize the important factors influencing the construction projects of the developing countries. In the previous study, we extracted the factors from the literature [9]. About 43 research articles that studied the factors and current relevant parts that affect the construction projects were identified. hence, 56 factors were identified and categorized out of 111, depending on the most common and related factors that were explained in the previous findings [9]. Next is the validation of the recognized factors obtained from the literature review to the construction projects in Yemen; this was done by carrying out the semi-structured email interviews. Meho [10] had reported that the email interview can be a suitable alternative in place of face-to-face and telephone interview because the availability to individuals is easy especially those that cannot interview because of the location or time. Moreover, it removes the challenges of fixing appointments to transcribe the interviews; this makes the interview to be easier with multiple participants [10]. In addition, an email interview gives the participants the opportunity to review and provide adequate responses. Also, this type of interview helps the careful presentation of equal materials and questions to every participant which can potentially yield excellent outcomes. Other benefits of email interview are cost saving of travelling and organization [11]. In order to stimulate the participants to take part in this study, they were contacted through the email invitations; the interview questions and consent forms were sent across to them through email. About 13 experts were chosen to rank and reconfirm the factors extracted from previous findings. These experts were from local construction industries situated in Yemen with about 15 years of working experience in their respective fields. Table 1 comprises the detailed information of all the participants. The identities of the participants were removed and the peculiar ID of identifying individual quotes was provided because of confidentiality.

Table1. Experts' profiles

Code	Organization Type	Profession	Years of Experience	Experience of construction project types
EXP1	Private	Civil engineer	Above 20	-Housing -Administrative and commercial -Infrastructure
EXP2	Government	Quantity survey	16-20	-Administrative and commercial - Infrastructure
EXP3	Private	Civil engineer	16-20	-Housing -Administrative and commercial

EXP4	Private	Civil engineer	Above 20	Industrial
EXP5	Government	Quantity survey	Above 20	-Administrative and commercial -Industrial -Infrastructure
EXP6	Government	Mechanical engineer	Above 20	-Housing -Administrative and commercial -Industrial
EXP7	Private	Architect	16-20	-Housing -Infrastructure
EXP8	International	Architect	16-20	-Infrastructure -Administrative and commercial -Industrial
EXP9	Private	Electrical engineer	Above 20	-Housing -Administrative and commercial -Industrial
EXP10	Private	Quantity survey	16-20	-Housing -Administrative and commercial
EXP11	Government	Civil engineer	Above 20	-Housing -Administrative and commercial -Infrastructure
EXP12	Private	Architect	Above 20	-Administrative and commercial
EXP13	Private	Civil engineer	16-20	-Housing -Administrative and commercial -Infrastructure

The interview questions were generated using previous studies. In order to identify the risks that affect construction projects, the anticipated responses were meant to authenticate and pinpoint the utmost vital risk factors that influence time, cost and quality of the construction projects in Yemen. For each question, the responses were weighted using a 7-point Likert scale (1 = Not at all Important, 2 = Low Importance, 3 = Slightly Important, 4 = Neutral, 5 = Moderately Important, 6 = Very Important, 7 = Extremely Important). Moreover, open-ended questions were provided within the interview survey to generate the lists of verification benchmark, obtain more factors other than the identified ones and obtain the indirect details from the participants. Notwithstanding, there were no new factors established from the interview survey, but the experts suggested that some factors must be merge as one factor; this includes war and political instability, exchange rate and inflation. Median rank was used as the basis of establishing the priority. The median was taken over mean because of its strength to the skewed distribution of the priority ratings. Based on the participants' views, the values allocated to individual factor were from a ratio scale of 1 to 7, where 1 and 7 indicate the lowest and highest importance, respectively.

3. Result and discussion

In Table 2, the greater score in the precedence was the most imperative factor based on the considered criteria. The results show that the participants recognised the greatest risk factors influencing the construction projects in Yemen to be inflation and exchange rate fluctuation, war and political instability, inadequate program scheduling, late handing over of the site, and insufficient resource availability. Conceivably, exchange rate fluctuation and inflation, war and political instability are the greatest risk factors whereby the mean was 6.08 and 5.77 because of the current situation in Yemen, which has affected the construction industry. Moreover, labour disputes and strikes, changes in laws and regulations, environment regulations procedures, public relationship, and culture difference did not play the important role for the construction projects in Yemen; thus, they have been neglected. The participants also recommended merging some factors which are: inflation and exchange rate fluctuation, war and political instability, defective work and quality of work.

Table 2. Means, standard deviation and rank of factors depending on of respondents' interviews.

Factors	Mean	S.D.	Rank	%
Turnover of skilled labour	4.77	1.36	23	68.13
Labour and equipment productivity	4.92	1.50	19	70.33
Accidents /safety	4.00	1.41	40	57.14
Defective materials	4.85	1.28	22	69.23
Damage to equipment	4.69	1.18	25	67.03
Damage to structure	3.62	1.26	47	51.65
Adverse weather conditions	3.38	1.26	48	48.35
Environment regulations procedures	2.69	1.33	54	36.26
Environmental hazards of the project	3.00	1.41	51	42.86
Force majeure/act of god	3.69	1.32	44	52.75
Difficulty to access the site	3.92	1.12	41	56.04
Different site conditions	4.85	1.46	20	69.23
Conflict of document	4.62	1.45	27	65.93
Defective design	4.77	1.01	24	68.13
Incomplete design scope	4.23	1.09	34	60.44
Unqualified designers	4.31	1.80	30	61.54
Design complexity	4.00	1.00	39	57.14
Inaccurate cost estimate	5.00	1.35	12	71.43
Contractors competition in bids	5.00	1.08	13	71.43
Inadequate program scheduling	5.69	1.32	4	81.32
Insufficient resource availability	5.46	0.88	6	78.02
Lack of Scope of work definition	4.15	0.99	37	59.34
Technology issues	4.62	0.77	28	65.93
Inflation/ Price fluctuation	6.00	0.71	2	85.71
Client's financial failure	4.23	1.74	35	60.44
Delayed payment on contract	5.00	1.29	14	71.43
Exchange rate fluctuation	6.08	0.76	1	86.81
Financial default of contractor	4.92	1.12	15	70.33
Labour disputes and strikes	2.92	1.19	52	41.76
Legal disputes	3.69	1.32	45	52.75
Delayed dispute resolutions	4.92	1.50	16	70.33
Owner demand changes and intervention	4.15	1.41	38	59.34
Requirement or delay for permits and approval	3.85	1.34	43	54.95
Changes in laws and regulations	2.69	1.55	53	38.46
Taxes and tax burdens	3.08	1.38	50	43.96
Conflict between actual quantities and contract quantities	4.92	1.80	18	70.33

IOP Conf. Series: Materials Science and Engineering **712** (2020) 012007 doi:10.1088/1757-899X/712/1/012007

Gaps between the Implementation and the specifications	4.62	1.89	29	65.93
Design changes	3.92	1.32	42	56.04
Quality of work	4.23	1.36	36	60.44
Defective work	4.31	1.32	31	6.54
Contractor competence	5.15	1.07	8	73.63
Change order negotiations	4.31	1.60	32	61.54
Corruption and bribery	4.92	1.55	17	70.33
Criminal acts	3.08	1.44	49	43.96
War and civil disorder	5.77	1.17	3	82.42
Government acts and intervention	3.69	1.49	46	52.75
Political instability	4.85	1.28	21	69.23
Economic instability	5.08	1.12	11	72.53
Construction Methods	5.08	1.55	10	72.53
Poor project planning & control	5.15	1.77	7	73.63
Poor communication between involved parties	5.08	2.06	9	72.53
Poor site management and supervision	4.62	1.56	26	65.93
culture difference	1.92	0.86	56	27.47
Public relationship	2.31	1.44	55	32.97
High performance or quality expectations	4.31	1.60	33	6.54
Late handing over of the site	5.54	1.13	5	79.12

Table 3 Risk factors influencing the construction projects in Yemen.

Risk category	Risk factors	Risk category	Risk factors
Environmental	Adverse weather conditions		Delayed payment on contract
	Environmental hazards of the project		Exchange rate fluctuation and Inflation
	Force majeure/act of god		Financial default of contractor
	Difficulty to access the site	Legal*	Legal disputes
	Different site conditions		Delayed dispute resolutions
Physical	Accidents /safety		Requirement or delay for permits & approval
	Defective materials		Taxes and tax burdens
	Damage to equipment	Political*	War and Political instability
	Labour and equipment productivity		Government acts and intervention
	Damage to structure		Economic instability
	criminal acts		corruption and bribery
Logistic*	Turnover of skilled labour	Construction*	Conflict between actual quantities and contract quantities
	Conflict of document		Gaps between the Implementation and the specifications
	Contractors competition		Defective work

	in bids		
	Inadequate program scheduling		Change order negotiations
	Insufficient resource availability		Construction Methods
	Lack of scope of work definition		Late handing over of the site
Design	Defective design	Management	Technology issues
	Incomplete design scope		Owner demand changes and intervention
	Unqualified designers		Contractor competence
	Design complexity		Poor project planning and control
	Design changes		Poor communication between involved parties
Financial*	inaccurate cost estimate		Poor site management and supervision
	Client's financial failure		High performance expectations

4. Conclusion
The identification of risk factors influencing the construction projects in Yemen is crucial it can help in achieving the set goals of projects which are time, cost and quality. Based on the review of literature which assisted in the identification of 111 factors influencing the construction projects in developing countries, 56 factors were selected using the highest frequency in the previous studies [9]. In order to validate and reconfirm the factors and select the most appropriate to Yemen construction industry, a semi-structured interview using 13 experts was employed. The validation result as provided in Table 3 with 48 factors are the risk factors influencing the construction projects in Yemen. These factors were grouped based on adoption from the previous studies [12-15]. Thus, the results of this study can give an improved understanding of the factors influencing the construction projects in Yemen. Additionally, the outcomes from this study can contribute to the construction projects in developing countries including the Yemen construction industry; the important risk factors that can influence the success of a project.

5. References
[1] A. Kassem M, Khoiry M A & Hamzah N 2019 Risk factors in oil and gas construction projects in developing countries: A case study. *International Journal of Energy Sector Management*. Vol. 13, No. 4, pp. 846-861.

[2] Goh C S & Abdul-Rahman H 2013 The identification and management of major risks in the Malaysian construction industry *Journal of Construction in Developing Countries*. Vol. 18, No. 1, pp. 19-32.

[3] Alaghbari W, Al-Sakkaf A A & Sultan B 2019 Factors affecting construction labour productivity in Yemen. *International Journal of Construction Management*. Vol. 19, No. 1, pp. 79-91.

[4] Bahamid R A & Doh S I 2017 A review of risk management process in construction projects of developing countries. *IOP Conference Series: Materials Science and Engineering*. Vol. 271, pp. 012042.

[5] Khoiry M A, Kassem M A & Hamzah N 2018 A pilot study on identify risk factors in oil & gas construction projects *Proceedings of Academic World 87th International Conference, Kuala Lumpur, Malaysia*. pp. 16-21.

[6] Ahmad S A, Issa U H, Farag M A & Abdelhafez L M 2013 Evaluation of risk factors affecting time and cost of construction projects in Yemen. *International Journal of Management*. Vol. 4, No. 5, pp. 168-78.

[7] Al-Fadhali N, Zainal R, Kasim N, Dodo M, Soon N K & Hasaballah A H A 2019 The desirability of Integrated Influential Factors (IIFs) Model of internal stakeholder as a panacea to project completion delay in Yemen. *International Journal of Construction Management*. Vol. 19, No. 2, pp. 128-136.

[8] Issa U H, Farag M A, Abdelhafez L M & Ahmed S A 2015 A risk allocation model for construction projects in Yemen. *Civil and Environmental Research*. Vol. 7, No. 3, pp. 78-88.

[9] Bahamid R A, Doh S I & Al-Sharaf M A 2019 Risk factors affecting the construction projects in the developing countries. *IOP Conference Series: Earth and Environmental Science*. Vol. 244, pp. 012040.

[10] Meho L I 2006 E- mail interviewing in qualitative research: A methodological discussion. *Journal of the American Society for Information Science and Technology*. Vol. 57, No. 10, pp. 1284-1295.

[11] Stacey K & Vincent J 2011 Evaluation of an electronic interview with multimedia stimulus materials for gaining in-depth responses from professionals. *Qualitative Research*. Vol. 11, pp. 605-624.

[12] Enshassi A & Mosa J A 2015 Risk management in building projects: owners' Perspective. *The Islamic University Journal (Series of Natural Studies and Engineering)*. Vol. 16, No. 1, pp. 95-123.

[13] Enshassi A, Mohamed S & Mosa J A 2008 Risk management in building projects in Palestine: Contractors' perspective. *Emirates Journal for Engineering Research*. Vol. 13, No. 1, pp. 29-44.

[14] Odimabo O 2016 Risk management system to guide building construction projects in developing countries: A case study of Nigeria. *Thesis Submitted to University of Wolverhampton*. pp. 1-313.

[15] Sharma S K & Swain N 2011 Risk management in construction projects. *Asia Pacific Business Review*. Vol. 7, pp. 107-120.

Acknowledgement

The support provided by Malaysian Ministry of Higher Education and Universiti Malaysia Pahang in the form of a research grant (RDU/UMP) vote number RDU190151 for this study is highly appreciated.

IOP Conf. Series: Materials Science and Engineering **712** (2020) 012008 doi:10.1088/1757-899X/712/1/012008

Numerical study of slope protected by rockfill and shredded tyre mixture gabion wall

A C Ainul Mardziah and A Amizatulhani

Faculty of Civil Engineering Technology, Universiti Malaysia Pahang, 26300 Gambang, Pahang, Malaysia

Abstract. Abundance of discarded scrap tyre in Malaysia demand diverse recycling products, in order to reduce its environmental pollutions and simultaneously the negative aftermaths to society and future generation. Hence, this paper is established to investigate the potential of tyre derived geo-material as rock fill alternative in conventional gabion type retaining wall, replacing a portion of the non-renewable rock fill. 4-tier scrap tyre gabion wall (STG wall) in terraced position with various scrap tyre ratio from total gabion rock fills volume are modelled to retain a 5.0m height of existing failed slope. The model are analysed by finite element method in Slope/W 2007 software (Morgenstern-Price analysis method) to assess overall stability of the gabion wall and factor safety (FOS) of the treated slope. From this study, higher FOS is acquired when STG wall incorporated into the failed slope compared to using conventional gabion wall.

1. Introduction

The debris volume at Malaysia landfills consistently increasing year by year consistent with rapid development and population growth. Waste disposal sourcing from construction, clinical (bio-medical), electronics, household, hazardous waste and scrap tyre are mostly end up in landfill [1]. It is found that the quantity of scrap tyre to be recycle is insignificant compared to the generation of scrap tyre [2]. At Port Klang, over hundreds of container cargos fill with scrap tyre have been shipped and left abandoned there at Northport and Westport [3]. Local authority mentioned that scrap tyre often mixed with other domestic waste in landfills, especially motorcycles tyre. The unusable tyre in open space attract disease-carrying pests, also has high possibility to pollute soil and groundwater. Imprudent practice of scrap tyre dumping has caused Sabah City Hall to conduct regular fogging activity to prevent the state landfills transform into mosquito habitat [4]. Improper way to decompose it may cause harm to human health and environment. Hence, ways are proposed to increase the recycling rate of scrap tyre while eliminating the hazardous effect. Recent studies proposed the application of scrap tyre as part of geotechnical solution. [5] has studied the compatibility of approximate 5% rubber crumb mixture and polypropylene short fibres towards concrete pavement. Scrap tyre mixed sand are also show potential as lightweight backfill material, replacing a portion of sand material, due to its mechanical properties [6–9]. Meanwhile, this paper focus on providing perspective into application of rock fill shredded scrap tyre mixture in gabion wall. The objectives of the study include assessment of the gabion wall made of rock fill and shredded scrap tyre effectiveness to rehabilitate slope failure also to study the effectiveness of percentage rockfill:tyre mixture ratio in protecting the failed slope.

IOP Conf. Series: Materials Science and Engineering **712** (2020) 012008 doi:10.1088/1757-899X/712/1/012008

2. Method

The studied slope data used in this paper are based on actual approximate 5.0m height slope failure event occurred in May 2016 nearby Hutan Pendidikan Bukit Gasing, Petaling Jaya, Selangor as shown in figure 1. The details of the failed slope and its subsoil properties are implied in this study.

(a) (b)

Figure 1. Slope failure at Bukit Gasing, Petaling Jaya, Selangor.

Two slope model are developed and assessed numerically in this paper, a 5.0m height of existing slope without any treatment and the slope model retained with a 4.0m height gabion wall. Engineering software, Geostudio Slope/W 2007 are utilised throughout the analysis, applying Morgenstern-Price type to analyse the slip circle of the slope and provide the global and local Factor of Safety (FOS).

2.1. Slope modelling details

Figure 2 shows slope model without any treatment. Meanwhile, figure 3 display slope model details retained with gabion wall in terraced position. Dimension used throughout the analysis are as per indicated. Considering the critical situation of continuous prolonged rainfall, full level of groundwater is opted in the slope models.

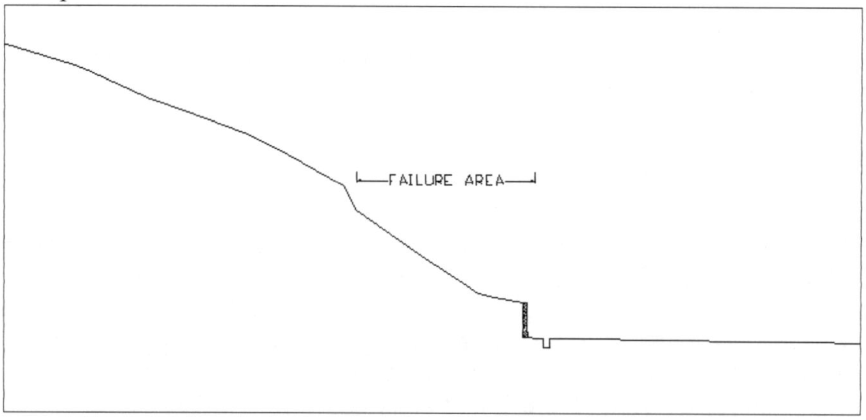

Figure 2. Existing slope model cross-section (without treatment).

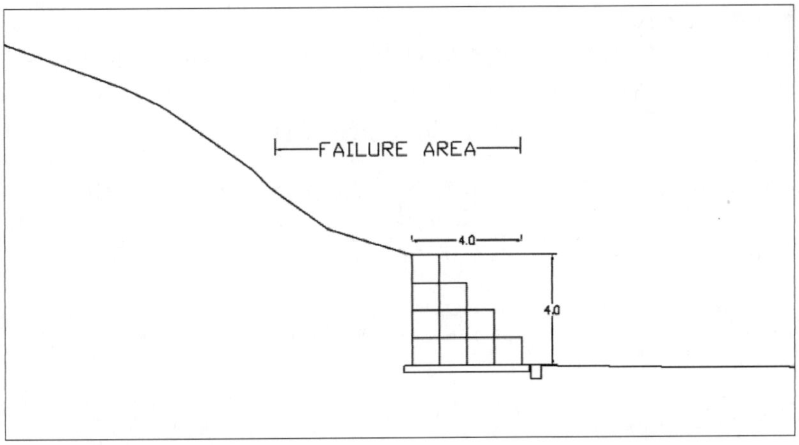

Figure 3. Treated slope model cross-section (with gabion wall).

2.2. Material size

The individual rock fill dimension ranging from 20mm greater than wire mesh opening and maximum dimension of 250mm [10]. According to [10], the wire mesh netting cage shall have minimum 2.70mm diameter with maximum mesh size of 100mm x 120mm, sufficiently coated with polyvinyl chloride (PVC) with minimum thickness of 0.55mm. No consideration is made for the wire mesh strength in analysis as per stated in BS 8002:1994. In analysis, tyre shreds of typical size ranging from 76mm to 305mm are utilised to replace a portion of rock fill volume inside the 1.0m x 1.0m x 1.0m standard size of wire mesh netting cage. Dimension of scrap tyre is assumed to have similar particle size distribution and the STG are assumed to be homogenous in the STG cage.

2.3. Material parameters

Available soil investigation (S.I) report are referred to determine the soil type and parameters. Table 1 summarize the Sandy SILT soil type parameter details.

Table 1. Soil parameters.

Soil layer	Unit weight, γ (kN/m^3)	Cohesion, c (kN/m^2)	Internal friction angle, θ (^{o})	Legend
SPT <20	18	1	29	
SPT 20-50	18	2	30	
SPT 50	19	4	32	

Mechanical properties of the rockfill shredded scrap tyre (STG) mixture used in this study are based on laboratoty test conducted. Unit weight value, γ of STG mixture are determined via Proctor Test while cohesion, c and internal friction angle, θ are obtained from Direct Shear Test. Granulated scrap tyre are purchased from a recycle tyre retread factory in Kuantan, Pahang for the laboratory tests. table 2 summarize the shredded scrap tyre parameters based on laboratory results.

Table 2. Soil parameters.

Model scrap tyre:rockfill mixture	Unit weight, γ (kN/m^3)	Cohesion, c (kN/m^2)	Internal friction angle, θ (^{o})
STG Wall of 0:100 ratio	16.48	1	30
STG Wall of 20:80 ratio	16.68	4	29

2.4. Rockfill shredded tyre mixture

A total of three (3) number of slope model are modelled and analysed in this paper. Table 3 lists the series of slope analysis with different mixture ratio.

Table 3. Soil parameters.

Model scrap tyre:rockfill mixture	Unit weight, γ (kN/m^3)	Cohesion, c (kN/m^2)	Internal friction angle, θ (^{o})
STG Wall of 0:100 ratio	16.48	1	30
STG Wall of 20:80 ratio	16.68	4	29

3. Results and discussion

3.1. Slope stability analysis results

The acceptability of a slope's stability is based on adequate Factor of Safety (FOS) against slope failure. As per Guideline for Slope Design, by Public Works Department Malaysia (see Figure 4), slope is considered stable if the Factor of Safety (FOS) is ≥ 1.5.

DESIGN COMPONENT	MODE OF FAILURE	MINIMUM FACTOR OF SAFETY	MAXIMUM PERMISSIBLE MOVEMENTS: VERTICAL	LATERAL	DIFFERENTIAL
1. Unreinforced Slopes	1.1 Local & Global Stability (cut & fill slopes) 1.2 Bearing (fill)	1.3 2.0	Analysis should be according to GEOTECHNICAL MANUAL FOR SLOPES (1984), GEO Hong Kong		
2. Reinforced or Treated Slopes (not on soft ground)	2.1 Local & Global Stability (cut & fill slopes) 2.2 Bearing (fill)	1.5 1.5			
3. Permanent Anchors	3.1 Tensile Resistance 3.2 Resistance at Soil Grout Interface 3.3 Creep/Corrosion	2.0 3.0	Geo Spec 1 (1989), GEO Hong Kong BS 8081		
4. Rigid Retaining Structures	4.1 Overturning 4.2 Sliding 4.3 Overall Stability 4.4 Bearing	2.0 1.5 1.5 2.0	15mm along face of wall Geoguide 1 (1983), GEO Hong Kong	15mm along face of wall	1 : 150 along face of wall
5. Reinforced Fill Walls/Structures	External Stability Internal Stability	BS 8006	± 5mm per metre height	± 15mm from reference alignment	1 : 100 along face of wall
6. Individual Foundation Piles (mainly under axial loads)	6.1 Shaft Resistance 6.2 Base Resistance	2.0 2.0	12mm along axis of pile at pile head at design load. 38mm or 10% pile size at pile head at twice design load. BS 8004		
7. Individual Foundation Loads (mainly under lateral & bending loads perpendicular to axis of pile)	Ultimate Lateral Resistance	2.5	12mm along axis of pile at pile head at design load. BS 8004	12mm perpendicular to axis of pile at design load	
8. Pile Group	Block Bearing Capacity	2.0	12mm at Working Load BS 8004		
9. Piles as Retaining Structures	As for 4, 6 & 7 above	As for individual foundation piles	As 4 above for rigid retaining structures BS 8004		
10. Embankment on Soft Ground	10.1 Bearing (short term) 10.2 Local & Global Slope Stability (long term)	1.4 1.2	7 years post construction settlement : (i) within 10m from bridge approach < 100mm (ii) road < 250mm		

Figure 4. Geotechnical Design Criteria for Slopes Design (Jabatan Kerja Raya Malaysia, 2010).

From the back analysis carried out in Geostudio Slope/W 2007 software, it is proven that the slope is failed with FOS 0.956≤1.5 (refer figure 5). Hence, 4.0m height of gabion wall is modelled and analysed. The results shown that the FOS value has slightly increased by 0.2 % applying 20% volume of scrap tyre derived geomaterial, comparing with conventional type of gabion wall. Table 4 summarize the analysis results.

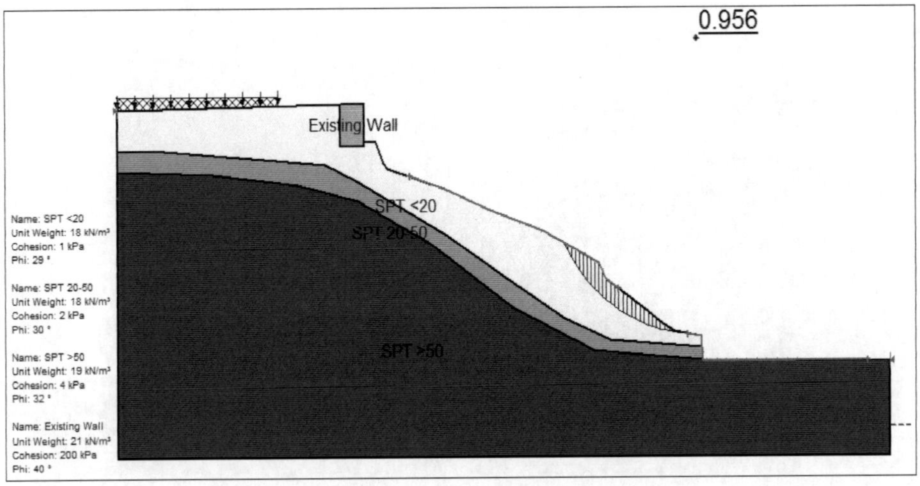

Figure 5. Existing failed slope.

Table 4. Series of slope models (Variance in scrap tyre:rockfill mixture ratio).

Model ID	Factor of Safety
1	0.956 ≤1.5 (Inadequate)
2	1.502 ≥1.5 (Satisfactory)
3	1.505 ≥1.5 (Satisfactory)

Meanwhile, the critical slip circle of slope is also observed. Figure 6 and figure 7 illustrate the output analysis from Geostudio Slope/W 2007. It can be seen that STG wall possess shallower slip circle compared to conventional gabion wall.

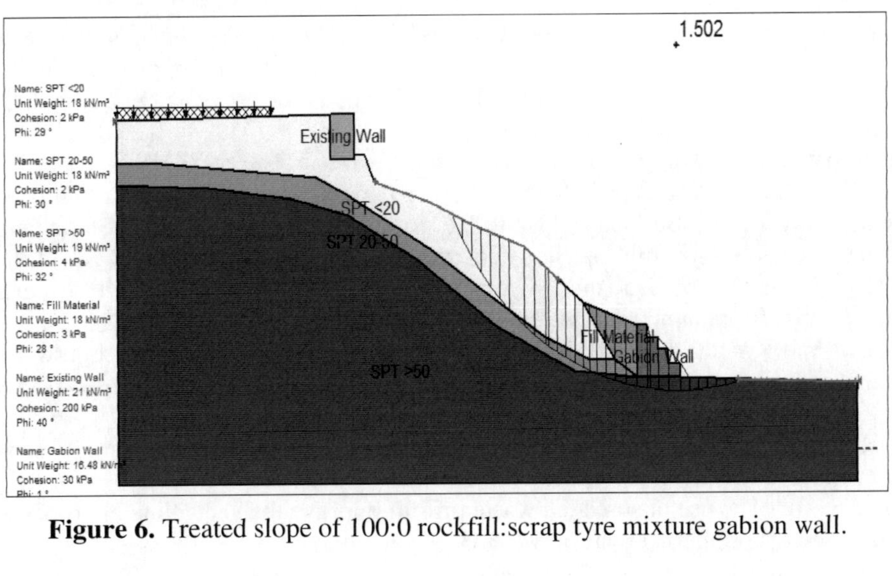

Figure 6. Treated slope of 100:0 rockfill:scrap tyre mixture gabion wall.

IOP Conf. Series: Materials Science and Engineering **712** (2020) 012008 doi:10.1088/1757-899X/712/1/012008

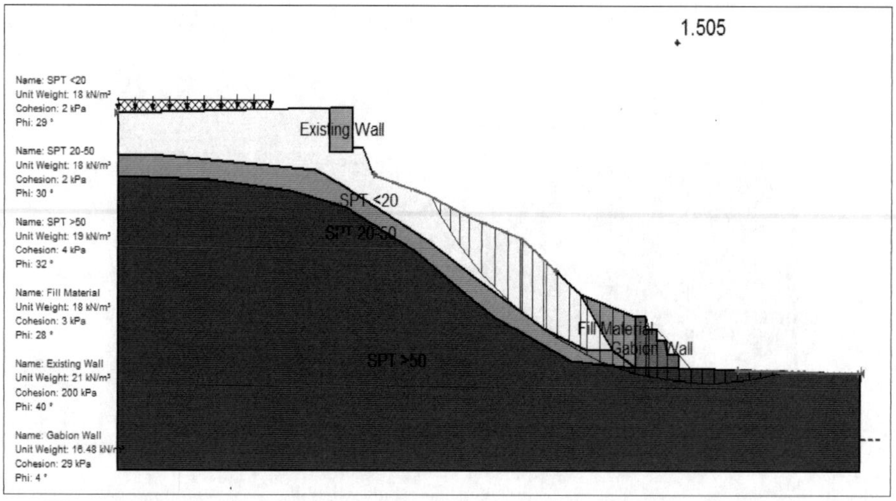

Figure 7. Treated slope of 80:20 rockfill:scrap tyre mixture gabion wall.

4. Conclusion

Numerical modelling and analysis of rockfill shredded scrap tyre mixture gabion type of retaining wall has been carried out. Slip circle and slope Factor of Safety (FOS) are studied and assessed, comparing the results with conventional gabion wall. From this study, STG wall resulted in higher FOS and caused shallower critical slip circle. Overall, scrap tyre derived geomaterial has potential as substitute material to replace a portion of non-renewable rockfill in gabion cages. However, further detailed analysis on the STG wall effectiveness shall be conducted, considering other parameters such as variance in soil parameters, scrap tyre:rockfill mixture ratio and retained heights. Small scale of laboratory experiments is also recommended to be conducted in order to verify the numerical analysis results.

5. References

[1] Performance Management and Delivery Unit 2015 *Solid Waste Management 2015 Final Report*

[2] Chemsain Konsultant 2011 *A study on scrap tyres management for peninsular Malaysia for National Solid Waste Management Department Ministry of Housing and Local Government* (Selangor: Chemsain Konsultant Sdn Bhd)

[3] Hemananthani S 2017 *Port Klang used as dumping ground* (The Star Online)

[4] Sario R 2015 *Sabah's RM700mil plan for environ-friendly landfills* (The Star Online)

[5] Hernandez-Olivares F, Barluenga G, Bollati M and Witoszek B 2002 Static and dynamic behaviour of recycled tyre rubber-filled concrete. *Cement. Concrete. Res.* **32** 10 p 1587–96

[6] Vanessa C, Linda G, Ase H, Joanne W and Krishna R 1996 Use of shredded tires as lighweight backfill material for retaining structures. *Waste. Manag. Res.* **14** 5 p 433–51

[7] Reddy S B and Krishna A M 2015Recycled tire chips mixed with sand as lightweight backfill material in retaining wall applications: an experimental investigation *Int. J. of Geosynthetic Gnd. Engr.* **1** 4 (Springer International Publishing) 2199-9279

[8] Sneha R C and Jiss K A 2018 Numerical modelling and analysis of retaining wall with crumb rubber as backfil*Int. Rsrsch. J. of Eng. And Tech.* **5** 4 p 4721–24

[9] Attom M F 2016 The use of shredded waste tires to improve the geotechnical properties of sands *Env. Geol.* **49** 4 (Springer Verlag) p 497–503

[10] Slope Engineering Branch 2010 *JKR Guidelines for Slope Design* (Public Works Department Malaysia) p 14

Acknowledgments

Authors expressing gratitude to all contributors who have been involved directly or indirectly in publishing the paper. A special thanks goes to Universiti Malaysia Pahang for providing financial support under grant of RDU1703305.

IOP Conf. Series: Materials Science and Engineering **712** (2020) 012009 doi:10.1088/1757-899X/712/1/012009

Performance of Coir Fiber Addition for Clay as a Sub-Grade for Pavement Design

R P Munirwan[1], Munirwansyah[1], Marwan[1], P J Ramadhansyah[2] and V Kamchoom[3]

[1]Department of Civil Engineering, Faculty of Engineering, Universitas Syiah Kuala, 23111 Banda Aceh, Indonesia
[2]Faculty of Civil Engineering Technology, Universiti Malaysia Pahang, 26300 Gambang, Pahang, Malaysia
[3]Department of Civil Engineering, King Mongkut's Institute of Technology Ladkrabang, Thailand

Abstract. Clay soil behavior often becomes problematic with building construction, it is about the shrink-swell behavior in clay when influenced by water content. Coconut husk fiber (coir fiber), in general, is an industrial waste which is still infrequent to be re-used. This study aims are to determine the effect of coir fiber addition for mechanical stabilization of clay soil in terms of CBR (California Bearing Ratio) value. Soil samples used for laboratory tests were collected from Ulee Glee area of Pidie Jaya Regency. According to AASHTO classification, the soil category is A-7-5 (25) while for USCS classification, the soil is OH (Organic High). The percentage of added coir fiber was 0%, 0.2%, 0.4% and 0.6% of the dry weight of the soil with a coir fiber length of 2 and 3 cm. Two treatments of clay-coir fiber mixing method were applied namely direct mixing and mixing by layers. The results of natural soil compaction test obtained that OMC (Optimum Moisture Content) value and dry soil weight ($\gamma_d$max) was 26.8% and 1.34 gr/cm^3 respectively. The highest CBR results were obtained for clay soil with 0.4% coir fiber 3 cm (direct mix) with CBR value = 17.7%. Furthermore, the lowest CBR value is 10% for percentage of 0.2% with coir fiber length 2 cm (by layer). In general, high organic clay soil with coir fiber mixture addition is able to increase the CBR values if compared to CBR of natural soil which is 8.15%. Thus, the use of coir fiber in this study is able to improve soil bearing capacity which is useful for construction material in the site.

1. Introduction

In civil engineering works, the ground must have sufficient bearing capacity both as a ground construction and as a construction material. As a construction material, generally, the soil used as embankment material is taken from the borrow area, which is generally in the form of cohesive soil such as clay. The problem of soil characteristics for construction may because of environmental conditions. Tropical season of dry and wet seasons give a great influence on soil behavior. Especially if dealing with expansive clay soil. Clay has properties that are very sensitive to the influence of water content and have high shrink-swell behavior properties.

Several latest researches to improve the soil strength by addition of chemicals and fiber materials have been done so far. [1] studied the effect of palm oil clinker concrete pile with foamed concrete pile as a soft soil floating foundation. The used of lime-clay stabilization for mechanical properties improvement of expansive soil was conducted by [2]. [3] studied the shear strength of soft clay with the reinforcement by encased lime bottom ash column. The result shows that soft clay shear strength improves by the installation of a single column of encapsulated lime bottom ash.

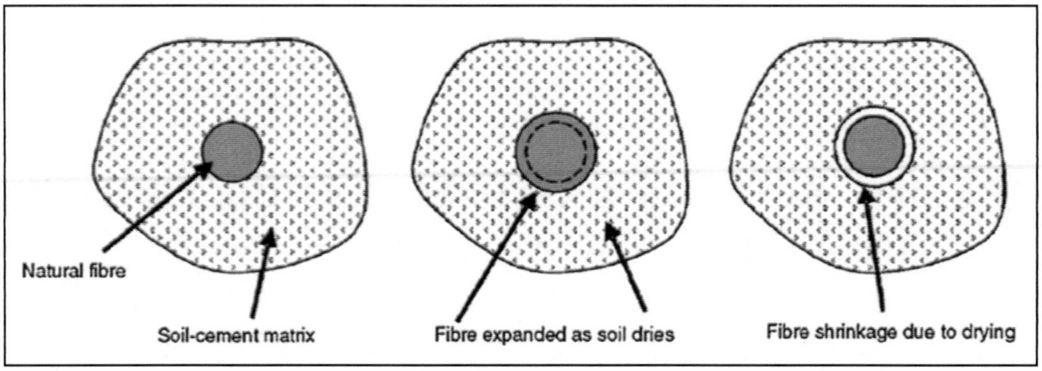

Figure 1. Influence of water absorption on natural fiber

Natural fiber used for soil improvement were already implemented by several previous research. Bagasse fiber mixed with hydrated lime was studied by [4] for expansive soil treatment. The result obtained that compressive strength of expansive soil increase by the increment of fiber-lime to the soil. Moreover, [5] found the improvement of clay soil geotechnical properties mixed with cement-rubber fiber which containing 6% of cement and up to 7.5% rubber fiber that can be used as civil engineering work material. [6] studied the effect of lime-rice husk ash and waste plastic fiber for engineering properties of silty soil. They found that the mixture ranges from 0.4-08% of dry mass were able to increase the value of compressive, tensile, and shear strength of silty soil properties.

Table 1. Coir Fiber Characteristics [11]

Characteristics of granulated	Results
Average length of the granulated one	10-20mm
Average thickness	0.1mm
Amount (percentile in weight)	0.5-0.7%
Ph	5.4
Electric Conductivity	1.8dS/m
Capacity of cationic exchange	92
Relation C/N	132
Specific mass	70g/L
Water retention	538ml/L
Capacity of aeration	45.5%
Porosity	95.6%

Coconut based stabilization agents were already applied for civil engineering material construction stabilization. The powder of coconut shell was used by [7] for modified asphalt binder. Further previous researches about coir fiber influence to soil were experimented by [8]-[10]. The research of [8] uses a nano-modification of coir fiber for treated with lime-marine clay in terms of shear strength parameter. The result shows good reaction between coir fiber and lime-marine clay for the value of shear strength, durability and also increase the tensile strength of the soil. Moreover, cracking of soil can also be prevented by using coir fiber as reinforcement according to [9], it was because of multifilament natural content and high lining content of the fiber.

A research article of [10] investigated comprehensive research for improvement of compressibility, swelling and strength parameter for black cotton soil with coir fiber stabilization. Furthermore, [11] conducted review research of coconut shell and fiber utilization for road construction. Their research result showed that coconut shell and fiber were able to improve the asphalt mixture of engineering properties when mixed with modified bitumen.

(a) (b)

Figure 2. Clay-Coir Fiber Samples for CBR Test, (a) top view of soil specimen for CBR test (b) soil specimen with coir fiber mixing stabilization

According to [12], there are two types of fibers used for soil improvement namely natural and synthetic fibers. Coir fiber, sisal, pal fiber, jute, flax, barely straw, bamboo, and cane are the example of natural fibers. Furthermore, polypropylene, polyester, polyethylene, glass, nylon, and steel fibers can be classified as synthetic fibers. Coir fiber, as a natural fiber is also one of the industrial wastes has also being considered as one of the clay mixtures to increase the bearing capacity and strength of the clay. Coir fiber is a fibrous material of matured coconut that has a thickness of about 3.5-5 cm with moisture content approximately 0.8-3.2% [13] and is the outermost part of the coconut fruit. Coir fiber is containing cellulose, lignin, pectin, tannin, and other water-soluble substance [12]. Coir composition is 43% of cellulose, 46% of lignin, 0.25% hemicellulose [13]. Good coir fiber basically has properties that are durable, friction-strength, not easily broken, resistant to water and also resistant to fungi or pests. Long-lasting coir fiber service life is around 4-10 years [12]. In one coconut can be obtained an average of 0.4 kg of coir where the coconut fiber contains almost 30% fiber. The properties and characteristics of coir fiber were shown in [11] as can be seen in Table 1. The influence of natural fiber for stabilization of soil can be imagined as in Figure 1. The absorption of water by natural fibers has a vital outcome on their adhesion [13]. In general, from previous studies, coconut fiber was able to increase soil strength but not significantly. With the right mixing method and good quality of coconut fiber, it is expected to maximize the use of coconut fiber in stabilizing clay.

Clay used in this study was collected from Ulee Glee Pidie Jaya area. Reasons for taking soil samples in Ulee Glee is because the soil was used for embankment construction of buildings and roads in Pidie Jaya Regency and its surroundings. Visually, this type of cohesive soil is basically not good for use in road embankments or buildings because in general cohesive soils have high shrink-swell behavior and low bearing capacity. The previous study regarding soil characteristic in Pidie Jaya area was also conducted and presented in several up to date researches. [14] evaluated the uplift force value of red clay soil in Pidie Jaya. The result of optimum moisture content of Pidie Jaya clay uplift force and swelling pressure was 2.466 kg and 0.868 cm respectively. Furthermore, [15] learned about expansive soil used as road sub-grade in Pidie Jaya. The article obtained that untreated clay soil of Blang Dalam – Pidie Jaya free swelling is 1.141% with swelling pressure for 23.18 kN/m^3.

The location of Pidie Jaya Regency which is an earthquake-prone area requires the use of good quality construction materials so that it can reduce the possibility of construction failure and achieve the construction age in accordance with the plan. Previous earthquake disaster in Pidie Jaya on December 2016 has caused about 104 human loss, more than 2,474 building need to be reconstructed, approximately 10 km roads and 50 bridges destroyed [16]. Pidie Jaya located along and close to Sumatera crust and some local faults like Lhokseumawe and Samalanga-Sipopok faults or known as (SAM-SIP) faults [17]. Better plan to prevent accident and destruction are important in earthquake disaster zone construction as [18] conducted a risk analysis research in Aceh Province.

2. Material and Methodology

The soil used in this study is a classification of A-7-5 (25), which is clay with high plasticity according to the classification of AASHTO, and OH (Organic High), namely organic clay with medium to high plasticity according to USCS classification with a specific gravity (SG) 2.63. Visually this clay soil is a red color, has a sticky character when squashed by hand. Though, for dry water conditions, the soil feels very stiff and hard while in the wet condition the soil is weak and expanded.

From laboratory test results for physical properties of the soil, the liquid limit (LL) and plastic limit (PL) of the soil were obtained at 54.15% and 33.73% respectively. Grain size distribution of # 200 pass filter analysis test obtained at 96.78%. All the physical parameter and mechanical parameter obtained were based on laboratory experiment of American Society of Testing and Materials (ASTM) standard.

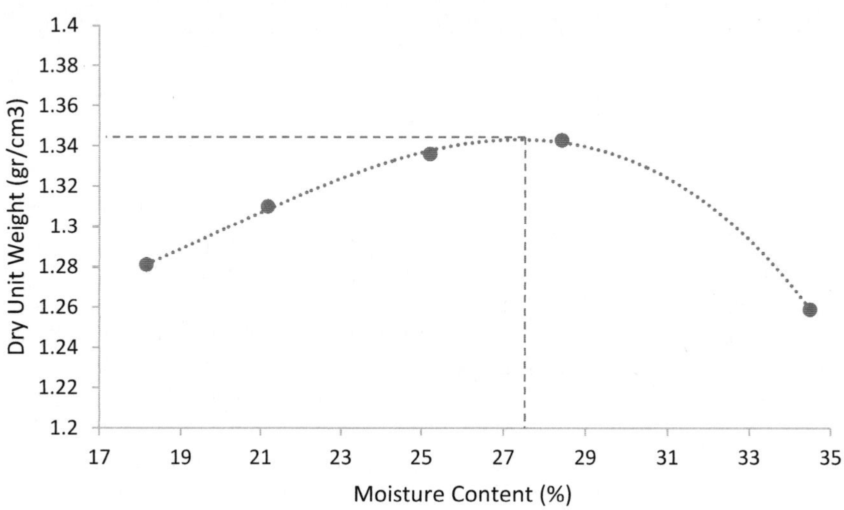

Figure 3. CBR result of high organic clay soil – coir fiber stabilization

Standard Proctor tests were performed to determine the optimum water content (OMC) value as in Figure 3. Total of 39 soil samples was prepared for CBR testing as can be seen in Table 2. Two lengths of coir fiber were arranged manually namely 2 cm and 3 cm. Coir fiber was dried under room temperature for 3 days. Manual sort and cut were conducted to obtain required length of coir fiber. The percentage between the mixture of soil and coir fiber was determined for 4 conditions, namely 0%, 0.2%, 0.4%, and 0.6%. The selected percentage was based on some previous study such as from [19] as they use 0%, 0.5%, 1.0%, and 1.5% of coir fiber in soil treated. The specimens were made using CBR molds by compaction method (ASTM D 1883-73) with a soil weight per test object approximately 4500 grams. Compaction test results (proctor standard) obtained the maximum dry unit weight (γ_{dmax}) 1.345 gr / cm^3 with optimum water content (OMC) was 27.5 % as can be seen in Figure 3. The stabilization mixing treatment used were direct mixing between soil and coir fiber and by mixing with layers of fiber in one sample specimen. The direct mixing treatment is mixing the clay with coir fiber before compaction. This treatment method was very common in soil stabilization techniques. On the other hand, layers treatment is coir fiber was added during soil specimen preparation of compaction test. The coir fiber was added after each layer of compaction test. This treatment method was trying to adopt simple site work method in soil compacting while contractor

IOP Conf. Series: Materials Science and Engineering **712** (2020) 012009 doi:10.1088/1757-899X/712/1/012009

pours the stabilization agent with specific amount after every layer compacted and continue compacting next layer with similar action. This treatment is suited for existing road construction techniques. The purpose of the difference in treatment of this soil-coir fiber mixture is to see the effect of the mixing treatment method on CBR values. List of test specimen designs and sample results of test objects can be seen in Table 2 and Figure 2.

Table 2. Soil Sample Designs for CBR test

Fiber Length	Mix Percentage (%)	Stabilization Method		Total
		Direct Mix	Layers	
2 cm	0	-	-	3
	0.2	3	3	6
	0.4	3	3	6
	0.6	3	3	6
3 cm	0.2	3	3	6
	0.4	3	3	6
	0.6	3	3	6
Total Sample				39

3. California Bearing Ratio Result

Detail of the CBR test result presented in Figure 4. The results showed that the addition of mixture variations in coir fiber percentage to Ulee Glee Pidie Jaya high organic clay soil affected the CBR value. The mixture between clay and coir fiber can generally increase the CBR value of high organic clay soil compared to the CBR value of natural clay soil sample (0% mix). The CBR value of natural soil without any addition is 8.15%, while all types of coir fiber mixtures with organic clay increase CBR values better than natural soil as shown in Figure 4. The highest CBR value of 17.70% was obtained from coir fiber 3 cm length with direct mixing treatment, followed by coir fiber 3 cm of layered treatment (16.15%). Additionally, the lowest CBR value was obtained in coir fiber 2 cm length with direct mixing treatment for 11.78% and layer treatment of 2 cm coir fiber length which is 10%.

The CBR testing results of 2 cm length coir fiber direct mixed treatment with a mixed variation of 0% obtained a CBR value of 8.15%. The most effective mixture was obtained at 0.4% addition of coir fiber, which is an increase in CBR value of natural soil by 93.62%. Whereas for the layer treatment mixing method for 2 cm length, the best CBR value was also for 0.4% mixture with an increase up to 66.38% from natural organic clay soil. Furthermore, the most effective percentage of coir fiber length 3 cm direct mixture treatment for CBR was obtained in 0.4% mixture for either direct mixed or layered treatment methods. The highest CBR value was 17.70% in the percentage of coir fiber 3 cm (mixed) 0.4% with an increase in CBR value more than double of natural CBR value of the soil that was 117.18%. Interestingly, the CBR value were decrease for 0.6% addition of coir fiber even still above CBR natural soil. It may because the quantity of coir fiber already too much compare to soil weight. The increment percentage of CBR value compared to natural high organic soil is presented in Table 3.

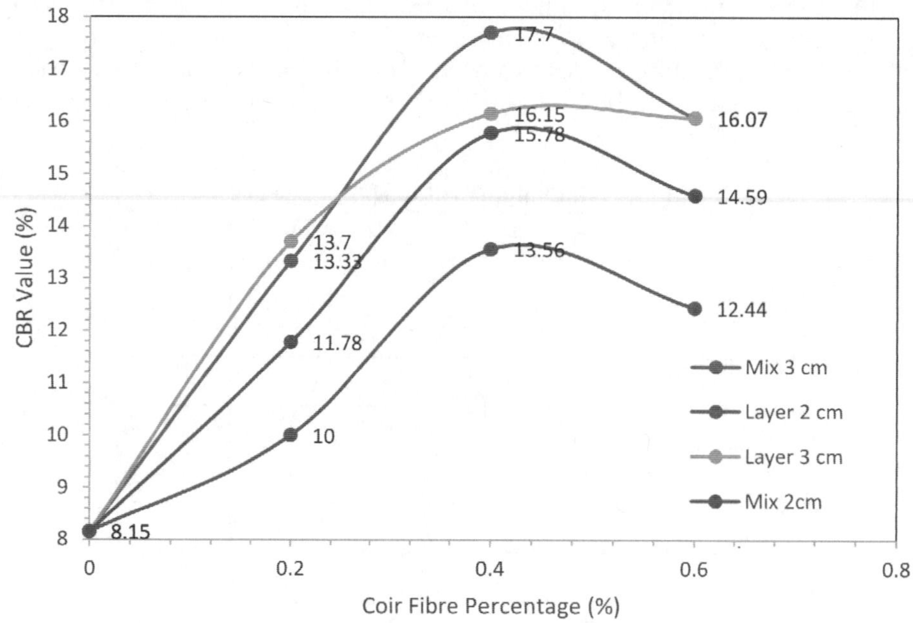

Figure 4. CBR result of high organic clay soil – coir fiber stabilization

Table 3. Coir fiber – clay soil CBR increment percentage from natural soil

Mix Percentage	2 cm (mix)		2 cm (layer)		3 cm (mix)		3 cm (layer)	
	CBR (%)	Increment Percentage from natural soil	CBR (%)	Increment Percentage from natural soil	CBR (%)	Increment Percentage from natural soil	CBR (%)	Increment Percentage from natural soil
0	8.15		8.15		8.15		8.15	
		44.55		22.70		63.56		68.10
0.2	11.78		10.00		13.33		13.70	
		93.62		66.38		117.18		98.16
0.4	15.78		13.56		17.70		16.15	
		79.02		52.64		97.18		95.46
0.6	14.59		12.44		16.07		15.93	

4. Conclusion

Based on finding results as explained previous, there are some highlights that can be considered for conclusions as follow:

- Clay which was stabilized with coir fiber by mixed treatment has higher CBR values compared to coir fiber with layered treatment.
- Clay-coir fiber with a length of 3 cm has higher CBR value compare to coir fiber with a length of 2 cm.
- The direct mixing treatment CBR results produce a higher value than the layered treatment method.
- However, the direct mixing method need more effort from contractors during application in the site if compared with layered treatment

IOP Conf. Series: Materials Science and Engineering **712** (2020) 012009 doi:10.1088/1757-899X/712/1/012009

5. References

[1] Sulaeman A, Fulazzaky M A, Haroen M, Bakar I, 2018 *Field Test Results of Palm Oil Clinker Concrete Pile and Foamed Concrete Pile for Floating Foundation in Soft Soil*, KSCE Journal of Civil Engineering Vol. 22, Issue 7 pp. 2232-2240

[2] Munirwansyah, Munirwan R P, 2016 *Lime-Clay Stabilization to Modified the Characteristic of Mechanical Properties and Reduce the Swelling Sub Grade,* The International Conference of Engineering and Science for Research and Development, Vol.1 (1)

[3] Hasan M, Yee K H, Pahrol M FA, Hyodo M, 2019 *Shear Strength of Soft Clay Reinforced with Encased Lime Bottom Ash Column (ELBAC),* International Journal of GEOMATE, Vol. 16 (57) pp. 62-66

[4] Dang L C, Fatahi B, Khabbaz H, 2016 *Behaviour of Expansive Soils Stabilized with Hydrated Lime and Bagasse Fibres,* Procedia Engineering, Vol.143 pp. 658-665

[5] Yadav J S, Tiwari S K, 2017 *Effect of Waste rubber Fibres on the Geotechnical Properties of Clay Stabilized with Cement,* Applied Clay Science, 149 pp. 97-110

[6] Muntohar A S, Widianti A, Hartono E, Diana W, 2013 *Engineering Properties of Silty Soil Stabilized with Lime and Rice Husk Ash and Reinforced with Waste Plastic Fiber,* Journal of Materials in Civil Engineering, Vol. 25 pp. 1260-1270

[7] Buhari R, Zabidi N F, Abdullah M E, Abu Bakar S K, Kamarudin N H K, 2016 *Rheological Behaviour of Coconut Shell Powder Modified Asphalt Binder,* Jurnal Teknologi, 78: 7-5, pp. 15-20

[8] Anggraini V, Asadi A, Farzadnia N, Jahangirian H, Huat B B K, 2016 *Reinforcement Benefits of Nanomodified Coir Fiber in Lime-Treated Marine Clay,* Journal of Materials in Civil Engineering, Vol.28 Issue 6

[9] Boddu R, Hong M, Deng Y, Chen F, Garg A, Bordoloi S, Kamchoom V, 2016 *Influence of Physical and Biochemical Composition of Three Cellulose Fibers on Cracking of Soil,* Proceeding of the 8th International Congress on Environmental Geotechnics, Vol. 3 pp. 348-355

[10] Sivakumar Babu G L, Vasudevan A K, Sayida M K, 2008 *Use of Coir Fibers for Improving the Engineering Properties of Expansive Soils,* Journal of Natural Fibers, Vol. 5 Issue 1 pp. 61-75

[11] Ting T A, Ramadhansyah P J, Hassan N A, Yaacob H, and Jayanti D S, 2015 *A Review of Utilization of Coconut Shell and Coconut Fiber in Road Construction,* Jurnal Teknologi (Sciences & Engineering) 76:14, pp. 121-125

[12] Hejazi S M, Sheikhzadeh M, Abtahi S M, Zadhoush A, 2012 *A Simple Review of Soil Reinforcement by Using Natural and Synthetic Fibers,* Construction and Building Materials, Vol.30 pp.100-116

[13] Laborel-Preneron A, Aubert J E, Magniont C, Bertron A, 2016 *Plant Aggregates and Fibers in Earth Construction Materials: A Review,* Construction and Building Materials, Vol.111 pp.719-734

[14] Munirwansyah, Munirwan R P, Sungkar M, and Fachrurrazi F, 2019 *Predicting Red Clay Swelling Behaviour of Jangka Buya-Pidie Jaya Earthquake Zone in Aceh,* AIP Conference Proceedings 2059 (1)

[15] Munirwansyah, Munirwan R P, 2017 *Stabilization on Expansive Soil for Road Subgrade for Geotechnic Disaster Approach,* International Journal of Disaster Management, Vol.1 (1) pp. 8-19

[16] Munirwansyah M, Munirwan R P, and Yunita H, 2018 *Geotechnical Engineering Aspect Related to Pidie Jaya – Aceh Earthquake Disaster and Mitigation* International Journal on Advanced Science Engineering Information Technology Vol. 8 No. 3 pp 870-875

[17] Muzli M, Umar M, Nugraha A D, Bradley K E, Widiyantoro S, Erbas K, Jousset P, Rohadi S, Nurdin I, Wei S, 2018 *The 2016 Mw 6.5 Pidie Jaya, Aceh, North Sumatra, Earthquake: Reactivation of an Unidentified Sinistral Fault in Region of Distributed Deformation*, Seismological Research Letters 89 (5), pp. 1761-1772

[18] Munirwansyah, Irsyam M, Munirwan R P, Yunita H, and Usrina M Z, 2018 *Geotechnical Approach for Occupational Safety Risk Analysis of Critical Slope in Open Pit Mining as*

Implication for Earthquake Hazard, IOP Conference Series: Materials Science and Engineering Volume 352 Conference 1

[19] Anandhamurugan A, Karuppasamy K, and Jagan S, 2017 *Study on the Stabilization of Soil Using Coir Fibers,* International Journal of Advanced Research in Basic Engineering Sciences and Technology, Vol. 3 (4), pp. 101-110

Acknowledgments

The result showed in this article are part of ongoing research at Soil Mechanics Laboratory of Syiah Kuala University (Unsyiah) which partially supported by Indonesia Ministry of Research and Technology and Lembaga Penelitian dan Pengabdian Masyarakat (LPPM) Syiah Kuala University Banda Aceh Indonesia under Lektor Research Scheme 2019 No. 204/UN11.2/PP/PNBP/SP3/2019 in 8 February 2019. The authors gratefully acknowledge their support.

IOP Conf. Series: Materials Science and Engineering **712** (2020) 012010 doi:10.1088/1757-899X/712/1/012010

Liquefaction Potential Analysis of Reusep Prestress Bridge in Pidie Jaya due to 6.4 Mw Earthquake

R P Munirwan[1], M Munirwansyah[1], K Jamaluddin[1], H Gunawan[1] and A Z Mubarak[2], P J Ramadhansyah[3] and D Youventharan[3]

[1]Civil Engineering Department, Faculty of Engineering, Universitas Syiah Kuala, 23111, Banda Aceh, Aceh, Indonesia
[2]Mechanical Engineering Department, Faculty of Engineering, Universitas Syiah Kuala, 23111, Banda Aceh, Aceh, Indonesia
[3]Faculty of Civil Engineering Technology, Universiti Malaysia Pahang, 26300 Gambang, Pahang, Malaysia

Abstract. Pidie Jaya is one of the districts in Aceh Province – Indonesia. On December 7, 2016, Pidie Jaya experienced a shallow 6.4 Mw earthquake which destroyed some vital buildings and also human loss. Based on preliminary field investigation, some geological damages like soil movement, soil cracks, and liquefactions occur because of the earthquake. Some liquefaction phenomena scattered in the area of Ulim, Panteraja, Meunasah Balek, Manohara Beach and Sagoe Trienggadeng. An evidence of liquefaction in Pidie Jaya area was displayed in this research based on several literature review. Furthermore, these paper aims to identify liquefaction possibilities in Trienggadeng Reusep Prestress Bridge based on a geotechnical investigation of N-SPT and some seismic data by using Kishida method. The surrounded area of the bridge was first evaluated in terms of geological setting, tectonics and seismic activity related to 6.4Mw of the previous earthquake. Furthermore, it was found from the soil profile that the studied area has a potency of liquefaction because of several saturated sand layers and high groundwater level. Effective overburden pressure of sand layers was calculated and plotted with N-SPT value to determine the possibility of liquefaction. The result shows that liquefaction potential shows high possibility for sand dominant layers both from bor-log 1 in 17.5 m and 25 m depths, and bor-log 2 in the depth of 17 m and 26.5 m.

1. Introduction

The liquefaction analysis potential was already studied in detail particularly since Niigata 1964 [1]. Research in liquefaction-related has been introduced in many ways for liquefaction potency estimation such as liquefaction histories and soil grain size conditions. Alluvial soil deposits liquefaction analysis by using an empirical method from the Nation Center for Earthquake Engineering Research was conducted by [2]. The research was conducted along the Nile river which already experienced liquefaction during October 1992 earthquake. There is also some previous liquefaction analysis in Indonesia which were conducted by [1], [3], and [4]. Assessment of liquefaction potency based on laboratory test was conducted by [1] while [3] try to predict the liquefaction probability by using shear wave velocity parameters. Moreover, [4] conducted another laboratory-based test to predict liquefaction according to the value of soil grain size and relative density. Specifically, liquefaction in Aceh was also analyzed by [5] and [6] by using Standard Penetration Test data. Aceh Province, where soil profile which has sand dominant layer shows the high possibility of liquefaction due to earthquake load as the Aceh location which closes to earthquake sources. Moreover, the shrink-swell behavior of soil also became a problem in Pidie Jaya soil as studied by [7].

Liquefaction damage to the structures due to earthquake disaster in Adapazari, Turkey was identified by [8]. The collapse of the building and construction during an earthquake to the effect of liquefaction potential are caused by structural defects poor quality of construction material. Research related to Pidie Jaya earthquake and the number of live and material loss was also conducted by [9] and [10]. The Pidie Jaya earthquake affected the geotechnical damage for up to 35 km radius mostly parallel to the seashore with approximately 101 people killed, more than 800 people wounded and economic loss around $139 million US [9]. As shown in figure 1, Aceh province surrounded by several faults which can be identified as an earthquake source. The main inland fault is known as Great Sumatran Fault. Moreover, according to [10], Samalanga-Sipopok fault (also known as Pidie Fault), was the source of the December 7^{th,} 2016 of 6.4 Mw earthquakes.

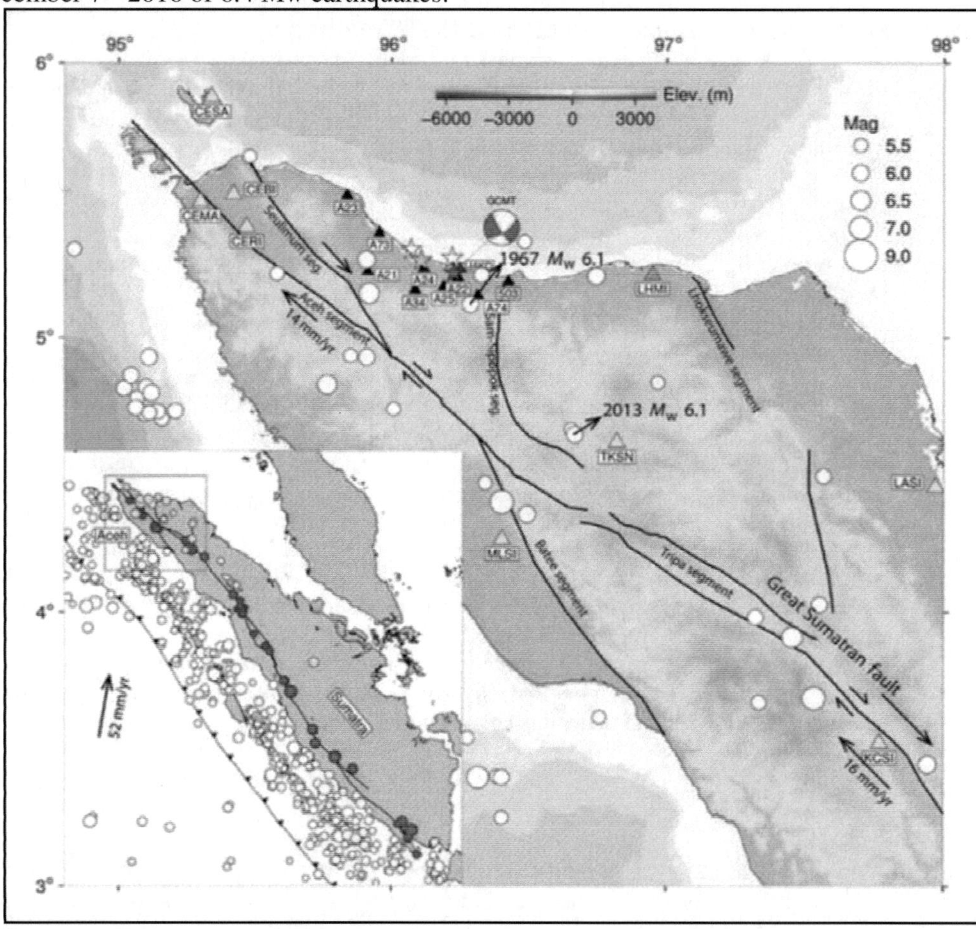

Figure 1. Aceh Faults along Sumatera [10]

Earthquake disasters often cause damage to buildings located far from the source of the earthquake. Previous study by [11] and [12] considering the earthquake load effect to civil engineering construction for design purposes. The difference in the level of damage to buildings indicates that the condition of the soil layer has a significant influence on the characteristics of seismic waves. Soft soil layers may amplify certain frequencies of earthquake vibrations. This condition may become extreme in areas that are based on thick soft soils [1], as a result, there is a change in material stiffness at the boundary between bedrock and soft soil that can cause amplification.

(a) (b)

(c) (d)

Figure 2. Liquefaction evidence in Pidie Jaya and the surrounding area, (a) boiling sand in local resident, (b) sand boil appears after the earthquake (c) sand boil shown up between tile crack (d) another sand boil evidence in resident area [14]

Some liquefaction phenomenal happened with the evidence of sandy like soil boiled in the west coast of Pidie Jaya along the shorelines such as in Manohara Beach [9]. The identified liquefaction zone spread along Ulim, Panteraja, West Panteraja, Meunasah Balek, Manohara Beach, and Sagoe Trienggadeng. Figure 2 explained some liquefaction evidence in Pidie Jaya and the surrounding area after the earthquake disaster from direct photograph in several resident areas surrounding Pidie Jaya after the earthquake disaster [13]. Reusep prestresses bridge, as a research location, was built in 2017 and located around Sagoe Trienggadeng at 5°14'21" N and 96°9'44" E. This bridge is 20 m long and 7 m wide. The soil profile below the bridge can be seen in figure 2. From the earthquake data in Table 1, it can be explained that Aceh Province lies in an active earthquake area, especially Banda Aceh which is an area that often experiences earthquakes. These earthquakes spread earthquake waves to the surface of the earth resulting in dynamic cyclic forces or regular alternating shear forces. Therefore, a study was conducted to find out the potential liquefaction. Only earthquake above Mw 5.0 was presented in Table 1. The raw data were obtained from Indonesia Agency of Meteorology, Climatology, and Geophysics in 2018 and had been sorted and presented as in Table 1.

Table 1. Recapitulation of recorded earthquake events in Aceh, from Indonesian Agency for Meteorological, Climatological, and Geophysics, Aceh 2018

No	Date			Latitude (N)	Longitude (E)	Depth (m)	Magnitude
	Year	Month	Day				
1	2004	December	28	5.11	96.162	30	5.2
2	2019	April	01	7.61	94.58	10	5.3
3	2017	February	15	5.19	96.16	27	5.4
4	2004	December	26	4.47	96.34	30	5.5
5	2016	December	07	5.25	96.24	15	6.5
6	2010	May	09	3.61	95.84	30	7.2
7	2008	February	20	2.77	94.31	16	7.5
8	2016	March	02	4.91	94.31	16	7.8
9	2005	March	28	2.09	95.50	30	8.6
10	2004	December	26	3.30	95.98	30	9.0

2. Analysis of Liquefaction

Kishida (1969) methods of liquefaction analysis were based on earthquake histories around the world for saturated sand [11]. This method was already used for predicting preliminary result of liquefaction potential. Furthermore, the method explained the relationship between N-SPT values and effective vertical stress (σ') with three types of liquefaction possibilities. The liquefaction potential can be classified as a high possibility, medium possibility and low possibility of liquefaction.

Table 2. Soil Parameters based on SPT (Cohesionless Soil: Fairly Reliable) [15]

SPT N-Value		**0 to 4**	**4 to 10**	**10 to 30**	**30 to 50**	**>50**
Compactness		very loose	loose	Medium	dense	very dense
Relative Density, Dr (%)		0 to 15	15 to 35	35 to 65	65 to 85	85 to 100
Angle of Internal Friction, ϕ (°)		<28	28 to 30	30 to 36	36 to 41	>41
Unit Weight (moist)	pcf	<100	95 to 125	110 to 130	110 to 140	>130
	kN/m^3	<15.7	14.9 to 19.6	17.3 to 20.4	17.3 to 22.0	>20.4
Submerged unit weight	pcf	<60	55 to 65	60 to 70	65 to 85	>75
	kN/m^3	<9.4	8.6-10.2	9.4 to 11.0	10.5 to 13.4	>11.8

The liquefaction analysis in this study is the liquefaction that occurs due to cyclic loading, the liquefaction that usually occurs in water-saturated sand soils that experience earthquake loads. The vibration may result the saturated sand loses it carrying capacity. This means that only soil layers with water-saturated sand were observed. Moreover, the sand layer must be below the groundwater level. Due to the limitation of soil parameter data for liquefaction analysis, engineering estimation based on N-SPT was used. Previous research in the correlation between N-SPT result with soil properties was showed by [15] and [16] as shown in Table 2. The N-SPT values were obtained from soil profile in figure 3 and the effective vertical stress was calculated and depend on soil type. Additionally, the unit weight and submerged unit weight of soil was needed to obtain soil vertical stress. The soil parameters used in the calculation were based on estimation by using table 2.

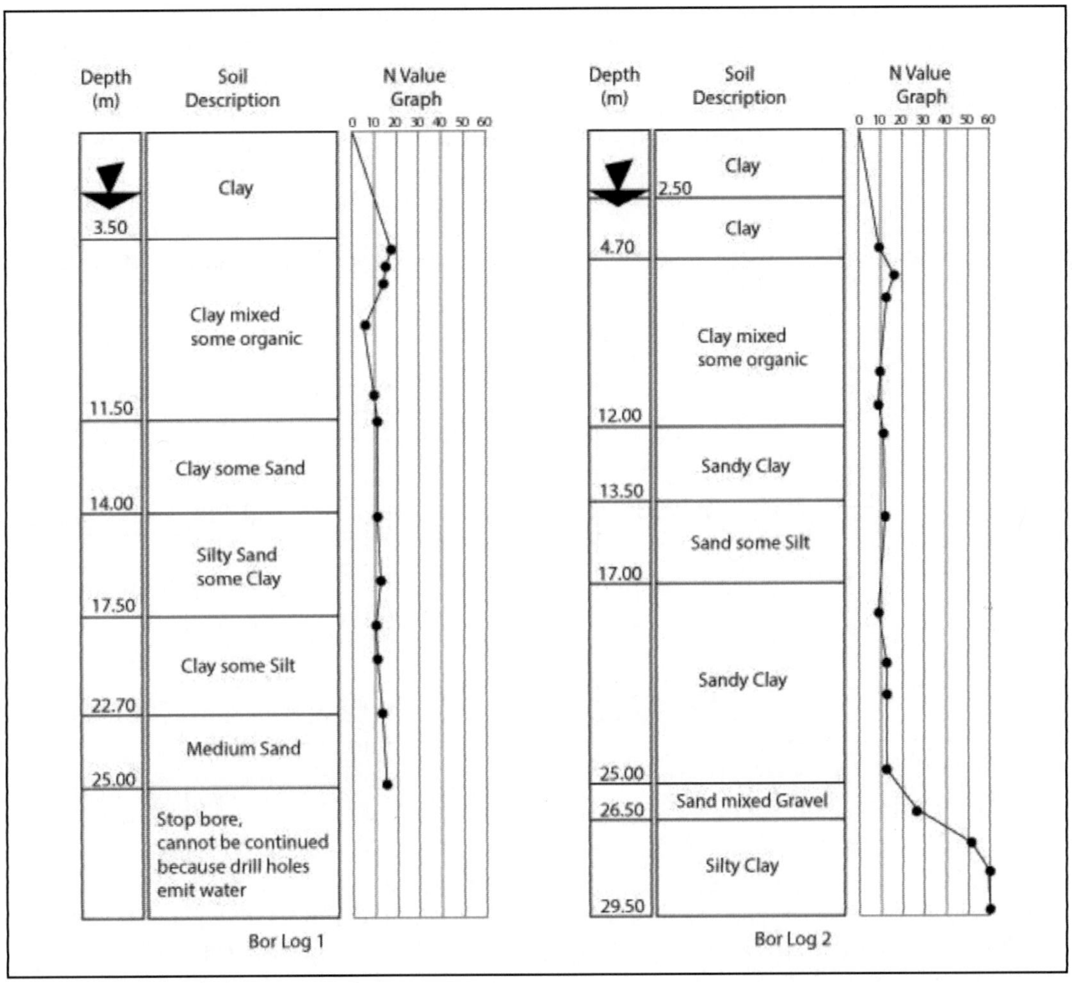

Figure 3. Soil Profile of Reusep Prestress Bridge from N-SPT Test

As can be seen in figure 3, there is more than 25.00 m depth of soil profile according to bor log 1 and up to 29.50 m depth of soil profile for bor log 2 with dominant soils are clay and sand. The groundwater level of the site is approximately 2.50 m below ground level. Sand dominant layers from bor-log 1 are in the depth of 14.00 – 17.50 m for silty sand with some clay (N-SPT average, 11) and 22.70 – 25.00 m for medium sand with N-SPT average is 14. Moreover, sand with some silt in 13.50 – 17.00 m with N-SPT average is 11 and sand mixed gravel (Average N-SPT, 23) in 25.00 – 26.50 m presented in bor-log 2. The potency of liquefaction may occur in layer of silty sand with some clay and medium sand for bor log 1 and in layers sand with some silt and sand mixed gravel for bor log 2. All liquefaction prediction layers were saturated layers and domination with sand soil as mention by [13].

3. Result and Discussion

Calculation results and discussion about the potential of liquefaction occurring in water-saturated sand layer at Prestress Bridge in Reusep Pidie Jaya can be explained in term of possibility level. From soil unit weight and soil depth observed, vertical stresses can be determined, both total and effective stress which then used as parameter for obtaining liquefaction potency. This article result was also another extended research result of [17] of determining soil characteristic and behavior under earthquake load in Pidie Jaya area. Depth and thickness of the layer under analysis were identified from the soil layer profile (drill log) as Figure 3. Furthermore, there are four saturated sand layers were identified based on soil profile in figure 4. These sands dominated layers were then analyzed for liquefaction potential. From bor log 1 it can be seen that at the depth of 14.00 - 17.50 m the soil is silty sand with some clay and at the depth of 22.70 - 25.00 m the soil is medium sand. Furthermore, according to bor log 2, there is sand with some silt layer at 13.50 - 17.00 m depth and sand mixed with gravel at 25.00 - 26.50 m depth.

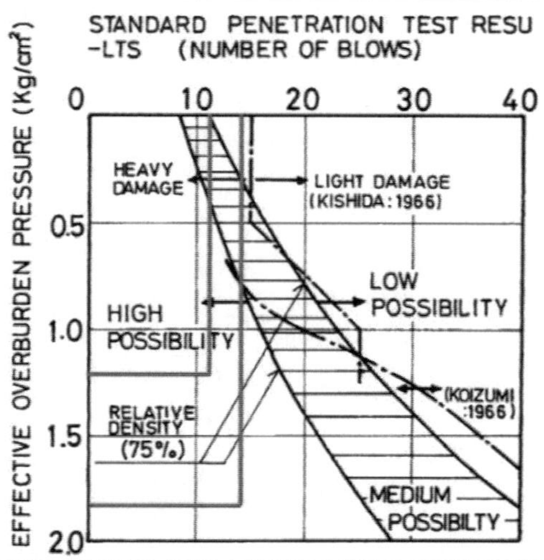

(a)

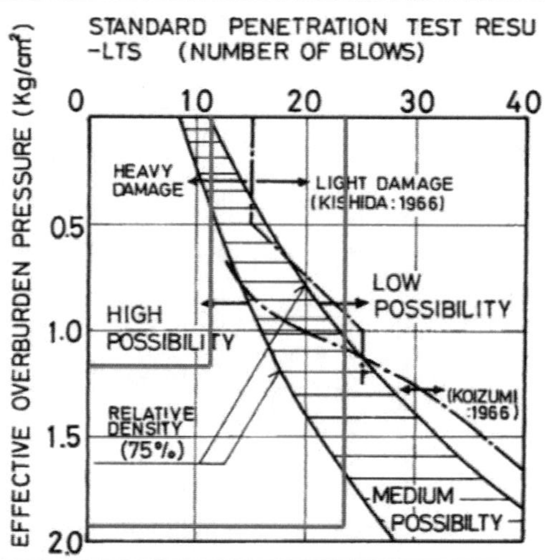

(b)

Figure 4. Probability of Liquefaction for Saturated Sand, (a) Bor log 1, (b) Bor log 2

By using N SPT value and correlation with Table 2, the effective overburden pressure (σ'_{vo}) calculation from bor log 1 for the depth of 17.5 m and 25.0 m are 118.02 kN/m^3 (1.20 kg/cm^2) and 179.21 kN/m^3 (1.83 kg/cm^2) respectively. Moreover, from bor log 2, σ'_{vo} obtained at 17.0 m depth and 26.5 m depth are 109.6 kN/m^2 (1.12 kg/cm^2) and 188.9 kN/m^2 (1.92 kg/cm^2) respectively. Based on effective overburden pressure of sand layers and average N-SPT value, then plotted to figure 4 to determine the possibility of liquefaction. Figure 4 is the probability of liquefaction for the saturated sand method proposed by [11]. The result shows that all plotted data were on the left side of the figure which means that liquefaction potential shows were high possibility for sand dominant layers both from bor-log 1 in 17.5 m and 25 m depths, and bor-log 2 in the depth of 17 m and 26.5 m.

4. Conclusion

The result of liquefaction analysis from this article can be classified as a preliminary result. Further soil parameter analysis from the laboratory is necessary for more appropriate results. The potency of liquefaction result is related to site condition as already described by [9]. Due to earthquake history and liquefaction potency in Pidie Jaya, for future construction development, liquefaction analysis is highly recommended for detail engineering design. Furthermore, the limitation of this liquefaction analysis was based on several factors including few site soil parameters. As a result, comprehensive physical soil parameter analysis from laboratory tests was important for more detail result of liquefaction analysis. For a large construction in areas known to have high liquefaction potential, firstly, it is best to treat the soil with stone columns or other soil liquefaction treatment such as grouting and compaction to improve soil strength and reduce the risk of liquefaction.

5. References

[1] Hakam A, Ismail F A, Fauzan, 2016 *Liquefaction Potential Assessment Based on Laboratory Test,* International Journal of Geomate, Vol. 11, Issue 26 pp. 2553-2557

[2] Ibrahim K M H I, 2014 *Liquefaction Analysis of Alluvial Soil Deposits in Bedsa South West of Cairo,* Ain Shams Engineering Journal

[3] Hartantyo E, Brotopuspito K S, Sismanto, Waluyo, 2013 *Predicting the Liquefaction Phenomena from Shear Wave Velocity Profiling: Empirical Approach to 6.3Mw, May 006 Yogyakarta earthquake,* Procedia Earth and Planetary Science

[4] Hakam A, 2016 *Laboratory Liquefaction Test of Sand Based on Grain Size and Relative Density,* Journal of Engineering and Technological Sciences, Vol. 48, No. 3 pp. 334-344

[5] Munirwansyah, Yunita H, Munirwan R P, 2017 *Liquefaction Potential Study due to Earthquake Based on SPT-N in Aceh Province (in Indonesia: Kajian Potensial Likuifaksi Akibat Gempa Berdasarkan Data SPT-N di Wilayah Provinsi Aceh),* Simposium II UNIID, Vol.2 (1), pp. 457-463

[6] Munirwan R P, Gunawan H, 2012 *Liquefaction Potential Evaluation in Krueng Raya Aceh Besar Shoreline Aceh Province (in Indonesia: Evaluasi Potensial Likuifaksi Pesisir Pantai Krueng Raya Aceh Besar Provinsi Aceh),* Jurnal Teknik Sipil Unsyiah, Vol. 1 (2), pp. 131-142

[7] Munirwansyah, Munirwan R P, Sungkar M, Fachrurrazi, 2019 *Predicting Red Clay Swelling Behaviour of Jangka Buya-Pidie Jaya Earthquake Zone in Aceh,* AIP Conference Proceedings Vol. 2059, Issue 1.

[8] Mollamahmutoglu M, Kayabali K, Beyaz T, and Kolay E, 2003 *Liquefaction-related Building Damage in Adapazari during the Turkey Earthquake of August 17, 1999,* Engineering Geology, Vol. 67, pp. 297-307

[9] Munirwansyah, Munirwan R P, and Yunita H, 2018 *Geotechnical Engineering Aspect Related to Pidie Jaya – Aceh Earthquake Disaster and Mitigation* International Journal on Advanced Science, Engineering and Information Technology Vol. 8 No. 3 pp 870-875 http://dx.doi.org/10.18517/ijaseit.8.3.4189

[10] Muzli M, Umar M, Nugraha A D, Bradley K E, Widiyantoro S, Erbas K, Jousset P, Rohadi S, Nurdin I, Wei S, 2018 *The 2016 Mw 6.5 Pidie Jaya, Aceh, North Sumatra, Earthquake: Reactivation of an Unidentified Sinistral Fault in Region of Distributed Deformation,* Seismological Research Letters, 89 (5), pp. 1761-1772

[11] Munirwan R P, Munirwansyah, 2019 *Escape Hill as Geotechnical Quick Response Method in Facing Upcoming Tsunami Disaster,* IOP Conference Series: Earth and Environmental Science Vol. 273(1)

[12] Munirwansyah, Irsyam M, Munirwan R P, Yunita H, and Usrina M Z, 2018 *Geotechnical Approach for Occupational Safety Risk Analysis of Critical Slope in Open Pit Mining as Implication for Earthquake Hazard* IOP Conference Series: Materials Science and Engineering Volume 352 Conference 1.

[13] Kishida H, 1969 *Characteristics of Liquefied Sands During Mino-Owari, Tohnankai and Fukui Earthquakes,* Vol. 9, Issue 1 pp. 75-92
[14] PUSGEN 2016 Laporan Gempa Pidie Jaya Provinsi Nanggroe Aceh Darussalam Indonesia [Pidie Jaya Earthquake Report Nanggroe Aceh Darussalam Province Indonesia] Indonesia
[15] Kumar R, Bhargava K, Choudhury D, 2016 *Estimation of Engineering Properties of Soils from Field SPT Using Random Number Generation,* INAE Letters, Vol. 1, Issue 3-4, pp. 77-84
[16] Rahman M M, 2017 *Foundation Design Using Standard Penetration Test (SPT) N-value,* Working Paper
[17] Munirwansyah, Munirwan R P, Sungkar M, Melinda Z, 2019 *The Effect of Soil-Structure Interaction on Multi-Storey Building Resonance and Dynamic Shear Modulus for Pidie Jaya Aceh Earthquake,* IOP Conference Series: Materials Science and Engineering Volume 523.

Acknowledgment

Authors would like to thank you for Universitas Syiah Kuala under Indonesia Ministry of Research and Technology and Lembaga Penelitian dan Pengabdian Masyarakat (LPPM) Syiah Kuala University Banda Aceh Indonesia for Lektor Research Scheme 2019 No: 204/UN11.2/PP/PNBP/SP3/2019 on 8 February 2019 for partly funded of this research. Moreover, authors also thank for any supports and appreciate to soil mechanics laboratory staffs and assistants for all their help.

The Influence of Steel Slag as Alternative Aggregate in Permeable Concrete Pavement

A A G Nadiatul Adilah[1], S Y Ayman Mohammed[1], P J Ramadhansyah[1], O Rokiah[1] and M R Hainin[1,2]

[1]Faculty of Civil Engineering Technology, Universiti Malaysia Pahang, 26300 Kuantan, Pahang, Malaysia
[2]Faculty of Engineering, School of Civil Engineering, Universiti Teknologi Malaysia, 81310 Johor Bahru, Malaysia

Abstract. This study identifies the pervious concrete with a high direct runoff on the pavement incorporating steel slag. This study is progressing in two stages i.e. 1ststage is obtaining the best mix design by using cubic concrete samples with dimension of 100mm height based on 7 days compressive strength test. Indeed, different proportion of aggregate and ratio were tested in order to select the highest compressive strength which has reached 10.9 MPa in the first 7 days of curing. The cylindrical concrete sample with the dimension of 200 x 100 mm was used in the 2nd stage. In addition, steel slag was used in the 2nd stage besides the granite aggregate. In fact, three different percentages of steel slag are involved; 50% Steel Slag and 50% Granite; 30% Steel Slag and 70% Granite; and 70% Steel Slag and 30% Granite. Hydrologically, 13.8 mm/s is the highest Infiltration rate that pervious concrete has reached which has been recorded throughout permeability test. The rainfall intensity which plays an important role on the pavement was identified. Thus, intensity distribution frequency curve was developed using MSMA and compared within filtration rate results of pervious concrete. Mix design ten (M10) has the ability to perform efficiently during highest rainfall intensity meanwhile strength needed to be increased since 10.9 MPa in the first 7 days of curing is not sufficient.

1. Introduction

Nowadays, flood is the most significant disaster in Malaysia that effects to the social and economic of the population. Pahang River is one of the rivers that contributes flood problem. Heavy rainfall can cause the excess of runoff rise to the high-water levels and causing the area to be flooded. Floods are the most significant natural disasters, which affect 4.9 million people and inflict damage worth of several million every year in Malaysia. About 29,720 km^{2}or 9% of the land area of the country is prone to flooding [1]. Pekan town is located on the banks of the Pahang River, regularly suffers both economic damages and physical destructions caused by the floods [2]. Flood occurs in Pekan due to drainage problem since it cannot cattle the quantity of water when storm-water runoff increases [3]. Pervious concrete or (thirsty concrete) is commonly used as an initially drainage tool to soak up most of runoff. This relieves the intensity of storm water drains and can minimize the expansive storm water from structures. Also, it assists to decrease storm water infrastructure costs. By infiltrating water, porous concrete can absorb nearby floor water sources and water surround root structures for timber and vegetation, which would not have obtained any water from an impermeable floor [4]. The use of porous concrete can be used in water sensitive urban sketch and structure to assist size the quantity of city runoff and pollutants flowing into natural waters [5]. Transportation is playing an essential position in any development of a community seeing that the rate motion made via either vehicle or humans represents energetic and active in the community. However, the presence of high depth direct runoff on roads is performing as boundaries against vehicles. Indeed, this direct runoff is paralyzing the mobility of any vehicle especially throughout evacuation time during flood session[6]. This study focused on investigation of the mechanical and hydrological properties of pervious concrete by compressive strength and permeability tests respectively. Indeed, it consists of 2 stages; 1st stage is obtaining preliminary results via 48 cubic concrete samples in dimensions of 100mm height based on 7 days compressive strength test. Indeed, various ratios of aggregate size are tested to select the highest compressive strength to be used in 2nd stage that consist of 12 cylindrical concrete samples in dimension of 200mm height and 100mm in diameter. In this stage, steel slag was use in 3 percentages as following:

a) 30% Steel Slag and 70% Granite
b) 50% Steel Slag and 50% Granite
c) 70% Steel Slag and 30% Granite

Infiltration rate is obtained through out permeability test. Oppositely, the Intensity Distribution Frequency IDF curve is performing by using MSMA 2nd edition for Rumah Pam Pahang Tua Station at site 3533102.

2. Materials

2.1. Cement

The cement manufactured according to the standards ASTM CI50. The physical and chemical properties of Portland cement are illustrated in Table1.

Table 1. Cement Properties [5]

Property	Test Result	Product Data	ASTM C 150	Standard
Physical & Mechanical Properties				
Normal Consistency	23.10 %	-		ASTM C 187
Blaine Fineness	361.3 m^2/kg	-	280 m^2/kg	ASTM C 204
Setting Time				ASTM C 191
Initial Set	1 hrs., 29 min	≥ 60 min	> 45 min	
Final Set	2 hrs., 15 min	-	< 6 hrs., 15 min	
Specific Gravity	3.15	-	3.14 - 3.16	ASTM C 188
Soundness	-	≤ 10		
Chemical Properties				
Loss on Ignition	-	≤ 5%		
Insoluble Residue	-	≤ 5%		
Sulphate (SO_3)	-	≤ 4%		
Chloride Content	-	≤ 0.1%		

2.2. Granite Aggregate

In this research, Granite Aggregates was collected from the Civil Engineering laboratory at Universiti Malaysia Pahang- Gambang campus.

2.3. Steel Slag

Steel slag is produced as a by-product during the oxidation of steel pellets in an electric arc furnace. This by-product that mainly consists of calcium carbonate is broken down to smaller sizes to be used as aggregates in asphalt and concrete. They are particularly useful in areas where good-quality aggregate is scarce [7].

3. Sample Preparation

Sieving were performed for granite and steel slag aggregates in order to separated them into different groups which are 20,14,12.5,9.5 and 4.75mm.This study has been carried out on 60 samples as stated in Table 2.

Table 2. Sample Distribution

Sample Type	Mould Type	No. of samples	Dimension
Cube	Plastic	48	100*100
Cylindrical	Steel	12	200*100

3.1 Testing Procedures

In 1st stage, the 16 mixtures were casted in plastic cubic moulds meanwhile in 2nd stage the 4 mixtures casted in cylindrical steel moulds. All mixtures were demoulded after 24 h, labelled and weighted for various testing. Then the samples were cured in a bath at room temperature, according to AS1012.8.1-2000 as shown in Figure 2. In 1st stage, all 3 samples of each mixture were prepared for 7 days compressive strength test. In 2nd stage, from each batch, 6 samples were prepared for permeability testing and compressive strength test.

3.2 Compressive Strength

Compressive strength of the pervious concrete was determined on cube and cylinder specimens that were cured in water curing and tested for 7 days as per ASTM C39M-18 as shown in Figure 1 and 2. All the specimens for compressive strength test have been stored in the curing tank until the testing day. 3 specimens have been tested for each curing age to obtain the average compressive strength. It was performed by using 2000kN UTM machine. The compressive strength test has been conducted according to the standard steps. The value of Compressive strength was calculated from the maximum loading acting on cross sectional area of cubes and cylinder by using this equation 1.

$$\text{Compressive Strength } \sigma = \frac{P}{A} \qquad (1)$$

Figure 1. Seven (7) Days curing

Figure 2. Compressive Strength Test

3.3. Permeability Test

One of the hydrological properties for the pervious concrete is the infiltration rate which is the speed of water passing through the porous concrete. The interconnected void of pervious concrete allows water to penetrate to hardened concrete. The relationship between strength and porosity is inversely proportional where highly porous combinations commonly yields decrease strength and vice versa. According to Abu Bakar et al. [8] PCPC with void ratios between 15% and 25% produce strength values greater than 13.8 MPa (2000 psi) with a permeability of about 200 L/m²/min (480 in./hr.). Permeability values up to 600 L/m²/min (1440 in./hr.) were reported for PCPC mixtures with a void ratio exceeding 20%. In terms of the permeability ability, the falling head test implemented based on the standard BS 1377-5:1990 and ASTM D2435–04. The falling head equipment test will be constructing in the University Malaysia Pahang laboratory for the pervious concrete. As shown in the Figure 3.

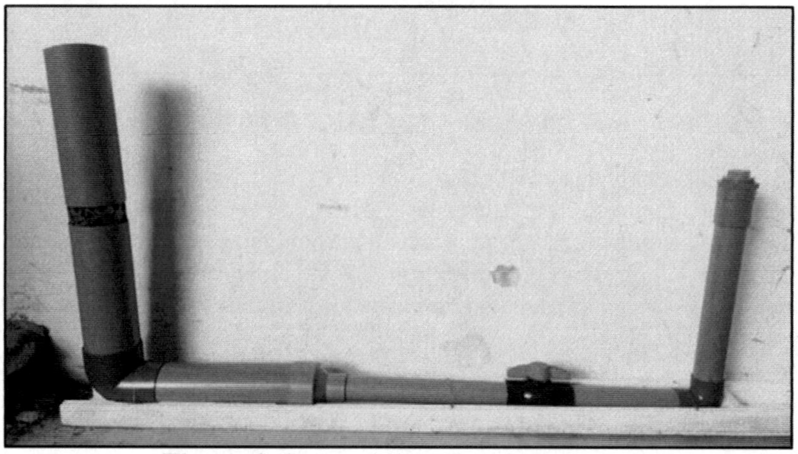

Figure 3. Permeability Test Instrument

3.4 Intensity Duration Frequency (IDF)

The data obtained is analysed with by using Equation 2. Intensity Duration Frequency (IDF) curves is developed. This study selects Rumah Pam Pahang Tua Station since it is located nearby Pahang River in Pekan which has experienced flood event during the past years located in Figure 5.

$$i = \frac{\lambda T^{\kappa}}{(d+\theta)} \qquad (2)$$

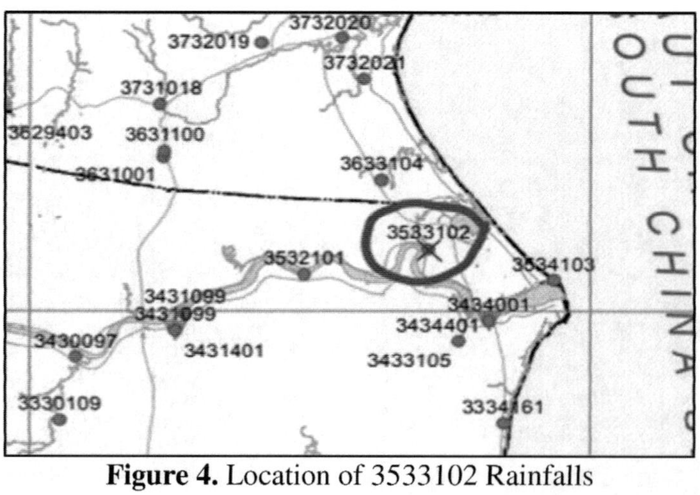

Figure 4. Location of 3533102 Rainfalls

Intensity Duration Frequency (IDF) curve is graphical representation gives the expected rainfall intensity of a given duration. Rainfall Intensity (mm/hr), Rainfall Duration (how many hours it rained at that intensity) and Rainfall Frequency (how often that rain storm repeats itself) are the parameters that make up the axes of the graph of IDF curve. An IDF curve is created with long term rainfall records and collected at a rainfall station [9]. All calculations are by MSMA 2nd Edition through Average Recurrence Interval (ARI) of 5, 10, 20, 50and 100.

4. Result and discussions

4.1. Preliminary Result

The effect of various coarse aggregate sizes on the compressive strength of pervious concrete was investigated. In the 1st stage, 20, 14 12.5, 9.5 and 4.75mm of granite aggregate were used. Based on Figure 6, the preliminary result which consists of 16 mixtures is illustrated. There are two mix design categories on basis of granite aggregate size which are single-sized aggregate and multi-sized aggregate. Average results of this stage are tabulated in Table 3 and 4. As noticed in table 4, ratio C: A of the PC is constant for all mixtures except M10 and M12 were 1:3.6 and 1:4.5 respectively. Also, mixtures 13, 14, 15 and 16 have an average granite aggregate density of 1333 kg/m^3 with 20mm and 14mm granite aggregate size. Also, the average weight of PC sample is ranged 1.7-2.1 Kg. Although all mentioned identical features, the compressive strength results are various Thus, the proportions of granite aggregate sizes dosage have an impact on the mechanical properties. However, upon a closer inspection of the results in Figure 6, it is observed that 16 mixtures of pervious concrete have different value of compressive strength due to various sizes and volume of aggregate 7 days compressive strength for pervious concrete mixtures ranged from 0.602 MPa to 10.9 MPa. Indeed, 10th mix has achieved the highest strength compression after 7 days curing in water at room temperature compared to the 16 mixtures. As known, 10.9 MPa is 2 thirds of the ultimate compressive strength which is indicated to be almost 17 MPa. Meanwhile, the previous studies of the control mix have reached the maximum 7 days compressive strength of 6.45 MPa [10].

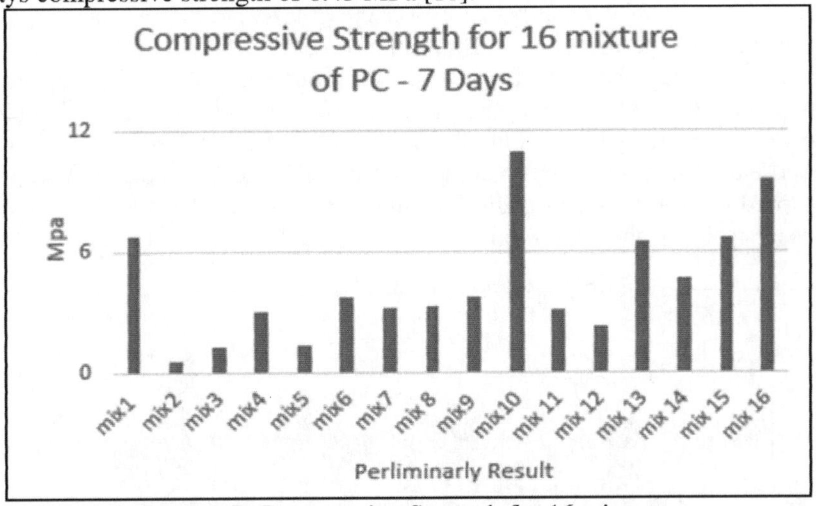

Figure 5. Compressive Strength for 16 mixtures

Table 3. Compression Strength Single-Sized Aggregate

Mix	Density Kg/m^3			Ratio		Compressive Strength (MPa)
	Cement	Aggregate	Water	Cement Aggregate ratio	w/c	
M1	347	1438	101	1:3.2	0.29	6.701
M2	270	1003	84	1:3.7	0.31	0.602
M3	355	1046	115	1:3	0.32	1.275
M4	359	1661	102	1:4.6	0.28	3
M5	355	1084	113	1:3.1	0.31	1.4

IOP Conf. Series: Materials Science and Engineering **712** (2020) 012011 doi:10.1088/1757-899X/712/1/012011

Table 4. Compressive Strength Multi-Sized Aggregate

Mix	Density Kg/m^3			Ratio		Compressive Strength (MPa)
	Cement	Aggregate	Water	Cement Aggregate ratio	w/c	
M6	311	1245	126	1:4	0.41	3.7
M7	312	1245	84	1:4	0.27	3.2
M8	311	1244	102	1:4	0.33	3.257
M9	310	1238	97	1:4	0.31	3.8
M10	440	1600	133	1:3.6	0.30	10.9
M11	440	1760	132	1:4	0.3	3.102
M12	360	1620	108	1:4.5	0.3	2.322
M13	334	1338	118	1:4	0.35	6.48
M14	334	1335	105	1:4	0.31	4.7
M15	334	1333	105	1:4	0.31	6.7
M16	334	1333	105	1:4	0.31	9.6

4.2. Mechanical and Hydrological Properties

The permeability results are various in the value of the infiltration rate since the range from 5 mm/s to 14 mm/s. Furthermore, 7 days compressive strength results were different due to different ingredients involved. Indeed, the first bar is representing the control mix for permeability which has achieved the lowest infiltration rate as 5 mm/s which because of using granite aggregate with no presence of steel slag; meanwhile, the highest permeability infiltration has reached the maximum of infiltration rate of 14 mm/s while the percentage of steel has reached 70 percentage of the mixture. It is also worth mentioning that even compressive strength results of the control mix have reached 10.9 MPa; meanwhile, with the first existing steel slag in the pervious the strength has increased to 11.8 MPa due to the high resistance that steel slag possesses. Indeed, the moderated permeable concrete has achieved in terms of the speed of water pass through the sample has occurred while the mixture consists of equivalent values of steel slag and granite on the basis of weight.

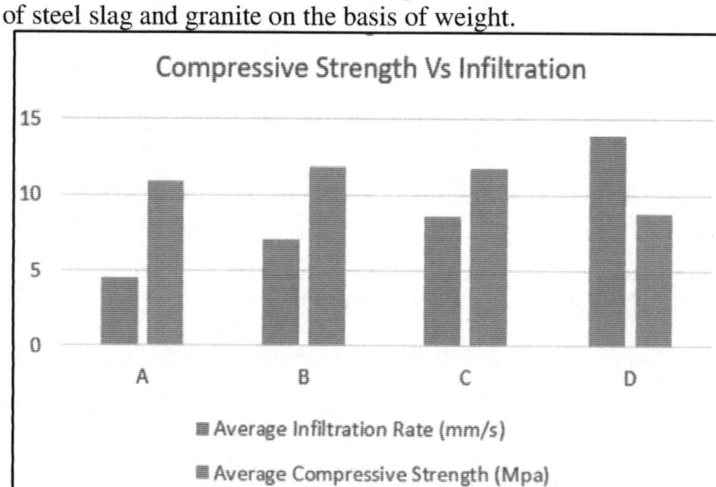

Figure 6. Compressive Strength vs Infiltration Rate

4.3. IDF Curve Developed

Referring to Table 5 and Figure 8, It is noticeable that the data collected from Rumah Pam Pahang Tua Station site 3533102 has been identified in the range of duration of 5 to 4320 minutes. Indeed, results were various; however, it is observed that while the duration of rainfall is shorter the higher value obtained. For example, in 5 ARI, for 5 minutes the value obtained is 173.3 mm/hr meanwhile for 4320 the value of intensity has decreased to 6.7 mm/hr. It is worthy to be highlighted that the maximum value of rainfallintensity gained is for 100 ARI is 327.11 mm/hr. Based on MSMA 2nd edition, 327.11 mm/hr considered as heavy rain.

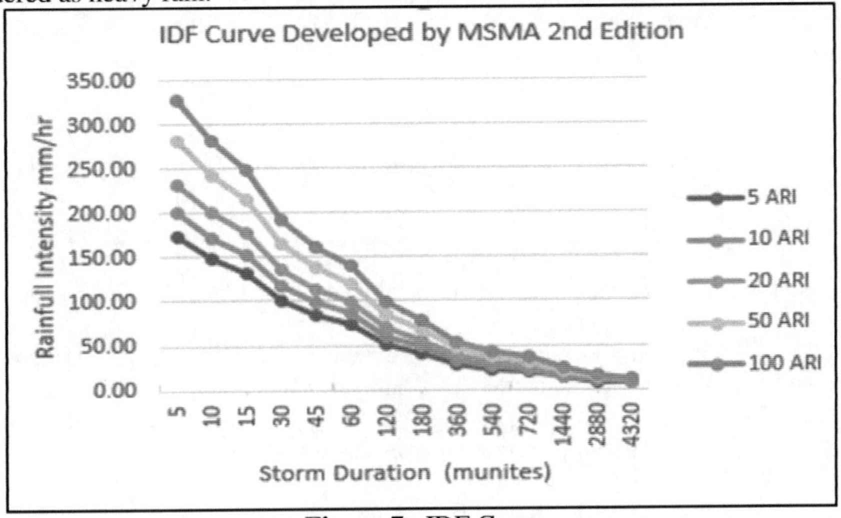

Figure 7. IDF Curve

Table 5. Intensity value by MSMA 2012

Duration	Average Recurrence Interval ARI (mm/hr)				
(minutes)	5	10	20	50	100
5	173.3	200.8	232.6	282.4	327.1
10	148.8	172.4	199.7	242.5	280.8
15	131.9	152.7	176.9	214.9	248.9
30	101.6	117.7	136.4	165.6	191.8
45	84.93	98.38	114	138.4	160.3
60	74.04	85.76	99.33	120.6	139.7
120	51.87	60.08	69.59	84.51	97.89
180	41.63	48.22	55.86	67.83	78.57
360	28.25	32.72	37.90	46.02	53.31
540	22.41	25.96	30.07	36.52	42.30
720	19.00	22.00	25.49	30.95	35.85
1440	12.72	14.73	17.06	20.72	24.00
2880	8.49	9.84	11.39	13.83	16.02
4320	6.70	7.76	8.99	10.92	12.65

5. Conclusions

To sum up, there are 3 outcomes which are stated clearly below regarding the compressive strength and permeability of the pervious concrete. Result of the IDF curve developed as well as the necessary comparison between intensity of rainfall and infiltration rate. Based on the preliminary result, the mix control has reached 10.9 MPa; meanwhile, with the first existing steel slag in the pervious the strength has increased to 11.8 MPa due to the high resistance that steel slag possesses. However, the more up to 50% of steel slag aggregate involved in the mixture much more strength obtained since 30% and 50% of steel slag have increased the strength of pervious concrete compared to mix control. The permeability results are various in the value of the infiltration rate since the range from 5 - 14 mm/s. The highest permeability infiltration rate has achieved the maximum of infiltration rate of 14 mm/s while the percentage of steel has reached 70 percentage of the mixture. It is worth to be said that the critical value of the rainfall intensity for 100 ARI obtained from the IDF curve is almost 1.1 mm/sec. Meanwhile, the lowest infiltration rate obtained in this study is 4.5 mm/s which is adequate to full fill the requirement of the rainfall intensity obtained from MSMA 2nd edition regarding at site 3533102 critical zone during flood session in Pekan.

References

[1] Annazirin E, Mardhiyyah S and Wan Zawiah W Z 2012 Preliminary Study on Bayesian Extreme Rainfall Analysis : A Case Study of Alor Setar, Kedah, Malaysia vol. **41**, no. 11, pp. 1403–1410

[2] Abdul M and Mohamed G 2013 Mitigation of Climate Change Effects through Non-structural Flood Disaster Management in Pekan Town , Malaysia *Procedia - Soc. Behav. Sci.*, vol. **85**, pp. 564–573

[3] Government of Malaysia Department of Irrigation and Drainage Urban Stormwater Management Manual for Malaysia MSMA 2nd Edition. .

[4] Gelong X, Weiguo S, Xujia H, Zhifeng Y, Jing W, Wensheng Z and Xiaoli Ji 2018 Investigation on the properties of porous concrete as road base material, *Constr. Build. Mater.* **158** pp. 141-148.

[5] Hainin M R, Ramadhansyah P J, Chan T H, Norhidayah A H, Nazri F M, Ichwana 2017 Mater. Sci. Forum. 889 pp. 265-269

[6] Othman M H and Hamid H A 2014 Impact of Flooding on Traffic Route Choices *SHS Web Conf.* **02** pp. 1–9, 2014.

[7] Wang G 2010 Determination of the expansion force of coarse steel slag aggregate," *Constr. Build. Mater.* **24**(10) pp. 1961–1966.

[8] Abu Bakar B H, Ramadhansyah P J and Megat Azmi M J 2011 Mag. Concrete Res. 63(5) pp. 313-320.

[9] Noor M, Ismail T, Chung E S, Shahid S and Sung J H 2018 Uncertainty in rainfall intensity duration frequency curves of Peninsular Malaysia under changing climate scenarios. *Water* **10**(12) pp. 1750

[10] Ibrahim A, Mahmoud E, Yamin M and Chowdary V 2014 Experimental study on Portland cement pervious concrete mechanical and hydrological properties. *Constr. Build. Mater.* **50,** pp. 524–529.

Acknowledgment

The authors gratefully acknowledge the University Malaysia Pahang for granting this research project which the vote number is RDU170344.

IOP Conf. Series: Materials Science and Engineering **712** (2020) 012012 doi:10.1088/1757-899X/712/1/012012

Ambient noise analysis for characterizing sub-surface dynamic parameters

B Setiawan[1], T Saidi[2], A Yuliannur[2], U Polom[3], P J Ramadhansyah[4] and M I Ali[4]

[1]Program Studi Teknik Geologi, Fakultas Teknik, Universitas Syiah Kuala, Indonesia
[2]Jurusan Teknik Sipil, Fakultas Teknik, Universitas Syiah Kuala, Indonesia
[3]Leibniz Institute for Applied Geophysics, Stilleweg 2, 30655 Hannover, Germany
[4]Faculty of Civil Engineering Technology, Universiti Malaysia Pahang, 26300 Kuantan, Pahang, Malaysia

Abstract. Ambient noise analysis of horizontal to vertical spectral ratio (HVSR) method has been used widely to provide reliable estimates of the site fundamental frequency and to constrain the inversion of near-surface shear wave velocity. The present paper focuses on the site measurement using the aforementioned analysis by means of the HVSR method for characterizing sub-surface dynamic parameters in the City of Banda Aceh, Indonesia. A Guralp CMG-6TD broadband seismometer was used in this study to cover a wide frequency range from 0.033 Hz to 50 Hz in standard operation. The instrument was deployed at two different sites (i.e. Location#1 of Blang Padang and Location#2 of Stadion Dirmutala) in the City of Banda Aceh for at least 2 hours for ambient noise recording. This continuous of 2 hours' microtremor time series was separated into 30 minutes record from which the site fundamental frequency and shear wave velocity of the measured site were deduced. The later sub-surface dynamic parameter was validated using another technique of reflection seismic. This investigation suggests the fundamental frequency of 0.45 Hz at Location#1 and of 0.65 Hz at Location#2. The estimated shear wave velocity of the top 30 m, $V_{s,30}$ of this investigation is 165 m/s at Location#1 and 156 m/s at Location#2. Both the site fundamental frequency and shear wave velocity are important for infrastructure design in the high seismic region of Banda Aceh, Indonesia.

1. Introduction

The velocities of seismic waves are much faster in rocks than in soils. Since the same amount of kinetic energy will propagate in rocks and soils, the amplitude of the seismic waves will strongly expand when the waves pass from rock through the soil layer. This amplitude expansion causes that the ground displacement at soil sites is usually greater than at rock sites. In some conditions, the ground shaking duration at soil sites also tends to be longer than at rock sites, which is also the result of a frequency transformation. These large ground displacement and long seismic duration are dangerous to any infrastructures founded on it [1,2,3]. Two sub-surface dynamic parameters, which are well accepted to contribute to the structural damage during a seismic event, are site fundamental frequency and the near-surface shear wave velocity profile. Ambient noise analysis of horizontal to vertical spectral ratio (HVSR) method has been used widely to provide reliable estimates of these site fundamental frequency and near-surface shear wave velocity [3,4,5]. Characterization of the sub-surface dynamic parameters based on single microtremor measurement method (sMSM) was carried out at two sites in the city of Banda Aceh, Nothern Sumatra, Indonesia. There are several advantages of the sMSM, i.e. simple, non-destructive, low cost and very feasible for sub-surface investigation in an urban area. This sMSM for seismic related studies have been applied by [6,7,8,9,10]. Empirically, there is a robust relationship between the structural seismic damages of buildings and their near-surface geological setting [11,12,13]. In the case of the structural frequency matches the fundamental frequency of the ground where the structure is founded, resonance will occur [1,2]. It is demonstrated in 1985 Mexico earthquake [1,3], 1995 Kobe earthquake [2], and 1999 Izmit earthquake [14]. Thus, characterizing the ground dynamic parameters is important in order to mitigate the seismic disaster and understand how the structure behavior during a seismic event.

2. Seismic source zones around the city of Banda Aceh

Two main seismic source zones of the tectonic subduction and the great Sumatran fault have been identified around the city of Banda Aceh. The tectonic subduction zone is the north-northeast oriented convergence subduction zone between the Indo-Australian plate and the Eurasian plate with the speed rate of up to 60 mm per year [15]. This zone has triggered several strong seismic events ($\geq$ 8.0), i.e. 1861 Nias earthquake, 2004 Sumatra-Andaman earthquake, and 2005 Simeulue earthquake [16,17]. These earthquake's slips are along the megathrust sections between the Indo-Australian and Eurasian plates (Figure 1). The 2004 and 2005 earthquake events were among the great seismic events to be studied using advanced seismological techniques. The later seismic source zone recognized around the city of Banda Aceh is the Great Sumatran Fault (GSF), as shown in Figure 2. The city of Banda Aceh is potentially exposed to significant seismic hazard of this major right-lateral strike-slip fault. The GSF is capable to cause up to M=7.9 earthquake [18]. Historically M=7.7 earthquake was occurred along this GSF in 1892 near the city of Sibolga (±570 km southeast of Banda Aceh) (cf. [17]). Furthermore, the city of Banda Aceh is built on up to 200 m thick alluvium sequences [19]. Any structures founded on thick alluvium poses a very high seismic vulnerability [1,3]. Therefore, sub-surface dynamic parameters study of Banda Aceh-Indonesia is urgently required for the city of Banda Aceh as the city is at risk for severe earthquake [20]. Understanding Banda Aceh's ground surface behavior is essential for developing the city's infrastructures.

3. Methods

The following sections describe the deployed equipment, data acquisition process, and data analysis used in this study, which is briefly explained, as follow:

3.1. Equipment

The seismometer used for the microtremor measurement in this study is the Guralp CMG-6TD/Broadband (0.033 Hz to 50 Hz in standard operation), as shown in Figure 3. It is a digital three-axis seismometer, which can record the north/south, east/west and vertical components of ground movement simultaneously. This broadband seismometer is perfectly suited for quick and easy installation in medium noise sites. The instrument offers an integrated 24-bit digitiser and a configurable output and ultra-lightweight (3 kg) and waterproof characteristics. Detailed main features of this seismometer are summarized in Table 1. The site installation of this equipment is shown in Figure 4.

IOP Conf. Series: Materials Science and Engineering **712** (2020) 012012 doi:10.1088/1757-899X/712/1/012012

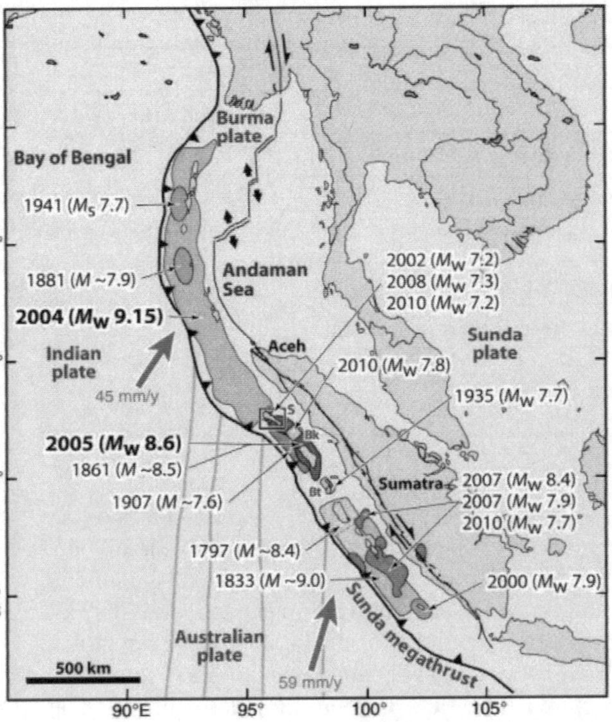

Figure 1. Tectonic setting of the Sumatran subduction zone showing major recent and historic plate boundary earthquake ruptures and their magnitudes (from [21], and references therein). Black lines = faults, grey lines = fracture zones.

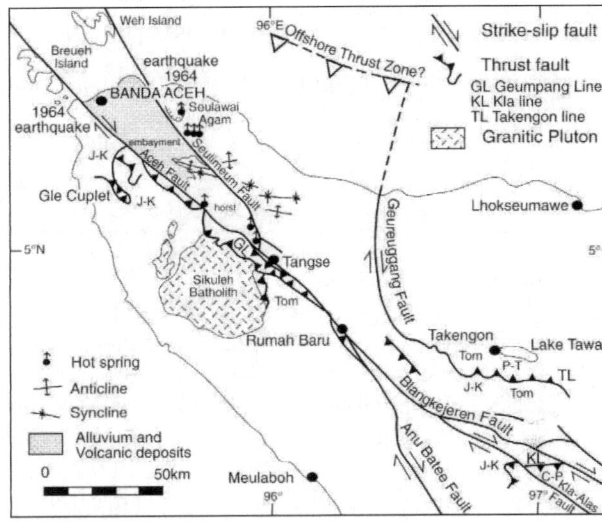

Figure 2. Faults around the city of Banda Aceh [22].

Figure 3. Guralp 6TD/Broadband seismometer, break out cable, GPS antenna, and battery for power supply

Figure 4. Illustration of installation of the deployed equipment

IOP Conf. Series: Materials Science and Engineering **712** (2020) 012012 doi:10.1088/1757-899X/712/1/012012

Table 1. Several important characteristics of the used seismometer

Parameter	**Description**	**Remarks**
Velocity output	30 s (0.03 Hz) to 100 Hz	Guralp CMG-6TD integrated digitiser
Output sensitivity	2400 V/ms^{-1} (2*1200 V/ms^{-1}) differential output	
Self-noise	-172 dB (Relative to 1[m/s^{-1}]^2 Hz^{-1})	
Power supply	11 to 28V DC, typically 7 mA @12 V	
Power consumption	0.93W without GPS or Ethernet at 12V DC	
Dimensions	153 mm diameter, 173 mm height (excluding handle)	
Weight	3.0 kg	
Temperature range	-20 ... +65 °C	

3.2. Data acquisition

The field ambient noise data were recorded from 15:40 pm on 20 April 2019 to about 01:00 am on 21 April 2019. The general weather during the data acquisition was calm with no strong winds or rain. Measurement locations were at near downtown of the city of Banda Aceh with Latitude of 05.549396 deg N and Longitude of 95.312956 deg E, as shown in Figure 5a (called Location#1) and at eastern of the city of Banda Aceh with Latitude of 05.567211deg N and Longitude of 95.340452 deg E, as shown in Figure 5b (called Location#2). Precaution was carried out prior to the data acquisition to avoid man-made holes (i.e. infills) and other underground structures. The field microtremor data collection at Location#1 was carried out from 22:45 PM on 20 April 2019 until about 02:00 AM on 21 April 2019. The field activity at Location#2 was started from 15:40 PM until 18:00 PM on 20 April 2019. At least 2 hours of ambient noise data were recorded at both Locations#1 and #2. A summary of the field data acquisition is presented in Table 2.

3.3. Data analysis using horizontal vertical spectral ratio (HVSR)

3.3.1. HVSR curve

The HVSR method is used to obtain the site fundamental frequency since introduced in 1989 by Nakamura [6]. The method focuses on the analysis of the spectral ratio of the Fourier amplitude spectrum of the horizontal (H) over vertical (V) components of the ambient noise particle movement velocity from which the HVSR curve and fundamental frequency of the measured site are estimated. Guidelines of data analysis using the HVSR method and interpretation of the result can be found in [23]. Several researchers i.e. [6,7,24] have successfully applied the HVSR method to compute the site fundamental frequency.

3.3.2. Inversion of the shear wave velocity

The measured HVSR curve and the estimated site fundamental frequency were used to obtain the sub-surface shear wave velocity profile of the measured site [25]. A tentative stratigraphic constraint is used in the inversion process by setting the layer depths, compression wave velocity, shear wave velocity, poison ratio, and density, in the neighborhood algorithm inversion code of Dinver [26] and [27]. The input parameters for the inversion adopted in this study are presented in Table 3 based on the studies [28] and [29]. Subsequently, the input parameters jointed with the obtained HVSR curve were used to generate at least 5,000 (i.e. 5,193) shear wave profile models, from which the best 20 shear wave velocity profile models are extracted and selected to propose the shear wave velocity profile of the measured site. The best 20 models are associated with the lowest 20 misfit inversions from the measured HVSR curve.

(a)

(b)

Figure 5. Ambient noise measurement sites at (a) near the downtown of the city of Banda Aceh, and (b) eastern of the city of Banda Aceh (yellow pins). The inserted small general maps show the locations on the Krueng Aceh river delta plain in the north of Sumatra.

Table 2. A summary of the field data sheet.

Parameter	Location#1 (Blang Padang)	Location#2 (Stadion Dirmutala)
Latitude	05.549396 deg	05.567211deg
Longitude	95.312956 deg	95.340452 deg
Sensor type	Guralp CMG-6TD	Guralp CMG-6TD
Sample frequency	100 Hz	100 Hz
Weather conditions	No wind; no rain; Approx. temperature 27 degrees	Weak wind; no rain; Approx. temperature 32 degrees
Ground type	Concrete pavement	Concrete pavement
Artificial ground/sensor coupling	None	None
Urbanization	Open field	Basketball field
Continuous noise sources	None	None

Table 3. Input parameters for the inversion

Layer No.	Depth (m)	Compression wave velocity (m/s)	Poisson's ratio	Shear wave velocity (m/s)	Density (kg/m^3)
Layer#1	1 – 50	200-500	0.2 – 0.5	100-350	1800-2000
Layer#2	20 – 100	200-800	0.2 – 0.5	150-500	1800-2000
Layer#3	70 – 300	200-800	0.2 – 0.5	150-500	1800-2000
Layer#4	150 – 400	400-1000	0.2 – 0.5	150-500	1800-2000
Layer#4	>400	700-5000	0.2 – 0.5	300-800	1800-2000

Note: Layer#4>Layer#3

4. Results

The HVSR curve, estimated site fundamental frequency, and developed shear wave velocity profile of the measured sites are presented. A brief discussion of the result is elaborated, also.

4.1. HVSR curve and site fundamental frequency

The results of the HVSR technique by means of site fundamental frequency of the measured sites which are Location#1 at Blang Padang and Location#2 at Stadion Dirmutala of Banda Aceh-Indonesia are presented. The HVSR curves of the two investigation sites are shown in Figures 6a and 6b. The site fundamental frequency obtained from the HVSR analysis at Location#1 estimates a frequency of 0.45 Hz. The site fundamental frequency obtained from the HVSR analysis at Location#2 is estimated at 0.65 Hz and indicates a secondary peak at nearly 2.7 Hz, which is not consistent over the four functions.

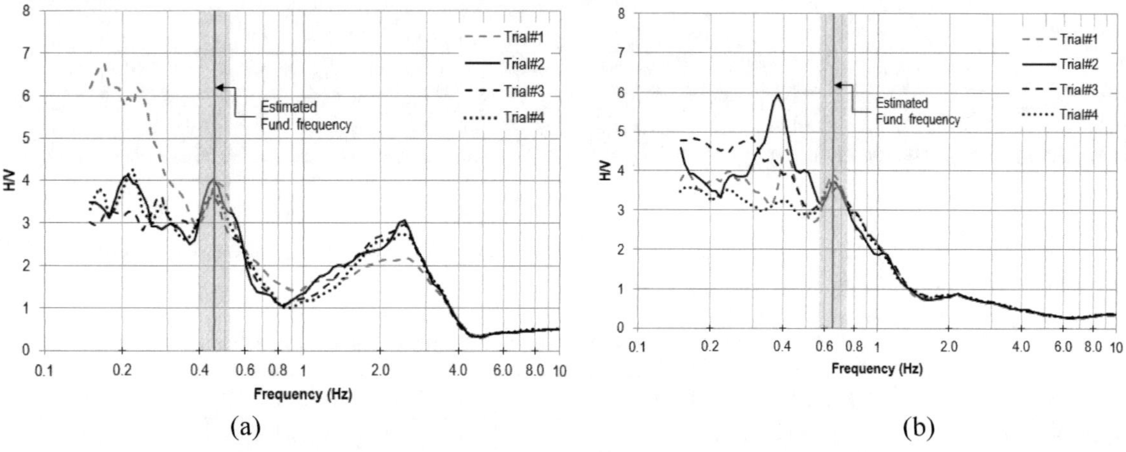

Figure 6. HVSR curves and estimated site fundamental frequencies at (a) Location#1, and (b) Location#2

4.2. The shear wave velocity profile

The inverted shear wave velocity-depth profiles using the neighborhood algorithm at both Locations #1 and #2 are presented in Figures 7a and 7b. The average of the four inversion trials for both Locations #1 and #2 are shown by a blocked mean function as red lines in Figures 7a and 7b. For Location#1, the simplified shear wave velocity-depth function at this location is gradually increasing from the ground surface to about 190 m in depth below the ground level (bgl), where a sharp increase to 600 m/s is visible (Figure 7a). Below 260 m bgl the velocity decreases to 480 m/s. The strong scattering of the inversion results in the interval 190-260 m probably indicates instability of the inversion algorithm (more about this in the discussion). For Location#2, the shear wave velocity inversion results are in a small scattering range up to 300 m bgl and show a gradually increase of the simplified shear wave velocity from 150 m/s at the surface to 350 m/s 300 m bgl.

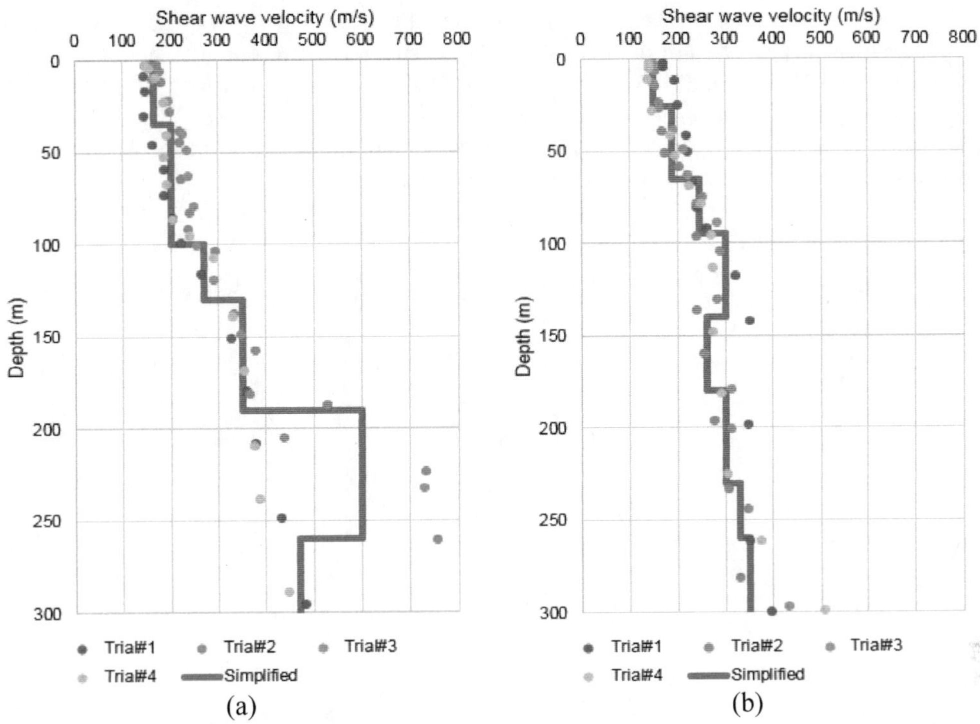

Figure 7. Inversion result by the neighborhood algorithm method of the shear wave velocity profiles at (a) Location#1 and (b) Location#2

4.3. Discussion

The studies [30,31,32] have investigated the relationship between the HVSR curve peak(s) and the resonance frequency of the shear wave. The results of the investigation can be summarized as follows: 1) In the case of a high-impedance contrast at the investigation site, the HVSR of the body S-waves always reveals a peak around the fundamental shear wave frequency; 2) In the case of a site consisting of a horizontally stratified media, the HVSR presents also peaks at the harmonics of the fundamental frequency; 3) In the investigation site case of a high-impedance contrast composed of a horizontally stratified media, the amplitude of the HVSR peak of the lowest frequency represents the shear wave amplitude amplification at the site. These findings are very useful for interpreting this study HVSR results. The HVSR curves at Location#1 indicate two peaks, which suggests a complex sub-surface layer structure at the site. Furthermore, the inversion shows unstable results in the depth range 190-260 m bgl. This complex sub-surface structure at Location#1 suggested by this study (Figure 7a) is also reported by Polom [29] (pers. communication, 2019), as shown in Figure 8a. The two peaks of 0.45 Hz and 2.7 Hz of the HVSR curve suggest the existence of two different impedance contrasts at different depths. In the case of this study, the first peak can be correlated to a deep stiff layer at nearly 18 m depth and the second one to a shallow stiff layer at nearly 140 m in depth [23,33,34]. The results of the two study locations suggests that the site fundamental period of the city of Banda Aceh is of $\approx$1.53 s to $\approx$2.22 s regarding the site fundamental frequency range of 0.45 and 0.65 Hz at the two locations. Therefore, seismic site amplification is suggested for standard reinforced concrete buildings >50 meters height in the city of Banda Aceh (see [35] and [36]). This general suggestion should be treated carefully as the frequency of the building is strongly depends on the building detailed structure. Recently, it is well accepted that a further amplification will occur when the building period is equal to or close to the ground period. The inverted top 30-meter shear wave velocity, $V_{s,30}$ models of this study are validated using [29]. In 2005 shear wave velocity measurements by the shear wave reflection seismic method were carried using a Geometrics GEODE recording system at both Locations #1 and

#2 in the frequency range 20-120 Hz [29]. The comparison is shown in Figure 8. Even though the range of measurement frequencies of the two methods are strongly different, a very good agreement is presented for Location#2 (Figure 8b). This study suggests $V_{s,30}$ of 165 m/s instead of [29] of 174 m/s at Location#1. The Vs,30 at Location#2 is estimated of 156 m/s instead of [29] of 162 m/s. Assuming a criterion of a quarter wavelength for the smallest structure resolution, i.e. 91 m for Location#1 and 60 m for Location#2 the resulting differences are in a very good agreement.

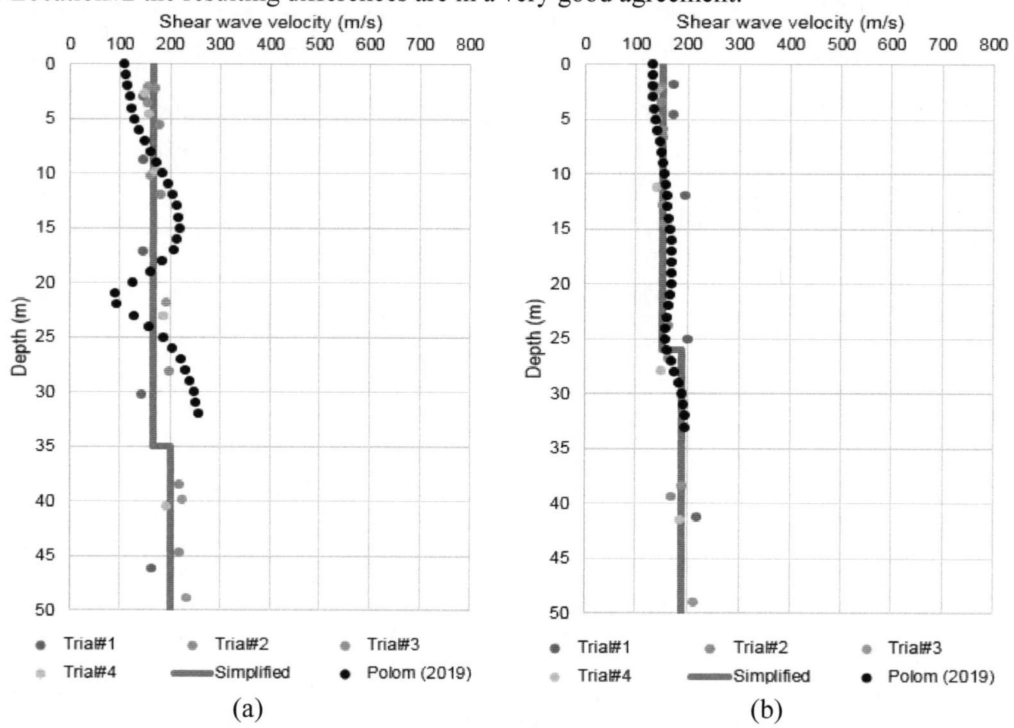

Figure 8. Comparison of the top 30-meter shear wave velocity, $V_{s,30}$ between this study and [28/26] at (a) Location#1 and (b) Location#2

5. Conclusion

Various studies have demonstrated the advantages and applicability of single point microtremor measurement and analysis for characterizing sub-surface dynamic parameters i.e. the site fundamental frequency and near-surface shear wave velocity. This single microtremor measurement and analysis have been carried out at two different locations in the city of Banda Aceh, which is highly vulnerable to seismic hazard. This study suggests site fundamental frequencies of 0.45 Hz and 0.65 Hz at the measured sites. This study also proposes $V_{s,30}$ of 165 m/s at Location#1 and of 156 m/s at Location#2. Therefore, it can be concluded that single microtremor measurement can provide easy, affordable and fast to obtain sub-surface dynamic parameters in the region of Banda Aceh city.

6. References

[1] Booth E D., Pappin, J.W., Mills, J.H., Degg, M.R and Steedman, R.S 1986 The Mexican earthquake of 19th September 1985. A field report by EEFIT, Institution of Structural Engineers.

[2] Brebbia, C.A., Beskos, D.E and Kausel, E 1996 The Kobe earthquake: geodynamical aspects. Computational Mechanics Publications, Southampton, ISBN:185312 4303, p. 165.

[3] Finn, W. D. L and Wightman, A 2003 Ground motion amplification factors for the proposed 2005 edition of the National Building Code of Canada. The National Research Council Canada (NRC) Research Press Web.

[4] Parolai, S., Facke, A., Richwalski, S.M and Stempniewski, L 2005 Assessing the vibrational frequencies of the Holweide Hospital in the city of Cologne (Germany) by means of ambient seismic noise analysis and FE modeling. Nat. Hazards, 34, 217–223.

[5] Setiawan, B., Jaksa, M. B., Griffith, M. C and Love, D 2018 Seismic site classification based on constrained modeling of measured HVSR curve in regolith sites. Soil Dyn Earthq Eng, 110: 244–261, DOI: 10.1016/j.soildyn.2017.08.006.

[6] Nakamura Y 1989 A method for dynamic characteristics estimation of subsurface using microtremor on the ground surface. Q Report Railw Tech Res Inst. 30(1):25–30.

[7] Mokhberi M, Davoodi M and Haghshenas E 2007 The application of the H/V spectral ratio technique for estimating the site characterization in the south of Iran. Soil Dyn Earthq Eng. 300–8. DOI: 10.1061/41102(375)37.

[8] Fah D, Wathelet M, Kristekova M, Havenith H, Endrun B, Stamm G, Poggi V, Burjanek J and Cornou C 2003 Using ellipticity information for site characterisation. Report for Network of Research Infrastructures for European Seismology, Sixth Framework Program, EC Project Number: 026130, 54 pp.

[9] Harutoonian P, Leo C, Doanh T, Castellaro S, Zou J, Liyanapathirana D, Wong H and Tokeshi K 2012 Microtremor measurements of rolling compacted ground. Soil Dyn. Earthq Eng. 23–31. DOI: 10.1016/j.soildyn.2012.05.006.

[10] Di Stefano P, Luzio D, Renda P, Martorana R, Capizzi P, D'Alessandro A, Messina N, Napoli G, Todaro S and Zarcone G 2014 Integration of HVSR measures and stratigraphic constraints for seismic microzonation studies: the case of Oliveri (ME). Nat Hazards Earth Syst Sci. 2:2597–637. DOI: 10.5194/nhessd-2-2597-2014.

[11] Graves, R. W., Pitarka, A and Somerville, P. G 1998 Ground motion amplification in the Santa Monica area: Effects of shallow basin edge structure. Bull. Seismol. Soc. Am. 88: 1224-1242.

[12] Setiawan, B 2017 Site Specific Ground Response Analysis for Quantifying Site Amplification at A Regolith Site. Indonesian Journal on Geoscience, 4 (3), p.159-167. DOI: 10.17014/ijog.4.2.159-167

[13] Setiawan, B., Jaksa, M. B., Griffith, M. C and Love, D 2018 Estimating bedrock depth in the case of regolith sites using ambient noise analysis. Eng Geol, 243: 145 – 159, DOI: 10.1016/j.enggeo.2018.06.022.

[14] Ozcep, F., Karabulut, S., Ozel, O., Ozcep, T., Imre, N and Zarif, H 2014 Liquefaction-induced settlement, site effects and damage in the vicinity of Yalova City during the 1999 Izmit earthquake, Turkey. J. Earth Syst. Sci. 123 (1) 73 - 89.

[15] McCaffrey, R., Zwick, P., Bock, Y., Prawiradirdjo, L., Genrich, J., Stevens, C., Puntodewo, S and Subarya, C 2000 Strain partitioning during oblique plate convergence in northern Sumatra: Geodetic and seismologic constraints and numerical modeling, J. Geophys. Res., 105(B12), 28,363–28,376.

[16] Lay, T., Kanamori, H., Ammon, C., Nettles, M., Ward, S., Aster, R., Beck, S., Bilek, S., Brudzinski, M., Butler, R., DeShon, H., Ekstro¨m, G., Satake, K and Sipkin, S 2005 The great Sumatra-Andaman earthquake of December 26, 2004, Science, 308, 1127–1133.

[17] Sørensen, M and Atakan, K 2008 Continued earthquake hazard in Northern Sumatra: Potential effects of a future earthquake. - EOS, Transactions, American Geophysical Union, 89 (14), 133-134.

[18] Petersen, M., Dewey, J., Hartzell, S., Mueller, C., Hansen, S., Frankel, A and Rukstales, K 2001 Probabilistic seismic hazard analysis for Sumatra, Indonesia and across the southern Malaysian peninsula, Tectonophysics, 390, 141-158.

[19] Siemon, B., Ploethner, D and Pielawa, J 2006 Hydrogeological Reconnaissance Survei in the Province Nanggroe Aceh Darussalam Northern Sumatra, Indonesia Survei Area: Banda Aceh / Aceh Besar 2005, Report C 1, BGR (Federal Institute for Geosciences and Natural Resources).

[20] Irsyam, M., Widiyantoro, S., Natawidjaja, D., Meilano, I., Rudyanto, A., Hidayati, S., Triyoso, W., Hanifa, N., Djarwadi, D., Faizal, L and Sunarjito, S 2017 Peta Sumber & Bahaya Gempa

Indonesia tahun 2017. Pusat Penelitian & Pengembangan Perumahan dan Permukiman, Badan Penelitian & Pengembangan, Kementerian Pekerjaan Umum & Perumahan Rakyat, ISBN: 978-602-5489-01-3.

[21] Meltzner, A.J., Sieh, K., Chiang, H.-W., Shen, C.-C., Suwargadi, B.W., Natawidjaja, D.H., Philibosian, B and Briggs, R.W 2012 Persistent termini of 2004- and 2005-like ruptures of the Sunda megathrust. Journal of Geophysical Research: Solid Earth, 117(B4):B04405. http://dx.doi.org/10.1029/2011JB008888

[22] Barber, A.J and Crow, M.J 2005 Chapter 13: Structureand structural history. -In: Barber, A.J., Crow, M.J. &Milsom, J.S. (eds.) (2005): Sumatra: Geology, Resources and Tectonic Evolution. 290 pp., Geological Society, London.

[23] SESAME 2004 Guidelines for the implementation of the H/V spectral ratio technique on ambient vibrations measurements processing and interpretation. SESAME European Research Project, WP12 –Deliverable D23.12 European Commission – Research General Directorate, Project No. EVG1-CT-200-00026.

[24] Picozzi M, Strollo A, Paraloi S, Durukal E, Ozel O, Karabulut S, Zschau J and Erdik M 2009 Site characterization by seismic noise in Istanbul, Turkey. Soil Dyn Earthq Eng. 469–82. DOI: 10.1016/j.soildyn.2008.05.07.

[25] Setiawan, B., Jaksa, M., Griffith, M and Love, D 2018 Passive noise datasets at regolith sites. Data in Brief, 20, 735-747. DOI: 10.1016/j.dib.2018.08.055

[26] Sambridge, M 1999 Geophysical inversion with a neighborhood algorithm I. Searching a parameter space. J. Geophysical Res. 103:4839-4878.

[27] Wathelet, M., Jongmas, D and Ohremberg, M 2005 Direct inversion of spatial autocorrelation curves with the neighborhood algorithm. Bull. Seismol. Soc Am. 95(5): 1787 - 1800.

[28] Polom, U., Arsyad, I and Kumpel, H 2008 Shallow shear-wave reflection seismics in the tsunami struck Krueng Aceh River Basin, Sumatra. Adv. Geosci., 14, 135–140.

[29] Polom, U 2019 Personal communication

[30] Caserta, A. 2011 Seismic site effects (Data analysis and modelling). Doctoral Thesis. Department of Geophysics, Faculty of Mathematics and Physics, Charles University in Prague, Czech Republic, 74 pp.

[31] Bonnefoy-Claudet, S., Cornou, C., Bard, P-Y and Cotton, F 2004 Nature of noise wavefield. Final Report, WP08, Deliverable D13.08, Site Effects Assessment Using Ambient Excitations (SESAME), European Comission – Research General Directorate, Project No. EVG1-CT-2000-00026 SESAME, 50 pp.

[32] Bonnefoy-Claudet, S., Cotton, F and Bard, P.-Y 2006 The nature of noise wavefield and its applications for site effects studies. A literature review. Earth-Science Reviews, 79:205 - 227.

[33] Toshinawa, T., Taber, J. J and Berril, J 1997 Distribution of ground motion intensity inferred from questionnaire survey earthquake recordings and microtremors measurements – a case study in Christchurch New Zealand during the Arthurs Pass Earthquake. Bull. Seismol. Soc Am. 87: 356-369.

[34] Gueguen P, Chatelain J-L, Guilier B, Yepes H, and Egred J 1998 Site effect and damage distribution in Pujili (Ecuador) after the 28 March earthquake. Soil Dyn Earthq Eng. 17:329–334.

[35] IS 1893:2002 Indian standard code of practice for earthquake resistant design.

[36] American Society of Civil Engineers 2005 Minimum design loads for buildings and other structures: SEI/ASCE 7-05. Virginia: American Society of Civil Engineers

Acknowledgments

The authors wish to acknowledge Universitas Syiah Kuala for providing a research funding (Contract No. 523/UN11/SPK/PNPB/2019) on the date of 8 February 2029. Also, the authors are grateful to the Faculty of Engineering of Syiah Kuala University for their support.

IOP Conf. Series: Materials Science and Engineering **712** (2020) 012013 doi:10.1088/1757-899X/712/1/012013

Site-specific ground response analysis at a site in the affected area of the 2016 Pidie Jaya earthquake

H Yunita[1], N Al Huda[1], T Saidi[1], A Yuliannur[1], B Setiawan[2], P J Ramadhansyah[3] and M I Ali[3]

[1]Jurusan Teknik Sipil, Fakultas Teknik, Universitas Syiah Kuala, Indonesia
[2]Program Studi Teknik Geologi, Fakultas Teknik, Universitas Syiah Kuala, Indonesia
[3]Faculty of Civil Engineering Technology, Universiti Malaysia Pahang, 26300 Kuantan, Pahang, Malaysia

Abstract. Analytical models have demonstrated that they are able to simulate reasonably well the seismic motions at the ground level. The most widely used model is the equivalent linear approach. This equivalent linear model was used to compute the free-field response of Meureudu-Pidie Jaya, Aceh Indonesia's soft soils during the 2016 Pidie Jaya earthquake. The model computes the ground response of horizontally layered soil deposits subjected to transient and vertically propagating shear waves through the one-dimensional soil column. Each soil layer is assumed to be homogeneous, visco-elastic and infinite in the horizontal extent. The equivalent linear estimation of soil properties is taken to express the nonlinearity of the soil's shear modulus and damping values. These values are assumed to be a function of shear strain amplitude and determined by an iterative process that must be consistent with the level of the effective strain induced in each sub-layer. Starting with the highest shear modulus and a low damping value, the shear modulus and the damping ratio of each sub-layer are modified. The modification is based on the applicable relationship between both properties and the shear strain. The calculation is repeated until strain-compatible modulus and damping values converge within a tolerance of 1%. This research reveals the ground motions of Pidie Jaya's soils. The results of the analysis are presented.

1. Introduction

A site-specific ground response analysis has to be taken into consideration for seismic hazard assessment. It has been well established that rock-based earthquake motions can be amplified on soft soil sites and cause structural damage, such as in the 1985 Mexico earthquake, the 1988 Armenian earthquake [1], the 1989 Loma Prieta earthquake in California [2], and the 1951 Adelaide earthquake [3,4]. Analytical models for a site-specific ground response analysis demonstrated that they are able to simulate reasonably well the soil behaviour due to dynamic loading. The widely used approaches are the equivalent linear technique which is included the EERA computer program [5]. The EERA (Equivalent-linear Earthquake Response Analysis) program was developed from the basic principles of the SHAKE program [6] which has been one of the most commonly used computer programs in geotechnical earthquake engineering since it became available in 1972. EERA was selected for this study because the program takes full advantage of the latest development of FORTRAN 90 and the Windows platform. EERA is not a stand-alone program. It is an add-on program embedded in

[2,*] Corresponding author: bambang.setiawan@unsyiah.ac.id

IOP Conf. Series: Materials Science and Engineering **712** (2020) 012013 doi:10.1088/1757-899X/712/1/012013

Microsoft Excel. The unavailability of the site-specific ground response analysis at a site in the affected area of the 2016 Pidie Jaya earthquake has become the motivation of this paper. A study by [7] found a greater seismic event than the 2016 Pidie Jaya earthquake at a deep potential seismogenic depth around the Pidie Jaya region. Furthermore, [7] suggested considering this potential larger event for seismic hazard evaluation of this region. Three actual seismic time histories and a synthetic of Pidie Jaya earthquake's time history were used. The equivalent linear approach was employed for this analysis. Several outcomes from these site-specific ground response analyses i.e. peak ground acceleration; stress and strain at each layer; amplification at the ground surface or the surface of each layer; Fourier amplitude; and the response spectrum of the soil are presented.

2. The 2016 Pidie Jaya earthquake and site characteristics

On 7 December 2016, 05:03:33 AM local time an Mw 6.5 earthquake shocked the northern coastal area of Aceh around the Pidie Jaya region (hereafter, Pidie Jaya earthquake). The Indonesian Agency for Meteorology, Climatology, and Geophysics (BMKG) stated the epicenter of this seismic is at 5.29° S and 96.22° E and the depth of the event is 15 km. The earthquake caused heavy damage on structure and killed and injured many people within a radius of ~35 km from the epicenter. To justify the fault geometry and associated tectonics, [7] deployed many seismometers to locate the aftershocks of the event. More than 300 events with magnitudes larger than 0.50 were recorded. The results are shown in Figures 1a and 1b which was suggested that the seismic has re-activated a small fault of Panteraja fault.

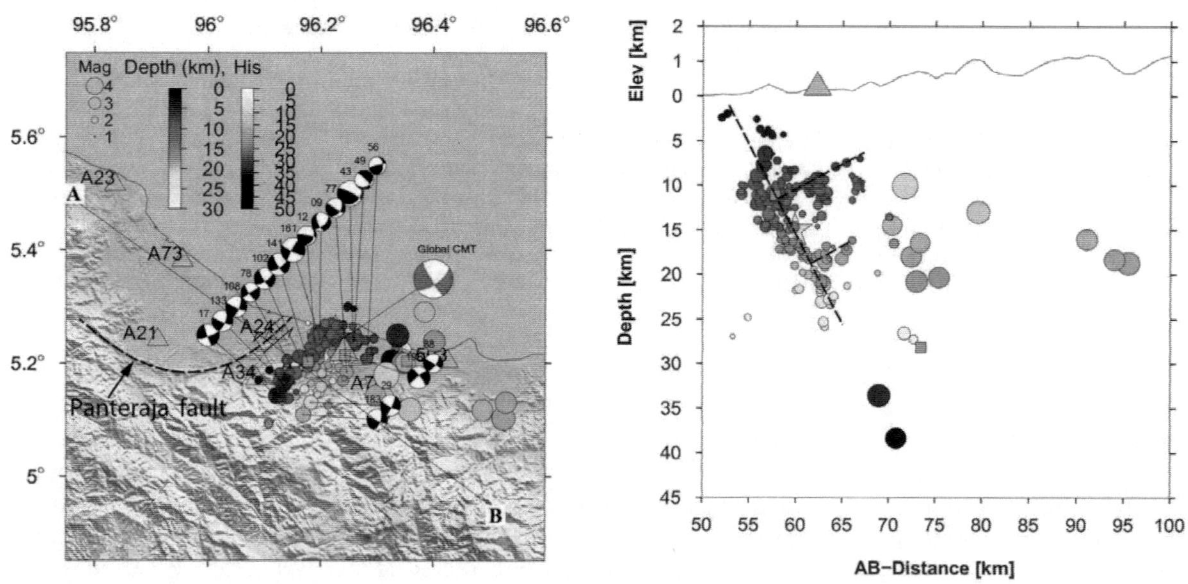

(a) (b)

Figure 1. (a) Map of aftershocks distribution and focal mechanism of the 2016 Pidie Jaya earthquake, and (b) cross-section A-B [7].

The characteristics of the study site were developed based on the results of the site investigations up to 30.5 m depth at a site in the affected area of the 2016 Pidie Jaya earthquake. Additional study results by [7] were used to estimate the sub-surface profile at a greater depth than 30.5 m. The HVSR curves at two different sites in the affected area are illustrated in Figure 2. These HVSR analyses suggested a relatively shallow bedrock at the measured sites.

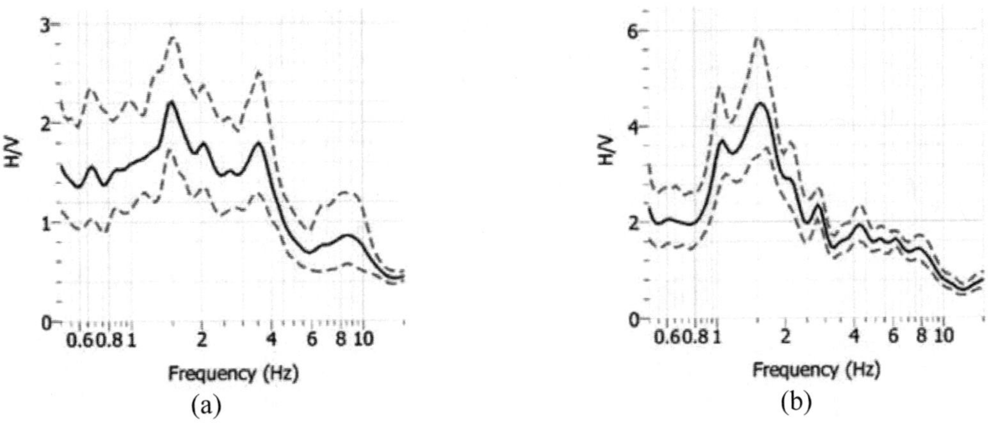

Figure 2. HVSR analysis at two different sites at Pidie Jaya [7].

3. Methods

A sequence of steps (Figure 3) is followed to interpret the earthquake motions at the stable ground surface or bedrock to account for their effects on the soil profile at any specific site. There are three parameters to be defined, which are earthquake input motions, soil profile, and dynamic soil characteristics i.e. strain dependent modulus reduction and damping behaviour.

3.1. Acceleration time histories

In this study, four seismic motions were used. These four historical seismic events are outlined in Table 1. Three seismic motions are actual of historical seismic events of the 2012 Simeulue II earthquake, the 2013 Mane-Geumpang earthquake, and the 2013 Bener Meriah earthquake. The last seismic motions of the 2016 Pidie Jaya earthquake is synthetic time histories generated using EXSIM of [8]. All the acceleration time histories of the four past events are shown in Figure 4.

Table 1. Four seismic events used in this study.

Event	Epicentre coordinates	Magnitude (ML)	Depth & distance from Pidie Jaya (km)
7:43:11 on 11-04 2012 Simeulue II	N 0.82 – 92.42 E	8.1	24 & 600
05:22:42 on 22-01-2013 Mane-Gempang	N 5.49 – 95.21 E	6.0	11 & 70
14:37:03 on 02-07-2013 Bener Meriah	N 4.70 – 96.61 E	6.2	10 & 90
05:03:33 on 06-12-2016 Pidie Jaya	N 4.70 – 96.61 E	6.5	15 & 10

3.2. Soil profile

One-dimensional (1D) sub-surface profile is developed for the site response analysis. The developed 1D profile i.e. soil type, layer thickness, unit weight, shear wave velocity is shown in Table 2. This 1D profile was developed using a borehole sunk at a site in the affected area of the 2016 Pidie Jaya earthquake. The estimation of the shear wave velocity of the 1D was developed from the mechanical

IOP Conf. Series: Materials Science and Engineering **712** (2020) 012013 doi:10.1088/1757-899X/712/1/012013

cone penetration test (CPT) and standard penetration test (SPT) ([9]-[18]). This detailed knowledge of the subsurface characteristics is important in the construction of the profile.

Table 2. Simplified soil profile input for ground response analysis

Note	Layer Number	Soil Material Type	The thickness of layer (m)	Total unit weight (kN/m^3)	Shear wave velocity (m/sec)
	1	Clay	4	21.31	120
	2	Sand	13	23.99	180
	3	Sand	13	23.48	220
	4	Sand	15	23.46	350
	5	Sand	10	23.26	500
Bedrock	6			22.80	800

3.3. Dynamic characteristics of shear modulus and damping curves

The default modulus reduction and damping curves provided by the EERA model have worked well in most applications [19]. The present study adopts these default curves to represent each typical material behaviour during strain since there was no laboratory testing to determine these curves.

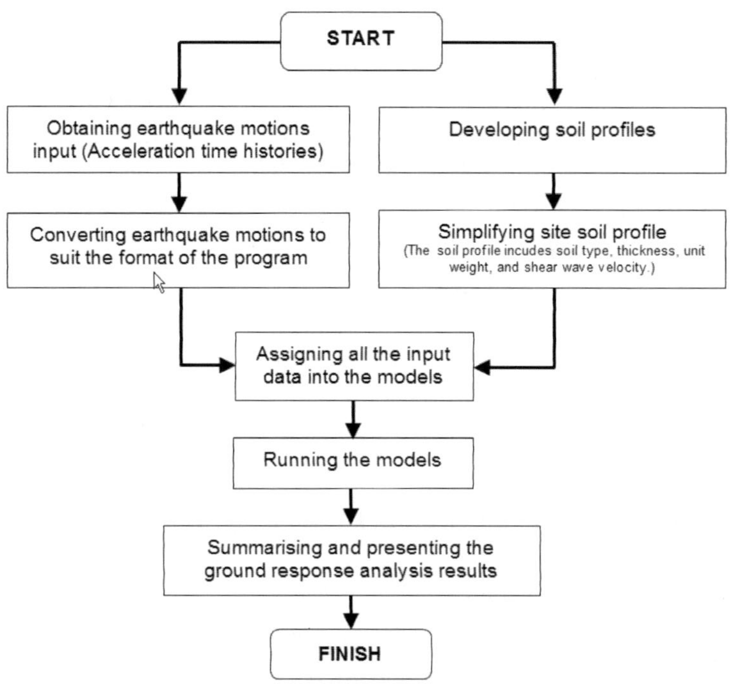

Figure 3. A sequence of steps for site-specific ground response analysis

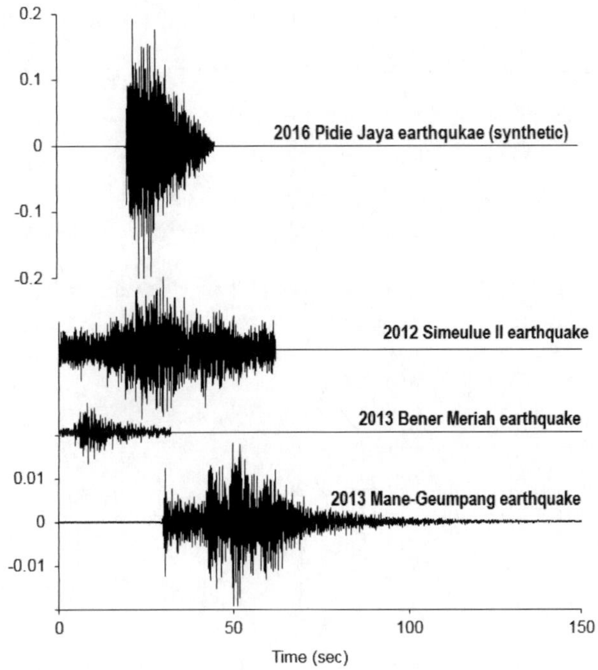

Figure 4. Time histories for ground response analysis

4. Results

The site-specific ground response analysis produces the following results: peak ground acceleration (PGA), response spectrum, fundamental site frequency, and site amplification. A summary of the site-response analysis outputs using the EERA is presented in Tables 3 and 4 below.

Table 3. PGA, maximum spectral acceleration and maximum spectral velocity results of site-specific ground response analysis

Parameters	Seismic event			
	2012 Simeulue II	**2013 Mane-Geumpang**	**2013 Bener Meriah**	**2016 Pidie Jaya**
PGA (g)	0.09	0.11	0.03	0.56
Max spectral acceleration (g)	0.35	0.51	0.10	1.95
Max spectral velocity (cm/s)	15.08	24.61	8.92	149.49

4.1. Peak ground acceleration (PGA), spectral ground acceleration and spectral velocity

The peak ground acceleration (PGA) results are 0.09g for 2012 Simeulue II earthquake, 0.11g for 2013 Mane-Geumpang earthquake, 0.03g for 2013 Bener Meriah seismic event, and 0.56g for the 2016 Pidie Jaya earthquake. As shown, the PGA varies across the seismic events. As expected, the highest PGA was caused by the 2016 Pidie Jaya seismic event (0.56g) and the lowest PGA of about 0.03g was triggered by the 2013 Bener Meriah earthquake. Similar results are shown in the max spectral acceleration and maximum spectral velocity. The highest spectral acceleration and velocity were caused by the 2016 Pidie Jaya seismic event of 1.95g and 149.49 cm/s, respectively. The lowest spectral acceleration and velocity were triggered by the 2013 Bener Meriah earthquake of 0.1g and 8.92 cm/s, consecutively. The spectral acceleration curves of all analyses are presented in Figure 5.

IOP Conf. Series: Materials Science and Engineering **712** (2020) 012013 doi:10.1088/1757-899X/712/1/012013

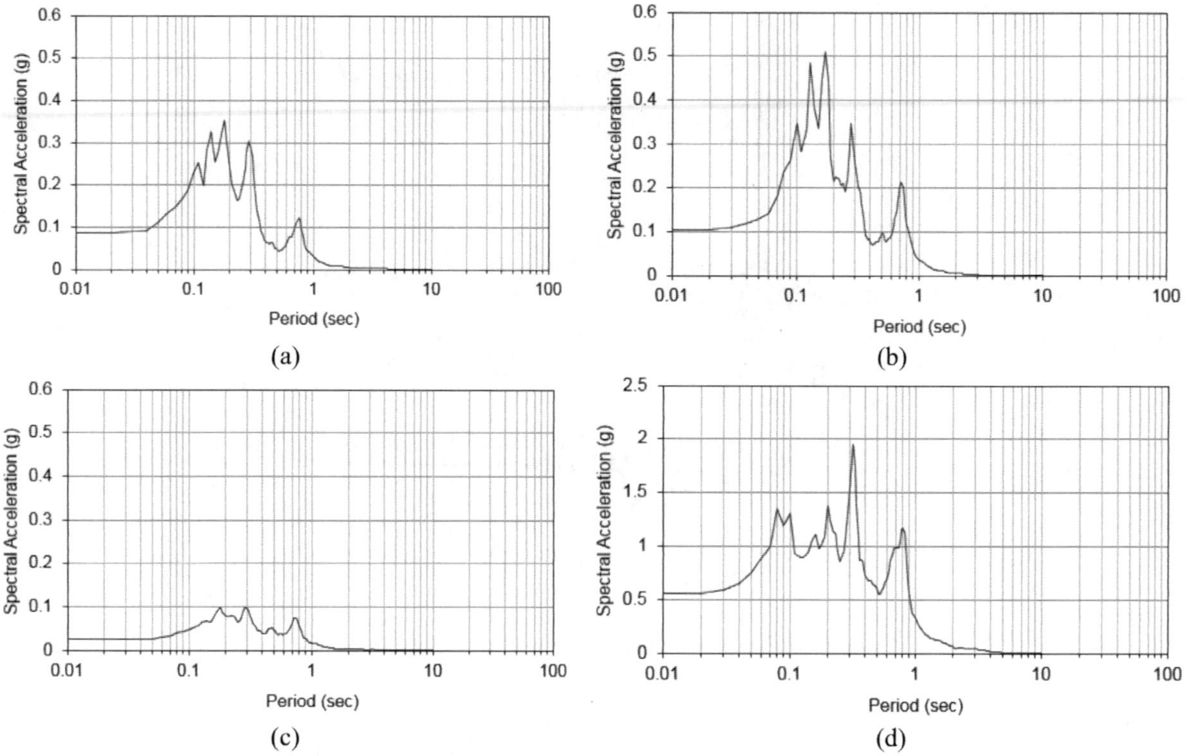

Figure 5. Spectral acceleration (a) 2012 Simuelue II, (b) 2013 Mane-Geumpang, (c) 2013 Bener Meriah, and (d) Pidie Jaya earthquakes

4.2. Site fundamental frequency and amplification

A summary of the site-response analysis outputs of the fundamental frequency and estimated amplification at ground level using the EERA is presented in Table 4. The estimated fundamental frequency of all studied sites is between 1.2 to 1.4 Hz. The typical estimated fundamental frequency is shown in Figure 6. The estimated amplification at the ground level at the investigated site is between 3.1 and 5.4. PGA profiles and amplification profiles of site response analysis are shown in Figure 7.

Table 4. Estimated fundamental frequency and amplification at ground level results of the site-specific ground response analysis

Parameters	**Seismic event**			
	2012 Simeulue II	**2013 Mane-Geumpang**	**2013 Bener Meriah**	**2016 Pidie Jaya**
Fundamental frequency (Hz)	1.4	1.4	1.4	1.2
Amplification at ground level	5.4	5.3	3.3	3.1

4.3. Discussion

In this study, the 2016 Pidie Jaya earthquake is estimated to produce PGA of 0.56g with a simplified average spectral acceleration (SA) of 1.0g at a period of 0.1 to 0.6s. These estimated PGA and simplified average SA of this study are in a reasonably well agreement with the USGS estimations [20] of 0.45g for PGA and ≈0.9g for peak SA, as shown in Figures 8a & 8b.

Another output of this study is estimated site fundamental frequency, which is important for building seismic resistant design [21]. Generally, the building fundamental frequency, f_B can be calculated using an Equation 1 [22].

$$f_B = \frac{10}{N} \qquad \text{Eq. 1}$$

Most buildings at the affected area of the 2016 Pidie Jaya earthquake area are single storey house and up to 3 storey shophouses, therefore the affected area of the Pidie Jaya earthquake's building frequency is estimated between 3.0 and 10 Hz (N is the number of building storey). In general, this study suggests a fundamental frequency of 1.2 to 1.4 Hz, which is only can significantly amplify the medium rise to high-rise buildings. However, detailed site fundamental frequency investigation using passive noise data, as shown in [25], is recommended as highly sub surface spatial variability around the study site is observed [7], [23], [24].

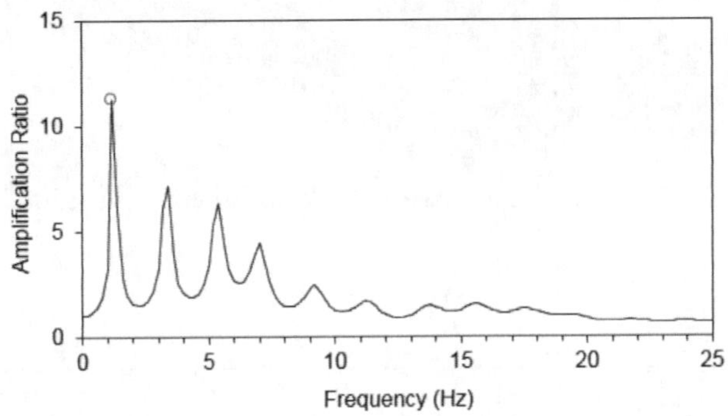

Figure 6. Typical estimated site fundamental frequency.

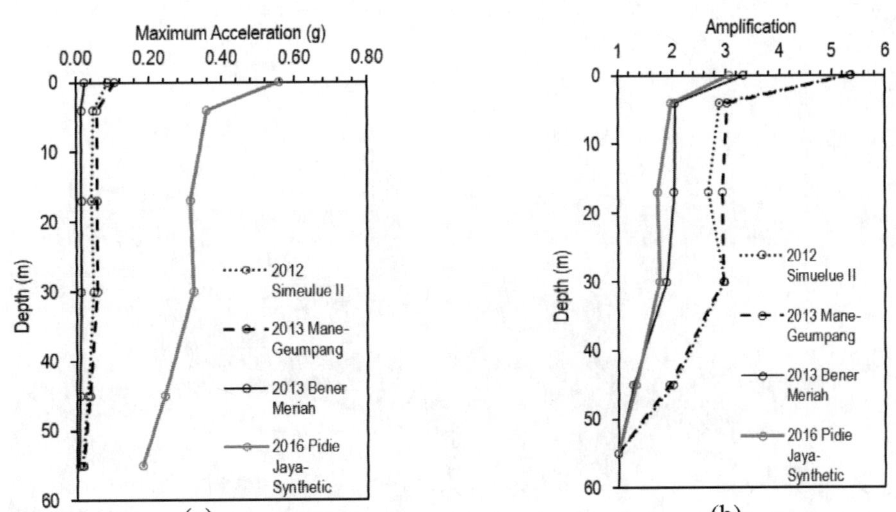

Figure 7. (a) PGA profiles of site response analysis, and (b) Amplification profiles of site response analysis.

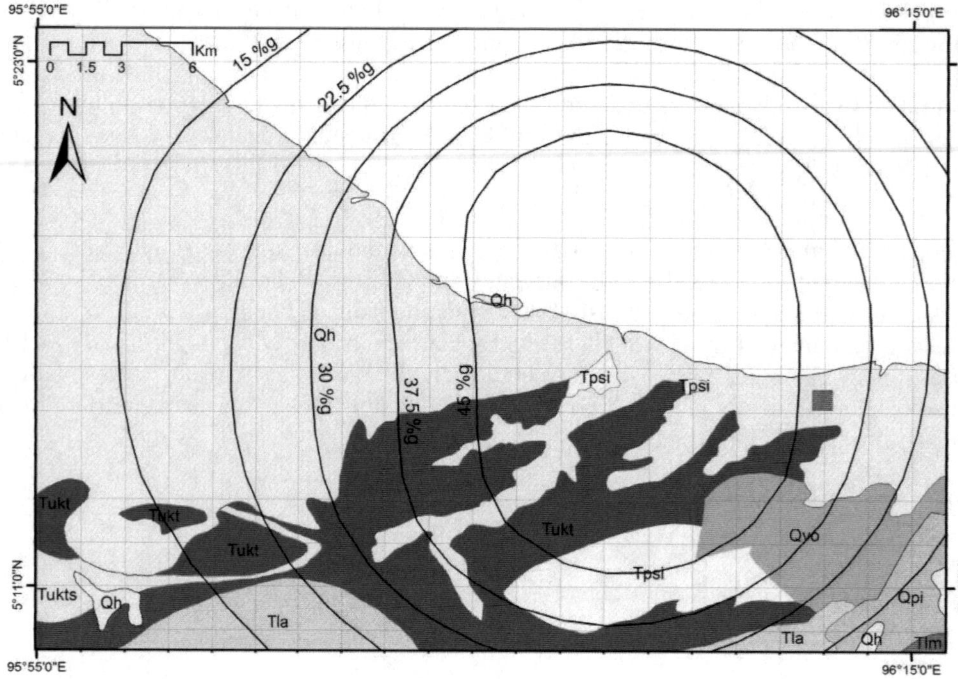

(a)

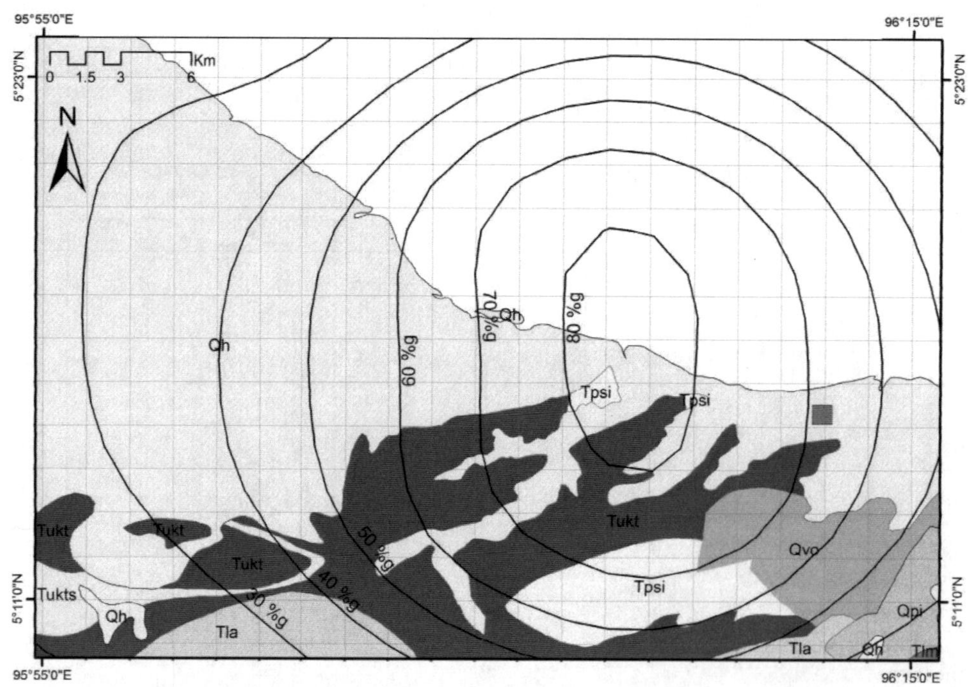

(b)

Figure 8. Estimated (a) PGA and (b) peak spectral acceleration (PSA) of Pidie Jaya earthquake by USGS [20] incorporated geological characteristics [23] and [24]. The red box is the location of the investigated site.

IOP Conf. Series: Materials Science and Engineering **712** (2020) 012013 doi:10.1088/1757-899X/712/1/012013

5. Conclusion

A case study involving ground response analysis at the affected area of the 2016 Pidie Jaya earthquake was presented. This site response analysis involved the input parameters of four earthquake motions, a 1D developed soil profile and three types of modulus and damping curves. The results of these ground response analyses show that the PGA is up to 0.56g. The estimated PGA amplification at the ground level is up to 5.4. Average spectral acceleration with a 5% ratio of critical damping in the range of period from 0.1 and 0.60s is about 1.0g. A comparison of PGA and peak spectral acceleration of this study with USGS estimation confirms the reasonably good agreement.

6. References

[1] Chávez-García, F. J., Jaramillo, H. M., Cano, M. G. and Ortega, J. J. V 2018 Vulnerability and Site Effects in Earthquake Disasters in Armenia (Colombia). I—Site Effects, Geosciences, 8 (254) 22pp. doi:10.3390/geosciences8070254.

[2] Hanks, T. C. and Krawinkler, H 1991 The 1989 Loma Prieta earthquake and its effects: introduction to the special issue, Bulletin of the Seismological Society of America, 81 (5), 1415-1423.

[3] Setiawan, B 2017 Site Specific Ground Response Analysis for Quantifying Site Amplification at A Regolith Site. Indonesian Journal on Geoscience, 4 (3), p.159-167. DOI: 10.17014/ijog.4.2.159-167.

[4] Setiawan, B., Jaksa, M. B., Griffith, M. C., and Love, D., 2018. An investigation of local site effects in Adelaide-South Australia: learning from the past. Boll. di Geofisica Teorica ed Applicata, 59(1): 27–46, DOI: 10.4430/bgta0218.

[5] Bardet, J. P., Ichii, K., and Lin, C. H., 2000. EERA a computer program for Equivalent-linear Earthquake site Response Analyses of layered soil deposits: Department of Civil Engineering, University of Southern California.

[6] Schnabel, P. B., Lysmer, J., and Seed, H. B., 1972. A computer program for earthquake response analysis of horizontally layered sites, Earthquake Engineering Research Center EERC Report 72-12. Berkeley, California: University of California.

[7] Muzli, M., Umar, M., Nugraha, A.D., Bradley, K.E., Widiyantoro, S., Erbas, K., Jousset, P., Rohadi, S., Nurdin, I., and Wei, S., 2018. The 2016 Mw 6.5 Pidie Jaya, Aceh, North Sumatra, Earthquake: Reactivation of an Unidentified Sinistral Fault in a Region of Distributed Deformation, Seismological Research Letters, 89(5): 1761-1772, doi: 10.1785/0220180068.

[8] Motazedian, D. and Atkinson, G., 2005. Stochastic finite fault modeling based on a dynamic corner frequency. Bull. Seism. Soc. Am. 95:995-1010.

[9] Mayne, P. W., and Kulhawy, F. H., 1982. K_0-OCR relationships in soil. Journal Geotechnical and Geoenvironmental Engineering, 22.

[10] Barrow, B. L., and Stokoe, K. E. I., 1983. Field investigation of liquefaction sites in Northern California. University of Texas Austin.

[11] Robertson, P. K., 1990. Soil classification using CPT. Canadian Geotechnical Journal, 27 (1), 8.

[12] Hegazy, Y. A. and Mayne, P. W., 1995. Statistical correlation between Vs and cone penetration data for different soil types. International Symposium on Cone Penetration Testing.

[13] Andrus, R. D., Stokoe, K. H., and Chung, R. M. 1999. Draft guidelines for evaluating liquefaction resistance using shear wave velocity measurements and simplified procedures: U.S. Department of Commerce.

[14] Andrus, R. D., Zhang, J., Ellis, B. S., and Juang, C. H., 2003. Guide for estimating the dynamic properties of South Carolina soils for ground response analysis: Clemson University.

[15] Mayne, P. W., 2006. In-situ test calibrations for evaluating soil parameters. The Overview Paper on In-situ Testing -Singapore Workshop, Singapore.

[16] Mayne, P. W., 2007. National Cooperative Highway Research Program (NCHRP) synthesis 368, cone penetration testing (a synthesis of highway practice). Washington D. C., USA: Transportation Research Board.

[17] Dikmen, U., 2009. Statistical correlations of shear wave velocity and penetration resistance for soils. J. Geophys. Eng. 6(1):61–72, DOI:10.1088/1742-2132/6/1/007.
[18] Tsiambaos, G. and Sabatakakis, N., 2010. Empirical estimation of shear wave velocity from in situ tests on soil formations in Greece. Bull. Eng. Geology & Environ., 70:291–297, DOI 10.1007/s10064-010-0324-9.
[19] Lasley, S.J., Green, R.A., and Rodriguez-Marek, A., 2014. Comparison of Equivalent-Linear Site Response Analysis Software, Proceedings of the 10th National Conference in Earthquake Engineering, Earthquake Engineering Research Institute, Anchorage, AK, 10 pp.
[20] USGS, (2019). M 6.5 - 14km WNW of Reuleuet, Indonesia, Accessed by 12 June 2019, https://earthquake.usgs.gov/earthquakes/eventpage/us10007ghm/shakemap/pga
[21] Mucciarelli, M., Masi, A., Gallipoli, M. R., Harabaglia, P., Vona, M., Ponzo, F., and Dolce, M., 2004. Analysis of RC Building Dynamic Response and Soil-Building Resonance Based on Data Recorded during a Damaging Earthquake (Molise, Italy, 2002), Bulletin of the Seismological Society of America, Vol. 94, No. 5, pp. 1943–1953.
[22] Mihaylov, D. and El Naggar, M. H., 2014. Soil response change and building fundamental resonances during earthquake shaking, Second European Conference on Earthquake Engineering and Seismology, Istambul, 25-29 August 2014, 10 pp.
[23] Bennet, J., Bridge, D., Cameron, N., Djunuddin, A., Ghazali, S., Jeffery, D., Kartawa, W., Keats, W., Rock, N., Thomson, S., and Whandoyo, R., 1981. Peta Geologi Lembar Banda Aceh, Sumatra Skala 1:250.000, Pusat Penelitian dan Pengembangan Geologi, Bandung.
[24] Keats, W., Cameron, N., Djunuddin, A., Ghazali, S., Harahab, H., Kartawa, W., Ngabito, H., Rock, N., Thomson, S., and Whandoyo, R., 1981. Peta Geologi Lembar Lhokseumawe, Sumatra Skala 1:250.000, Pusat Penelitian dan Pengembangan Geologi, Bandung.
[25] Setiawan, B., Jaksa, M., Griffith, M., and Love, D., 2018. Passive noise datasets at regolith sites. Data in Brief, 20, 735-747. DOI: 10.1016/j.dib.2018.08.055.

Acknowledgments
The authors wish to acknowledge Universitas Syiah Kuala for providing a research grant with a contract no. 521/UN11/SPK/PNPB/2019 date 8 February 2019. Also, the authors are grateful to the Faculty of Engineering of Syiah Kuala University for their support.

IOP Conf. Series: Materials Science and Engineering **712** (2020) 012014 doi:10.1088/1757-899X/712/1/012014

Selection of Contractor by Using Analytical Hierarchy Process (AHP)

P Z Razi, N I Ramli, M I Ali and P J Ramadhansyah

Faculty of Civil Engineering Technology, Universiti Malaysia Pahang, 26300 Gambang, Pahang, Malaysia

Abstract. Construction projects is progressing rapidly in Malaysia and have been as one of an important factor in contributing the boost of country's economic. Well known as a sectors with involvement of multiple players, thus, contractor selection is an important aspect in construction sector especially for client or parties involve that need to hire the best contractor to manage and complete their project within time-line and bearable cost couple with comprised quality. Furthermore, without a suitable method use in selection of contractor, it will affect the completion of whole project. This paper aiming in make use of the analytical hierarchy process (AHP) as a decision support model to select contractor. The AHP allows constructing decision as hierarchies and each criterion can be evaluated through weighted determined by the expert in construction field. The objectives of this study is to develop the main criteria used by client in selection of contractors also identify the weighted criteria by using AHP method from current practice in Malaysia. Besides, this method is one of the decision-making that is necessary to eliminate the risks of project failure due to poor contractor's performance.

1. Introduction

The decline in the construction industry occurred around 2005 to 2006 with a rate of negative 5.1%. There was a reduction in the number of projects and many bankrupts' contractors [1]. In Malaysia, issues faced for selection of contractor where they do not emphasizing an important aspect in the delivery of construction projects where it linked to project success, in term of time schedule, cost, and quality. Besides, the overall project quality and owner satisfaction is relevant to the contractor performing the work. Contractor need to understand the procedures for obtaining government, private projects or tenders. Many of them are blacklisted because they cannot afford financial risk and responsibility given to complete the projects, also demand in price from chosen contractors when come from closed tender. From observation, it is found that the contractors with insufficient financing where most of them do not have sufficient capital to finance their undertakings. Then, lack of experience and skills in technical or through management in construction phase which contractor unable to complete the project given according to agreed costs and time scheduled. Also, their quality performance for previous project that give them positive or negative impact. However, this study will identify the best criteria or factors that are important during selection of contractors using Analytical Hierarchy Process (AHP) where a theory of measurement through pair wise comparisons and relies on the judgments of experts to derive priority scales was applied.

2. Literature Review

In this challenging era, to get successful project construction we need to manage the flow or process in an effective manner. The demands from clients, competition, and regulatory agencies have been growing rapidly [2], a failure to properly manage them can lead to problems for the entire project and construction team. Various studies have shown that overall project quality and owner satisfaction is directly related to the contractor performing the work [3,4]. Selection of a contractor deemed as major challenge for many construction firms and there is a need for employing systematic tools prior to select the best alternative among different contractors [5]. The selection of a proper construction contractor increase chances of successful completion of a construction project [6]. The selection of contractors is an important aspect in the delivery of construction projects and is linked to project success, in the terms of schedule, cost, and quality [7]. To choose capable construction contractor is one of the important aspect faced by the client or parties involve who wishes to achieve successful complete projects outcomes. This type of tasks is challenging, because construction industry is rapid growth and competitive around the world. [8] Agree that the probability of construction failure is quite high for individual contractors, and it is important for project owners to confront and manage these risks if they wish to achieve good project results. In the other words, we need to find the decision making tools where we can seek the best of the satisfactory options to be the best solution to the problem [9]. However, failure to make effective decision leads to poor, ineffectual and wrong decisions [10].

3. Research Methods

This study employing quantitative analysis a multi-criteria decision making (MCDM) namely the Analytical Hierarchy Process (AHP). Figure 1 below depicts the research flow process. Based on the figure, this study first scrutinizing the literature review on the "factors for the selection of contractors". Drawing largely upon factors of pre-selection of contractors developed by [11], a hierarchy of the selection was further developed. Analytical hierarchy process (AHP) is a decision aiding method developed by Professor Saaty back in 1980's [12] and this method is still being in practised up during this era. Regardless of aiding in decision making, AHP also assists stakeholders especially in solving complex and conflict multicriteria and subjective issue by facilitating to prioritize and rank the associated problem [13].

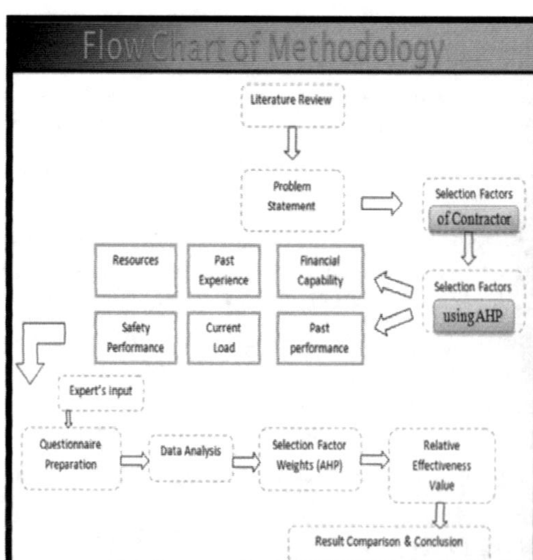

Figure 1. Research Methods Flowchart

AHP is designed based on the pair-wise comparison scale and as shown in **Table 1.** It relies on the judgement of expert from scale one (1) to nine (9).

Table 1. Scale of Relative Importance for Pair-Wise Comparison [12]

Numerical Rating	Verbal Judgement of Importance
9	Extremely importance
7	Very strongly importance
5	Essential or Strong importance
3	Moderate Importance
1	Equal Importance
2,4,6,8	Intermediate value

The pairwise comparison of various factors generated are organized into a square matrix as depicted in **Eq. 1** below.

$$A = aij = \begin{Bmatrix} a11 & \dots & a1n \\ \vdots & \ddots & \vdots \\ an1 & \cdots & ann \end{Bmatrix} \quad \textbf{Eq. 1}$$

The principal eigenvalue and the corresponding normalized right eigenvector of the comparing matrix give the relative importance of the various factors being compared. The elements of the normalized eigenvector are termed weights with respect to the factors or sub-factors and ratings with respect to the alternatives. Equation (3) showed formula of each matrix that needs to be normalized.

$$Aw = \lambda\ max.W \quad \textbf{Eq. 2}$$

[12] demonstrated that λ_{max} = n is a necessary and sufficient condition for consistency. Inconsistency may arise when λ_{max} deviates from *n* due to varying responses in the pairwise comparisons. Therefore, [12] proposed a method to measure the inconsistency by first estimating the consistency index (CI). The CI is defined in Eq. (3). Then, to obtain the consistency ratio (CR), the CI is divided by the random consistency index (RI) in Eq. (4), value of RI as tabulated in **Table 2.** The CR value should not greater than 0.1 otherwise the pairwise comparison result should be rejected.

$$CI = \frac{(\lambda\ max - n)}{(n-1)} \quad \textbf{Eq. 3}$$

$$CR = \frac{CI}{RI} \quad \textbf{Eq.4}$$

Table 2. Random Consistency Index (RI)

Number of Factors	3	4	5	6	7	8
Random Index (RI)	0.58	0.90	1.12	1.24	1.32	1.41

As this kind of multi-criteria decision making method requires only expert judgement, thus, sampling size is not a big issue, according to [14], if an expert is well versed in his field, his decision is sufficient unless political expediency requires that several judges from different constituencies are necessary. This study employing the purposive sampling methods with the pair-wise questionnaire only considered construction expert with ten (10) years of construction experience. Distribution of the questionnaire conducted by using face to face method, taking into account that this method may unfamiliar to the respondent.

4. Results and Discussion

4.1. Demographic Data

The respondents profile is tabulated in **Table 3** below. From the table, only respondents with more than 10 years of construction experience are taken into account for further analysis. This leads to only the opinion of twelve (12) construction experts who will be analyzed to the next level. Data in the table depicts that respondent with 10 to 14 years of experience is 58% while respondent with over 15 years is 42%. While the percentage breakdown for profession shows Civil and Structural Consulting Engineer recorded the highest percentage with 33.33%, Archictect and Public Authority at 25% respectively, while the rest belongs to senior site project manager at 16.67%.

Table 3. Respondent Profile

Profession	Experiences (Years)			
	1 to 5 years	6 to 9 years	10 to 14 years	Over 15 years
C&S Consultant Engineer	*2	*1	2	2
Architect	-	-	2	1
SiteProject Manager	-	-	1	1
Public Authority	-	-	2	1

* data is not accounted for because it does not comply with inclusion criteria expert with > 10 years experiences.

4.2. AHP Results and Discussion

Table 4 below depicts results of the AHP analysis results into six (6) factors that have been listed as factors that contribute in the selection of contractors. The analysis illustrates that the priority vector (PV) as the relative importance weighted for each factor, while prioritized ranking (PR) has arranged in descending in order to facilitate the discussion of each factors. Obviously in the table, financial capability (0.478) monopolizes the factor compared to the other five factors. Next is followed by past performance (0.252), past experience (0.148), resources (0.073), current workload (0.038) and finally the safety performance (0.006) factors.

Table 4. Factors PV and PR

Factors (F)	Priority Vector (PV)	Prioritized Ranking (PR)
F1 (Financial Capability)	0.4776	1
F2 (Past Performance)	0.2518	2

F3 (Past Experience)	0.1483	3
F4 (Resources)	0.0735	4
F5 (Current Workload)	0.0376	5
F6 (Safety Performance)	0.0066	6

The financial capability of a contractor company is indeed an important element in the selection criteria of the contractor prior to being procured. The finding of financial capabilities correspond to previous study [15] whom conduct a risk assessment related to the choice of contractors and found that among the important criteria to be assessed is the contractor financial capabilities. Similarly, study by [16] ranked financial capability as the top final contractor selection factor by using the analytical hierarchy process. In like manner, [17] found the top knotted of critical success factors for Malaysian contractors in international construction projects are the contractor's experience, effectiveness in decision making and contractor's cash flow. Past performance report of a contractor strongly encourages the client to decline or select the contractor. The contractor's past performance will also stimulate the success of a construction project. Nevertheless, the report should undergo a transparent and fair system without any personal interest. Among the significant factors in the evaluation of past performance reports listed by [18] are through a transparent and fair procurement system. Interestingly, this study found that the respondents evaluated safety performance as an unimportant factor in the selection of contractors. This is in line with the results obtained by [16] whom reported safety performance as the last factor to be assessed for the selection of contractor. On the other hand, the respondent also ignored the factors of current workload or burden borne by the contractor. A possible explanation for this finding is that, if the contractor is competent enough, the current workload is not a major problem for the prospective contractors. This is largely dependent on the maturity level of contractor companies in managing and allocating their in hands projects without affecting their construction performance level.

5. Conclusion

This study aims to identify which factors is prioritized during selection of a contractor. The results provides some empirical findings that illuminates client perspectives in making fast decision during selection of a contractor. It appears that financial capability (0.478) dominates the ranking, followed by past performance (0.252), past experience (0.148), resources (0.073), current workload (0.038) and finally the safety performance (0.006) factors. This study is conducted on the basis of extracting expert judgement quantitatively. Future studies can consider to employ a qualitative method in the form of depth interview pertaining the obtained factors in this study, or adopting a mixed method approach.

References

[1] CIDB 2005 CIDB Annual Report Kuala Lumpur, Malaysia.
[2] Crowley B.L.G and Hancher D.E 1995 Risk Assessment of Competitive Procurement, *ASCE* June 230–237
[3] Cheung S.O. Wong P.S.P Fung A.S.Y and Coffey W.V 2006 Predicting project performance through neural networks International Journal of Project Management **24** 3 207–215.
[4] Maloney W.F. 2002 Construction Product/Service and Customer Satisfaction Journal of Construction Engineering Management **128** 6 522–529.
[5] Demirkesen S. and Bayhan H.G. 2019 Subcontractor Selection with Choosing-By-Advantages (CBA) Method IOP Conference Series: Materials Science and Engineering **471** 022020.
[6] Alhazmi T. and McCaffer R. 2000 Project Procurement System Selection Model Journal of Construction Engineering and Management **126** 3 176–184.

[7] Hatush Z. and Skitmore M. 1994 Contractor Selection using Multicriteria Utility Theory: An Additive Model Building and Environment **2–3** 33 105–115.
[8] Kangari R. and Bakheet M. 1994 Linguistic Analysis of Risk in Project Planning Computing in Civil Engineering.
[9] Whitney S.N. 2003 A new model of medical decisions: Exploring the limits of shared decision making Medicial Decision Making **23** 4 275–280.
[10] Drucker P.F. 2001 An excerpt from The Essential Drucker, First. Claremont, California: HarperCollins Publisher.
[11] Almeida M. 2016 Pre-Qualification Of Contractors For High-Rise Building Projects In Philippines: A Selection Method in Construction Management using Analytic Hierarchy Process (AHP) 2016 as a Tool in Decision Making DLSU Research Congress **4**.
[12] Saaty T.L. 1988 What Is the Analytic Hierarchy Process ? Mathematical Model for Decision Support **48** 109–121.
[13] Ishizaka A. and Labib A. 2009 Analytic Hierarchy Process and Expert Choice: Benefits and limitations ORInsight **22** 4 201–220.
[14] Saaty T.L. and Özdemir M.S. 2015 How Many Judges Should There Be in a Group ? Annals of Data Science **3–4** 359–368.
[15] Valitova I. Trofimova T. Simankina T. and Stein E 2018 Risk assessment related to the choice of contractors performing the job MATEC Web Conference **170** 01068.
[16] Fong P.S.W. and Choi S.K.Y 2000 Final contractor selection using the analytical hierarchy process Construction Management and Economics **18** 5 547–557.
[17] Derrick J-Z T. and Farid Ezanee M.G. 2011 Critical Success Factors for Malaysian Contractors in International Construction Projects using Analytical Hierarchy Process EPMM **September** 127–138.
[18] Zainal R. Hamza G. Manap N and Syahrom N.S 2019 Significant Ways to Improve the Performance of Medium and Large Local Contractors in the Jigawa State Nigeria International Conference on Built Environment and Engineering (IConBEE2018) **3018**.

Selection of Method in Construction Industry by using Analytical Hierarchy Process (AHP)

P Z Razi, N I Ramli, M I Ali and P J Ramadhansyah

Faculty of Civil Engineering Technology, Universiti Malaysia Pahang, 26300 Gambang Kuantan

Abstract Method in construction industry is very important in construction industry. The type of method used in construction may affect the quality of the project. Three distinct method that will be explore in this paper are traditional method, design and build and industrial building system (IBS). The traditional procurement method, known for its separate entity by the client, architect and contractor. The architect will responsible for design work, while contractor will responsible for construction work. Meanwhile, the design and build, whereby the contractor accepts responsibility for some or all of the design finally the IBS is when the construction built using pre-fabricated components. This paper aiming in identifying the criteria and suitability for selecting different kind of construction delivery method in construction by using the multi-criteria decision making (MCDM) namely the Analytical Hierarchy Process (AHP) method. Results provides some empirical finding which contractor for Grade 7 is suitable employing the industrial building system(IBS) method while the traditional method is appropriately for the Grade 1 contractor.

1. Section heading

The choice of the appropriate method in a construction project is indeed an important decision and is a major factor that greatly influences the productivity and efficiency of the construction project in addition to the large impact aspect given by the method of construction on productivity, quality, and cost [1]. The method of construction is generally known as one of the ways in which it is used to alter the raw material and is exhibited in a product form [2]. Rendering to [3], any sophisticated programming and management of a system that was introduced in the construction project, it will be worthless if the method of construction used does not giving any optimal impact in terms of time, cost, quality and safety. Construction professionals including engineers, (structural or mechanical) expertise was needed during the early stage of a project. Choosing the wise construction method largely crucial during the progressive delivery of a construction developments which often deeds as an actual obstacle to expending the expertise and acquaintance of all project companions meritoriously in the design and inception of the project [4]. In addition, extensive attitude, perception and commanding divergences amidst single project and project groups causing an isolating between design stage and development project growth. Remarkable differences of design and build contract if compare to traditional method is that the clients will have lacking in command and power when comes to the design issues and resist for any major modification. Henceforth, the contractor need to responsible for both design and construction of the project. This causing some clients and contractor having less interest on the leverages offered by design and build method. On the other hand, for the traditional method, client will assign different parties such as architect to performed design work. Reflecting to the issue, within

Malaysia context, the number of projects procured under design and build method decreasing from March 2012 to March 2014 [5]. In similar vein, prior to finish the project in timely manner, associated risk that arise from the verdict made by the contractor for design and construction work should cover by a liability. This is extremely contrast to traditional method, which risk appear from the design stage will fully responsible by the architect [6]. Most small contractors still choose to adopt conventional systems or traditional methods as they are familiar with the facilities offered by conventional systems compared to IBS technology. They argue that the IBS system is more suitable to be applied to large-scale construction projects. [6]. Among the constraints and challenges identified in implementing IBS for construction projects are the financial problems and constraints faced by small contractors who have made it difficult for them to build a manufacturing plant that is in desperate need of high and demanding capital investment. Detrimental the issues, lacking of knowledge and expertise owned by civil engineers in Malaysia in use of the structure analysis for the pre-fabricated design are among obstacles identified by prior study[6]. Given so many choices of methods available in the industry, thus this study is conducted in identifying and analysing within the three method using a systematic decision making tools namely the AHP. This paper will aid construction practitioners that are focussing within the contractor scope in evaluating the feasibility within the conventional, design and build or industrialised building system.

2. Literature Review

Construction industry player has been widely practiced globally and locally in a traditional method (TM). Therefore, the process of the installation of the timber or plywood formwork and steel reinforcement for the building at site is by fabricating components that called as a conventional method [7]. On the other hand, in conventional method have two parts. For the first part is structural system that the frame is in cast in-situ such as column, slab and beam. Hence, cast in-situ process is going into four action which are install the steel reinforcement bar, erection of the timber of formwork and scaffolding, using of pouring fresh concrete and disassemble of formwork and scaffolding [8]. Design and Build (D & B) procurement method of delivery are recognized as the ultimate substitute for most of public and private projects in many countries [9]. D & B is seen as one of the alternative procurement approaches that promote a single entity or consortium that takes sole responsibility for the project, couple with the fixed pricing offer [10]. The popularity of the D&B procurement system was driven by the separation of the design phase from the construction team project [9]. The point is due to the fact that construction projects require a more integrated design and build (D&B) process moreover on-time project delivery by having a privilege consortium entity that only takes on the sole responsibility of the particular project. The D & B system became popular and referred to in Malaysia beginning in the 1980s and 1990s where the Public Works Department (PWD) played their main role. Typical in the D & B project implementation process is the end user who initially identified the project and applied for a budget. Industrialized Building System (IBS) is a construction system built using pre-fabricated components. Component assemblies are systematically made using sophisticated machines, mould systems and mechanical equipment. These components are made out of the construction site (off-site) and upon completion of the manufacture, will be sent to the construction site for installation. IBS is characterize as a construction system where each component is manufactured in a factory or a construction plant or outside of the construction site, then placed and installed into the building structure by requiring only minimal additional site work [9]. The progress of this system is seen as very economical, reducing construction time while reducing construction workers.

3. Research Methods

This study that aiming in identifying the best method to adopt within the Malaysia's contractor scope. In determining which method works best for each contractor, multi-criteria decision making (MCDM), AHP tools are used to assist and facilitate construction practitioners in making quick decisions. AHP was first introduced by Professor Saaty in the 1980s [10], and many studies especially in the field of construction management adopt this method. Among them include [11-13]. For the context of this study, hierarchical and alternative structures are built first as shown in **Figure 1** below.

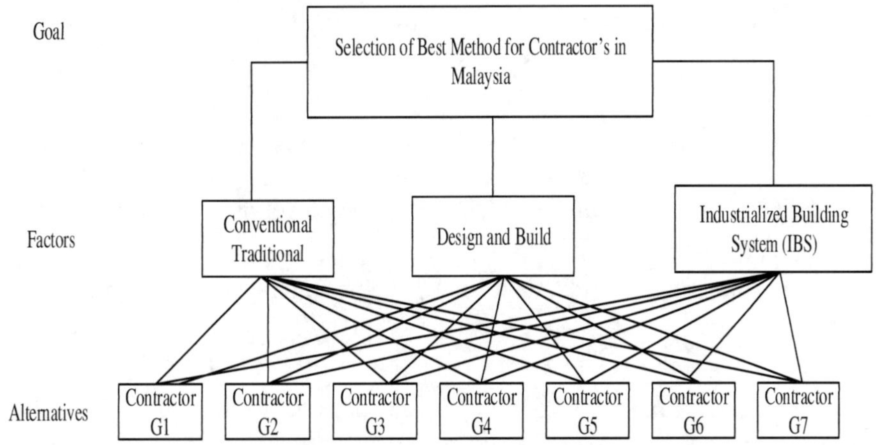

Figure 1. Decision Hierarchy Structure

AHP requires a hierarchy structures to develop first that composed of the goal, factor and optionally any alternatives within the context of the goal. On the other hand, the methods adopted in this study is presented in flowchart diagram shown in **Figure 2.** Uniquely, the AHP method employing the pair-wise questionnaire with nine (9) linguistic scale. The pairwise questionnaire distributed to all local and active construction practitioners. Upon getting their feedback, screening made for any unattended question and if there are any, will revert back to the respondent. Once confirmed, the analysis begin and all the data converted to square matrix form, normalizing the data and finally the data being validate by checking the respondent judgement consistencies and inconsistencies.

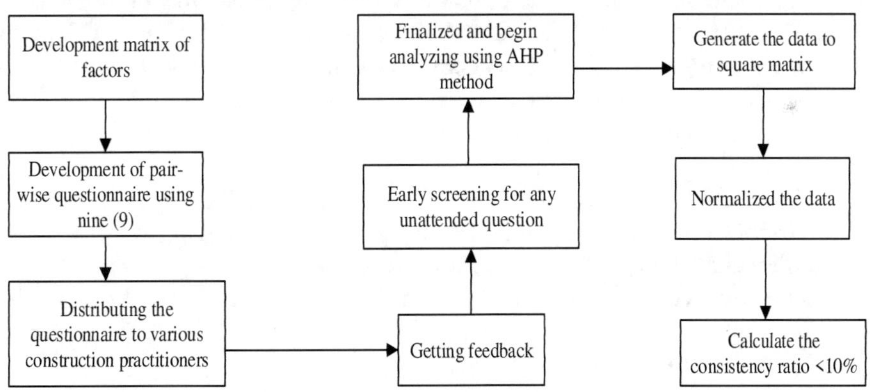

Figure 2. Research Methos Flowchart

In the AHP method all judgements are recorded in a matrix of pairwise comparisons $A = (a_{ij})_{n \times n}$ in which the dimension of matrix n means that we compared n parameters. The element of the matrix a_{ij} denotes a pairwise comparison of parameter i with parameter j, we gave the inverse comparison (comparing parameter j with parameter i). So the reciprocal value:

$$a_{ij} = 1/a_{ij} \qquad (1)$$

We can employ the eigenvector method to derive the parameter weights from the matrix of pairwise comparisons A, which means that we must solve the equation:

$$Aw = \lambda_{max} w \quad (2)$$

where λ_{max} is the maximal eigenvalue of matrix *A*. For every matrix of pairwise comparisons *A* we must also calculate the consistency ratio, which measures the level of inconsistency between pairwise comparisons:

$$CR = \frac{CI}{RI} \quad (3)$$

where $CI = \frac{\lambda_{\max} - n}{n-1}$ is the consistency index, *n* is the size of matrix *A* and *RI* is the average consistency index. We assumed that if $CR \leq 0.10$, then the inconsistency level of matrix *A* is still acceptable.

In the case of group decision making where m is the number of decision makers, we aggregate the individual judgments into one joint judgment ${a_{ij}}^{group}$ applying the geometric mean method:

$${a_{ij}}^{group} = \sqrt[m]{\prod_{k=1}^{m} {a_{ij}}^{k}} \quad (4)$$

Where ${a_{ij}}^{k}$, $k = 1, ..., m$ are the individual judgements of *m* decision makers. [14] Addressed that the geometric mean method Equation (4) is the only appropriate method for aggregating individual judgments into group judgments as it satisfies some necessary axiomatic conditions like preserving reciprocity. For the analysis to be reliable, the acquired data is require a consistency ratio of less than 10% or ($CR \leq 0.1$). On the other hand, a systematic software program namely the 'Expert Choice' is further utilized during the data analysis process. The Expert Choice software is a multi-objective decision support tool based on the Analytic Hierarchy Process (AHP), a mathematical theory first developed at the Wharton School of the University of Pennsylvania by one of Expert Choice's founders. The AHP is a powerful and comprehensive methodology designed to facilitate sound decision making by using both empirical data as well as subjective judgments of the decision-maker(s) [15]. The AHP assists with the decision making process by providing decision-makers with a structure to organize and evaluate the importance of various objectives and the preferences of alternative solutions to a decision [15]. Following are the steps proposed by [15] in AHP and Expert Choice :

- Brainstorm and structure a decision problem as a hierarchical model.
- Set the type and mode of pair wise comparisons or data grid functions.
- Group enable the model.
- Import data to Expert Choice from external databases.
- If applicable, pair wise compare the alternatives for their preference with respect to the objectives, or assess them using one of the following: ratings or step functions, utility curves, or entering priorities directly.
- Pair wise compare the objectives and sub-objectives for their importance to the decision.
- Synthesize to determine the best alternative.
- Perform sensitivity analysis.

4. Results and Discussion

4.1. Demographic Respondent Analysis

This section is explanatory from **Figure 3.** To **Figure 6.** Figure 3. Illustrates the respondents by gender. From the total 30 responses received, 21 responses (70%) received from male and 9 responses (30%) received from female. Figure 4 below shows the composition of respondents by profession. From the total 30 responses received, 8 responses (29%) received from site engineer and designer, 8 responses (28%) from engineer and 4 responses (14%) received from site engineer. On the other hand, figure 5 illustrates the respondent experiences in construction industry. From the total 30 responses received, 13 responses (44%) received from less than 5 years experiences, 6 responses (20%) received from the 5-10 years experiences, 7 responses (23%) received from the 11-15 years experiences and 4 responses (13%) received from more than 15 years. While figure 6 depicts the organization of the company. It shown that, 4 responses (13%) received from the government (public authority), 4 responses (14%) received from the consultant and 22 responses (73%) from the contractor.

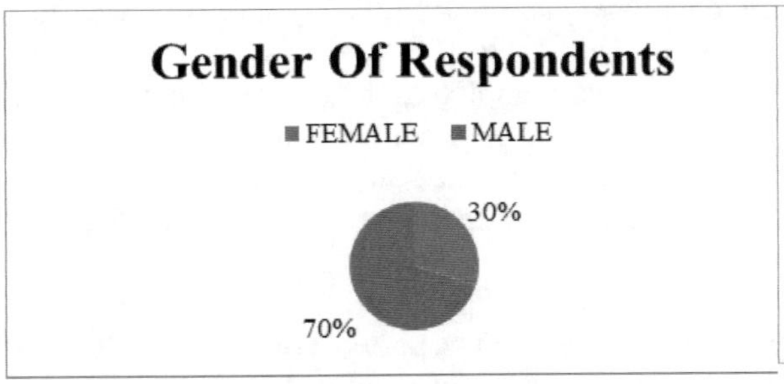

Figure 3. Respondent's by Gender

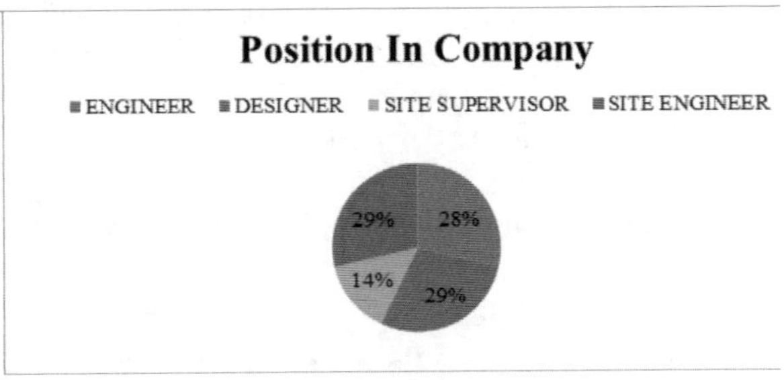

Figure 4. Respondent's Position

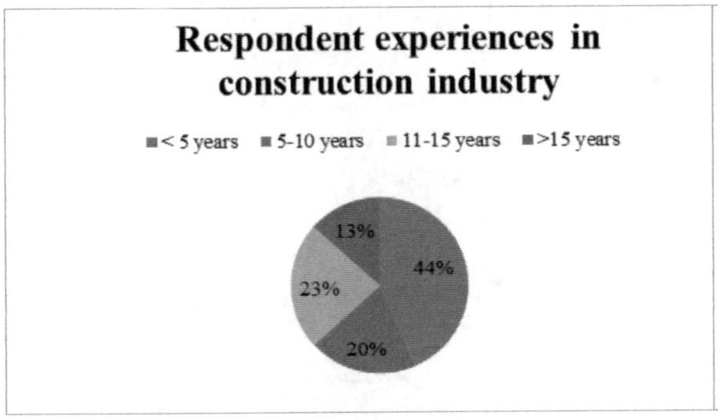

Figure 5. Respondent's Experiences

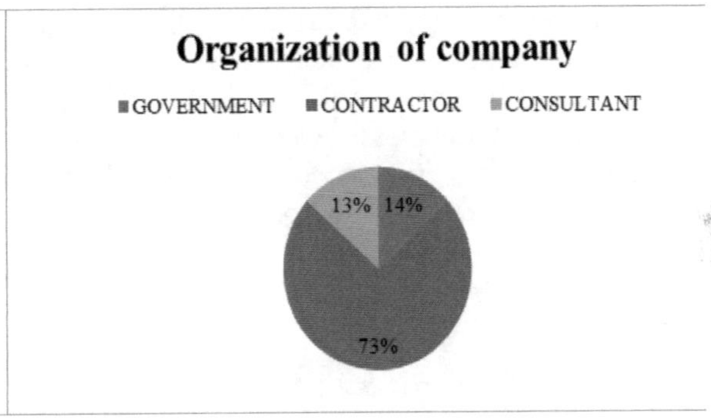

Figure 6. Organization of the company

4.2. AHP Results and Discussion

Figure 7 below depicts the results on the prioritization of the selection for construction method derived upon analysis using the 'Expert Choice' program.

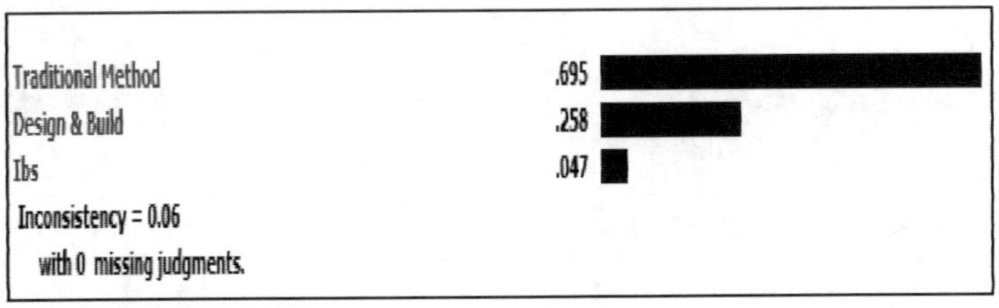

Figure 7. Priorities of selection of construction method

Apparently, the results achieved inconsistencies at 0.06 less than 0.10 (10%) as per recommendation by [10], [16]. Analysis of AHP analysis shows that contractors are more likely to choose conventional building systems rather than D&B and IBS. This is in line with current thinking that contractors prefer to choose conventional building systems. Results shows that traditional method weighted the most prioritized at 0.695 rather than proposing D&B weighted at 0.258 and IBS system weighted at 0.047. A possible explanation for this might due to shuffle of building system from conventional to IBS is not motivated by financial constraint borne by the contractor. Additionally, for decades, most contractors have been disclosed and educated in conventional building systems coupled with the large number of foreign workers employed inexpensively in Malaysia [17].

IOP Conf. Series: Materials Science and Engineering **712** (2020) 012015 doi:10.1088/1757-899X/712/1/012015

The delay in IBS implementation and acceptance in the construction industry in Malaysia is largely driven by the lack of knowledge of IBS in the industry. In addition, the lack of exposure to IBS technology has contributed to the structure analysis and design of the poorly installed components in the manufacturing plant [18]. It further leads to problems when assembling the components resulting to difficulty prior to fitting the installation process [19]. According to [20] many small contractors are in hesitant to embrace IBS system and choose to remain employed the conventional method. The rationale of this lies in the fact that, contractors are formerly adapted with the conventional system and they presume that the available technology system are appropriate and suitable for them in handling with small scale projects. In similar vein, among the constraints and challenges identified in implementing IBS for construction projects are the financial problems and constraints faced by small contractors who have made it difficult for them to build a manufacturing plant that is in desperate need of high and demanding capital investment. Eventually, it was also reported some contractors in Malaysia is affected from poor efficiency and financial performance to commit in IBS [21]. **Figure 8** below illustrates the sensitivity analysis of each construction method with respect to the alternatives i.e. the grade of contractors. From the figure, observable remark can be made that contractor G7 dominates in embracing the IBS technology and design and build rather than the conventional method. Thus, it can be presume that with the higher financial capabilities borne by the contractor (G7), the higher their willingness to adopt new technology than the conventional method. Additionally, low-grade contractors (G1), their level of readiness to adopt the latest technology is also relatively low. This can be seen from the diagram where G1 contractors embrace more on conventional methods rather than design and built and IBS.

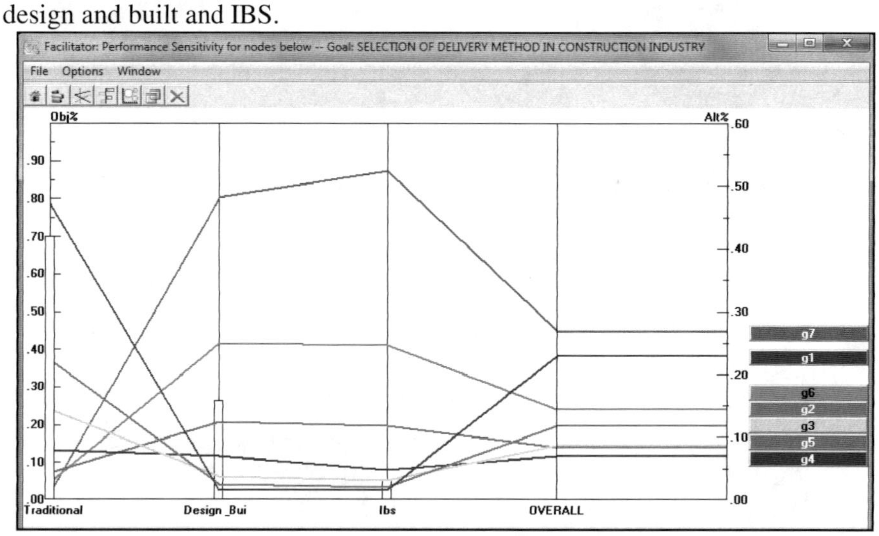

Figure 8. Sensitivity Analysis of Alternatives with respect to the goal

5. Conclusion

Based on the overall result, the conclusion drawn as follows. First, the traditional method is captured the most prioritized ranking for the method used in construction industry is due to the fact that traditional procurement method is the project's client and have full influences toward the overall process of the project. This also contributes to the increase on quality and functionality of the project. Other than that, best quality of project is guarantee when the building contract attached with the bill of qualities and project specification. Traditional procurement method is hassle-free type of procurement while in develop any project. The effect of variation may reduce the number of contract conflict. In addition, it having balanced allocation of risk between project's client and contractor compare to design and build method. It also known as tried procurement method because it have been tested in the long term period and is a very familiar procurement method. On the other hand, for the design and build method cost factors are key factors that need to be addressed in the design and build contract.

However, other factors such as duration, team reputation and quality should still be considered. The basis in making the decision for awarding the contract is based on the lowest bid provided by the contractor. However, the consideration of other criteria's should also be taken into account, this due to the fact that cost factors is commonly dynamic in nature. The selection of best value provides consideration of both cost factors and other subjective factors such as project management, quality control, and team reputation. These factors has gain popularity among project contracting building team due to its ability to consider all relevant factors affecting the design proposal's. It can be concluded that, in determining the best method selection, cost factors need to be in line with other subjective factors such as project management, quality control, and project team reputation. These subjective factors are always taken into consideration by the project team because of its ability to consider all the factors associated with design proposals. As a driver to the movement in activating IBS involvement in the construction industry, the government should implement IBS-related forums periodically. In addition, academics and industry practitioners can also actively cooperate in the form of information and experience sharing, the development of new techniques and advice on the promotion and implementation of IBS. Finally, an online portal may also be available to disseminate information on IBS internationally, such as products produced and processes related to IBS.

References

[1] Randolph Thomas H. Maloney W.F. Horner R.M.W. Smith G.R. Handa V.K. and Sanders S.R. 1990 Modeling construction labor productivity *Journal Construction Engineering Management* **116** 4 705–726.

[2] Nam C.H. and Tatum C.B. 1988 Major characteristics of constructed products and resulting limitations of construction technology *Construction Management and Economics* **6** 2 133–147.

[3] Illingworth J.R. 2000 *Construction Methods and Planning*, First. London, UK: CRC Press.

[4] Kamarul Anuar M.K. Mohamed Nor Azhari A. and Mohd Nasrun M.N. 2010 IBS survey:Drivers, barriers and critical success factors in adopting industrialised building system (IBS) construction by G7 contractors in Malaysia *Journal of Engineering Science and Technology* **9** 4 490–501.

[5] CIDB 2014 CIDB Report Construction Industrial Development Board (CIDB) Kuala Lumpur, Malaysia, 2014.

[6] Ling T.M. 2014 Project Procurement Method: The Conflicts in Construction Projects Procured under Design and Build Method *Thesis* Universiti Malaysia Pahang.

[7] Eric L. and Kamar K. 2012 Industrialised Building Systems (IBS): Strategic Outlook of Manufactured Construction in Malaysia *Journal of Architectural Engineering* **18** 69–74.

[8] Alinaitwe H.M. Mwakali J. and Hansson B. 2016 Assessing the degree of industrialisation in construction – A Case of Uganda *Journal of Civil Engineering Management* **12** 3 221–229.

[9] CIDB 2003 Industrialised Building System Seminar - Towards Industrialisation of Malaysian Construction Industry Kuala Lumpur, Malaysia.

[10] Saaty T.L 1988 What Is the Analytic Hierarchy Process ? *Mathematical Model for Decision Support* **48** 109–121.

[11] Sepasgozar S. and Davis S. 2018 Construction Technology Adoption Cube: An Investigation on Process, Factors, Barriers, Drivers and Decision Makers using NVivo and AHP Analysis *Buildings* **8** 6 74.

[12] Singh R.P. and Nachtnebel H.P. 2016 Analytical hierarchy process (AHP) application for reinforcement of hydropower strategy in Nepal *Renewable and Sustainable Energy Reviews* 55 43–58.

[13] Pan N.F. 2008 Fuzzy AHP approach for selecting the suitable bridge construction method *Automation in Construction* **17** 8 958–965.

[14] Saaty T.L. and Özdemir M.S. 2015 How Many Judges Should There Be in a Group ?*Annals Data Science* **1** 3–4 359–368.

[15] Barfod M.B. 2014 Graphical and technical options in Expert Choice for group decision making

Denmark.
[16] Saaty T.L. 2008 Decision making with the analytic hierarchy process *International Journal of Services Sciences* **1** 1 83.
[17] Ariff H. Salit M.S. Ismail N. and Nukman Y. Use of Analytical Hierarchy Process (AHP) for Selecting The Best Design Concept *Jurnal Teknologi* **49(A)** Disember 1–18.
[18] Abd.Rahman A.B. and Omar W. 2006 Issues and Challenges in the Implementation of Industrialised Building System in Malaysia *Proceedings of the 6th Asia-Pacific Structural Engineering and Construction Conference (APSEC 2006)* **September** 5–6.
[19] Nick B. and Ron W. Drivers 2007 Constraints and the Future of Offsite Manufacture in Australia *Construction Innovation* **Special Edition**.
[20] Hamid Z. Kamar K. Zain M. Ghani K. and Rahim A. 2010 Industrialized Building System (IBS) in Malaysia: The Current State and R&D Initiatives *Malaysia Construction Research Journal* **2** 1 1–13.
[21] CIDB 2011 The Current State of Industrialised Building System (IBS) Construction in Malaysia: Drivers, Barriers and the Way Forward *Proceedings of 1st IBS Roundtable Workshop* Kuala Lumpur Malaysia.

Selection of Best Consultant by using Analytical Hierarchy Process (AHP)

P Z Razi, N I Ramli, M I Ali and P J Ramadhansyah

Faculty of Civil Engineering Technology, Universiti Malaysia Pahang, 26300 Gambang, Kuantan, Malaysia

Abstract Consultant selection is a complex problem involving qualitative and quantitative multi criteria decision. In this study, an AHP-based consultant selection model is applied to develop and assist in decision making process to resolve the consultant selection problem in choosing the most favourable consultant's criteria combination. The Analytic Hierarchy Process (AHP) is an established decision method used to synthesize judgements and select the best alternative. When the objectives and scopes of study are set, then the data collection are made through two methods which are through literature review and pair-wise questionnaire survey. Information obtained from the questionnaire is interpreted into a simpler form and analysed by using the Expert Choice software. Results shows that financial capability of the consultant prioritize the decision of the selection while dominating the sub-criteria for selection is the profit made by the consultant within the last 3 years.

1. Introduction

Malaysia has a large construction industry of over RM 146 billion. Construction sector contributed to Malaysia's GDP at RM13695 Million [1]. The highest percentage share was contributed by civil engineering projects which recorded 42.6%. This was followed by the construction of non-residential building at 28.4%, followed by 24.2% in construction of residential buildings, and finally special trades at 4.8%. From the figure, private sectors dominates at 56.2% equivalent to RM20.5 billion while public sector 43.8% proportionate to RM16 billion [1]. Being a primary national economic contributor, the construction industry is facing prolonged problems such as time overrun, cost overrun, poor safety and poor quality [2]. Construction projects are known with the involvement of various parties from the industry itself as well as the support from the government for approving the development, financial budget and so forth. Prior to accomplish successful project, all parties involved directly or indirectly should work closely with aligning their respective areas of work and having collective coordination coupled with bilateral communication within private and public sector that involves for the project. It is concede that there are many factors that drive success for a construction project, primarily the mega projects which including financial, construction quality, and the progress level of the project itself [3]. Preceding to transform Malaysia into developed world by 2020, good teamwork among clients, consultants and contractors to implement and complete construction projects in a timely manner will be relish by in Malaysia's construction practitioners[4]. The design consultants typically consist of engineers, civil engineers and structures, mechanical and electrical engineers and other specialist designers, including interior designers, landscape architects, lighting experts, city planners, and so on. In the context of the construction project, the consultants specialize in their respective fields of expertise. Consultants act as facilitators who fulfil the client's request and also act as the facilitator of the construction project. In addition, they also act as advisers to clients on matters pertaining to design and legal requirements and by-laws that are closely related to the construction project. In the context of cost and contract negotiations, commonly a quantity surveyor acts as a cost consultant and contract for a project that covers the requirements of the deed and contract of the project. The need to choose the right and accurate consultants for the project is a very important yet risky challenge for any construction client's. Hence the right choice should be based on a decision-making that is analytic[3]. It is vital for the client/owner to select the best consultant in their project. Hence, client satisfaction towards consultant's work is very important in order to be able to survive in the marketplace. Based on the research findings, the construction cost would be five times higher to form a new construction client's rather than maintaining existing clients. It is therefore ultimate concerns for multiple construction players to be prepared as in order for them to be remain and survive in nowadays global marketplace, they should continuously attempt to elevate their achievement prior to be endure and sustain in nowadays comprehensive marketplace [5]. Successful project is led by the best selection of suitable, qualified project team, and experience project manager with good leadership skills. The criteria are most often related to the candidate's personal skills and experience in handling similar or multi projects. This study aims in providing an effective selection and recruitment of consultants on different projects for the contribution success of a project in terms of time, cost and quality. By employing a multi-criteria decision making, it is important for the selection of this consultant to be based on elements categorized into a price-related element and not closely related to the price. [6,7]. Consultant selection can be viewed as an intricate indefinite grouping problem, in which decisions are purely judge based to the consultant's qualification criteria, consultant's performance criteria and the field expertise judgment. Furthermore, multi-criteria decision making that are inter-related on the choice of consultant's is largely reliant on the ambiguity and uncertainty symbolic nature of construction projects couple with individual judgments of the construction expert's [7]. In earlier study by [8], also justified that when involving a development, the consultant selection process needs to consider the several elements that will influence the selection results as well as the decision-making process requiring the participation of various parties looking on numerous point of view. The most important issue in the process of consultant selection is to develop an effective method to select the best one [9]. As mentioned above, this problem is a group decision making. Hence, in this research all issues and problems associated to the consultant will be discussed and the best criteria for selecting consultants will be weighted by using Analytical Hierarchy Process (AHP) method.

2. Literature Review

Prior studies have reported lacking in performance of a construction project is mainly derived by the factors of faulty in design or issues related to technical that was primarily caused by the consultant. This is much largely dependent on the limited ability of public authority to evaluate consultant competence, especially aspects related to design and execution of projects [10]. One of the problems that arose from the issues is that the client have to face for a time overrun eventually causing delay. It is a big issues related to project management consultants (PMC) as surveys was being conducted among them [11]. Time overrun can be due to the changes of design schedule and government restraints. By the perspective of the clients, even though the construction projects were planned and organized perfectly, it will still face the risk of delays. Although many tools such as computer modelling technologies, Building Information Modelling (BIM) which created to assist the process of site management, delays keep occurring in construction projects [12]. In addition, insufficient cost also may contribute to the construction delay due to the inflation and capital equipment. The latest technological developments are indispensable in a construction project. The latest technology that has begun to be accepted slowly in the construction industry is the Building Information Modelling (BIM) system. BIM, which is basically a 3D processed model, combines architectural, engineering and construction (AEC) information and is capable of replacing and simplifying design procedure. However, the use of BIM has a risk where it requires the justification of acoustic consultants to ensure that the change in information complies with the prototype or the standards permitted to meet current requirements [13]. Building failure is not a rare issues as it can occur due to the type of materials used, the design method adopted, the environmental impact of the problem, the construction method used and the type of building to be built. Building failures usually occur when the tested material is more geared towards its strength limit, so the material fails to cover the pressure applied to it. As an example, a new multi-purpose stadium in Eastern state of Malaysia which was constructed by an international construction firm was collapsed a major part of the roof construction. The major cause such stadium collapsed has been identified due to failure of design and materials used, coupled with urgency during construction and lack of supervisory just to reach on completion within the prescribed period. This is definitely the fault of design consultants as they do not consider the wind load while designing it and lack of supervision throughout the designing and construction process.

3. Research Methods

This study entails the use of multi-criteria decision making (MCDM) tools namely the Analytical Hierarchy Process (AHP) which was first introduced by Professor Saaty [14]. AHP is employed to elucidate multifaceted choice argument, engage a multi-level hierarchy structure of goals, criteria, sub-criteria, and options [15]. A methodology in obtaining the optimal consultant criteria is firstly, by identifying the criteria for consultant selection. The process involved in AHP basically involves the assessment of criteria and sub-criteria by relevant experts in their respective field. Among the key contributions given by AHP is that it can determine the most dominant criteria and sub-criteria for each given goal. The first stage in AHP is to build a hierarchical framework that consists of goals, criteria and sub-criteria. Next, the AHP model will be developed through a pair-wise questionnaire that will be assessed by the experts. Subsequently, the results of the questionnaire will be computed and calculated the geometric mean before it is formed to the matrix structure and the weight of each criteria is obtained. Specifically, the AHP method, can determine to the extent on which responses are consistent or inconsistent through the consistency index. The framework of consultant's selection are shown in the model presented in **Figure 1** below.

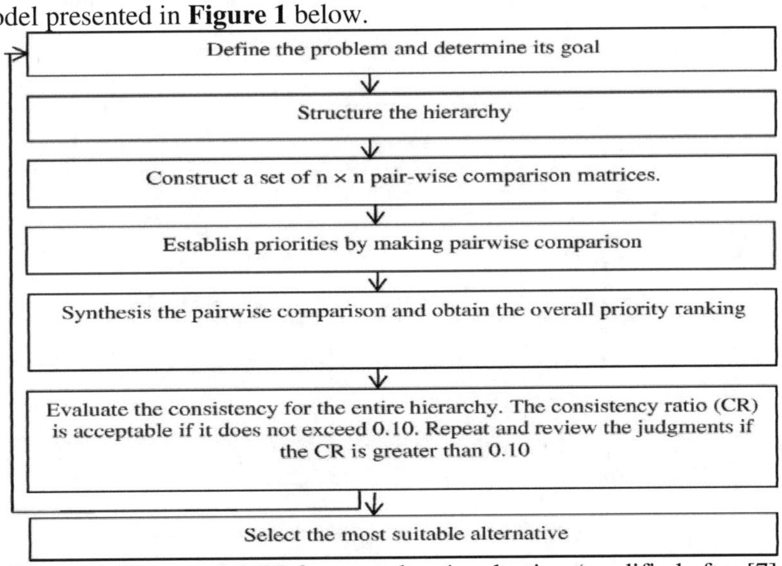

Figure 1. Proposed AHP for consultant's selection (modified after [7]

To create a judgement in a structured manner and to produce preferences, the judgement requisite to be fragmented by succeeding steps below and also simplified in **Figure 2**.

i) Specify a research problem and decide the choice of understanding desired.
ii) Arrange the judgement hierarchy from the uppermost is the aim or the goal, followed by the criteria from an extensive perceptions subject to the goal and finally the sub-criteria belongs to the criteria.
iii) Create a set of pair-wise comparison matrices. Every criteria in the higher hierarchy being measured to the lower sub-criteria respectively.

iv) Every weighted is measure to determine the most dominant criteria and sub-criteria. Weighted obtained for every criteria is known as local weight, while the multiplication of weighted obtained for criteria and sub-criteria produces the global weight.

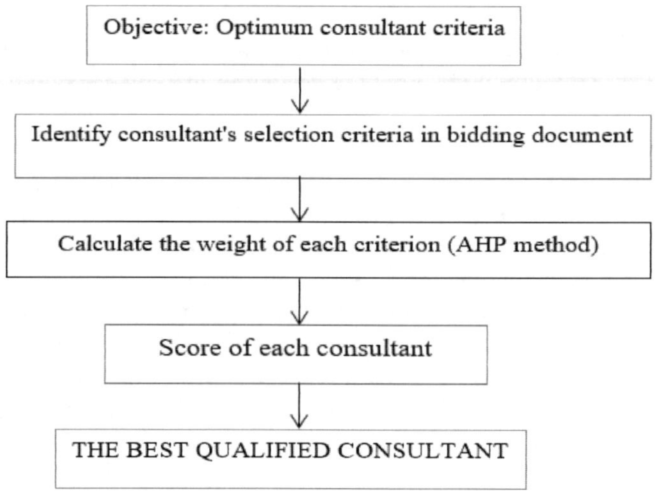

Figure 2. AHP in decision making

Within the context of this study, hierarchical structures that contains goal, criteria and sub-criteria is shown in **Figure 3** below.

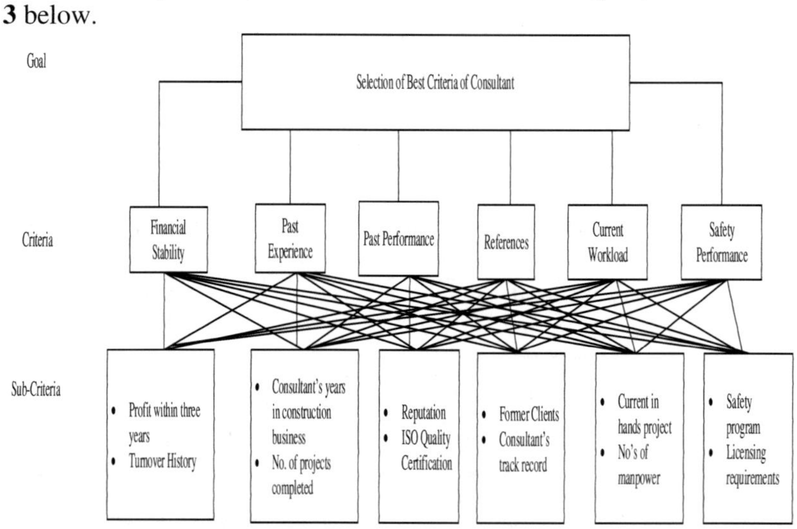

Figure 3. Decision Hierarchy Structure

AHP is used to develop relation measures from isolated and repeated paired comparisons [16]. Pairwise comparisons are the central theory within the utilization of AHP method. Pairwise assessments include choosing which criteria is much more imperative to the others criteria and if one is determined important, on what scale of qualitative judgement is the importance. The pairwise comparisons are then incorporate over the use of definite algebra, and preferences for every criteria. Every criteria are then normalized to aggregating of one, and the dominant criteria is ranked as the top best preferences [17]. **Table 1** below depicts number of criteria being compared, in n(n -1)/2. This is due to the reciprocatory values and the transverse criteria are proportionate for a homogeneousness.

IOP Conf. Series: Materials Science and Engineering **712** (2020) 012016 doi:10.1088/1757-899X/712/1/012016

Table 1. Pairwise compare the criteria and sub-criteria

j \ i	Objective 1	Objective 2	Objective 3	Objective 4	...	Objective n
Objective 1	1	i_1j_2	i_1j_3	i_1j_4	...	i_1j_4
Objective 2	i_2j_1	1	i_2j_3	i_2j_4	...	i_1j_4
Objective 3	i_3j_1	i_3j_2	1	i_3j_4	...	i_1j_4
Objective 4	i_4j_1	i_1j_2	i_1j_3	1	...	i_1j_4
...	...	...	...	...	...	...
Objective n	i_4j_1					$i_nj_n=1$
i_aj_b : (a=1,..n; b=1,..n) = values (see table 1.)						

AHP is very well known for its simple concepts. In similar vein, despite being able to provide the dominant criterion among others, it is an option because its ability to give precision in the results through the consistency index analysis offered. This consistency index is calculated on each criterion, and in the event of inconsistency, the consistent ratio can be calculated and if necessary, the pair-wise comparison can also reconsidered [18]. Generally, as suggested by Prof. Saaty, the consistency index (CI) should be less than 0.1 (10%) so that analysis can be adopted and analysed further [14,19-21]. If the CI is too high, it clearly shows that the respondent's decision is inconsistent and dubious in its accuracy. This further explains that the analysis of the consistency index is very useful in determining the probability of an error during the expert judgment [22]. In contrast, the decision making can be further examined, conceding that the values of consistency index and consistency ratio are sufficient. The AHP incorporates an effective technique for checking the consistency of the evaluations made by the decision maker when building each of the pairwise comparison matrices involved in the process. By reducing complex decisions to a series of pairwise comparisons, and then synthesizing the results, the AHP helps to capture both subjective and objective aspects of a decision. In addition, the AHP incorporates a useful technique for checking the consistency of the decision maker's evaluations, thus reducing the bias in the decision making process.

4. Results and Discussion

4.1. Demographic Respondent Analysis

This section is explanatory from **Figure 4** to **Figure 7.** Figure 4 illustrates the respondents by gender. From the total 30 responses received, 18 responses (60%) received from male and 12 responses (40%) received from female. Figure 5 below shows the respondent's experience in construction. It appears that only one respondent is having construction experience more than 10 years, while nine respondents having 5-6 years of experience, the others ten respondents are having experience 3-4 years and 1-2 years respectively. On the other hand, figure 6 depicts the respondent's position. It shows that, 11 respondent's monopolized in the survey is a Design Engineer, 10 respondents is a site engineer, 2 holds a position as Executive and Project Coordinator respectively, and 1 respondent hold position as Site Supervisor, Manager, Architect, Quantity Surveyor, and Structural Engineer accordingly while Figure 7 illustrates the company/organization of the respondents's. Apparently, 47% (14) of the respondents from the contractor organization, 20% (6) is the consultant, 13% (4) is the client, 13%(4) from the government or public authority while the remaining 7% (2) is from the architect firm.

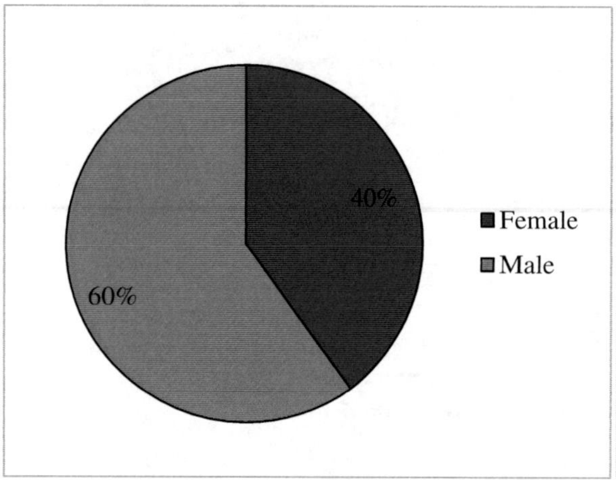

Figure 4. Respondent's by Gender

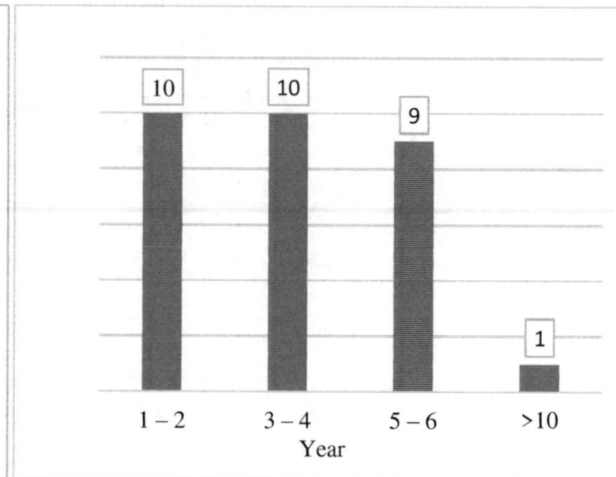

Figure 5. Respondent's Experiences

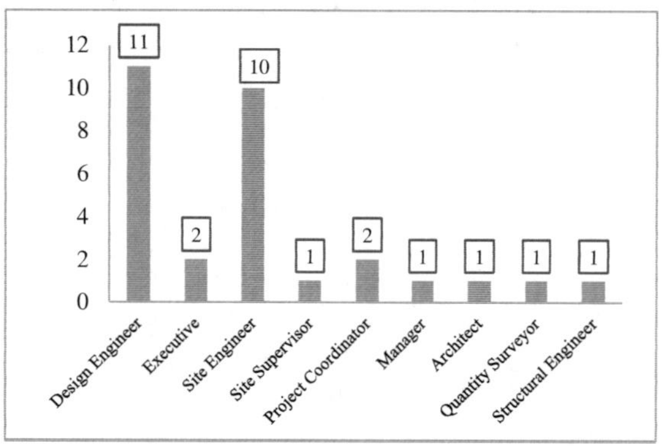

Figure 6. Position of the Respondent's

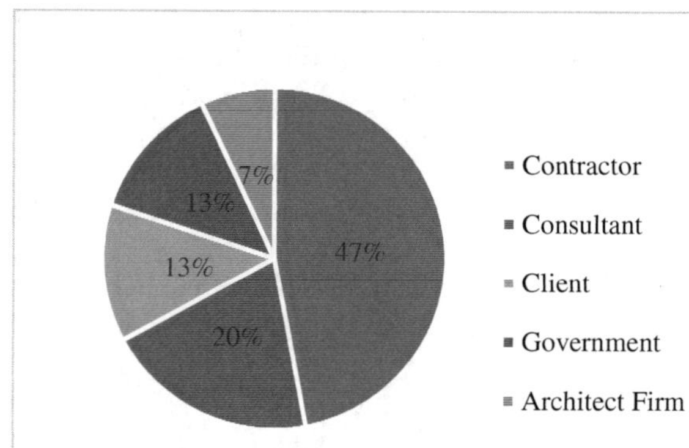

Figure 7. Company/Organization of the Respondents's

4.2. AHP Results and Discussion

Figure 8. below depicts the results on the prioritization of criteria for the selection of consultant by analysing by using the 'Expert Choice' program.

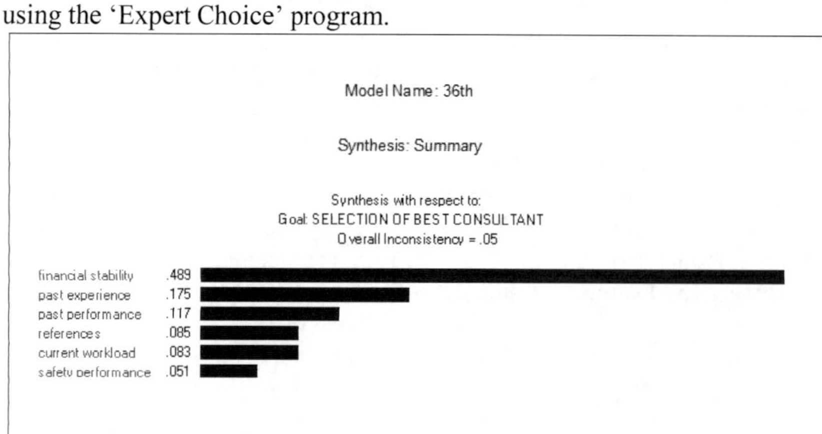

Figure 8. Priorities of criteria selection of consultant

Apparently, from the figure 8, the results achieved consistencies at 0.05 (5%) and is less than 0.10(10%) as per recommendation by [14,19]. Financial stability is dominating the prioritization weighted at 0.489, followed by past experience of the consultant weighted at 0.175, past performance of the consultant (0.117), references by others weighted at 0.085, consultant workload weighted at 0.083 finally safety performance of the consultant at 0.051. The findings of financial stability is cohere with a report by [23]. The importance of the consultant's selection is closely related to the consultant financial position and stability[23]. Figure 9 on the other hand, depicts the results for the prioritization of the sub-criteria selection of the consultant. It shows from the figure that, top prioritization of sub-criteria in selecting the consultant is the profit made by the consultant within the last 3 years weighted at 0.291, No's of projects completed weighted at 0.199, reputation of the consultant weighted at 0.156, consultant track record 0.075, consultant's current project 0.061, turnover history 0.058, safety program 0.046, consultant's year in construction business 0.033, ISO quality certification 0.026, manpower 0.020, former clients 0.019 finally licensing requirements 0.015.

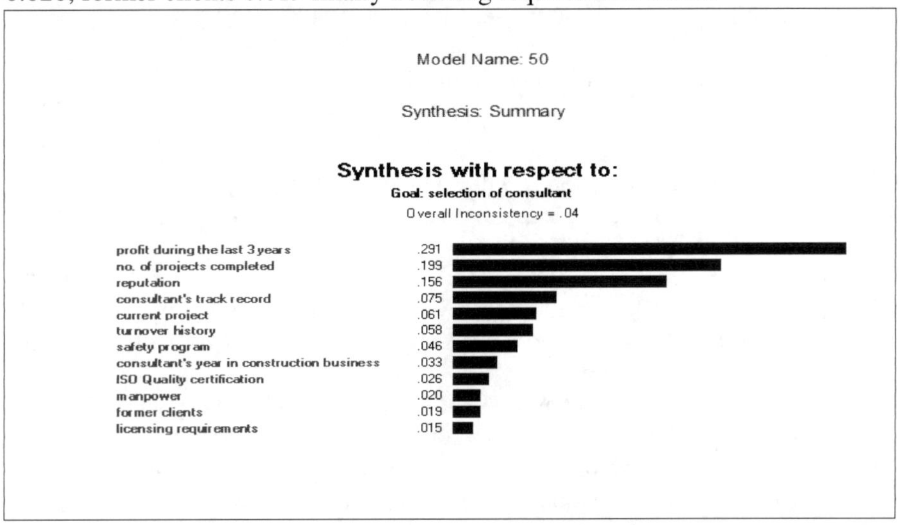

Figure 9. Priorities of sub-criteria selection of consultant

The finding confirms that in the construction sector where the function of a consultant is indispensable, among the criteria identified in the project consultant selection is the experience of the consultant. This shows that job perceptions and past experience are very valuable in the success of a project. This is acknowledged as the consultant is the party to be referred by each project team in the event of any technical or non-technical problems. In addition, it can be concluded that data analysis and results may differ if analysed to other construction clients. This is influenced by the different cultural organizational factors of the client. In order to solve the problem of these differences, proper selection of consultants with great criteria needs to be built. According to this paper's findings, the results obtained can be support by a few evidences from the previous research. It is widely accepted view that, at a minimum, performance measures of a project are based on time, cost and quality[24]. Furthermore, in an observation-based study by [25], in order to achieve a completed project that meets the owner's quality expectations, all parties to a project must acquire an understanding of those expectations, incorporate them into the contract pride and other contract documents to the extent possible, and commit in good faith to carry them out. This is supported by [26], stated the effectiveness and quality of the consultant's work is taken into account. In placing more emphasis, [10] states that the competence, commitment and attitudes of design consultants strongly influence the quality and cost of built facilities. Thus, selecting the right design consultants is of utmost importance for any construction client.

According to a study by [27] the cost variance was the most common technique used to measure design performance. It is not only confined to the tender sum, but the overall cost that a project incurs from inception to completion, which includes any costs arise from variations, modification during construction period and the cost arising from the legal claims, such as litigation and arbitration. In advance, other study claimed that time path and budget were considered more important for consultants that were also involved as project manager [28]. These studies have found the similar results where price is the professional fee charged by the consultant. Which states that in many cases, price is the main determinant in consulting selection [29].

5. Conclusion

The success of any construction project is much dependend on the selection of the team-work or parties involved during the construction stage, including the consultant. This study esteem the practise of multi-cireteria decision making (mcdm), the Analytical Hierarchy Process (AHP) for the selection of consultant. Results shows that financial capability of the consultant prioritize the decision of the selection while dominating the sub-criteria for selection is the profit made by the consultant within the last 3 years. Results presented will be an aided tools specifically the client for choosing the right consultant while not risking the construction performance.

References

[1] DOSM 2018 Department of Statistic Malaysia Construction Report Putrajaya, Malaysia.

[2] Nahmens I. and Ikuma L.L.H. 2009 An empirical examination of the relationship between lean construction and safety in the industrialized housing industry *Lean Construction Journal* **2009** 1–12.

[3] Ha T.T. Le Hoai L. and Lee Y.D. 2015 A Fuzzy AHP Model for Selection of Consultant Contractor in Bidding Phase in Vietnam *KICEM Journal of Construction Engineering and Project Management* **5** 2 35–43.

[4] Fadhlin A. Chiet C.V. Anuar K. and Tiean Shen T. 2004 An Overview on the Growth and Development of the Malaysian Construction Industry *Workshop on Construction Contract Management*, Johor Bahru Malaysia.

[5] Cheng J. Proverbs D.G. and Oduoza C.F. 2006 The satisfaction levels of UK construction clients based on the performance of consultants: Results of a case study *Engineering Construction and Architectural Management* **13**, 6 567–583.

[6] San Cristobal J.R. 2012 Critical Path Definition Using Multicriteria Decision Making: PROMETHEE Method *Journal Management Engineering* **29** 2 158–163.

[7] Singh D. and Tiong R.L.K. 2004 A Fuzzy Decision Framework for Contractor Selection *Journal of Construction Engineering and Management* **131** 1 62–70.

[8] Russell J.S. and Skibniewski M.J. 1988 Decision Criteria in Contractor Prequalification *Journal of Management and Engineering* **4** 2 148–164.

[9] Mahmoud M. *et al.* 2009 A formal framework for scenario development in support of environmental decision-making *Environmental Modelling and Software* **24** 7 798–808.

[10] Sporrong J. 2011 Criteria in consultant selection: Public procurement of architectural and engineering services," *Australasian Journal of Construction Economics and Building* **11** 4 59–76.

[11] Dadzie J. Abdul-Aziz A.R. and Kwame A. 2012 Performance of Consultants on Government Projects in Ghana : Client and Contractor Pespective *International Journal of Business and Social Research* **2** 6 256–267.

[12] Głuszak M. and Leśniak A. 2015 Construction Delays in Clients Opinion – Multivariate Statistical Analysis *Procedia Engineering* **123** 182–189.

[13] Philipp N.H. 2013 Building information modeling and the consultant: Managing roles and risk in an evolving design and construction process *Proceedings of Meetings on Acoustics* **19**.

[14] Saaty T.L. 1988 What Is the Analytic Hierarchy Process ? *Mathematical Model for Decision Support* **48** 109–121.

[15] Olakitan Atanda J. 2019 Developing a social sustainability assessment framework *Sustainable Cities and Society* **44** September 237–252.

[16] Vaidya O.S. and Kumar S. 2006 Analytic hierarchy process: An overview of applications *European Journal Operation Research* **169** 11–29.

[17] Wind Y. and Saaty T.L. 1980 Marketing Applications of the Analytic Hierarchy Process *Management Science* **26** 7 641–658.

[18] De Boer L. Van Der Wegen L. and Telgen J. 1988 Outranking methods in support of supplier selection *European Journal of Purchasing and Supply Management* **4** 2–3 109–118.

[19] Saaty T.L. 2008 Decision making with the analytic hierarchy process *International Journal of Services Sciences* **1** 1 83.

[20] Saaty T.L. and Vargas L.G. 2012 The possibility of group choice: Pairwise comparisons and merging functions," *Social Choice and Welfare* **38** 3 481–496.

[21] Saaty T.L. and Özdemir M.S. 2015 How Many Judges Should There Be in a Group ? *Annals Data Science* **1** 3–4 359–368.

[22] Belton V. and Stewart T.J. 2002 *Multi Criteria Decision Analysis*, Two. London, UK: Kluwer Academic Publisher.

[23] NCOSS 2015 Selecting a consultant NSW Council of Social Service(NCOSS) Woolloomooloo, New South Wales.

[24] Barkley B.T. and Saylor J.H. 1994 *Customer-Driven Project Management: A New Paradigm in Total Quality Implementation*. Singapore: McGraw-Hill.

[25] Ganaway N.B. 2006 *Construction Business Management: A Guide to Contracting for Business Success*, First. Oxford, United Kingdom: Elsevier.

[26] De Caluwe L. and Stoppelenburg A. 2004 Developing Criteria for Effectiveness of Consultant's Work S*econd International Conference on Management Consulting of MC Division of the AoM* **June.**

[27] Salter A. and Torbett R. 2003 Innovation and performance in engineering design *Construction Management and Economics* **21** 6 573–580.

[28] Walenberg B. 2010 The Effect of Consulting Approach and Context Characteristics on Perceived Relevance of Effectiveness Criteria of Consultants in The Netherlands *Thesis* Universiteit Van Tilburg.

[29] Ling Y.Y. Ofori G. and Low S.P. 2003 Evaluation and Selection of Consultants for Design-Build Projects *Project Management Journal* **34** 1 12–22.

IOP Conf. Series: Materials Science and Engineering **712** (2020) 012017 doi:10.1088/1757-899X/712/1/012017

The Neo-Deterministic Seismic Hazard map (NDSHA) of Sumatra compared with official 2010 and 2017 derived from PSHA method

I Irwandi[1], M Umar[1], J Khaizal[1], M Asrurifak[2], F Usama[3] and M Ridwan[4]

[1]Universitas Syiah Kuala, Darussalam Banda Aceh, Indonesia, 23111
[2]PuSGeN (Pusat Studi Gempa Nasional), Bandung, Indonesia, 40393
[3]Coffey, A Tetra Tech Company, Wellington, New Zealand, 6011
[4]Housing and Settlement Research Centre (Puskim-PUPR), Bandung, Indonesia, 40393

E-mail: irwandi@unsyiah.ac.id

Abstract. Typical seismic hazard problem lies on determination of ground motion characteristics associated with future earthquakes. Strong ground motion characteristics such as Peak Ground Acceleration (PGA) is important information for engineers in order to provide earthquake resistant building. The Indonesian government has provided update of maps of PGA by released a series official PSHA (Probabilistic Seismic Hazard Assessment) maps in 1978, 2002, 2010 and 2017. On the other hand, the NDSHA (Neo-Deterministic Seismic Hazard Assessment) method has been developed and successfully applied to Sumatra. Because of the availability of numerical PGA data for the official PSHA 2010 and 2017, this paper discusses the comparison of PGA between derived from NDSHA and that of official PSHA updated in 2010 and 2017. The spatial resolution of the PGA digital map of PSHA is 0.01 degree and NDSHA is 0.1 degree. Due to the different resolution, a program and a script are developed to incorporate PSHA data into the NDSHA data points by performing interpolation procedure. The comparison results show the PGA of PSHA 2017 is significantly greater than PGA PSHA 2010. However, the PGA values of the 2017 PSHA map are lower than that of PSHA 2010 such as along the Sumatran fault in Central and South Sumatra. The PGA values estimated by using standard NDSHA with 10 Hz cut-off is higher than those of the computed PGA from PSHA at PE (Probability of Exceeding) of 2% in 50 year for official map released in 2010 and 2017. For comparison with PE of 2% in 50 year, the near field NDSHA gives the PGA value greater than that of the PSHA; and for the far field sources. This means the NDSHA calculation for Sumatra gives reasonable results based on the adopted magnitude distance threshold. This is due to the fact that the standard NDSHA is based on the realistic physical simulation of the seismic wave propagation. The comparison between the PGA computed based on the proposed enhanced version of NDSHA and PSHA shows that the updated version gives the same values at near field and lower values at the far sites; whereas the standard one gives a higher value in NDSHA, which in turn gives a higher PGA value for the near and far field sources.

1. Introduction

Possible magnitudes and impacts of earthquakes in an area are among unanswered questions for geoscientists. Meanwhile, we have to construct earthquake-resistant buildings on earth where we live. The earthquake experts in Indonesia have attempted to produce official seismic hazard maps. In 1983, the initial seismic hazard map formally adopted by the Indonesian government which is published in a

IOP Conf. Series: Materials Science and Engineering **712** (2020) 012017 doi:10.1088/1757-899X/712/1/012017

book titled "*Peta Hazard Gempa Indonesia dalam Peraturan Perencanaan Tahan Gempa Indonesia untuk Gedung 1983*" or Indonesian Earthquake Hazard Map in Earthquake resistant building planning 1983. The map classified the 8 levels of seismic hazard zone which based on the work of Beca and Ferner [1]. In 2002 the Ministry of Public Works issued a revised earthquake hazard map based on the calculation of PSHA (Probabilistic Seismic Hazard Assessment) [2]. Following the 2004 earthquake and tsunami, the government carried out further studies to anticipate future destructive earthquake by releasing new seismic hazard map in 2010 [3] [4] which included the new study of active Great Sumatra Faults (GSF) system [5] [6] [7]. After 5 years of the publication of the 2010 earthquake hazard map, there were several earthquakes occurred in the previously considered safe area therefore the government plan to issue and update the seismic hazard map in 2016. However, it was surprising that the Pidie Jaya earthquake [8] on December 7, 2016 which was destructive and took many casualties occurred in areas that had not been defined as active fault zones previously. Based on these events, the team needs to evaluate to include more detailed active fault map and the release of the new map was delayed until mid-2017 [9]. The official earthquake map issued by the Indonesian government in 2010 and 2017 uses the PSHA procedure as shown in figure 1a. The PSHA estimate the strong ground motion from GMPE (Gound Motion Prediction Equation) [10] [11] [12] [13]. On the other hand, the NDSHA method has been developed and applied in several countries [14] [15] with the procedure shown in Figure 1b. The NDSHA calculate the strong ground motion from synthetic seismogram [14] [16]. The technique for producing the synthetic seismogram has been verified with field observation of BMKG accelerometer and it's respond spectral [17]. The seismic hazard maps have a PGA map, so it is interesting to investigate the difference in PGA values among the three maps discussed in this paper.

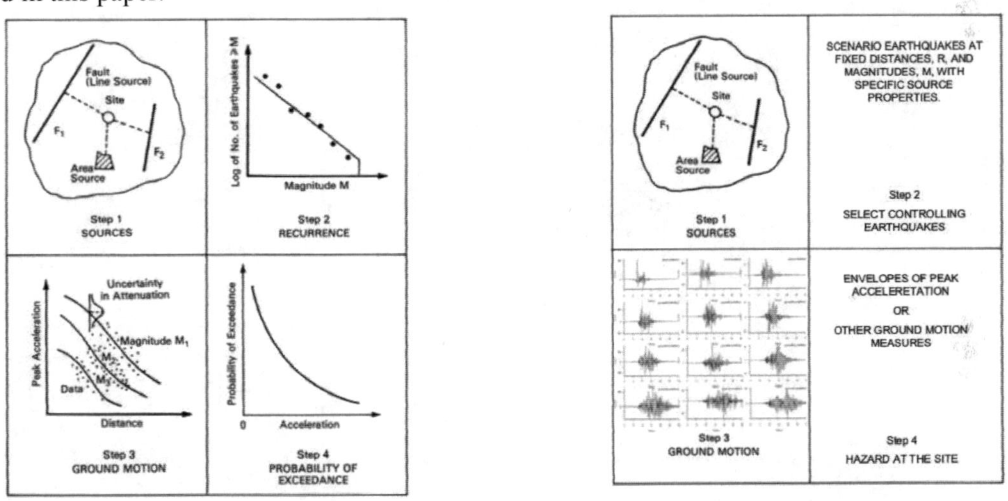

(a) PSHA (b) NDSHA

Figure 1. The comparison procedure in performing PSHA and NDSHA [13] [17]

2. Methods

The discussion of this paper only focuses on the comparison of the PGA maps in Sumatra. Therefore, there is no complicated computing process but the numerical data in map form is used. The software used to manage and visualize the data into map form, as well as results on the comparison.

2.1 Numerical Map of PGA Data

To difference of the PGA values generally can be noticed by comparing two or more maps visually. However, the visual comparison has limitation due to the limited number associated with number is known. By using numerical comparisons the difference of the PGA values can be calculated more accurately because it directly compares the values of each related point. Therefore, this study requires seismic hazard map information in the form of text format containing latitude, longitude, and PGA values.

As discussed above, there are several official maps of PSHA 1983 [19] [2], 2010 and 2017. However, the 1983 and 2002 PSHA maps are only available in the form of images of pdf. Fortunately, numerical data of official maps of PSHA 2010 and 2017 were obtained. The PSHA 2010 data is obtained from Fauzi Usamah in 2012 and the PSHA 2017 data was obtained from M. Asrurifak in the framework of research collaboration in 2018. Both numeric maps have been replotted as shown in Figure 2b and Figure 2c for PSHA 2017 and 2010, respectively. The maps are in accordance with the image map in the official book of earthquake hazard. Actually, for the PSHA map, we have more than one PE (possibility of exceeding) PGA maps, such as PE of 2% and 10% in 50 years. However, we limit the discussion to the PSHA map with a PE of 2% in 50 years. Other than comparing PSHA map released in different years, comparisons with other methods such as NDSHA were also carried out as shown in Figure 2a. NDSHA numerical data was obtained from the results of studies at the University of Trieste in 2016.

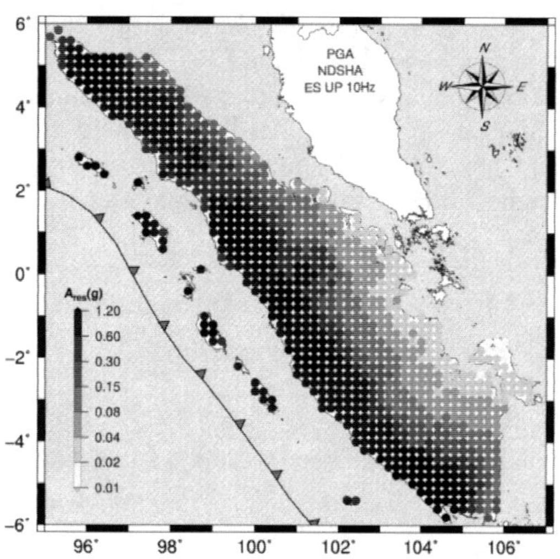

(a) NDSHA

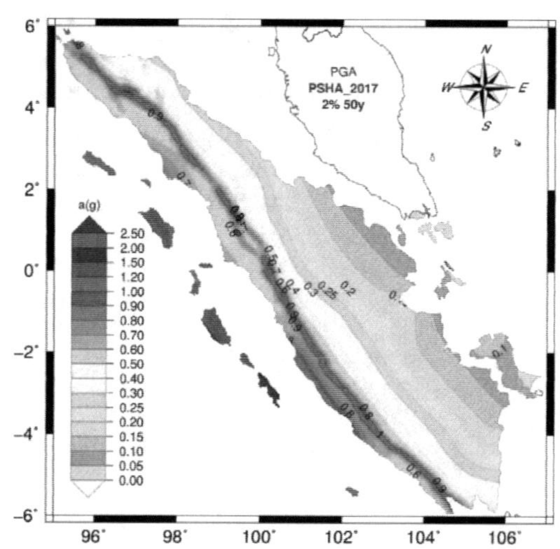

(b) PSHA 2017

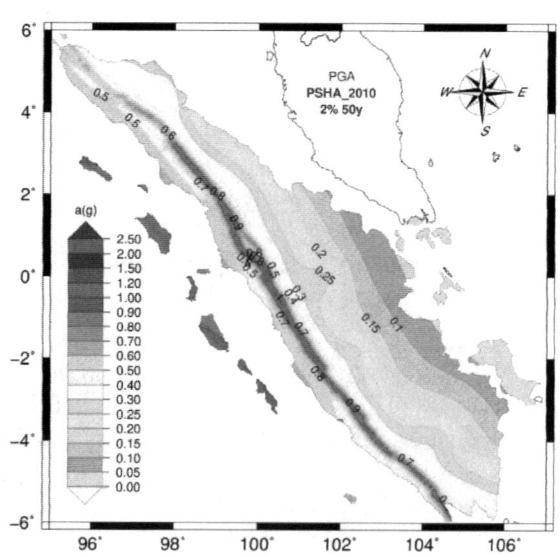

(c) PSHA 2010

Figure 2. PGA maps that compared each other in this paper and the results of the plotting are shown in:

(a) The PGA NDSHA map plotted using points. Each point represents a PGA value obtained from many realistic simulation models of synthetic seismograms.
(b) Map of PGA PSHA 2017 which is officially issued by PuSGeN in the form of the book “Peta Sumber dan Bahaya Gempa Indonesia Tahun 2017”
(c) Results of the 2010 PSHA of Sumatra are officially issued by *Tim Revisi Peta Gempa (TIM 9).*

2.2 Software and Tools
Most researchers use proprietary software with friendly GUI facilities. However, we concern to copyright, we prefer to use open source software. We performed the whole procedure in Linux Ubuntu 16.04 operating system which is equipped with various other support tools from GNU. Final plotting process is by using GMT software [20]. All works use bash shell-based scripts so that automation processes can be perform in input, process, and output. The aim of the script is to simplify routine data processing for certain data format and the script that is created still allows settings through passing parameters. Following command was used to produce image 2a.

```
haz2map.sh -Caccg.cpt -rel -z0.23 shta9nf2res.fin -sub sunda3.gmt
north.sh ESUP.ps diff.lab -tno -bWeSn -Gblue diff.lab -png
```

PGA data from PSHA has a relatively higher resolution of 0.01 degree compared to the PGA data obtained from NDSHA (only has 0.1 degree resolution). The PSHA procedure calculates the PGA by using GMPE so that it does not require long computing time even for very dense observation points. Meanwhile for NDSHA, PGA value is calculated based on realistic physical seismic wave propagation that is high computational cost so that the point is made relatively less dense in order to avoid long computational time. Due to the different resolution between NDSHA and PSHA, a program is created to interpolate PSHA point to point NDSHA that can be compared between one to another. The interpolation program was made by using Gfortran [21].

2.3 Comparison Schematics
Each version of the earthquake map, the PSHA 2010, the PSHA 2017 and the NDSHA, produces many different parameters for different purposes. For example for PSHA, one produced maps for PE of 2% and 10% for 50 years. Comparing all existing maps produce multiple differential maps and is more complicated to discuss. For NDSHA which is not a probability base, we only produces one map with MCE Input. Therefore, 3 maps were selected to be compared as shown in Figure 3 and labelled as the corresponding images. The comparison procedure of PGA maps is represented by a line that connects the PGA map which produces 4 lines (Figure 3). Especially for differences in PSHA 2017-2010, two maps were prepared differently. While the NDSHA comparison with PSHA 2010 and PSHA 2017 uses point plotting style mapping because it follows the lowest map resolution.

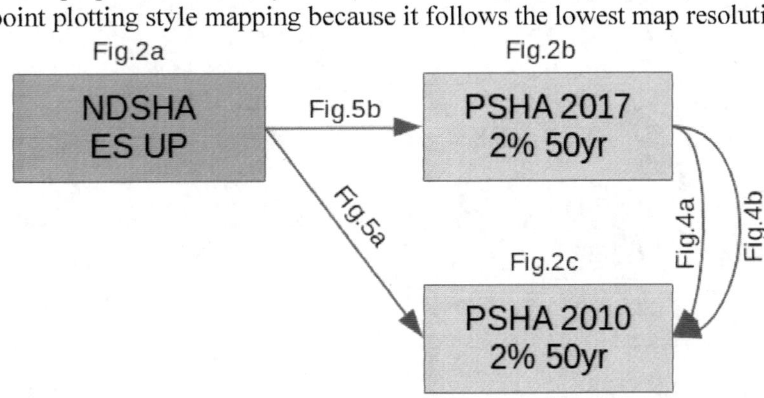

Figure 3. Schematics of comparison PGA and associated figure discussed in this paper

3. Results and Discussion
3.1 The Different 2017 to 2010 Official PSHA
The discussion is starting from official PGA PSHA for different updated times in 2017 and in 2010 with 2% PE in 50 year as shown in Figure 4. The PGA for PE of 10% in 50 years is not discussed and PGA value which is smaller than PGA for PE of 2% in 50 year. In view of the worst scenario side, it is better to discuss the PGA for PE of 2% in 50 years. In figure 4, the PGA PSHA 2017 seems to be larger than the 2010 PGA PSHA except for several spots such as the central part of Sumatran fault and the central part and the southern part of Sumatra. Likewise, with the archipelago in the forearc section,

there has been an increase in PGA. There is also an increase of PGA values in the Bangka Island which had been considered as a seismically safe zone. The increase in PGA is very significant in the northern Sumatra fault in the province of Aceh.

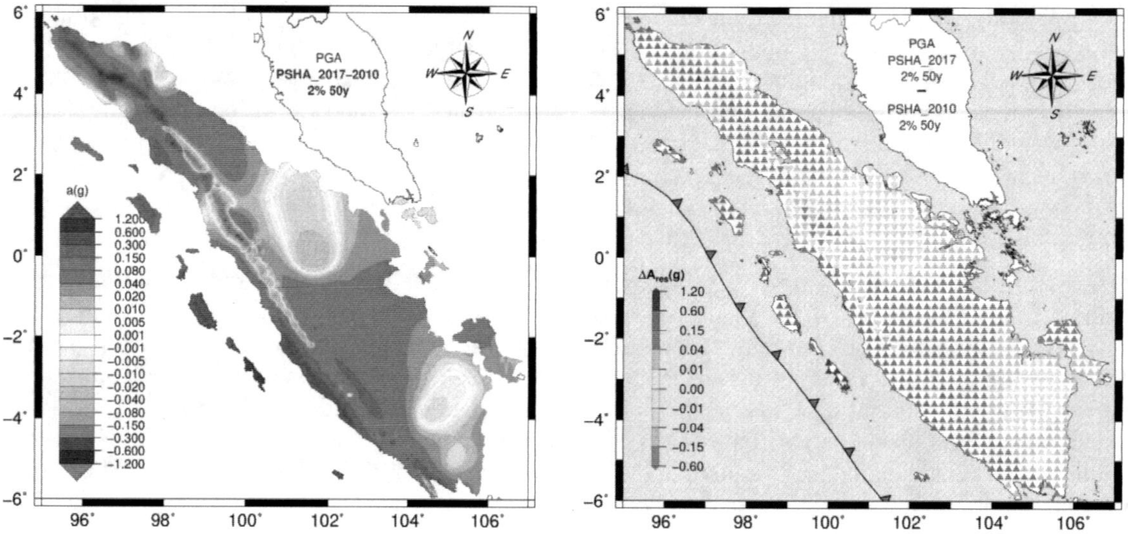

(a) area plotting style (b) point plotting style

Figure 4. The differential map of PGA from official PSHA map 2017 and 2010 with (a) area plotting style (b) point plotting style

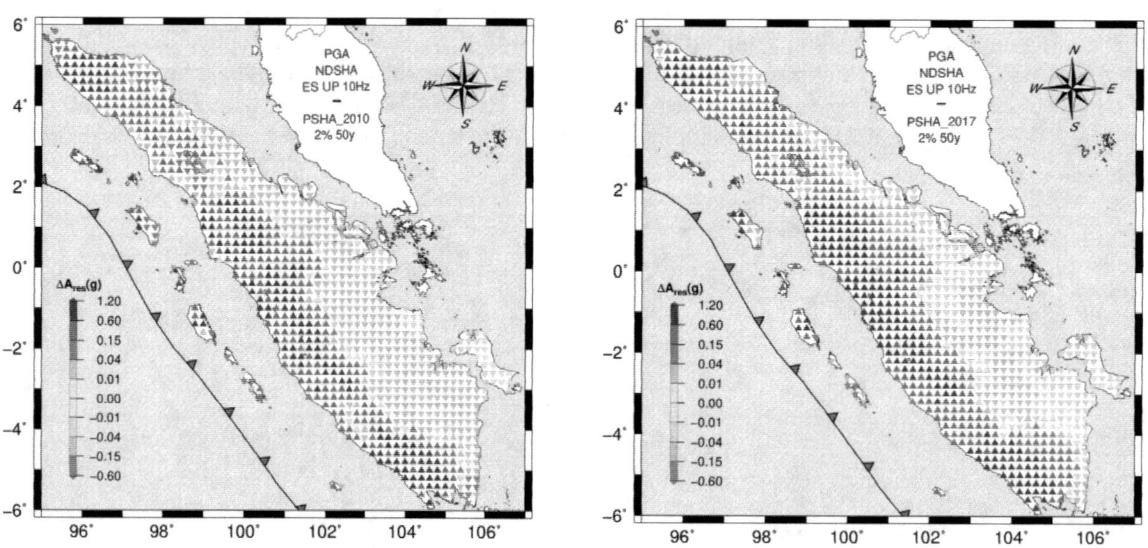

(a) NDSHA – PSHA 2010 (b) NDSHA – PSHA 2017

Figure 5.The differential map of PGA of NDSHA with official PSHA map for (a) 2010 (b) 2017

3.2 The Different NDSHA to 2017 and 2010 Official PSHA

The second part of the discussion is the comparison among different methods. The PGA NDSHA was compared to those of PSHA 2010 and PSHA 2017 for PE of 2% in 50year shown in Figure 5a and 5b, respectively. In general, the PGA NDSHA value is greater than the both PGA of the 2010 and the 2017 PSHA for the location close to Sumatran fault and smaller to areas far from the Sumatran fault.

This can be interpreted that the NDSHA PGA value is attenuated significantly in distance compared to the PSHA PGA values. Both differential maps, 2010 update map (figure 5a) and 2017 updated maps (figure 5b), visually look similar in the same the scale bar. The only different area is around Lake Toba. For the Banda Aceh city, the value of PSHA PGA is higher than the value provided by NDSHA for both differential maps.

4. Conclusion

The value of the 2017 PGA PSHA is greater than the 2010 PGA PSHA except for a number of spots such as in the central part of the Sumatran fault and the central and southern part of Sumatra. Significant increase of the PGA occurs in the Sumatra fault which is in the northern part of the province, precisely. The NDSHA PGA value is greater than the PSHA PGA, especially those close to the Sumatran fault, which is valid for areas far from the Sumatran fault, which can be interpreted as NDSHA is dampened more quickly by increasing distance compared to the PGA PSHA values. On the other hand, this shows that the NDSHA method provides more realistic results than the PSHA method because the calculation of PGA is based on realistic physics simulation of the seismic wave propagation were close to the fault will have a very large PGA compared to PGA produced by areas far from the sources.

5. Reference

[1] Beca, Carter, Hollings, and Ferner, "Indonesian Earthquake Study," *Unpublished,* vol. 6, 1978.

[2] Standar Nasional Indonesia (SNI), *Tata Cara Perencanaan Ketahanan Gempa untuk Bangunan Gedung (SNI 03-1726-2002),* 2002.

[3] Irsyam, M., Sengara, W., Aldimar, F., Widiyantoro, S., Triyoso, W., Natawidjaja, D.H., Kertapati, E., Meilano, I., Asrurifak, M., Suhardjono and Ridwan, M, "Summary of Study Development of Seismic Hazard Maps of Indonesia for Revision of Hazard Map in SNI 03-1726-2002," *Indonesia Seismic Hazard Map Revision Team,* 2010.

[4] Gempa, Tim Revisi Peta, "Peta Hazard Gema Indonesia 2010 Sebagai Acuan Dasar Perencanaan dan Perancangan Infrastruktur Taha Gempa Indonesia," 2010.

[5] Natawidjaja., Hilman, D., and Triyoso, W, "The Sumatran Fault Zone-from Source to Hazard," *Journal of Earthquake and Tsunami,* vol. 1, no. 1, pp. 21-47, 2007.

[6] Tabei, T., Kimata, F., Ito, T., Gunawan, E., Tsutsumi, H., Ohta, Y., Yamashima, T., Soeda, Y., Ismail, N., I, Nurdin., and Sugiyanto, D, "Geodetic and Geomorphic Evaluations of Earthquake Generation Potential of the Northen Sumatran Fault," 2015.

[7] Muksin, U., Bauer, K., Muzli, M., Ryberg, T., Nurdin, I., Masturiyono, M., and Weber, M, "AcehSeis Project Provides Insight into the Detailed Seismicity Distribution and Relation to Fault Structures in Central Aceh, Northern Sumatra," *Journal of Asian Earth Science,* vol. 171, pp. 20-27, 2019.

[8] Muzli, M., Umar, M., Nugraha, A.D., Bradley, K.E., Widiyantoro, S., Erbas, K.,Jousset, P., Rohadi, S., Nurdin., Irwandi., and Wei, S, "The 2016 Mw 6.5 Pidie Jaya, Aceh, North Sumatra, Earthquake: Reactivation of an Unidentified Sinistral Fault in A Region of Distributed Deformation," *Seismological Research Letters,* vol. 89, no. 5, pp. 1761-1772, 2018.

[9] (PuSGeN), Tim Pusat Studi Gempa Nasional, "Peta Sumber dan Bahaya Gempa Indonesia," *Kementerian Pekerjaan Umum dan Perumahan Rakyat,* 2017.

[10] Petersen, M.D., Dewey, J., Hartzell, S., Mueller, C., Harmsen, S., Frankel, A., and Rukstales, K, "Probabilistic Seismic Hazard Analysis for Sumatra, Indonesia and across the Southern Malaysian Peninsula," *Tectonophysucs,* vol. 390, no. 1, pp. 141-158, 2004.

[11] Harmsen, S, "USGS Software for Probabilistic Seismic Hazard Analysis (PSHA)," *USGS Draft Document,* 2007.

[12] Zuccolo, E., Vaccari, F., Peresan, A., and Panza, G.F, "Neo-Deterministic and Probabilistic

Seismic Hazard Assessment: a Comparison Over the Italian Territory," vol. 168, 2011.

[13] Irwandi, I., Muzli., Muksin and Jamaluddin, K, "GMPE based Shakemap Generation of Peak Ground Motion and Intensity Maps for Pidie Jaya Earthquake," *Journal of Physics,* vol. 1120, p. 012092, 2018a.

[14] Panza, G.F., Mura, C.L., Peresan, A., Romanelli, F., and Vaccari, F, "Seismic Hazard Scenarios as Preventive Tools for a Disaster Resilient Society," *Advances in Geophysics,* vol. 53, pp. 93-165, 2012.

[15] Peresan, A., Zuccolo, E., Vaccari, F., and Panza, G.F, "Neo-Deterministic Seismic Hazard and Pattern Recognition Techniques: Time-Dependent Scenarios for North Eastern Italy," *Regional Seismic Hazard and Seismic Microzonation Case Studies,* vol. 169, no. Advanced Seismic Hazard Assesment, 2012.

[16] Irwandi, I., Idris, Y., Muksin., Jamaluddin, K., Zulfakriza, Joni., and Muzli, "Realistic Shakemap M6.5 Pidie Jaya Earthquake 7 Desember 2016 Based on Modal Summation Technique," in *Earth and Environmental Science*, 2019.

[17] Irwandi, Irwandi., I, Rusydy., Muksin, U., and Rudyanto, A, "Response Spectra Analysis of the Modal Summation Technique Verified by Observed Seismometer and Accelerometer Waveform Data of the M6.5 Pidie Jaya Earthquake," in *Materials Science and Engineering*, 2018c.

[18] Irwandi, "Advantages of Realistic Model Based on Computational Method: NDSHA versus Standard PSHA," *Earth and Environmental Science,* vol. 56, 2017.

[19] PU (Departemen Pekerjaan Umum), Ditjen Cipta Karya, Direktorat Masalah Bangunan, *Peraturan Perencanaan Tahan Gempa Indonesia untuk Gedung,* 1983.

[20] Wessel, P., and Smith, W.H, "Free Software Helps Map and Display Data EOS," *Transactions American Geophysical Union,* vol. 72, no. 41, pp. 441-446, 1991.

[21] A. Maintainer, "GFotran: Compiling a 1,000,000+ line Numerical Weather Forecasting System," in *In GCC Developers' Summit*, 2005.

[22] Irwandi, Irwandi., I, Rusydy., Muksin, U., and Rudyanto, A, "Synthetic Accelerogram of Modal Summation Technique and its Validation with the Pidie Jaya Earthquake Data," in *AIP Conference Proceedings*, 2018b.

Acknowledgments
The researchers are grateful to the developer of GNU package software under Ubuntu and GMT for computation and plotting the maps. This work is financially supported by Directorate of Research and Community Service, Ministry of Research, Technology and Higher Education with the project "Penelitian Terapan", contract No. 215/SP2H/LT/DPRM/2019.

IOP Conf. Series: Materials Science and Engineering **712** (2020) 012018 doi:10.1088/1757-899X/712/1/012018

Effect of Defect Width upon Burst Capacity of Composite Repaired Pipe

K E Leong[1], K S Lim[1], A S Sulaiman[2], S C Chin[1] and N Yahaya[3]

[1]Faculty of Civil Engineering Technology, Universiti Malaysia Pahang, 26300 Gambang, Pahang, Malaysia
[2]Faculty of Mechanical and Automotive Engineering Technology, Universiti Malaysia Pahang, 26300 Gambang, Pahang, Malaysia
[3]Faculty of Engineering, School of Civil Engineering, Universiti Teknologi Malaysia, 81310 Johor Bahru, Malaysia

Abstract. In recent years, FRP composite wrap repair is the most preferable pipeline rehabilitation system used in the pipeline industry. However, some issues concerning this repair method are not fully understood by the industry. Effect of defect geometries toward the efficiency of composite repaired pipe is one of the issues that concerned by the industry. Pipeline repair design codes and standards have been developed mainly focus on the defect depth and neglect other defect geometries such as defect length and defect width. Previous studies stated that defect geometries especially defect width should not be ignored in evaluating and designing pipe repair system. Therefore, the burst pressure of the composite repaired pipeline subjected to various defect widths was determined through this study in order to evaluate the effect of defect width upon the burst capacity of composite repaired pipeline. Finite element analysis was used to determine the burst capacity of the composite repaired pipe with rectangular shape of defect. There are three different widths were selected with constant defect length and depth. The base model of composite repaired pipe was developed and validated and then modified with the various defect widths in this study. The result shows that burst pressure for three different models vary with a percentage of 12.51% between the maximum burst pressure and minimum burst pressure. The stress contour plot extracted from the finite element analysis revealed that the area of highest stress (557.7MPa) is located around defect region. As the defect is getting wider and subsequently creating a bigger defect area, stress concentration is also getting larger at the defect region. With this, the composite repaired pipe tends to fail at lower pressure when the defect getting wider. Based on the results, the defect width is proven to affect the burst capacity of composite repaired pipe.

1. Introduction

Steel pipelines have been used as a basis element to transport oil and natural gas in large quantities over a long distance as it can resist high pressure of fluid and gases [1-3]. However, deterioration will happen on pipeline due to the critical environment and hence cause reduction of load bearing capacity of pipelines, wall thinning or cracks which might result in leakage of oil or gas and may damage the environment [4-6].

Corrosion is one of the mechanical damage on oil and gas pipelines. When the pipeline service duration increase, it will be affected by corrosion mechanism and the corrosion can occur in internal and external surface of pipeline [7-9]. The condition of pipeline contains corrosion defect will cause high stress concentration at the deepest point of flaw area. The pipeline may then burst at that point if

the operating pressure in the pipe has reached its maximum burst pressure [10]. Therefore, methods to repair corroded pipeline is necessary to ensure its safe operation. In recent years, many pipeline operators prefer to use Fibre Reinforced Polymer (FRP) repair system to restore the strength of damaged pipes. The corroded defect area on the transmission pipeline will be strengthening by wrapping a composite sleeve bonded by epoxy grout to the pipe. Repair of pipeline will be conducted based on the evaluation of the pipeline condition to determine whether repair is necessary or not [11,12]. However, there are some issues about the FRP composite repair method that are not fully understood by the industry include the effect of the defect geometries, hence defect width towards burst capacity of a composite repaired pipe is focused in this study [13-15].
Currently, standards such as ASME PCC-2 and ISO 24817 were used by industry to ensure the safety and effectiveness of FRP composite repair system [16,17]. These standards are considered to be conservative as they always overdesign repair system due to safety factory and premature replacement [18,19]. Assessment codes and standards are also essential in repairing pipe and many modifications had applied on the original equations of assessment codes to reduce conservativeness of design codes. For example, code ASME B31Gmod was modified from the code ASME B31G by the American Society of Mechanical Engineers [20]. This assessment code is only accounts the remaining pipe wall thickness, maximum defect length and outer diameter of pipe but neglect the defect width. In contrast, assessment code such as DNV-RP-F101 used defect width as one of the parameters to estimate the burst pressure of the damaged pipeline [21]. As stated by Ref. [22], the prediction of the burst pressure of flaws not only depends on the material properties, but also on the defect geometry. By considering the influence of defect width, the composite repair system may be less conservative and more realistic in using the adequate amount of material to rehabilitate the damaged pipelines. Therefore, detailed research and experiment needs to be conducted to investigate the effect of defect width in sustaining the maximum pressure of composite repaired pipes [23].

2. Methods

2.1. Stage 1: Development of Base Model

ABAQUS Finite Element (FE) modelling software was utilized to build models, create meshes and conduct FE calculations to simulate burst capacity of pipe. The base model used in this study contains a corroded steel pipe, epoxy grout and composite wrap. A 1200mm long of hollow steel pipe with outer diameter of 168.3mm and 7.11mm thickness of pipe wall was created. A metal loss defect with dimension of (100x100x3.555) mm located in the middle of the pipe was modelled. The epoxy grout was created to cover the defect region while the composite wrap was created as a thin shell layer with 168.3mm diameter and 300mm length. The components that modelled individually were assigned with relevant material properties obtained from works done by Ref. [15]. Interaction between different materials was created followed by applying boundary conditions on them after assembled the component into integrated structure for analysis. The analysis was set to run 500 seconds with 50MPa pressure, which means the pressure rate is 0.1MPa/s, loaded on the internal wall of pipeline. An optimum meshing size was applied on the structure. The result of burst pressure of from simulated FE model was compared with published experimental test data. The base model is considered validated with error margin less than 10% between the results [24].

Table 1. Material Properties.

Material Properties	Material		
	Steel pipe	**Putty**	**Composite**
Young's Modulus, E (GPa)	222	19	14.3 (Hoop)
			10.1 (Axial)
			5.5 (Radial)
Density (kg/m3)	7850	-	1659.2
Poisson's Ratio	0.3	0.35	-
Ultimate Tensile Stress (MPa)	557.7	20.01	241.28 (Hoop)
			169.43 (Axial)

2.2. Stage 2: Parametric Study

Three models with different defect widths of 168.3mm × 84.15mm (D × ½D), 168.3mm × 168.3mm (D × D) and 168.3mm x 336.6mm (D × 2D) were created by modifying the defect geometry in the validated base model. The parameter for this study to analyse is defect width which is the length of defect in hoop dimension. The simulation was conducted with the constant value of defect length, defect depth, material properties and boundary conditions. The ultimate burst pressure of each models obtained from finite element analysis was extracted to evaluate the effect of different defect width towards burst capacity of composite repaired pipeline.

3. Results and Discussion

The error margin between the burst pressure of experiment test (33MPa) and FE analysis (31.77MPa) for base model is 3.73% which means the base model is validated as it is less than 10%. This validated base model was used to modify with three defect width to do parametric study. After the internal pressure applied on the inner wall of the defective pipe model with defect geometry of 168.3mm x 84.15mm (D × ½D model), the results of the simulation is shown in Figure 1 with the stress contour plot of completed model and all individual components which are defective steel pipe, putty and composite wrap. It only shows hoop stress contour plot for steel since hoop stress is the greatest stress experienced by a pressurised pipe.

When the stress of the corroded area reaches the ultimate tensile stress, pipeline failure will occur as the load bearing capacity of the pipeline is lower than the ultimate stress can be withstand by the pipeline [25]. It was observed that the highest stress (557.7MPa) is concentrated at the both edges of the defect region along the axial direction. In contrast, the stress observed at the centre of the defect region was smaller than that. This high stress concentration region that is predicted by finite element analysis can be considered as the failure location of the pipe in experiment. The putty shows that the whole structure of putty experienced the highest tensile stress which is 20.01MPa. This shows that the epoxy grout has lower strength compared to other components and it will experiences failure first than others. On the other hand, 272.5MPa stresses was observed on composite wrap where the region have defect region underneath it and they were located upon the four corners of the defect region. The composite wrap also experienced 227.1MPa stress which is higher than the stress on the centre of the composite along axial direction. This region also at the same region mentioned before but it is on the edges of the defect region along hoop direction.

IOP Conf. Series: Materials Science and Engineering **712** (2020) 012018 doi:10.1088/1757-899X/712/1/012018

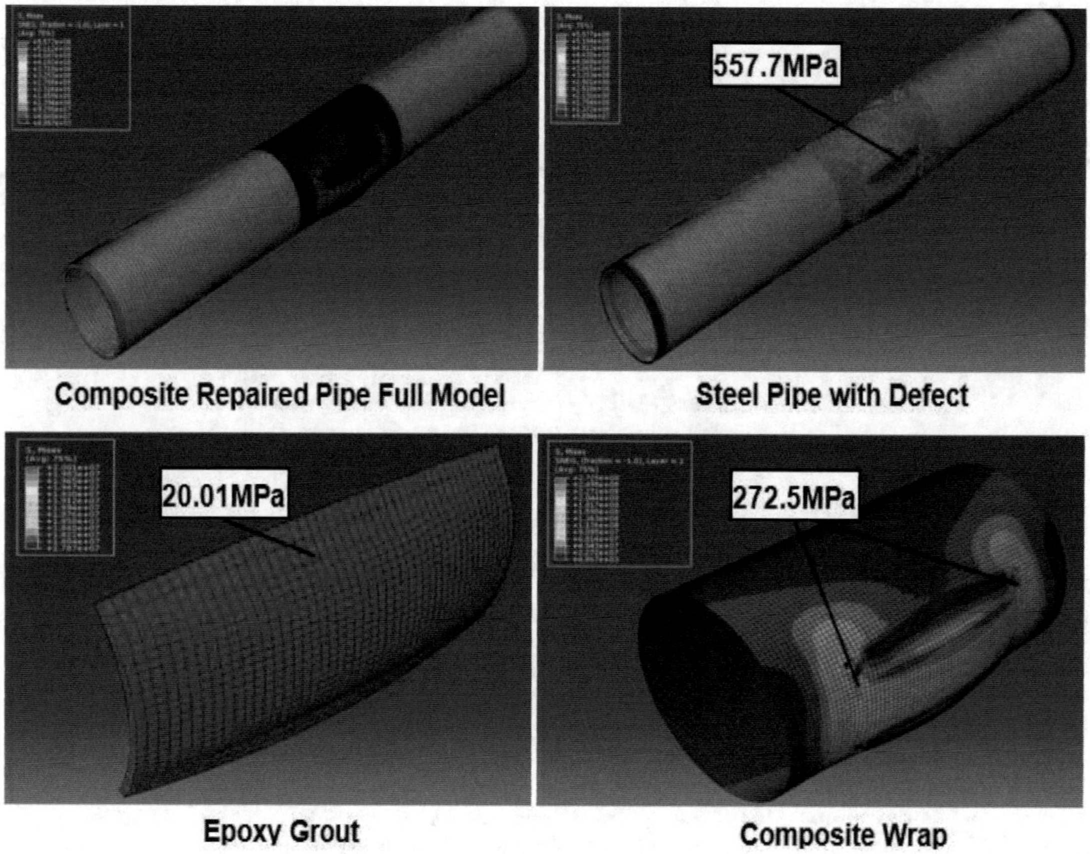

Figure 1. Stress contour plot of completed model with all components.

Since this study is focusing on analyse whether the burst pressure of composite repaired pipe is significantly affected by the changes on defect width, the stress sustained around the defect region with different width were analysed in more details. Figure 2 shows the stress contour plot of steel pipe component with defect geometries of 168.3mm × 168.3mm (D × D model). The highest stress of 557.7MPa was observed at both edges of the defect region along the axial direction. By comparing to the defective pipe with defect of 168.3mm × 84.15mm (D × ½D model), the area of highest stress concentration at the edges of the defect region in the pipe with defect of 168.3mm x 168.3mm (D × D model) is narrower. Besides that, a small area of 557.7MPa of stress along axial direction was observed to occur in the middle of the defect region. It was also observed that the areas between the edge and middle of defect region along the axial direction were sustained 479.3MPa of stress.

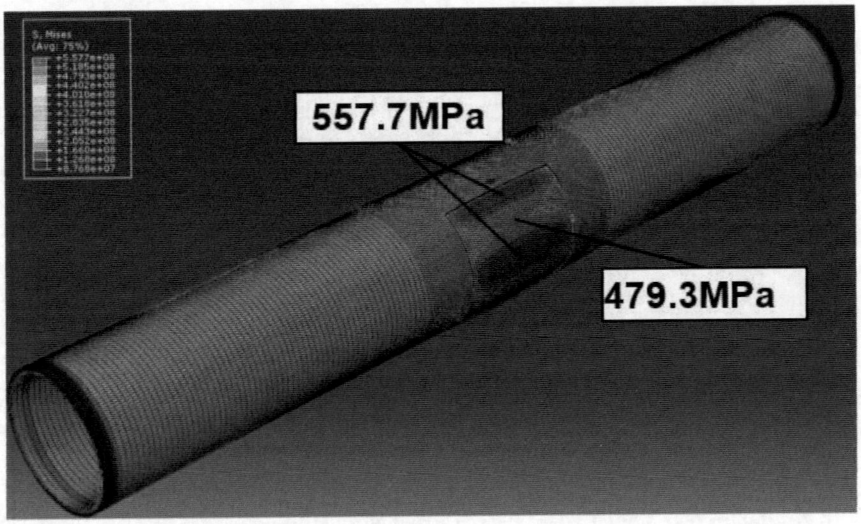

Figure 2. Stress contour plot of defective steel pipe with defect geometries (D × D model).

As the defect getting wider, the distribution of stress concentration on the defect region will be located in different location around the defect. We can see the difference in Figure 3 that shows the stress contour plot of defective steel pipe with defect geometries of 168.3mm x 336.6mm (D × 2D model). Both edges and middle of the defect region experienced the highest stress of 557.7MPa along axial direction but the area of stress concentration at the edges of defect was the narrowest amongst the three defective pipes. In contrast, the area of stress concentration in the middle of the defect was getting wider and there was another area of stress concentration of 557.7MPa stress occurred nearby the centre of the defect region. From the stress contour plot diagrams, we could see that the stress concentration area at the edges of the defect region is getting smaller along the axial direction when the defect width increasing. On the other hand, the highest stress was getting more concentrated in the centre of defect region as the area of 557.7MPa stress increasingly occur at the middle of the defect region as the area getting wider.

Figure 3. Stress contour plot of defective steel pipe with defect geometries (D × 2D model).

Figure 4 shows the result of finite element analysis on burst pressure of composite repaired pipe subjected to various defect widths. With the simulation of finite element modelling, the burst capacity of composite repaired pipe can be predicted based on the changing of defect width. The simulated burst pressure of D × ½ D model is 29.33MPa which is the highest burst pressure that sustained by the defective pipe amongst three models. The predicted burst pressure for D x D model and D x 2D model are 28.62Mpa and 25.66MPa, respectively. The predicted burst pressures in all cases were found to be slightly different. The variation between the maximum predicted burst pressure and the minimum predicted burst pressure is 12.51%. Based on these results, it can see that as the defect getting wider, the load bearing capacity of composite repaired pipe is getting lower. This is condition can be refer to previous studies such as ASME B31G states that the bigger the defect area on the corrode pipe, the lower the pressure of a leak or rupture to occur [26].

From the results of finite element analysis, when the defect width decreases, the area of the stress concentration in defect region is smaller compared to other models with bigger defect area. The low stress concentration in the defect area will be transmitted to the putty and composite wrap, thus the pipe will burst at higher pressure when the stress at the composite wrap reached ultimate tensile stress. In contrast, with longer defect width and larger defect area, higher stress will be experienced by the pipe as the area of stress concentration is getting wider and the stress distributed unevenly along the defect region. Therefore, the composite repaired pipe will burst at the low pressure.

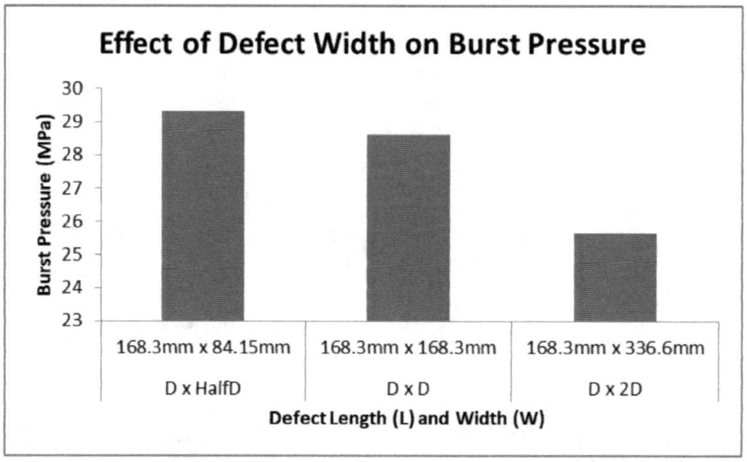

Figure 4. Effect of defect width on burst pressure of composite repaired pipe.

4. Conclusion

In this study, finite element models were created to determine the burst pressure of the externally corroded steel pipeline subjected to various defect widths. The influence of defect width in composite repaired pipe upon its burst capacity was evaluated through the finite element analysis. The result from finite element analysis shows that the changes in defect width will affects the burst capacity of composite repaired pipe. The result of variation between burst pressures shows 12.51%, which is more than 10%, means that the finding of this study has effect in assessing and designing of composite repair system in a more realistic and effective way. Since this research only conducts finite element analysis, it is suggests to perform experiment test to further validate the result of this study for the research to have more confident on simulation results. It also suggest to do more details on parametric study by adding more defect width with smaller difference into the finite element analysis in order to get a detailed trend of the burst capacity of composite repaired pipe influence by various defect width. This can be used to predict the burst pressure of pressurize pipe depends on its defect geometry after being repaired.

IOP Conf. Series: Materials Science and Engineering **712** (2020) 012018 doi:10.1088/1757-899X/712/1/012018

References

[1] Li Y, Li T X, Cai G W and Yang L H 2013 Influence of AC interference to corrosion of Q235 carbon steel *Corrosion Engineering, Science and Technology* **48(5)** 322-326.

[2] Mior Mohd Tahir S N F, Md Noor N, Yahaya N and Lim K S 2015. Relationship between in-situ measurement of soil parameters and metal loss volume of X70 carbon steel coupon. *Asian Journal of Scientific Research* **8(2)** 205-211.

[3] Ehsani M 2015 Repair of Corroded/Damaged Metallic Pipelines Using Fiber-Reinforced Polymer Composites *Rehabilitation of Pipelines Using Fiber-reinforced Polymer (FRP) Composites* ed V M Karbhari (United Kingdom: Elsevier), p 39-59.

[4] Alamilla J L, Espinosa-Medina M A and Sosa E 2009. Modelling steel corrosion damage in soil environment. *Corrosion Science* **51** 2628-2638.

[5] Lim K S, Yahaya N, Md Noor N, Mior Mohd Tahir S N F, Paik J K, and Mohd M H 2017 Effects of soil properties on the corrosion progress of X70-carbon steel in tropical region. *Ships and Offshore Structures* **12(7)** 991-1003.

[6] Abdel-Karim R, Nabil M, Reda Y and El-Raghy S 2018. Corrosion characteristics of ASTM A106 grade B carbon steel pipelines exposed to sodium sulfate solutions. *Materials Performance and Characterization* **7(1)** 480-494.

[7] Shamsuddoha M, Islam M M, Aravinthan T, Manalo A and Lau K T 2013 Characterization of mechanical and thermal properties of epoxy grouts for composite repair of steel pipelines. *Material and Design* **52** 315-327.

[8] Othman S R, Yahaya N, Md Noor N, Lim K S, Zardasti L and Rashid A S A 2017 Modeling of external metal loss for corroded buried pipeline. *Journal of Pressure Vessel Technology, Transactions of the ASME* **139(3)** 031702.

[9] Barkanov E N, Lvov I and Akishin P 2017 Optimal Design of Composite Repair Systems of Transmission Pipelines *Non-destructive Testing and Repair of Pipelines* eds Barkanov *et al* (Switzerland:Springer) p 387-398.

[10] Fekete G and Varga L 2011 Extension of pit corrosion effect on pipelines *Periodica Polytechnica Mechanical Engineering* **55(1)** 15-20.

[11] Haladuick S and Dann M R 2018 Decision making for long-term pipeline system repair or replacement *ASCE-ASME Journal of Risk and Uncertainty in Engineering Systems, Part A: Civil Engineering* **4(2)** 04018009.

[12] Liu X, Zheng J, Fu J, Ji J and Chen G 2018 Multi-level optimization of maintenance plan for natural gas pipeline systems subject to external corrosion. *Journal of Natural Gas Science and Engineering* **50** 64-73.

[13] Ma W F, Luo J H and Cai K 2011 Discussion about application of composite repair technique in pipeline engineering *Advanced Materials Research* **311-313** 185-188.

[14] Shamsuddoha M, Islam M M, Aravinthan T, Manalo A and Lau K T 2013 Effectiveness of using fibre-reinforced polymer composites for underwater steel pipeline repairs *Composite Structures* **100** 40-54.

[15] Lim K S, Azraai S N A, Yahaya N, Md Noor N, Zardasti L and Kim J H J 2019 Behaviour of steel pipelines with composite repairs analysed using experimental and numerical approaches *Thin-Walled Structures* **139** 321-333.

[16] Alexander C R 2009 Development of standards for composite repair systems *Pipeline Gas Technology Magazine* **October Edition** 1-4.

[17] Lim K S, Azraai S N A, Yahaya N, Zardasti L and Md Noor N 2017 Strength development of epoxy grouts for pipeline rehabilitation *Jurnal Teknologi* **79(1)** 9-14.

[18] Duell J M, Wilson J M and Kessler M R 2008 Analysis of a carbon composite overwrap pipeline repair system *International Journal of Pressure Vessels and Piping* **85** 782-788.

[19] Lim K S, Yahaya N, Valipour A, Zardasti L, Azraai S N A and Md Noor N 2018 Mechanical properties characterization and finite element analysis of epoxy grouts in repairing damaged pipeline *Journal of Pressure Vessel Technology, Transactions of the ASME* **140(6)** 061701.

[20] ASME International 2011 *ASME PCC-2-2011. Repair of Pressure Equipment and Piping* (New York: The American Society of Mechanical Engineers)
[21] Det Norske Veritas 2010 *Recommended Practice DNV-RP-F101.* (Norway: Det Norske Veritas)
[22] Cunha S B and Netto T A 2012 Analytical solution for stress, strain and plastic instability of pressurized pipes with volumetric flaws *International Journal of Pressure Vessels and Piping* **89** 87-202.
[23] Chiodo M S and Ruggieri C 2009 *Proc. Int. Conf. on ASME 2009 Pressure Vessels and Piping Conference (Prague)* Vol 6 (Czech Republic: ASME) p 1167-1176.
[24] Chan P H, Tshai K Y, Johnson M, Choo H L, Li S and Zakaria K 2015 Burst strength of carbon fibre reinforced polyethylene strip pipeline repair system – a numerical and experimental approach *Journal of Composite Materials* **49(6)** 749-756.
[25] Liu H, Khan F and Thodi P 2017 Revised burst model for pipeline integrity assessment. *Engineering Failure Analysis* **80** 24-38.
[26] ASME International 1991 *ASME B31G-1991. Manual for Determining the Remaining Strength of Corroded Pipelines: A Supplement to ASME B31 Code for Pressure Piping* (New York, USA: The American Society of Mechanical Engineers)

Acknowledgments
The author gratefully acknowledges the financial and technical support provided by Ministry of Education Malaysia (Grant No. FRGS/1/2018/TK07/UMP/02/8) and Universiti Malaysia Pahang (Grant No. RDU190150).

IOP Conf. Series: Materials Science and Engineering **712** (2020) 012019 doi:10.1088/1757-899X/712/1/012019

Influence of Desert Sand and Cooling Regime on the Compressive Strength of High Strength Concrete after High Temperatures

M H Zhang[1], H F Liu[1], S Tian[1], Y C Ma[1] and S I Doh[2]

[1]School of Civil and Hydraulic Engineering, Ningxia University, 750000, Yinchuan, Ningxia, China

[2]Faculty of Civil Engineering Technology, Universiti Malaysia Pahang, 26300 Gambang, Kuantan, Pahang

Abstract: An experimental investigation is conducted to analyze the influence of desert sand replacement ratio and cooling regime on the compressive strength of desert sand-based high strength concrete (DS-HSC). According to the design of experiment, six desert sand replacement ratios (0%, 20%, 40%, 60%, 80% and 100%) are employed to concrete mixture. DS-HSCs are subjected to two cooling regimes (air cooling and water cooling) and various temperatures (20℃, 400℃, 500℃, 600℃, 700℃ and 850℃, respectively) in this paper. The results indicate that the compressive strength of DS-HSC with air cooling is higher than that with water cooling. Moreover, the highest compressive strength appears with the desert sand replacement ratio of 20%. The compressive strength of DS-HSC gradually decreases with the increase of desert sand replacement ratio when the desert sand replacement ratio is more than 20%.

1. Introduction

Sand is a non-renewable resource and plays a significant role in building. The ordinary material of concrete has a huge demand every year in building domain, which leads to the excessive consumption of sand. It is important to find a new material to solve this problem. Plentiful desert sand place in China. If desert sand has a good workability in concrete, it will promote the development of local economy. Researchers have already performed some research on desert sand these years [1-5]. Dune sand from Tenggeli and Mu Us desert could be used as fine aggregate in concrete construction [6]. Properties of drying shrinkage cracking of concrete using dune sand and crushed sand were investigated [7]. It was found that the maximum value of compressive and tensile strength occurred when the dune sand replacement was 20%. Moreover, the compressive and tensile strength increased firstly and then decreased with increasing dune sand. The physical characteristics and mechanical performance test of concrete for pavement application containing Sahara dune sand were carried out [8]. It was found that the concrete for pavement application had a good workability with a mix of 60% dune sand and 40% crushed sand. The relationship of dune sand replacement ratio on mechanical properties of concrete was also surveyed [5, 9, 10]. These above indicate desert sand has a potential to be used in building domain.

Recent years, the requirement of high-rise building is huge due to the rapid development of economy. High strength concrete is widely used in building domain according to its characteristic of high durability, stiffness and strength. Meanwhile, the trend of high-rise building fire gradually grows with increasing the high-density population and quantity of high-rise building. Fires frequently cause the damage of building structure and threaten the safety of people. The pictures of high-rise building fire are displayed in figure 1. Some researcher have already focused on this status and performed some work. The mechanical properties of fiber high-strength concrete exposed to high temperature were performed [11-12]. The different grades of high-strength concrete and normal concrete were researched on their compressive strength, splitting tensile strength and flexural strength [13]. High-strength concrete exposed to high temperature lost its mechanical strength in a manner similar to that of normal concrete [14]. Moreover, some works were conducted to other properties of high-strength concrete. The influence of eccentric loading on mechanical property of high-strength concrete column, tube and bar was investigated [15-16]. It was found that eccentric loading significantly affected the axial stress-strain behavior of FRP-confined high-strength concrete. Some researchers concerned the effect of experimental condition on compressive properties of high-strength concrete. Curing method and curing time had a close relationship with mechanical property of high-strength concrete [17-18]. It was found that the compressive strength of high-strength concrete increased with increasing curing temperature and curing time. The cooling regime had a significant effect on mechanical properties of high-strength concrete [19]. It was illustrated that water cooling caused only a bit more deterioration in compressive strength than that of furnace cooling.

In this paper, the compressive strength of DS-HSC exposed to different high temperatures is investigated. The influence of desert sand replacement ratio and cooling regime on the compressive strength of DS-HSC exposed to high temperature is analyzed. In addition, the optimal value of desert sand replacement ratio is found.

Figure 1. High-rise building fire

2 Experimental methods

2.1. Materials

In this paper, the raw materials, used to produce DS-HSC element, contained: cement, fly ash, stone, mountain sand and desert sand. The 42.5R grade of Ordinary Portland cement, provided by Saima cement plant, was used in DS-HSC element. The stone was purchased from local stone plant as coarse aggregate. The maximum value of stone was 25mm. The apparent density, bulk density, void content and mud content of coarse aggregate were 2698kg/m^3, 1430kg/m^3, 47% and 0.78%, respectively. Desert sand from Mu Us desert and local mountain sand were employed as fine aggregate. The apparent density, bulk density and mud content of desert sand were 2624kg/m^3, 1400kg/m^3 and 0.14%, respectively, while those of mountain sand were 2636kg/m^3, 1570kg/m^3 and 0.6%, respectively. The fineness modulus of desert sand and mountain sand were 0.194 and 2.59, respectively, so desert sand was super-fine sand in accordance with Chinese National Standard JGJ52-2006[20]. The appearance of desert sand from Mu Us desert is displayed in figure 2. The fly ash of class I provided by local power plant was used to make elements and the ignition loss ratio was 2.8%. In this work, the water used was local running water. Polycarboxylate superplasticizer in power was used and the ratio of reducing water ranged from 25% to 30%. Moreover, the dosage of Na_2SO_4 was below 3%.

Figure 2. Appearance of desert sand from Mu Us desert

2.2. Mix proportion and element preparation

Mix proportion has a significant influence on workability of concrete in test. In this paper, the water-cement ratio and sand ratio were 0.28 and 0.35, respectively. Six desert sand replacement ratios (0%, 20%, 40%, 60%, 80% and 100%) were employed to make DS-HSC elements of grade C60. Moreover, the 10% content of fly ash was placed in all concrete elements. The mix proportion of DS-HSC is shown in Table 1. The name meaning of every group is expressed in Table 2.

According to GB/T50081-2002[21], raw materials of DS-HSC were successively placed into the agitator mixing different time, coarse aggregate and fine aggregate mixing 1min, Ordinary Portland cement and fly ash mixing 0.5min, water mixed with superplasticizer mixing 2min. After the mixing process, concrete mixture was put into molds and then vibrated 2min on the vibrator in order to shape and discharge bubbles of elements. In total, 216 cubic elements (100mm×100mm×100mm) were made in this paper. All elements were removed to the curing room (18~22℃, humidity of 95% above) maintaining 28 days after elements demolded.

Table 1. Mix proportion of DS-HSC.

Materials	Mass (kg/m^3)
Cement	643
Fine aggregate	553
Coarse aggregate	1024
Water	180

Table 2. The name meaning of every group.

Group	Content description
C1	0% desert sand and 10% fly ash
C2	20% desert sand and 10% fly ash
C3	40% desert sand and 10% fly ash
C4	60% desert sand and 10% fly ash
C5	80% desert sand and 10% fly ash
C6	100% desert sand and 10% fly ash

2.3. Testing

In order to research the mechanical property of DS-HSC on fire, DS-HSC at the age of 28 days was exposed to different high temperatures(20°C, 400°C, 500°C, 600°C, 700°C and 850°C). The heating process was conducted with a muffle furnace (10°C/min) and maintained DS-HSC element in target temperature for 1h to keep thermal stability of element. The specific heating process of DS-HSC is shown in figure 3. After the heating process, DS-HSC element was subjected to two cooling regime (air cooling, water cooling). In particular, DS-HSC element with water cooling were put into water(20°C) for 0.5h before placing room for 1 day. Finally, all DS-HSC elements were conducted to the compressive test until damage according to Chinese National Standard GB50081-2002[21]. The loading rate was 0.8~1.0MPa/s. The pictures of heating furnace and compressive machine are shown in figure 4.

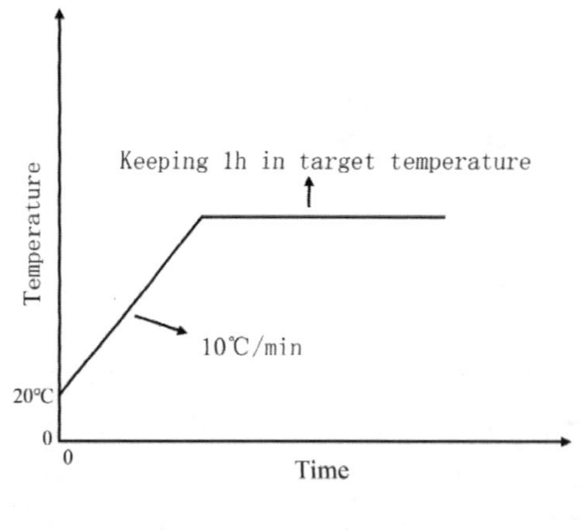

Figure 3. Curve of DS-HSC exposed to heating temperature

IOP Conf. Series: Materials Science and Engineering **712** (2020) 012019 doi:10.1088/1757-899X/712/1/012019

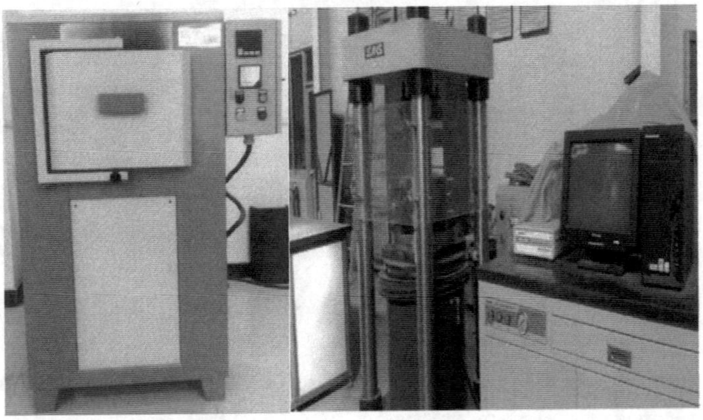

Figure 4. Muffle furnace and compressive machine.

3. Results and discussions

3.1. Physical characteristic

3.1.1. Appearance. DS-HSCs exposed to different high temperatures are subjected to air cooling and water cooling, respectively. The apparent characteristics of DS-HSC C6 exposed to high temperature with two cooling regimes are shown in figure 5. For DS-HSC with air cooling and water cooling, micro-cracks appear for DS-HSC element exposed to 400°C. Crack widens and increases its quantity with increasing temperature. The effect of temperature and cooling regime causes the difference of DS-HSC elements in colors. The colors of DS-HSC exposed to 400°C, 500°C, 600°C, 700°C and 850°C with air cooling are dark grey, bluish grey, grey with yellow little, off-white with yellow little and off-white, respectively, while these with water cooling are dark grey, bluish grey, off-grey, grey with yellow little and off-white, respectively.

DS-HSC element exposed to 850°C with water cooling arises the phenomenon of scaling while it does not arise for DS-HSC element with air cooling. It may be attributed to the scouring effect of stream. DS-HSC element exposed to 600°C above arises the spalling. This phenomenon can be explained as follow: DS-HSC element has a great compaction and produces evaporating pressure for DS-HSC element exposed to high temperature. The evaporating pressure of DS-HSC element does not have an efficient way to discharge due to the great compaction of DS-HSC element. Therefore, the evaporating pressure of DS-HSC element increases with increasing temperature until spalling.

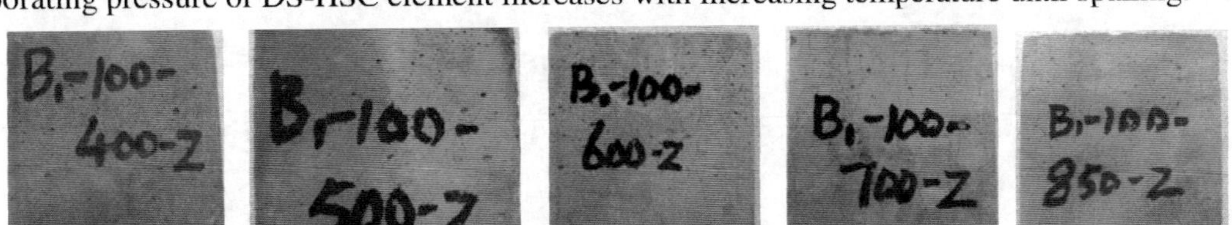

(a) air cooling

IOP Conf. Series: Materials Science and Engineering **712** (2020) 012019 doi:10.1088/1757-899X/712/1/012019

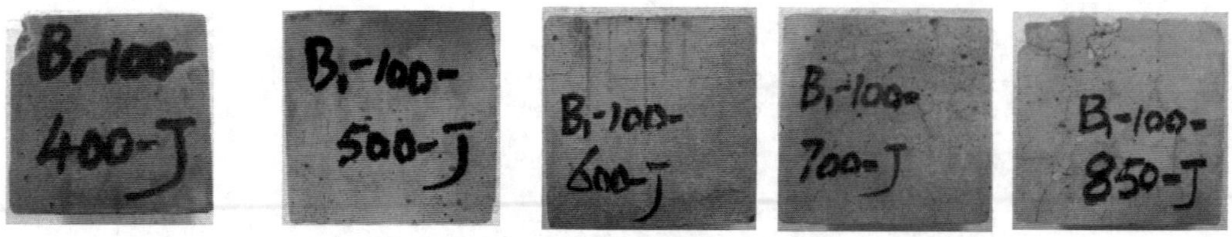

(b) water cooling

Figure 5. Apparent characteristics of DS-HSC C6 exposed to high temperature with two cooling regime.

3.1.2. Mass loss rate. All cubic elements were weighed before burning and after removing from furnace for placing 24h. The ratio of the difference in mass of DS-HSC before and after high temperature to the mass of DS-HSC before high temperature is defined as mass loss rate. The relationship between temperature and mass loss rate of DS-HSC is displayed in figure 6. The mass losses of DS-HSC are different levels under different temperatures and cooling regimes. As shown in figure 6, the mass loss rates of DS-HSCs with air cooling and water cooling gradually increase with increasing temperature. For instance, the mass loss rates of DS-HSC C2 exposed to 400°C, 500°C, 600°C, 700°C and 850°C with air cooling are 1.7%, 2.5%, 3.8%, 5.3%, 6.4%, respectively, while these with water cooling are 0.5%, 0.7%, 1.1%, 1.5%, 3.9%, respectively. This phenomenon is due to the evaporation of water and decomposition of C-S-H and $Ca(OH)_2$[22].

The relationship between desert sand replacement ratio and mass loss rate of DS-HSC exposed to 700°C is shown in figure 7. It can be found from figure 7 that the mass loss rate of DS-HSC with air cooling is higher than that with water cooling. This can be explained as follow: Water comes into DS-HSC when DS-HSC exposed to high temperature performs water cooling. Therefore, the DS-HSC can continue to hydration.

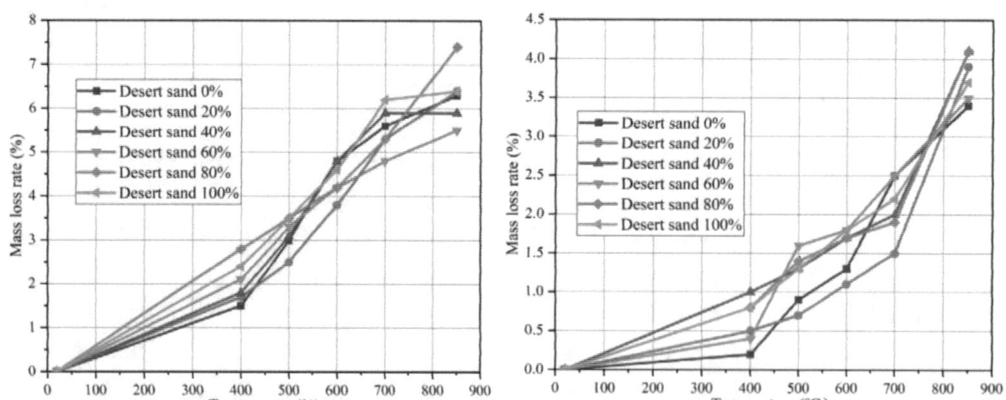

(a) air cooling (b) water cooling

Figure 6. Relationship of temperature and mass loss rate of DS-HSC

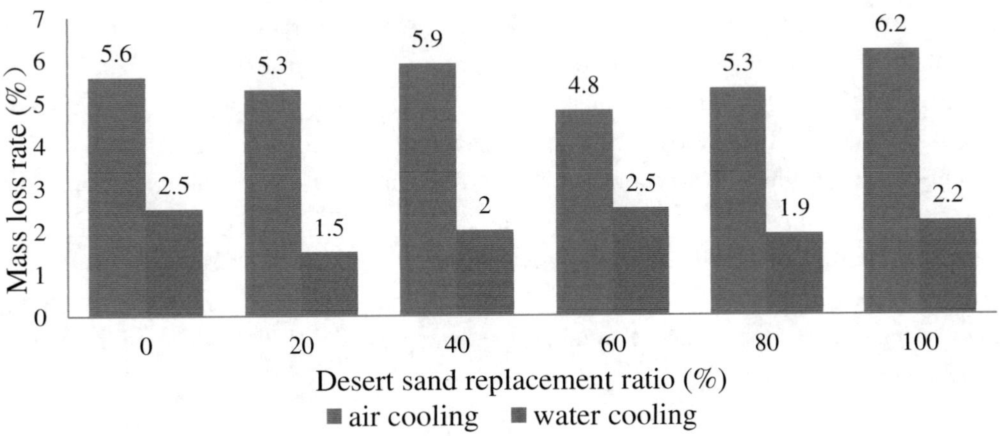

Figure 7. Relationship between desert sand replacement ratio and mass loss rate of DS-HSC exposed to 700℃.

3.2. Compressive strength

3.2.1. Desert sand replacement ratio. The relationship between desert sand replacement ratio and compressive strength of DS-HSC exposed to high temperature is shown in figure 8. As displayed in figure 8, no matter which cooling regime is employed, the compressive strength of DS-HSC increases with the desert sand replacement ratio to be the maximum value at 20% and gradually decreases after the desert sand replacement ratio is over 20%. This phenomenon can be explained as follow. Due to the characteristic of small particle for desert sand, desert sand improves the gradation of fine aggregate and then enhances the cohesive force between cementitious material and fine aggregate. Therefore, the compressive strength of DS-HSC increases with increasing the desert sand replacement ratio. At the same time, the strength of desert sand is lower than that of mountain sand. In the second region, as the desert sand replacement ratio increases, the effect of desert sand on cohesive force is smaller than effect of strength of desert sand on compressive strength of DS-HSC. Therefore, the compressive strength of DS-HSC decreases with increasing the desert sand replacement ratio.

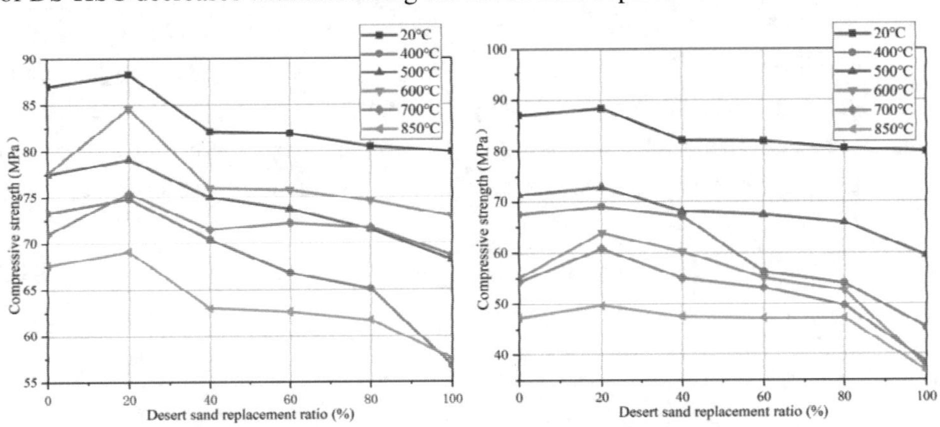

(a) air cooling (b) water cooling

Figure 8. Relationship of desert sand replacement ratio and strength exposed to high temperature

3.2.2. Cooling regime. The influence of cooling regime on the compressive strength of DS-HSC is measured in every desert sand replacement ratio and high temperature. Experimental results of DS-HSC with desert sand replacement ratio of 20% are taken for an example, which is plotted in figure 9. It is clearly displayed in figure 9 that the influence of cooling regime on compressive strength is significant for the DS-HSC exposed to high temperature. The compressive strength of DS-HSC exposed to high temperature with air cooling is higher than that with water cooling. This phenomenon can be explained as follow. The DS-HSC element with air cooling slowly finishes the process of thermal release. At the same time, DS-HSC element produces drastic thermal shock in the rapid process of water cooling, so the internal DS-HSC element with water cooling produces some residual stresses which increase cracks [23-24]. Therefore, the damage of DS-HSC element with water cooling is higher than that with air cooling.

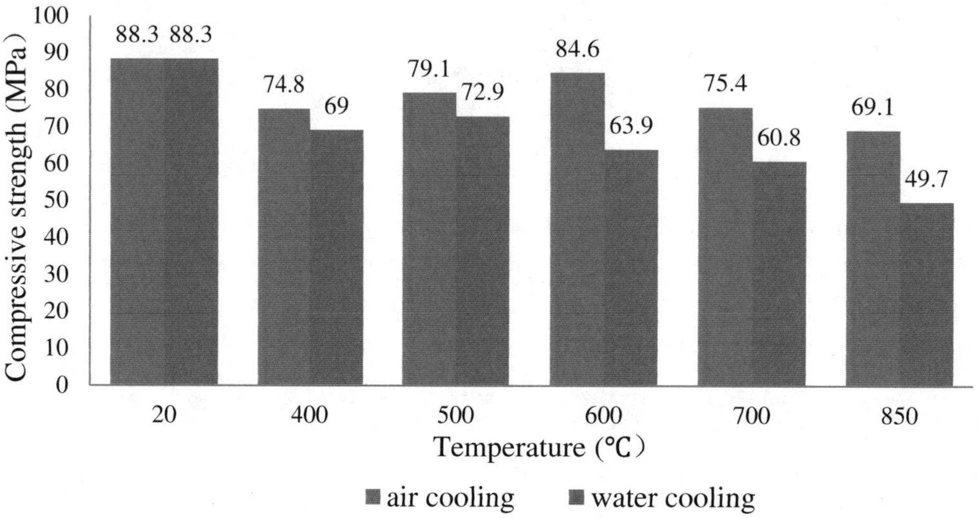

Figure 9. The influence of cooling regime on the compressive strength of DS-HSC with desert sand replacement ratio of 20%

4. Conclusions

This paper made 216 cubic DS-HSC elements (100mm × 100mm × 100mm) totally and carried out an experiment to investigate the influence of desert sand replacement ratio and cooling regime on the compressive strength of DS-HSC exposed to high temperature. According to experimental results, these conclusions can be found: When DS-HSC is exposed to high temperature, the compressive strength of DS-HSC increases with the desert sand replacement ratio to be the maximum value at 20% and gradually decreases after the desert sand replacement ratio is over 20%. Moreover, the compressive strength of DS-HSC exposed to high temperature with air cooling is higher than that with water cooling.

References

[1] Lopez-Querol S, Arias-Trujillo J, Gm-Elipe M, Matias-Sanchez A and Cantero B. 2017. Improvement of the bearing capacity of confined and unconfined cement-stabilized aeolian sand. *Constr. Build. Mater.*, **153**: 374-84.

[2] Liu H F, Ma J R, Wang Y Y and Ning J G. 2017. Influence of desert sand on the mechanical properties of concrete subjected to impact loading. *Acta Mech. Solida Sin.*, **30**(6): 583-95.

[3] Yacine A, Ratiba M, Abderrahmane G. 2016. Ceramic waste influence on dune sand mortar performance. *Constr. Build. Mater.*, **125**: 703-13.

[4] Jiang C, Zhou X, Tao G and Chen D. 2016. Experimental study on the performance and microstructure of cementitious materials made with dune sand. *Adv. Mater. Sci. Eng.*, **2016**.

[5] Luo F J, He L, Pan Z, Duan W H, Zhao X L and Collins F. 2013. Effect of very fine particles on workability and strength of concrete made with dune sand. *Constr. Build. Mater.*, **47**: 131-7.

[6] Zhang G, Song J, Yang J and Liu X. 2006. Performance of mortar and concrete made with a fine aggregate of desert sand. *Build. Environ.*, **41**(11): 1478-81.

[7] Euibae L, Sangjun P and Yongjic K. 2016. Drying shrinkage cracking of concrete using dune sand and crushed sand. *Constr. Build. Mater.*, **126**: 517-26.

[8] Khay S E E, Neji J and Loulizi A. 2011. Compacted dune sand concrete for pavement applications. *Proceedings of the Institution of Civil Engineers-Construction Materials*,, 164(2): 87-93.

[9] Bouziani T, Bederina M and Hadjoudja M. 2012. Effect of dune sand on the properties of flowing sand-concrete (FSC). *Int. J. Concr. Struct. Mater.*, **6**(1): 59-64.

[10] Al-Harthy A S, Halim M A, Taha R, and Al-Jabri K S. 2007. The properties of concrete made with fine dune sand. *Constr. Build. Mater.*, **21**(8): 1803-08.

[11] Behnood A and Ghandehari M. 2009. Comparison of compressive and splitting tensile strength of high-strength concrete with and without polypropylene fibers heated to high temperatures. *Fire Saf.*, **44**(8): 1015-22.

[12] Chen B and Liu J Y. 2004. Residual strength of hybrid-fiber-reinforced high-strength concrete after exposure to high temperatures. *Cem. Concr. Res.*, **34**(6): 1065-69.

[13] Li M, Qian C X and Sun W. 2004. Mechanical properties of high-strength concrete after fire. *Cem. Concr. Res.*, **34**(6): 1001-5.

[14] Chan Y N, Peng G F and Anson M. 1999. Residual strength and pore structure of high-strength concrete and normal strength concrete after exposure to high temperatures. *Cement. Concrete. Comp.*, **21**(1): 23-7.

[15] Pour A F, Gholampour A, Zheng J and Ozbakkaloglu T. 2019. Behavior of FRP-confined high-strength concrete under eccentric compression: tests on concrete-filled FRP tube columns. *Compos. Struct.*, **220**:261-72.

[16] Salah-Eldin A, Mohamed H M, Benmokrane B. 2019. Structural performance of high-strength-concrete columns reinforced with GFRP bars and ties subjected to eccentric loads. *Eng. Struct.*, **185**: 286-300.

[17] Wedatalla A M O, Jia Y and Ahmed A A M. 2019. Curing effects on high-strength concrete properties. *Adv. Civ. Eng.*, **2019**.

[18] Yazıcı H. 2007. The effect of curing conditions on compressive strength of ultra-high strength concrete with high volume mineral admixtures. *Build. Environ.*, **42**(5): 2083-89.

[19] Luo X, Sun W and Chan S Y N. 2000. Effect of heating and cooling regimes on residual strength and microstructure of normal strength and high-performance concrete. *Cem. Concr. Res.*, **30**(3):

379-83.

[20] Standard C. JGJ52-2006. 2006. Standard for technical requirements and test method of sand and crushed stone (or gravel) for ordinary concrete. *China Architecture & Building Press*, Beijing.

[21] Standard C. GB/T50081-2002. 2017. standard for test method of mechanical properties on ordinary concrete. *Standards Press of China*, Beijing.

[22] Savva A, Manita P, Sideris K K. 2005. Influence of elevated temperatures on the mechanical properties of blended cement concretes prepared with limestone and siliceous aggregates. *Cement. Concrete. Comp.*, **27**(2), 239-48.

[23] Wang Y Y, Liu F Q, Xu L F and Zhao H. 2019. Effect of elevated temperatures and cooling methods on strength of concrete made with coarse and fine recycled concrete aggregates. *Constr. Build. Mater.*, **210**, 540-7.

[24] Nadeem A, Memon S A and Lo T Y. 2014. The performance of fly ash and metakaolin concrete at elevated temperatures. *Constr. Build. Mater.*, **62**, 67-76.

IOP Conf. Series: Materials Science and Engineering **712** (2020) 012020 doi:10.1088/1757-899X/712/1/012020

Properties of Concrete Containing Crushed Palm Oil Clinker as Partial Fine Aggregate Replacement

M A Sulaiman[1], M I Ali[1], M Y Al-Amri[1], K Muthusamy[1,2], A M Albshir Budiea[3], N Nordin[4], Y Duraisamy[1] and R Othman[1]

[1]Faculty of Civil Engineering Technology, Universiti Malaysia Pahang, 26300 Gambang, Pahang, Malaysia
[2]Earth Resources and Sustainability Centre, Universiti Malaysia Pahang, 26300 Gambang, Pahang, Malaysia
[3]Faculty of Civil and Environmental Engineering, Universiti Tun Hussein Onn Malaysia, 86400 Batu Pahat, Johor Bahru, Malaysia
[4]IGeo Disaster Research Centre, Infrastructure University Kuala Lumpur, Unipark Suria, Jalan Ikram-Uniten, 43000 Kajang, Selangor Darul Ehsan.

Abstract. The growing demand for sand supply for construction industry use especially in concrete production drives the sand mining industry to flourish. However, excessive sand mining would affect the river environment, aquatic life and water quality. At the same time, the increasing quantity of palm oil clinker, which is a by-product at palm oil mill disposed at landfill, poses negative impact to the environment. The use of palm oil clinker waste as sand replacement in concrete production would be able to reduce the use of natural sand from the river and dumping of clinker waste at landfill. The current research investigates the fresh and mechanical properties of concrete containing palm oil clinker as partial fine aggregate replacement. A total five concrete mixes were produced. The control specimen was prepared by using 100% natural sand. Other mixes were formed by integrating crushed palm oil clinker as partial fine aggregate replacement ranging from 10%, 20%, 30% and 40%. All specimens were water cured up to 28 days. The specimens were subjected to slump test, compressive strength test, flexural strength test and water absorption test. The finding shows that suitable amount of palm oil clinker can be used as partial fine aggregate replacement in concrete production. Concrete containing palm oil clinker up to 40% replacement exhibits water absorption value lower than 10%, thus classified as good quality concrete.

1. Introduction

In Malaysia, the continuous growing construction trade and palm oil industry contributes towards the economic prosperity of the country. At the same time, these industries also cause negative impact to the environment. Development in construction industry has increased the demand for building material production especially concrete. The expanding concrete industry also pushes for larger supply of raw materials, which harvested from the environment. Sand is one of the raw materials obtained through mining activity at the river. The use of sand as fine aggregate in concrete plays the role of producing a more compact structure for strength and durability enhancement. Thus, the development in concrete production directly boosts the growth of sand mining industry. However, uncontrolled sand mining would affect the water quality and habitat of aquatic life [1, 2]. Continuous on going mining activity at one location would destroy the green river environment and affect other elements related to it namely fauna, flora and inhabitants. There are rivers in certain parts of the world are suffering the negative impact of extreme sand mining activity [3]. In addition, acute aggregate shortage problem happened in certain parts of the world [4]. Thus, discovering other alternative material especially from locally generated waste would reduce the harvesting local natural sand and also reduce environmental pollution due to waste disposal. In relation to that, Malaysia being one of the largest palm oil producers in the world palm oil industry contributing to the prosperity of economic also generates large amount of by-products, which disposed as waste. This industry has been flourishing ever since it begins to be commercialized in the end of 20th century. The Malaysian palm oil exported increased from less than 100 000t in 1960 to 16.05 MT in 2016 [5]. At the same time, the quantity of by-product generated from palm oil mills also increased along with the production growth. Approximately, 2.6 MT of solid waste is generated per year [6]. Palm oil clinker is a by-product, which formed in incineration chamber during the combustion of shell and fibre to generate power supply for the palm oil mill. It is thrown as waste material in the factory-dumping site. This rock like is available in abundance and have a small commercial value locally [6]. As the production of palm oil industry continues to increase, it is expected larger quantity of clinker would be produced. Thus, continuous dumping would lead to piling up of the waste at the allocated area creating unhealthy environment to the inhabitants [7]. Utilizing this readily available waste material as fine aggregate replacement in concrete production would contribute to lesser dependency of industry on natural sand supply as well as reduces usage of land to dump palm oil clinker waste. Thus, the present research discusses the effect of crushed palm oil clinker as partial fine aggregate replacement on the workability and mechanical properties of concrete.

2 Method of Research

2.1 Materials

Among the materials used to prepare concrete specimens in this research are cement, water, coarse aggregate, fine aggregate and palm oil clinker. Ordinary Portland cement (OPC) from a single source was used throughout the experimental work. Both coarse and fine aggregate were supplied from the local sources. Tap water was used for mixing and curing purposes. Palm oil clinker (POC) was obtained from a palm oil mill located in East Coast of Peninsula Malaysia. POC was collected from an open dumping area at mill as illustrated in Figure 1. POC were cleaned to remove the debris sticking on it and oven dried for 24 hours. After that, the clinkers were crushed and sieved passing sieve 2 mm. Only clinker passing 2mm were packed and stored for concrete mixing work. Figure 2 illustrates the palm oil clinker before and after crushing process.

Figure 1: Palm oil clinker waste collection

IOP Conf. Series: Materials Science and Engineering **712** (2020) 012020 doi:10.1088/1757-899X/712/1/012020

Figure 2: Palm oil clinker before and after crushing process

2.2 Sample Preparations and Testing

In this experimental programme, five concrete mixes consist various quantity of palm oil clinker as partial sand replacement were prepared as in Table 1. Control specimens (POC-0) were prepared by using 100% river sand. Another, four concrete mixes consists of 0%, 10%, 20%, 30% and 40% palm oil clinker by weight of sand were prepared. After casting process, all the samples were covered with wet gunnysack for 24 hours before demoulded. All specimens were water cured up to 28 days. All specimens were subjected to compressive strength test, flexural strength test and water absorption test. The effect of palm oil clinker content on concrete workability were determined through slump test that was carried out following the procedures in BSEN, 12350-2 [8]. The hardened concrete specimens were subjected to compressive strength test and flexural strength test in accordance to BSEN, 12390-3 [9] and BSEN 12390-5 [10] respectively. The water absorption test was carried in accordance to BS 1881 : Part 122 [11].

Table 1: Mix proportion of control specimen

Mixes	**Cement (kg/m^3)**	**Sand (kg/m^3)**	**Granite aggregate (kg/m^3)**	**Crushed Clinker (kg/m^3)**	**Water (kg/m^3)**
POC-0	350	980	850	-	245
POC-10	350	882	850	98	245
POC-20	350	784	850	196	245
POC-30	350	686	850	294	245
POC-40	350	588	850	392	245

3 Result and Discussion

3.1 Workability

Figure 3 and 4 shows that the slump decreases as the percentage of palm oil clinker content replaced becomes higher. The result indicates that concrete become less workable as the amount of palm oil clinker added increases. This implies the need for larger amount of water required to make the mixes more workable. The reduction of workability is due to the porous clinker particle unlike the solid sand particles. Past researcher, Neville [12] has pointed out that physical properties of the aggregate used have influence on the workability of concrete. Similar trend has been reported by Nazrin et al. [13] and Abutaha et al. [14] when clinker is used as partial fine aggregate replacement in lightweight concrete mix.

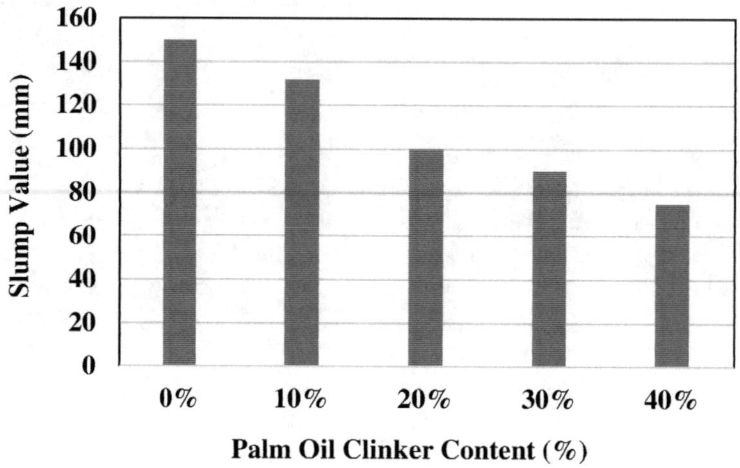

Figure 3: Slump test results

Figure 4: Effect of clinker on slump of concrete mix

3.2 Compressive Strength and Flexural Strength

Figure 5 and 6 illustrate the compressive strength and flexural strength results of the concrete specimens containing various percentage of crushed palm oil clinker. Generally, the strength of all specimens increases as the curing age become longer due to continuous presence of water promoting undisturbed hydration process. As a result, larger quantity of CSH gel was formed in concrete internal structure thus creating a more compact concrete with higher load sustaining capacity. One of the researchers, Neville [12] has highlighted the importance of continuous water supply during curing in enhancing the concrete strength. As for the effect of clinker on concrete strength, it is evident that the use of porous type of clinker as fine aggregate somehow affects the concrete strength. Inclusion of clinker, which absorbs water owing to its porous physical appearance, reduces the concrete mix workability. This cause difficulty during compaction works and also results in loss of adhesion due insufficient water for hydration process resulting in concrete with lower strength. As larger amount of clinker is added, the mix stiffer and hard to be compacted thus creates more voids that reduce concrete strength.

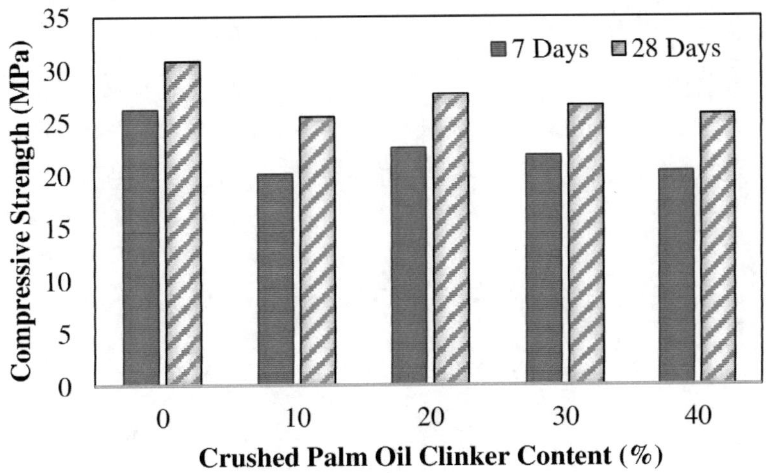

Figure 5: Compressive strength results

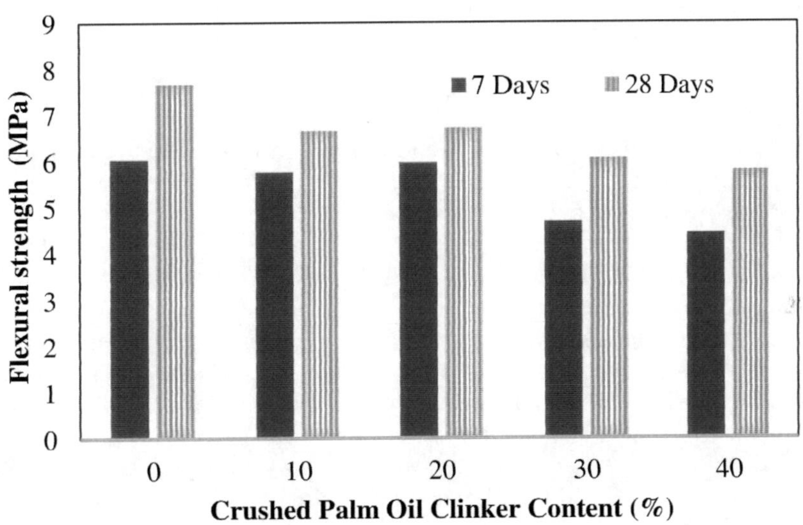

Figure 6: Flexural strength results

3.3 Water Absorption

Figure 7 illustrates the water absorption of concrete containing various POC content. Evidently, the water absorption of concrete increases as more palm oil clinker is replaced. Water absorption was increasing because POC is a porous material, which has the ability to absorb water better than river sand. This finding is in line with, indicates that the absorption capacity of concrete is high due to the porous nature of aggregate used. However, all concrete specimens containing crushed clinker up to 40% exhibit water absorption less than 10% enabling it to be classified as high quality concrete. According to Neville [12], concrete with water absorption below 10% is classified as high quality concrete.

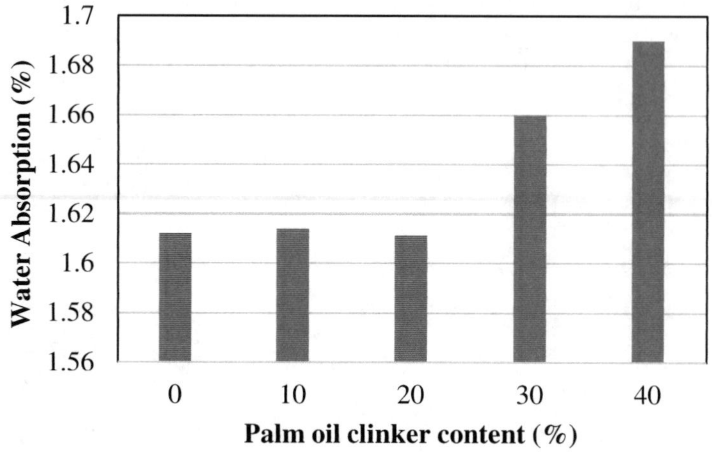

Figure 7: Water absorption test results

4 Conclusions

Based on the results obtained from the experimental work, it is concluded that the use of crushed palm oil clinker content as partial sand replacement effect the concrete properties. Increase in the quantity of palm oil clinker content, reduce concrete workability. The presence of palm oil clinker as partial replacement of sand in concrete reduces the compressive strength and flexural strength of concrete.

5 References

[1] Ashraf MA, Maah MJ, Yusoff I, Wajid A and Mahmood K 2011Sand mining effects, causes and concerns: A case study from Bestari Jaya, Selangor, Peninsular Malaysia *Scientific Research and Essays* **6(6)**1216-31
[2] Ramachandra TV, Vinay S and Chandran MS 2018 Quantification of annual sediment deposits for sustainable sand management in Aghanashini river estuary *Journal of environmental management* **206** 1263-73
[3] Ashish DK 2019 Concrete made with waste marble powder and supplementary cementitious material for sustainable development *Journal of cleaner production* **211** 716-29
[4] Patra RK and Mukharjee BB 2017 Influence of incorporation of granulated blast furnace slag as replacement of fine aggregate on properties of concrete *Journal of cleaner production* **165** 468-76
[5] Nambiappan B, Ismail A, Hashim N, Ismail N, Nazrima S, Idris NA, Omar N, Saleh KM, Hassan NA and Kushairi A 2018 Malaysia 100 years of resilient palm oil economic performance *Journal of Oil Palm Research* **30(1)**13-25
[6] Abutaha F, Razak HA and Kanadasan J 2016 Effect of palm oil clinker (POC) aggregates on fresh and hardened properties of concrete *Construction and Building Materials* **112** 416-423
[7] Muthusamy K, Mirza J, Zamri NA, Hussin MW, Majeed AP, Kusbiantoro A and Budiea AM 2019 Properties of high strength palm oil clinker lightweight concrete containing palm oil fuel ash in tropical climate *Construction and Building Materials* **199** 163-77
[8] B.S. EN, 12350-2: 2009 Testing fresh concrete: Slump-Test
[9] B.S. EN 12390-3: 2019Testing Hardened 646 Concrete. Compressive Strength Test
[10] B.S. EN 12390-5:2019 Testing hardened concrete–Part 5: Flexural strength Test
[11] B.S 1881-122-2011Testing concrete, Method for Determination of Water Absorption
[12] Brandt AM. 2005 Cement-based composites: materials, mechanical properties and performance *CRC Press*

[13] Ahmad Zawawi M. N, Muthusamy K. and Mohd Ali N 2016 Compressive strength of oil palm shell lightweight aggregate concrete containing clinker as sand replacement, *Proc. of The National Conference for Postgraduate Research*, Universiti Malaysia Pahang
[14] Abutaha F, Abdul Razak H and Ibrahim H 2017 Effect of coating palm oil clinker aggregate on the engineering properties of normal grade concrete *Coatings*. **7(10)** 175

Acknowledgement
The authors would like to acknowledge the help and co-operation received from the lecturers and technical staff of Universiti Malaysia Pahang in conducting the experimental work. The financial support received from Universiti Malaysia Pahang through grant RDU1703109 is highly appreciated.

IOP Conf. Series: Materials Science and Engineering **712** (2020) 012021 doi:10.1088/1757-899X/712/1/012021

Rainfall Trend Analysis using Box Plot Method: Case Study UMP Campus Gambang and Pekan

A A G Nadiatul Adilah[1,2], M Mohamad Zarif[2] and A Mohamad Idris[2]

[1]Researcher, Earth Resources & Sustainability Centre (ERAS), Universiti Malaysia Pahang, 26300 Kuantan, Pahang, Malaysia
[2]Faculty of Civil Engineering Technology, Universiti Malaysia Pahang, 26300 Kuantan, Pahang, Malaysia

Abstract. Climate change leads to changes in precipitation. This phenomenon has already begun to transform rainfall pattern in Malaysia. Rainfall data, temperature data, and evaporation data are collected and recorded monthly to display the relationship between rainfall, temperature, and evaporation to determine the pattern of hydrologic cycle. The relationship obtained, shows weather pattern at campus Universiti Malaysia Pahang (UMP) in Gambang and Pekan. By monitoring of the weather conditions, it can help in controlling the activity in the UMP. Besides, weather data is very important in our life as the rainfall is an important consideration in design runoff conveyance and erosion control. Changing trend in rainfall distribution also gives an effect on hydrological analysis especially related to historical rainfall record. Box-plot method was used to determine the rainfall trend of hydrologic cycle. It is shown that the rainfall data that was collected that the rainfall event differs every year, as an example for 2016, a total of 1982.9 mm was collected and year after it decreased sharply to 994.7 mm in 2017 and for 2018 raised slightly to 1020.9 for UMP Pekan. Whereas for UMP Gambang, 1072.8 mm in 2016, and also decreased slightly to 972.5 mm in 2017 and in 2018 increased to 1309.8 mm. It shows that each year trend of rainfall for each UMP campuses had change.

1. Introduction

The normal features of the climate of Malaysia are uniform temperature, extensive rainfall and high humidity. Rainfall being predominant form of precipitation causing stream flow, especially the flood flow in a river. The magnitude of precipitation varies with time and space [2]. Malaysia is observed to have a tropical climate, means the average temperature of the country, usually ranges from 21°C to 32°C and the humidity is range in between 70% to 90% [3]. The climate is affected by the northeast and southwest monsoons, tropical wind that alternative during the course of the year. The direction of the wind in this inter-monsoon season is variable and usually more than 10 knots [4]. Due to the seasonal rainfall in Malaysia, the probability of occurrence of rainfall amount is varied during the whole year [5]. The northeast monsoon blows from November to March and the southwest monsoon from May to September. The increase in temperature and rainfall will affect the water resources [6].
Daily human activity is influenced by weather conditions, monitoring of weather conditions can help in controlling the activity. The behavior of and pattern changes of the weather is not all the same in the UMP Gambang and UMP Pekan area and its nearest place. Differences in the magnitude of rainfall in various parts at a given time and variation of rainfall at a place in various seasons of the year are obvious and need no elebration [7]. It is important to monitor and study the pattern of weather at surrounding. Other than that, the weather consists of wind, humidity, temperature and precipitation. The pattern and trend of weather at Gambang and Pekan area can be identify by making analysis study

of hydrological data from hydrological station. The primary goal of this project is to conduct experimental studies to identify the trend of rainfall event at UMP Gambang and Pekan campus.

Studying the weather is a remote part of science. It will be able to study faraway places weather without ever having to go there. A weather station is a weather observation facility with tools and equipment to monitor atmospheric conditions to study the weather. It helps people measure wind, air pressure, rainfall, moisture and temperature. All data will be collected from the instruments used and compared with the weather data of the station in Kuantan. To search for patterns and trends, all weather data collected from the previous one week period will use and analyze it using graph, chart and averages. Data interpretation will be based on computational skills. The weather can be described from the graph or chart and how weather affects everyday life.

A rain gauge is an instrument that can be used in a given time period to measure the amount of rainfall received in the area. Rain gauge is a simple tool that can be installed and used. The reading amount of rainfall is in inches or millimeters. Rain gauges are the most worldwide used devices for in-situ point measurements of precipitation intensity and duration, especially for the Tipping-Bucket rain gauge since it can not only accurately measure rainfall intensity from low-to-intermediate level, but also recording remotely with reliability and suitability [8]. The rain gauge can be used to measured the amount of rainfall since it has a high level of accurancy in measuring the amount of rainfall [9].

2. Study Area

The rain gauge was set up at the field at KolejKediaman 2 (KK2) in Universiti Malaysia Pahang (UMP) Campus Gambang (Figure 1). The field at KolejKediaman 2 (KK2), Universiti Malaysia Pahang (UMP) is chosen due to the stability and compromise to fulfil the ideal criteria to set up a rain gauge. The ideal location of the site should be free from obstacle, building and steep slopes which will be effect the data collection of the weather station. Figure 2 shows rain gauge in KK2, Campus Gambang.

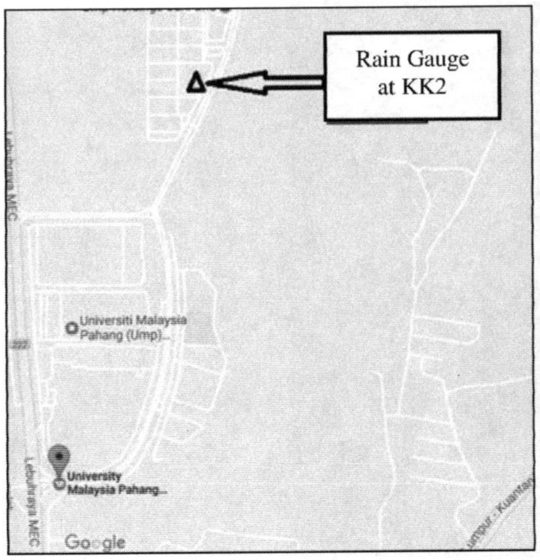

Figure 1. Location of rain gauge at Kolej Kediaman 2(KK2), UMP Campus Gambang

Figure 2. Rain gauge at Kolej Kediaman 2, UMP Gambang

There is also weather station that is set up at UMP Pekan located near to the kindergarten that is inside the UMP Pekan (Figure 3). This site is chosen due the criteria for the research which is to compare the rainfall event between UMP Pekan and UMP Gambang. Stationed here also is the rain gauge which located one meter apart from each other. Figure 4 shows location of weather station in Campus Pekan.

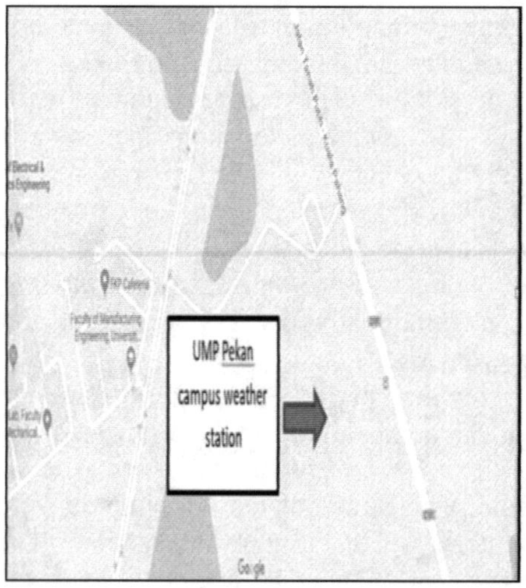

Figure 3. Location of weather station at UMP Pekan campus

Figure 4. Rain gauge at Kolej Kediaman 2, UMP Gambang

3. Methodology

The rainfall data from HOBO rain gauge is collected once a month from both rain gauge stations that situated in UMP Gambang and UMP Pekan campus, the data is collected by plugging in USB cable from the computer to the rain gauge in order to get the raw rainfall data. Figure 5 show the process of data collection from rain gauge at KK2, Campus Gambang.

Figure 5. Getting the data from the HOBO rain gauge in KK2 in UMP Gambang

At the same time with the planning and process to setting up and upgrading the weather station in Pekan, the weather station in Pekan is collected once a month. Figure 6 shows the data collected from weather station after transfer to PC and the data recorded from weather station. The Figure 7 shows the data recorded in the console from weather station.

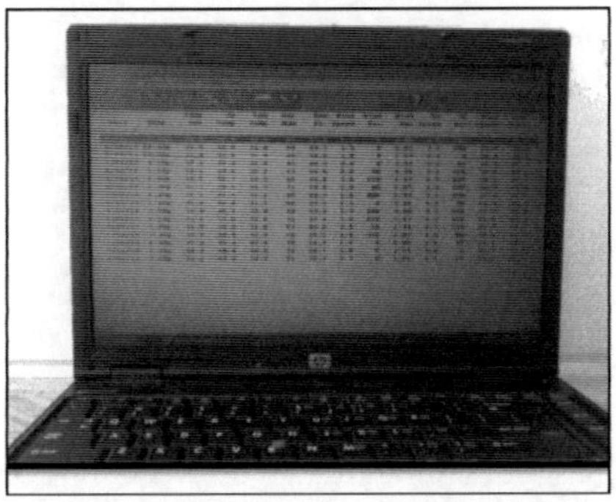

Figure 6. The data collected from weather station after transfer to PC

Figure 7. The data recorder from weather station

3.1 Box plot method

There are a lot of methods that can be used in order to measure the rainfall event, the box-plot method can also be used to determine the event of rainfall. According to [10], box plot is the form of summary of a given dataset which includes, the median, the inter quartile range and the computation and the meaning each of the values are described below.

3.1.1 The median. Median is the middle data observation in a ranked of any dataset and as a measure for central tendency of the data and the same as the 50th percentile of a data.

3.1.2. The inter quartile range. The box represents the middle 50% of the ranked data and is drawn from the lower quartile value to the upper quartile value which is the 25th until 75th percentile.

3.1.3. The outer range. The whiskers represent an outer range and are drawn as a vertical lines extending outward from the ends of the box. This whisker represents the maximum and minimum of said data for the sample. Figure 8 shows what the box-plot will represent the data when it is plotted.

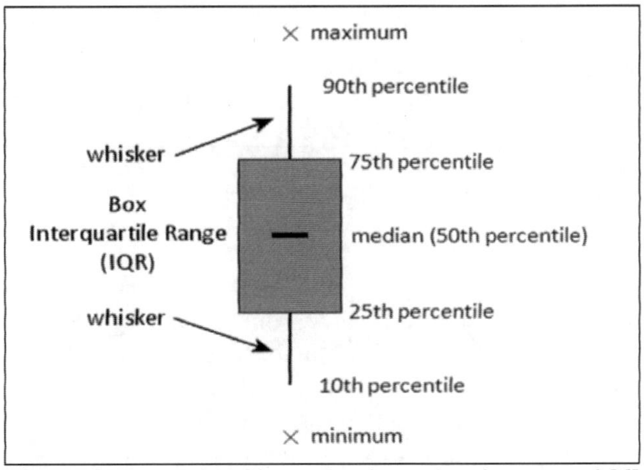

Figure 8. Example of box-plot data when interpreted [6]

After collecting the data from both UMP Gambang and UMP Pekan campus, the suitable method proposed to analyse the data is box-plot method. The purpose of using this method is the ability to be able to check the trend and skewness of the rainfall trend for UMP Gambang and UMP Pekan. The advantage of the box-plot method is that it is able compare multiple datasets side-by-side, as idealized in Figure 9. Important characteristics of each dataset (central tendency, skewness, dispersion, and extremes) are very easy to interpret and visualize [8].

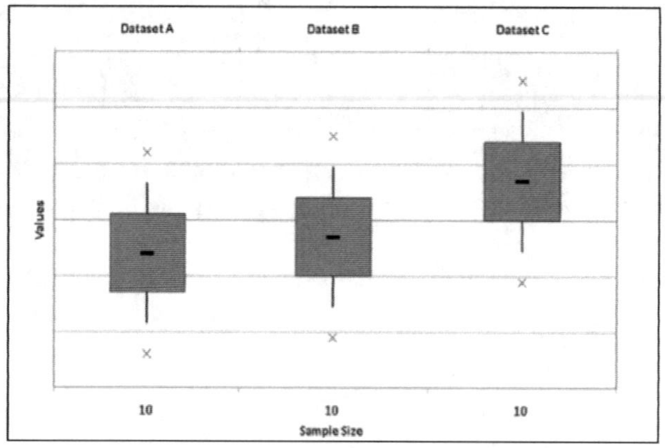

Figure 9. Example of datasets compared side-by-side [6]

From the box plot too, the trend or skewness can be determined by using the rainfall data, below in Figure 3.10 shows the idealized distribution example of rainfall data trend or skewness that can be determined by the shape of the box-plot that has been plotted by using dataset of rainfall data.

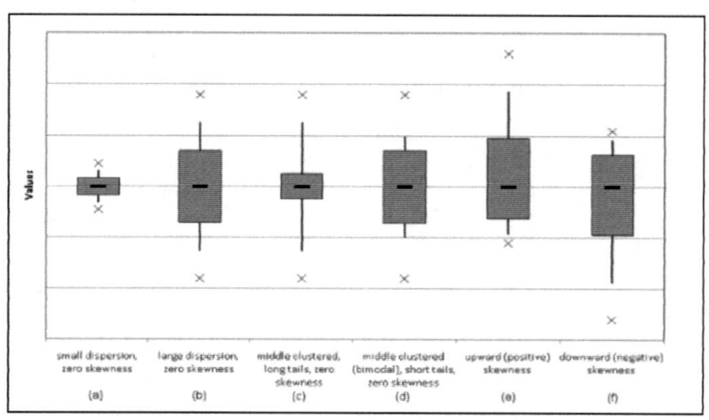

Figure 10: Idealized box-plot data distribution [6]

4. Results and Discussions

Based on the Table 1 and Table 2, it shows the raw rainfall data collected from Rain Gauge at KK2, UMP and also from UMP Pekan. The analysis data start from 2016 until March of 2019. The data which is from 2016 until early 2018 was carried out by past researches and is taken for further observation for this research. As for UMP Gambang, from 2016 to 2019, it shows that highest total rainfall collected is in 2018 with an average rainfall per month of 109.15 mm. And because 2019 is only taken until March, so there is not much data to be use and to compare with the other year in comparison. As for UMP Pekan, it is figured out that during year 2016 it collected the most rainfall data which is 1982.9 mm by the rain gauge. And as for the average rainfall per month for 2016 is 165.24 mm. It is shown that for UMP Gambang that rainfall data collected to be up and down as in 2016, it is 1072.8 mm, and the year after 2017, it decreased to 972.5 mm. It began to increase more the year after which is 1309.8 mm and as for 2019 which consists of only 3 months data, with 196.2 mm so far. And as for UMP Pekan, 2016 is the highest with 1982.8 mm and decreases sharply the year after in 2017 with 994.7 mm and increases slightly with a collected of 1020.9 mm and for 2019 with 82.9 mm so far. This can be shown that data collected between the two places varies differently.

Table 1. Data from rain gauge in UMP Gambang

Gambang		
Year	Total Rainfall (mm)	Average Rainfall per month (mm)
2016	1072.8	89.4
2017	972.5	81.04
2018	1309.8	109.15
2019	196.2	65.4

Table 2. Data from rain gauge in UMP Pekan

Pekan		
Year	Total Rainfall (mm)	Average Rainfall per month (mm)
2016	1982.8	165.24
2017	994.7	82.89
2018	1020.9	85.08
2019	82.9	27.63

Chart 1 below shows that the rainfall data are from year 2016 to 2017 for UMP Gambang. The highest total rainfall occurs at the month of October 2016 which is 296.0 mm and can be seen on the end of the whisker of the box-plot for year 2016. As for year 2017, the highest rainfall data collected is also in October 2017 with 248.6. As for Chart 2, it continues for UMP Gambang but for year 2018 and 2019. For 2018, the highest is in December 2018 with 266.6 mm. As for year 2019, the highest is in January 2019 with 108.8 mm, but it is only for 3 months worth of data are only taken in 2019 so it will not amount to much.

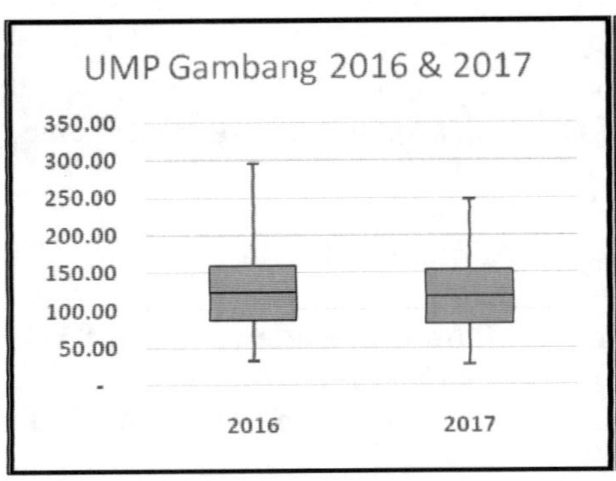

Chart 1. UMP Gambang 2016 & 2017

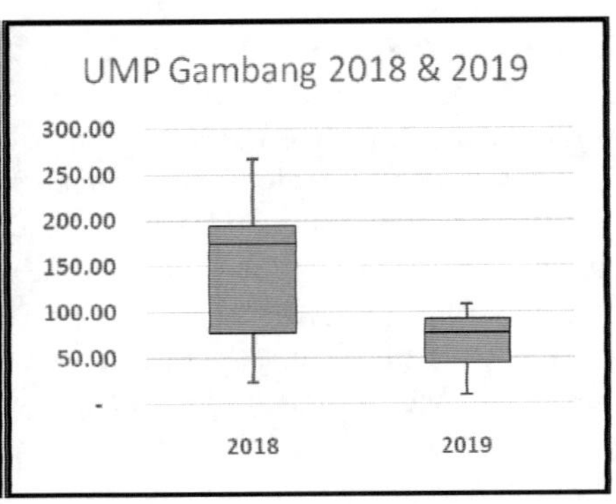

Chart 2. UMP Gambang 2018 & 2019

Chart 3 shows that the rainfall data are from year 2016 to 2017 for UMP Pekan. The highest total rainfall occurs at the month of November 2016 which is 991.4.0 mm and can be seen on the end of the whisker of the box-plot for year 2016. As for year 2017, the highest rainfall data collected is also in November 2017 with 389.2 mm. As for Chart 4, it continues for UMP Pekan but for year 2018 and

2019. For 2018, the highest is in December 2018 with 346.7 mm. As for year 2019, the highest is in January 2019 with 69.3 mm, but it is only for 3 months worth of data are only taken in 2019 so it will not amount to much.

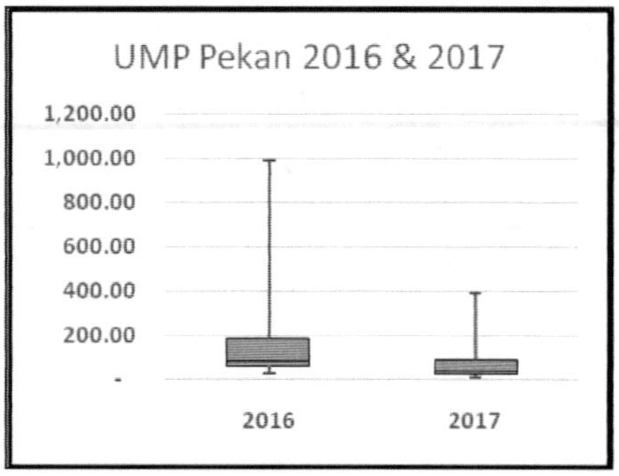

Chart 3. UMP Pekan 2016 & 2017

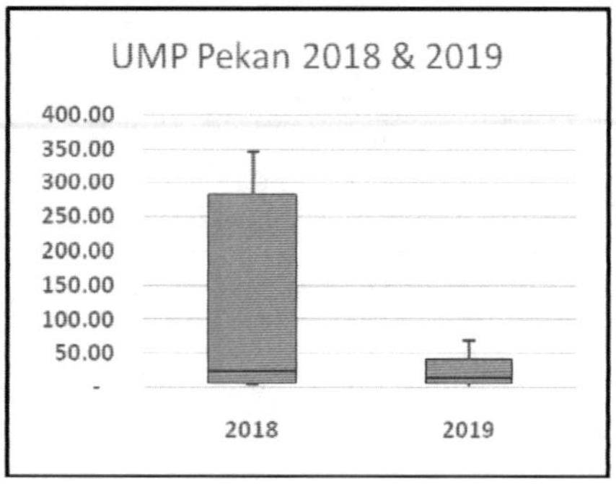

Chart 4. UMP Pekan 2018 & 2019

5. Conclusions

Weather at UMP Gambang campus shows that rainfall trend is varied throughout the year and would increase and decrease the following year. It can be concluded that near the end of each year which is around October to December, both UMP Gambang and UMP Pekan collected the highest rainfall data reading.

6. References

[1] Pettazi A, and Salson S 2012 Combining radar and rain gauges rainfall estimates using conditional merging: A case study. In Erad 2012-The Seventh European Conference On Radar In Meteorology and Hydrology

[2] Subramanya, K 2013 Engineering Hydrolog. 4th edition. New Delhi: McGraw Hill Education (India) Private Limited

[3] Tangang FT, Juneng L, Salimun E, Kwan MS, Loh JL and Muhamad H 2012 Climate change and variability over Malaysia: Gaps in science and research information.Sains Malaysiana pp.1355-66

[4] Ho MK and Yusof T 2012 Homogeneity tests on daily rainfall series in Peninsular Malaysia *Int. J. contemp. Math. Sciences* pp.9-12

[5] Suhaila J and Jemain A A 2009 A comparison of the peninsular of the rainfall patterns between stations on the East and the West coasts of peninsular Malaysia using the smoothing model of rainfall amounts Meteological Applications pp.391-401

[6] Wang K 2013 Engineering Hydrology 4th Edition Mc Graw Hill

[7] Subramanya K 2013 Engineering Hydrology 4th edition New Dehli: Mcgraw Hill Education (India) Private Limited

[8] Song, Y 2016 High Temporal Resolution Rainfall Information Retrieval from Tipping-Bucket Rain Gauge Measurements *Procedia Engineering* 1193-1200.

[9] Pettazi,A. & Salso, S 2012 Combining Radar and Rain Gauges Rainfall Estimates using Conditional Merging: A case study in Erad 2012 The Seventh European Conference on Radar in Meteorology and Hydrology

[10] Banacos P C 2011 Eastern Region Technical Attachment Box and Whisker Plots for Local Climate Datasets : Interpretation and Creation using Excel 2007 / 2010 *Interpretation A Journal of Bible and Theology* Vol.**1** pp. 2–20 doi: 10.1007/s00338-013-1085-2

Acknowledgment

The authors gratefully acknowledge the University Malaysia Pahang for granting this research project which the vote number is RDU170344.

IOP Conf. Series: Materials Science and Engineering **712** (2020) 012022 doi:10.1088/1757-899X/712/1/012022

Water Quality Status and Heavy Metal Contains in Selected Rivers at Tasik Chini due to Increasing Land Use Activities

A A G Nadiatul Adilah and H Nurul Nadia

Faculty of Civil Engineering Technology, Universiti Malaysia Pahang, 26300 Kuantan, Pahang, Malaysia

Abstract. This study investigates the current water quality levels in the rivers around Lake Chini namely Jemberau River and Chini River during the dry season of 2018 and the rainy season of 2019. There are thirteen types of water quality parameters and seven types of heavy metals have been made experimental and classified based on the Water Quality Index which is from the Department of Environment (DOE-WQI) and the National Water Quality Interim Standard, Malaysia (INWQS). There are chemical and physical parameters that have been analyzed such as pH, temperature, turbidity, Electrical Conductivity (EC), Suspended Solids (SS), Total Suspended Solids (TSS), Dissolved Oxygen (DO), Biochemical Oxygen Demand (BOD), Chemical Oxygen Demand (COD), Ammonia Nitrogen (NH3-N), Potassium (K), Nitrate (N) and Phosphorus (P) were analyzed at the sample site and laboratory experiments were also performed. In addition, there are seven types of heavy metals that have been tested in the laboratory to determine the concentration of heavy metals contained in the water samples from the Chini River and the Jemberau River which are Copper (Cu), Chromium (Cr) Cadmium (Cd), Ferrum (Fe), Zinc (Zn), Manganese (Mn) and Lead (Pb) were tested using laboratory tools named as Atomic Absorption Spectroscopy (AAS). Based on the results of the study conducted according to the Water Quality Index (WQI), the water quality at selected locations which are Jemberau River and Chini River located near Lake Chini was classified as Class III which means water in the river requires extensive treatment to ensure water quality return clean and safe to use as drinking water to the locals in the future. According to the results of the study, there are activities conducted around Lake Chini which have caused water pollution which involves the release of heavy metals. Among the activities identified as the main cause of water pollution in Lake Chini are iron mining, agriculture, illegal logging and waste disposal from residential areas such as the National Service Training Program (PLKN).

1. Introduction

Water is very important to human daily life in order to do their daily routine especially for drinking. Lakes are one of major water source in Malaysia as freshwater because they are often fed by rivers, rain and springs. The lake is very sensitive area because there are potential exposure pollutants from many sources. For example, pollution can flow through the water body of the lake from runoff and from the deposition of atmospheric. The limited water movement within the lake affects the level pollution in the lake environment. Moreover, high concentrations pollutants can reduce the biodiversity of the lake ecosystem and change the physical environment around the lake. In Malaysia, Tasik Chini which is located within the state of Pahang on the east coast of Peninsular Malaysia is the second largest natural lake in Malaysia after Tasik Bera. Surface area of the Tasik Chini covers 12,565 acres and it was surrounded by natural forest and aboriginal settlements. Tasik Chini has a unique shape that consists of 12 small lakes that local people called as “Laut” which is interconnected by natural channels. Tasik Chini has recreational value and ecological importance in terms of its biodiversity as this area richly endowed with biological resources and some 288 species of plants, 21 species of aquatic plants, 92 species of birds, and 144 species of freshwater fish.

Tasik Chini is an ecological area which is importance due to large biodiversity and it is also very important for local parties' economy. The mainstay economy is mainly forest-based and agriculture-based activities [1].These activities could be the source for pollution to the lake. There are logging, iron ore mining, rubber plantation, palm oil plantation and residential area. There is presence of heavy metals that come from iron ore mining activities which gives bad effect to water quality of Tasik Chini. Accumulation of metals and organic pollutants in the sediments may have long-term adverse effects on aquatic organisms [2]. Heavy metals concentration such as Lead (Pb), Iron (Fe), Manganese (Mn), Cadmium (Cd), Zinc (Zn) and Chromium (Cr) become increasing due mining activities. Heavy metals are a metallic element that has a relatively high density, specific gravity, or atomic weight and has toxic effects [3]. Meaning that, if there are unwell-operated for mining activity, it will causes increasing of heavy metal concentrations in water body because lake is a stagnant water area surrounding by land which is term water pollution refers to any types of aquatic pollution between two extremes of a highly productive body of water poisoned by toxic chemicals that eliminates living organisms. Water quality is among the most important environmental issues related to sustainable development, especially to ensure national drinking water safety [4].

The water quality pollutants, such as organic pollutants (permanganate index), nutrients (total phosphorus), total suspended solid, and chlorophyll-a were significantly influenced by the proportion of forest and grass land and the consumption of phosphate fertilizers in tributaries [5]. Water quality can be defined as the chemical, physical, biological, and radiological characteristics of water. This is a measure of water conditions over the requirements of one or more biotic species and or to any human purpose. This is most often used with reference to a set of standards for which compliance, generally achieved through water treatment that can be assessed. The most commonly used standards for assessing water quality are ecosystem health, human safety, and drinking water. The water quality parameters for the lake, measured according to Malaysia Department of Environment Water Quality Index (DOE-WQI) also, will classified by using National Water Quality Standard, Malaysia (NWQS).

1.1 Research Objectives

i) To evaluate the characteristics of each water quality parameters and to analyze the current status of water quality at Jemberau River and Chini River during rainy season in 2018 and dry season in 2019.
ii) To identify current heavy metals level in Jemberau River and Chini River.

1.2 Scope of Study

This research is conducted on October 2018 until February 2019. The location of study area is at Tasik Chini, Pahang. The scope of study in this research is about effect from land use activities to the water quality at Jemberau River and Chini River that are currently facing problem due to mining activity, logging activity and agricultural activity. All of this activity can contribute to the increasing of heavy metal concentration in the river and will give worse effect to the water quality at Tasik Chini. The increase in water demands causes more conflict between the human system and the river ecological system [6]. Figure 1 show the view of Jemberau River. The coordinate for this station is $03^{o}41.952''$ North and $102^{o}93.121''$ East as shown in Figure 2. Figure 3 show the view of Chini River. The coordinate for this station is $03^{o}45.091''$North and $102^{o}89.194''$ East in Kuantan, Pahang as shown in Figure 4.

IOP Conf. Series: Materials Science and Engineering **712** (2020) 012022 doi:10.1088/1757-899X/712/1/012022

Figure 1. Jemberau River, Tasik Chini

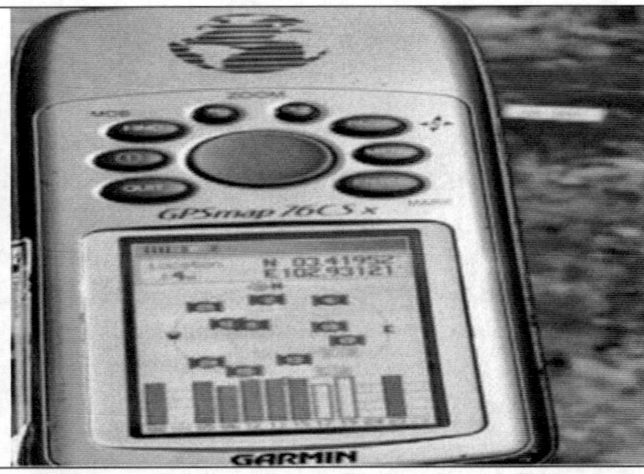

Figure 2. Coordinate for Jemberau River

Figure 3. Chini River, Tasik Chini

Figure 4. Coordinate for Chini River

2. Study Area

The lake is serves to refill underground water, affect the water quality of the groundwater catchment, and maintain the biodiversity and habitat of the area. Dam, reservoirs and lakes are important in the conservation of basic national resource-water [7]. The two largest freshwater lakes in Malaysia which are Tasik Chini and Tasik Bera are located on the same river basin known as Pahang River Basin.

Tasik Chini is the study area for this research. Tasik Chini is the second largest natural lake in Peninsular Malaysia which is located at Pahang. This lake plays vital role as a natural wetland ecosystem due to its presences that can decrease the velocity, frequency and level of floods and riverbank erosion. This lake is important for local people for providing the fish source and basic facility for transportation such as boating. Since Tasik Chini flows to the main river which is Pahang River, it is influence by the intertidal monsoon season.

In recent years, specific concerns about the impacts of climate change on water eutrophication, which causes global environmental challenges regarding the management of water resources, have been raised [8].Due to uncontrolled flowing-in of pollutants such as organics, nitrogen, phosphorus, receiving water bodies often experience algal blooming, floating, and deterioration of biodiversity, eventually negatively-affecting human life [9]. The worse effect from these changes were damages to the plant community, aquatic ecosystem and raise the sedimentations level in the lake.

3. Methodology

Water quality parameters for lake, are measured in accordance to Malaysia Department of Environment Water Quality Index (DOE-WQI) also, will categorized by using skill of the use of National Water Quality Standard, Malaysia (INWQS) [10].

IOP Conf. Series: Materials Science and Engineering **712** (2020) 012022 doi:10.1088/1757-899X/712/1/012022

Water samples are taken twice during the rainy season and dry season. Water samples for the rainy season were taken on October 2018, while samples for dry season were taken on February 2019. At each station, three bottles of 500 ml for water sample was collected at 10 cm below the surface area. There are two types of test carried out which are in-situ tests and laboratory tests. For in-situ tests, five parameters are measure such as temperature, pH, turbidity, electrical conductivity (EC), and Dissolved Oxygen (DO). While for laboratory test, 10 tests were conducted on Biological Oxygen Demand (BOD) test, Chemical Oxygen Demand (COD) test, Total Suspended Solid (TSS) test, Suspended Solid (SS), Nitrate test, Ammoniacal Nitrogen test, Phosphorus test, Potassium test, Hardness test and Heavy Metal test. Water samples are collected at 10cm below the surface water using HDPE bottles. The sample should be kept in the ice box at temperature 4°C and transported to the laboratory for analysis. The preservation technique of water sample was different based on the tests that to be carried out. The preservation techniques are shown in the Table 1 below:

Table 1. Preservation Techniques

Parameter	Container	Preservation	Maximum Holding Time
	Inorganic Tests		
Ammoniacal Nitrogen	P,G	Cool, 4°C $H^2 SO^4$ to pH <2	28 days
Biochemical oxygen demand		Cool, 4°C	48 hours
Chemical oxygen demand	P,G	Cool, 4°C $H^2 SO^4$ to pH <2	28 days
Hardness	P,G	HNO_3 to pH <2	6 months
Nitrate	P,G	Cool, 4°C	48 hours
Phosphate	G	Cool, 4°C	48 hours
Residue, non- filterable (TSS)	P,G	Cool, 4°C	28 days

Adopted from Environmental Protection Agency Guidelines for handling and preserving samples, P=plastic, G= glass

4. Results and Discussions

Table 2 shows in-situ and laboratory test results for sample collected at Jemberau River during rainy season in 2018 while Table 3 shows the results for dry season in 2019.

Table 2. Data collection and laboratory test for Jemberau River during rainy season 2018

PARAMETER	UNIT	SAMPLE VALUE
Temperature	°C	25.10
DO	mg/L	5.36
pH	-	5.08
EC	µS/cm	32.40
Turbidity	NTU	7.06
BOD	mg/L	0.20
COD	mg/L	29.00
NH3-N	mg/L	0.05
TSS	mg/L	3.00

IOP Conf. Series: Materials Science and Engineering **712** (2020) 012022 doi:10.1088/1757-899X/712/1/012022

Table 3. Data collection and laboratory test for Jemberau River during dry season 2019

PARAMETER	UNIT	SAMPLE VALUE
Temperature	°C	27.30
DO	mg/L	4.38
pH	-	5.76
EC	µS/cm	34.10
Turbidity	NTU	4.88
BOD	mg/L	8.06
COD	mg/L	30.50
NH3-N	mg/L	0.11
TSS	mg/L	17.00

Table 4 shows in-situ and laboratory test results for sample collected at Chini River during rainy season in 2018 while Table 5 shows the results for dry season in 2019.

Table 4. Data collection and laboratory test for Chini River during rainy season 2018

PARAMETER	UNIT	SAMPLE VALUE
Temperature	°C	28.60
DO	mg/L	2.49
pH	-	5.64
EC	µS/cm	35.60
Turbidity	NTU	45.60
BOD	mg/L	1.25
COD	mg/L	18.50
NH3-N	mg/L	0.10
TSS	mg/L	36.00

IOP Conf. Series: Materials Science and Engineering **712** (2020) 012022 doi:10.1088/1757-899X/712/1/012022

Table 5. Data collection and laboratory test for Chini River during dry season 2019

PARAMETER	UNIT	SAMPLE VALUE
Temperature	°C	27.30
DO	mg/L	1.82
pH	-	6.32
EC	µS/cm	31.30
Turbidity	NTU	17.70
BOD	mg/L	7.30
COD	mg/L	13.00
NH3-N	mg/L	0.50
TSS	mg/L	66.00

There are six parameters that are selected for WQI which are Dissolved Oxygen (DO), Biochemical Oxygen Demand (BOD), Chemical Oxygen Demand (COD), and Suspended Solid (SS). Ammoniacal Nitrogen (AN) and pH. In addition, the calculations are not done on their own parameters but on their sub-index which are known as SIDO, SIBOD, SICOD, SIAN, SISS and SIpH.
The formula used to calculate the Water Quality Index is shown below:

$$WQI=(0.22*SIDO)+(0.19*SIBOD)+(0.16*SICOD)+(0.15*SIAN)+(0.16*SISS)+(0.12*SipH) \quad \text{(Eq. 1)}$$

where;

SIDO	=	Sublndex DO (%saturtlon)
SIBOD	=	Sublndex BOD
SICOD	=	Sublndex COD
SIAN	=	Sublndex NH3-N
SISS	=	Sublndex SS
SipH	=	Sublndex pH

$0 \leq WQI \leq 100$

Figure 5 shows the Water Quality Index (WQI) for Jemberau River and Chini River. Based on DOE-WQI, the selected station at Jemberau River during rainy season in 2018 is classified in Class II, while Jemberau River during dry season in 2019 and Chini River during both rainy and dry season are classified in Class III. From above data, it is means that the water quality of Jemberau River and Chini River is different from the location of the sampling station.

IOP Conf. Series: Materials Science and Engineering **712** (2020) 012022 doi:10.1088/1757-899X/712/1/012022

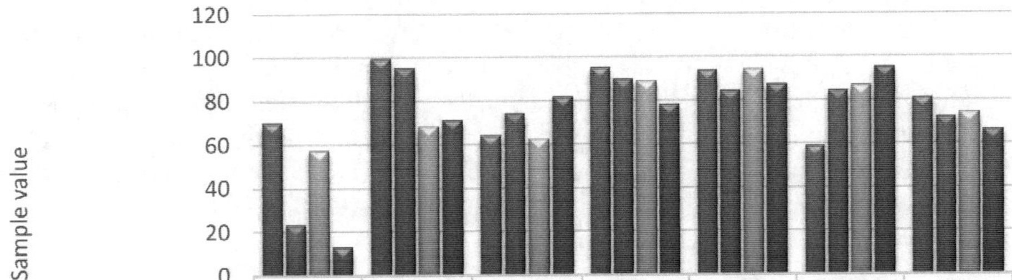

	SIDO	SIBOD	SICOD	SIAN	SISS	SIpH	WQI
Jemberau River (Rainy Season)	70.06	99.55	64.17	95.25	93.92	59.19	81.01
Chini River (Rainy Season)	23.25	95.11	74.5	90	84.61	84.45	72.28
Jemberau River (Dry Season)	57.38	68.52	62.59	88.95	94.22	86.79	74.49
Chini River (Dry Season)	13.04	71.56	81.81	78.08	87.5	95.14	66.68

Parameter

Figure 5. Water Quality Index at Jemberau River and Chini River during rainy season in 2018 and dry season in 2019

From Figure 6, it is shows that the water quality for all sampling stations at TasikChini during rainy season and dry season. According to DOE-WQI, station at Jemberau River during rainy season in 2018 is classified in Class II because the WQI value is 81.01 which is in the range 76.50-92.70, so that the water quality at this station is consider clean. While the other stations at Jemberau River during dry season in 2019 and Chini River during both rainy and dry season are classified in Class III. It is classified in Class III because of the WQI value for those rivers are in the range 51.90-76.50 which is the index range is classified as slightly polluted. The water pollution might be coming from the land use activities surrounding rivers such as mining activity, residential area and agricultural activity which generated by local people.

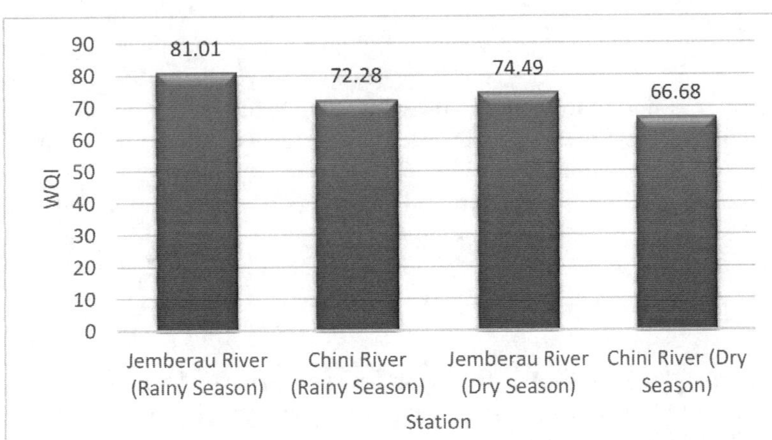

Figure 6. Water Quality Standard based on WQI at Jemberau River and Chini River during rainy season in 2018 and dry season in 2019

Table 6 shows the concentration in unit mg/L for seven parameters of heavy metals in sample obtained at Jemberau River and Chini River during rainy season and dry season. These heavy metal levels are determined by Atomic Absorption Spectroscopy (AAS) which are seven parameters carried out in laboratory by using AAS method are Chromium (Cr), Lead (Pb), Copper (Cu), Manganese (Mn), Iron (Fe), Cadmium (Cd) and Zinc (Zn). Based on the data tabulated above, the concentration of heavy metals in water samples is slightly higher according to permissible range of NWQS. This situation occurs because of land use activities surrounding the rivers such as agricultural, mining and also emission of waste from residential area. All of this activity can lead to increasing water pollution at Jemberau River and Chini River.

Table 6. Result of heavy metals using AAS method

Jemberau River	**Chromium, Cr (mg/L)**	**Lead, Pb (mg/L)**	**Copper, Cu (mg/L)**	**Cadmium, Cd (mg/L)**	**Iron, Fe (mg/L)**	**Zinc, Zn (mg/L)**	**Manganese, Mn (mg/L)**
October	-0.080	0.013	0.032	-0.021	0.492	0.005	0.069
February	-0.139	0.304	-0.031	0.454	-	-	-
Chini River	**Chromium, Cr (mg/L)**	**Lead, Pb (mg/L)**	**Copper, Cu (mg/L)**	**Cadmium, Cd (mg/L)**	**Iron, Fe (mg/L)**	**Zinc, Zn (mg/L)**	**Manganese, Mn (mg/L)**
October	-0.087	0.004	0.035	-0.023	1.57	0.00	0.080
February	-0.131	0.211	-0.036	0.454	-	-	-

5. Conclusions

Based on the analysis that has been made, station at Jemberau River and Chini River are classified as Class III which is according to index range of WQI, the station at Jemberau River and Chini River was classified as slightly polluted. In addition, the water sample collected from station Jemberau River is classified in Class III due to land use activity nearby the river such as mining activity that can cause surface runoff and contributes to heavy metals pollution. While, Chini River is also classified in Class III which is slightly polluted because there have the National Service Training Programme near the river and the effluent discharge from that activity flowing into the Chini River and cause the water pollution. In addition, Chini River also acts as outlet to Pahang River, so that it will give worse effect in the future if the river is not treating well from now. Based on the collected data for two different stations of water samples that has been analyzed, we can conclude that the main sources of water pollution are comes from nearby mining activity. From the in-situ analysis and laboratory experiment, there are varied reading of parameter concentration based on different location and seasons. Other than that, the land use activity at certain location give the most impact to the reading taken especially for heavy metals concentration. According to NWQS and DOE-WQI, since the rivers are classified as Class III, hence the water supply need an extensive treatment and it is not suitable for drinking uses.

6. References

[1] Ahmad H, Jusoh H, Idris M, Er AC, Selvadurai S, Toriman MI, Sharifah Mastura and Buang A 2013 Revitalizing ecotourism for a sustainable tasik chini biosphere reserve *Asian Social Science* **9** (14 SPL) pp 70–85

[2] Sun Z, Brittain JE, Sokolova E, Thygesen H, Saltveit SJ, Rauch Sand Meland S 2018 Aquatic biodiversity in sedimentation ponds receiving road runoff – What are the key drivers? *Science of the Total Environment* Elsevier BV **610–611** pp 1527–1535

[3] Jamshaid M, Khan AA, Ahmed K and Saleem M 2018 Heavy metal in drinking water its effect

on human health and its treatment techniques – a review *International J. of Biosciences (IJB)* 12(**4**) pp 223–240

[4] Gao B, Gao L, Gao J, Xu D, Wang Q and Sun K 2019 Simultaneous evaluations of occurrence and probabilistic human health risk associated with trace elements in typical drinking water sources from major river basins in China *Science of the Total Environment* Elsevier BV **666**

[5] Tian Y, Jiang Y, Liu Q, Dong M, Xu D, Liu Y and Xu X 2019 Using a water quality index to assess the water quality of the upper and middle streams of the Luanhe River northern China *Science of the Total Environment* Elsevier BV **667** pp 142–151

[6] Yan Z, Zhou Z, Sang X and Wang H 2018 Water replenishment for ecological flow with an improved water resources allocation model *Science of the Total Environment* Elsevier BV **643** pp 1152–1165

[7] Baharim NB, Yusop Z, Askari M, Ismail Y, Wan Muhd Tahir WZ, Othman Z and Zainal Abidin M.R 2012 Preliminary Results of Stratification Study in Sembrong 2nd Int. Con. on Water Resources (ICWR2012) pp 1–7

[8] Nazari Sharabian M, Ahmad S and Karakouzian M 2018 Climate Change and Eutrophication: A Short Review *Engineering Technology & Applied Science Research* **8** (6) pp 3668–3672

[9] Boelee E, Geerling G, Van Der Zaan B, Blauw A and Vethaak AD 2019 Water and health: From environmental pressures to integrated responses Acta Tropica Elsevier **193** (February 2018) pp 217–226

[10] Suhaimi O, Ahmad M and Lim EC 2012 The 12th World Lake Con.Seasonal influence on water quality and heavy metals concentration in Tasik Chini, Peninsular Malaysia

Acknowledgment
The authors acknowledge the Department of Irrigation and Drainage (DID) of the Ministry of Natural Resources & Environment, Malaysia for their assistance in providing the data used in this study. This research was supported by a grant of RDU180350 from Universiti Malaysia Pahang (UMP).

IOP Conf. Series: Materials Science and Engineering **712** (2020) 012023 doi:10.1088/1757-899X/712/1/012023

Nanoparticle in Asphalt Binder: A State-of-The-Art Review

P J Ramadhansyah[1], K A Masri[1], A H Norhidayah[2], M R Hainin[2], M W Muhammad Naqiuddin[2], Y Haryati[2], M K I M Satar[2] and A Juraidah[3]

[1]Faculty of Civil Engineering Technology, Universiti Malaysia Pahang, 26300 Gambang, Pahang, Malaysia
[2]Faculty of Engineering, School of Civil Engineering, Universiti Teknologi Malaysia, 81310 Johor Bahru, Malaysia
[3]Faculty of Civil Engineering, Universiti Teknologi MARA, 40450 Shah Alam, Selangor, Malaysia

Abstract. Increasing traffic volumes and the rising cost of bitumen making it necessary to improve the performance of asphalt binder through bitumen modification. Various types of modifiers have been employed as bitumen binders in order to improve the bitumen mixture, particularly with regard to its resistance to aging, cracks due to fatigue and thermal conditions, moisture-induced damage and permanent deformation. This review study summarizes the performance of nanoparticles as an asphalt modifier. Several relevant literatures were reviewed, which related to the matter associated with nanomaterials such as nano-silica, carbon nano-tubes and nano-titanium dioxide. Recent studies showed that the utilisation of nanoparticle gave great benefits in many ways.For instance, nano kaolin as a mineral clay promises non-hazardous effects to road users. This can result in a healthy environment and help towards increasing the application of green technology road construction.

Content

1. Nanotechnology in Asphalt Binder

Recently, increased traffic volumes, larger and heavier trucks, and higher tire pressures have aggravated the severity of the condition upon the pavement system, resulting in rutting, stripping and fatigue cracking [1]. Therefore, the concept of new technology, named nanotechnology has been introduced as an alternative to enhance the physical and rheological behaviours of asphalt binders, resulting in enhanced performance of the asphalt mixture [2]. Improvements in asphalt binder and mixture performance through other types of modifier at macro-scale and micro-scale have been achieved, yet it is interesting to explore what nanotechnology offers in terms of improving asphalt pavement performance. Therefore, the application of nanotechnology in the binder was widely explored and reported all around the world. Nanotechnology is considered to be the latest knowledge for pavement in order to develop a safe and sustainable pavement infrastructure. Although researchers, material producers, and engineers have explored the potential for many years, nanotechnology usage has been limited. Therefore, new efforts and exploration of the development of nanomaterials for pavement application that improve the nanoscale mechanical and physical properties as well as durability of the mixture were further investigated.

2. Nanomaterials in Asphalt Binder Modification

The previous study proved that various kinds of modifiers such as a polymer, crumb rubber, fibers are introduced to improve performances of asphalt binder. These performances include moisture susceptibility, permanent deformation, fatigue life, anti-ageing, and so on. From all these modifiers, polymer was the most widely used in binder modification. However, one of the major shortcomings of pure polymer modifiers is that most of the polymers are thermodynamically incompatibility with asphalt binder due to the large differences of density, polarity, molecular weight and solubility between the polymer and the asphalt binder. This can result in delamination of the composite during thermal storage, which is not readily apparent and adversely affects the material when it is used in construction. Therefore, more and more researchers concentrate on the introduction of nanomaterials to modify the asphalt binder due to the rapid development of nanotechnology and overcome this issue. The nanomaterial is described as a material with at least one dimension within 1 – 100 nm [3]. Based on previous research, the properties of nano-composites such as physics, chemistry and biology were significantly different from their initial ones [4]. Due to the unique characteristic of nanomaterials such as the tunnel effect of macroscopic quantum and surface effect were also found attributing to the natures of its large surface area and small particles size. It is noted that nanomaterials exhibited high-temperature sensitivity, high ductility, large surface area, high strain resistance, and low electrical resistivity [5,6]. There are various types of nanomaterials that commonly used in binder modification such as nano-silica, carbon nanotubes (CNTs), nano titanium dioxide (TiO_2) and nano clay. The following section discussed the application of nanomaterials in binder modification in term of their properties and performance.

3. Nano-silica

Silica is an abundant compound worldwide that is largely employed in industries to produce silica gels, colloidal silica, fumed silica and etc. [7]. While nano-silica is an inorganic material was produced mainly from silica precursors. There are few methods to produce nano-silica which are vapour phase method, sol-gel methods and other methods, however, among them, the sol-gel process method is considered as the most widely applied method for its mild condition and pure products [8]. Nano-sized silica is interesting because it is applied in emerging areas like medicine and drug delivery. Amorphous nano-silica is qualified as a nano-bio pesticide. The advantage of these nanomaterials resides in a low cost of production and the high-performance features. According to Yao *et al.* [9], nano-silica is also a material with a huge surface area, strong absorption, good dispersal ability, high chemical purity and excellent stability. Therefore, it is expected that the utilisation of nano-silica enhances the performance properties of the asphalt binder and mixture. Table 1 shows the main physical and chemical properties of nano-silica.

Table 1. Properties of nano-silica [4,9]

Color	White				
True density, g/cm^3	2.4				
Surface area, m^2/g	180-600				
Bulk density, g/cm^3	<0.10				
Purity, %	+99				
Chemical properties	SiO_2	Ti	Ca	Na	Fe
	>99%	<120 ppm	<70 ppm	<50 ppm	<20 ppm

3.1. Application of Nano-silica in Modified Binder

The study on the utilisation of nano-silica in binder modification was performed to improve the performance of asphalt binder in terms of physical, rheological and mechanical properties. A laboratory evaluation conducted by Md. Yusoff *et al.* [10] reported that the addition of nano-silica improves the resistance to rutting at high temperature and fatigue deformation at intermediate temperature as compared to the control mixture. The study was performed using polymer modified asphalt mixture (PMA) with 0%, 2% and 4% of nano-silica by weight of asphalt binder. Figure 1 shows the results of the resilient modulus test at 25 °C. The result shows that PMA mixed with 4% nano-silica shows least susceptibility to fatigue deformation with highest resilient modulus value. Similar trends can also be observed for aged samples. This indicated that the addition of nano-silica would improve resistance to fatigue deformation at intermediate temperatures, both for un-aged and aged samples. In addition, another finding was also can be observed at which as temperature increases, the difference in resilient modulus is more notable, with a decline in stiffness at temperature of 40 °C as shown in Figure 2. By considering the difference in the resilient modulus value at higher temperatures, this indicates that PMA mixed with 4% nano-silica is the least susceptible to rutting compared to control mix. These findings coincide with previous studies done by Yao *et al.* [9], and You *et al.* [11], who found that the rut depths of nanomaterial modified asphalt mixtures decrease compared to the control mixture, and a smaller rutting depth was observed for a greater percentage of nanomaterials in mixtures. Another finding on the improvement of rutting resistance can also be observed through dynamic creep test as shown in Figure 3. The result indicates that the addition of nano-silica can reduce the rutting depth by almost half as compared to the control mixture with 4% nano-silica mixed with PMA shows the lowest permanent deformation values.

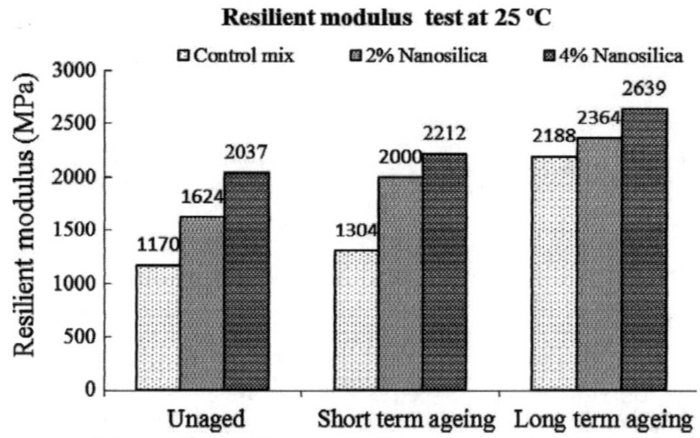

Figure 1. Resilient modulus test at 25 °C [10]

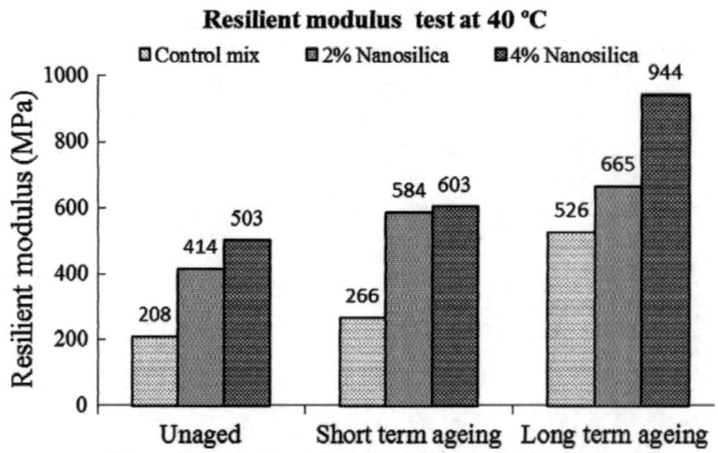

Figure 2. Resilient modulus test at 40 °C [10]

IOP Conf. Series: Materials Science and Engineering **712** (2020) 012023 doi:10.1088/1757-899X/712/1/012023

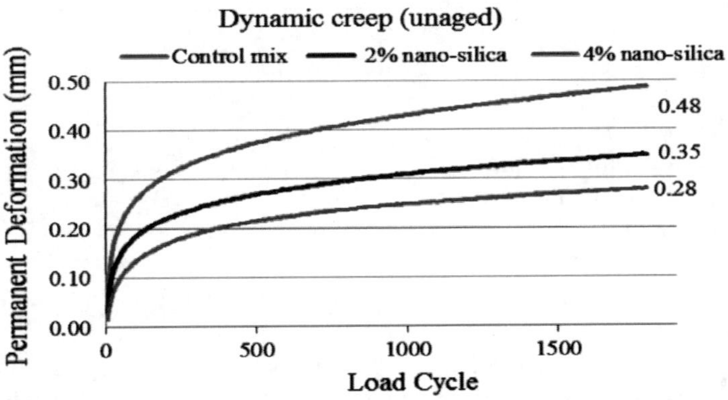

Figure 3. Dynamic creep test result for un-aged mixtures [10]

In 2015, Masri *et al.* [12] conducted research on the porous asphalt mixture incorporating nano-silica. This study was performed to investigate the effect of nano-silica modified binder on binder-drain down of porous asphalt. Binder drain-down was one of the main problems of porous asphalt because is having a high air void content which allowed the binder to flow through those air voids. This leads to the binder being drained off from the mixture. Different proportions of nano-silica ranging from 1 to 6% by weight of binder were used in this study. The nano-silica used was in colloidal form with the average size between 10 and 15 nanometres (nm). Table 2 tabulated the results obtained from the Cantabro and drain-down tests. The result shows that the Cantabro decreased as well as binder drain-down until the optimum percentage of nano-silica content which is 4% by weight of the binder. This indicates that the existence of nano-silica is capable of enhancing the performance of porous asphalt mixture in terms of abrasion resistance and binder drain-down. It can be concluded that the performance of anti-stripping property of nano-silica modified binder was significantly enhanced.

Table 2. Cantabro and drain-down test results for different proportions of nano-silica [12]

Nano-silica (%)	Cantabro (%)	Drain-down (%)	Drain-down (g)
0	17	0.324	3.564
1	13	0.277	3.047
2	12	0.222	2.442
3	10	0.197	2.167
4	9	0.186	2.046
5	11	0.234	2.574
6	14	0.285	3.135

4. Carbon Nanotubes (CNTs)

A carbon nanotube (CNT) is a one-atom-thick sheet of graphite rolled up into seamless hollow cylinders which have a diameter starting from one nanometre. CNTs were discovered by Iijima [13], who first reported the arc-discharge synthesis and characterisation of helical microtubules, formed by molecular-scale fibres with structures related to fullerenes. Since then, an enormous amount of research has been conducted in the area of CNT production and their applications due to the superiority of their electronic and mechanical properties [14]. CNTs are divided into two different types which are single-walled carbon nanotubes (SWCNTs) and multi-walled carbon nanotubes (MWCNTs), based on the number of grapheme sheets. However, as a modifier, MWCNTs are better than SWCNTs as they are stiffer, easier and cheaper to produce on a large scale [15]. Currently, several techniques are available to produce CNTs. The three common methods are arc discharge, laser ablation and chemical vapour deposition (CVD) [16]. Table 3 shows the properties and characteristics of MWCNTs.

Table 3. Properties of MWCNTs [17]

Property	Value
Colour	Black
Purity, (%)	>90
Outside diameter, nm	10 – 30
Inside diameter, nm	5 – 10
Length, μm	10 – 30
Density, g/cm^3	2.1
Surface area, m^2/g	>200

Dispersion of CNTs has been one of the largest challenges due to the aggregation of the nanotubes [18]. Due to the strong, cohesive forces of CNTs, it can easily cluster together when added to asphalt binder. Therefore, for modified asphalt binder with CNT particles, the mixing process is a key factor. Improvement of material properties can be achieved by a proper dispersion technique. Improper dispersion leads to nanotubes damage and size break down, which deteriorate the material properties. Shirakawa *et al.* [19] used asphalt emulsion as solvent to disperse CNT and stated in their studied that CNT disperses better in anionic and non-ionic emulsion than cationic emulsion. Microstructure analysis performed by Faramarzi *et al.* [20] using Scanning Electron Microscope (SEM) analysis show that CNTs has strong tendency for aggregation with formation of random network of contacting aggregate as shown in Figure 4. Meanwhile, the morphology structure of CNT modified binder is shown in Figure 5, which presents a continuous matrix with uniform distributing nanometre phase. Both figures were observed using the same scale reference which is at 5 μm. In this study two process of dispersion in asphalt binder were performed: simple mixing and wet mixing. In simple mixing, CNTs was manually added into binder and mixed using mechanical stirrer at a speed of 1550 rpm for 40 minutes to reach the required homogeneity. While wet mixing involved the dispersed of CNTs in the solvent by means of sonication process and high shear mixing. Kerosene was chosen as solvent because it was petroleum-based product, cheap and easily available. Based on the morphology structure of CNTs modified asphalt binder, simple mixing CNT modified asphalt binder shows some large accumulated particles as shown in Figure 5(a). Meanwhile, Figure 5(b) shows CNTs particle distribute uniformly with a few mass accumulated particle for wet mixing. This indicated that utilising the wet mixing can lead to a decrease in the accumulating tendency of CNTs and consequently the increase in the mechanical properties as compared to simple mixing. Since CNTs exhibit great mechanical properties, it is expected to produce significantly stronger and more improved bituminous composites than conventional reinforcing materials such as glass fibers, carbon fibers and so on.

Figure 4. SEM images of carbon nanotubes [20]

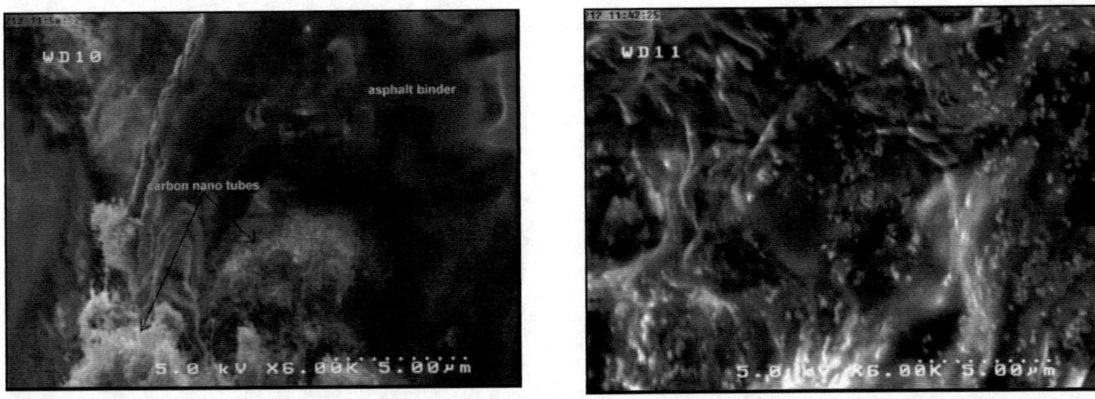

Figure 5. Simple mixing CNT-modified asphalt binder (left) and wet mixing CNT-modified asphalt binder (right) [20]

4.1. Application of CNTs in Modified Binder

One of the most desired properties of nanomaterials in road construction is their capability to confer a mechanical reinforcement to bituminous materials. While a large number of research studies have been carried out in the last decade on modification of asphalt binder by different nanoparticles, very few studies have been conducted in the area of CNTs modified binders and mixtures. Theoretically, the addition of CNTs in asphalt binder resulted in a reduction of penetration value and increased softening point due to increasing of binder stiffness. This was supported by Amin *et al.* [17], which found that the softening point value increased significantly with the increasing of CNTs content in the binder. This indicated that the modified binder with CNTs shows stiffer and lower temperature susceptibility. These characteristics are considered important especially when significant fluctuation in temperature occur during the day or in areas with heavy traffic. In addition, laboratory evaluation conducted by Amin *et al.* [17] reported that the addition of CNTs in binder enhanced the rheological properties of binder by increased the rutting factor, thus enhancing the pavement resistance to permanent deformation. Furthermore, rheological properties of binder in aging condition were also investigated in their study. A similar trend as un-aged condition was observed which it can be noticed that rutting factor significantly increased as amount of modifier increased. However, they found that modified binder with 0.5% of CNTs content or less did not cause a significant improvement. This finding was supported by Santagata *et al.* [21] where modifying asphalt binder with high percentages of CNTs (high than 0.5% of asphalt binder content) positively improve the rheological characteristics of the binder. These characteristic improvements lead to decrease asphalt pavement rutting in high temperature and thermal cracking at low temperature. Therefore, the selection percentage of modifier is one of the main parameters to produce the optimum performance of binder.

5. Nano-Titanium Dioxide (nano-TiO_2)

Titanium dioxide is the naturally occurred oxide of titanium and usually found in the form of rutile, anatase and brookite. TiO_2 has been known as a useful photocatalytic material that is attributed to the several characteristics includes high photocatalytic activity compared with other metal oxides, and it's compatible with conventional bituminous materials without changing any original performance. Its chemical formula is TiO_2 which it belongs to the family of transition metal oxides [22]. Nano-TiO_2 is produced from the oxidation of titanium precursor at a high temperature. It consists of 80% anatase and 20% rutile. Table 4 shows the main properties of nano-TiO_2. Generally, there are two main production methods to obtain nano-TiO_2 which are using sulfate process and chloride process. However, both methods have potential to cause negative impacts on environment. For instance, the sulfate process produced a large number of acidic wastes, while chloride process may result in a lot of solid, liquid and mineral wastes [23]. The characteristics of nano-TiO_2 are large surface area, small diameter and very low opacity as compared to ordinary TiO_2. Because of the unique properties, it is expected that nano-TiO_2 can improve the performance of modified asphalt binder and mixture.

Table 4. Properties of nano-TiO_2 [22,23,24]

Property	Value
Colour	White
Purity, (%)	>99.8
Particle size, nm	20-60
Density, g/cm^3	3.9
Bulk density, g/cm^3	0.08
Surface area, m^2/g	50-80

5.1. Application of Nano-TiO_2 in Modified Binder

Many kinds of research are carried out to investigate the properties of the modified binder with nanomaterial. For example, Buhari *et al.* [22] performed laboratory evaluation on physical and rheological properties of modified asphalt binder with nano-TiO_2 R15. The studies carried out using 2%, 4%, 6%, 8% and 10% of nano-TiO_2 content by weight of binder. The result shows that the addition of nano-TiO_2 has significantly reduced the penetration value and increased the softening point temperature. This demonstrated the improvement on the stiffness of the binder and also better temperature susceptibility as compared to control binder. In addition, viscosity value reduced as nano-TiO_2 content increased. This indicated that nano-TiO_2 modified binder has lower mixing and compaction temperature. Therefore, it can be considered that nano-TiO_2 modified binder is the environmental friendly because of less energy consumption during preparation of the mixture. Furthermore, the decreasing in viscosity value leads to improve the workability of the binder. In term of rheological properties, addition of nano-TiO_2 did not show positive improvement on the rutting factor ($G^*/\sin \delta$) as compared to control binder even though it is still in the same performance grade which is 64 °C. This finding was in contradictory with Fang *et al.* [25], where they stated that rutting factor and the elasticity of the nano-TiO_2 modified binder were increased. Thus the resistance to permanent deformation was enhanced, and the service life of modified asphalt mixture was prolonged. The studies performed by Shafabakhsh *et al.* [24] proved that the utilisation of nano-TiO_2 in bituminous material does not only improve the properties of binder but also in the mixture. The studies were carried out to investigate the properties of hot mix asphalt (HMA) with different percentages of nano-TiO_2 content (1%, 3%, 5% and 7% by weight of binder). The mechanical property of the mixture was evaluated in term of fatigue performance of both control mixture and nano-TiO_2 modified mixture. Based on the laboratory works, they found that nano-TiO_2 modified mixture tend to have longer fatigue life as compared to control mixture. This indicated that nano-TiO_2 reduced the generation and propagation rate of micro-cracks which can lead to high tensile strength in the modified mixture. Therefore, it can be concluded that utilisation of nano-TiO_2 can improve the fatigue life of mixture by prevents tensile and vertical cracks from being easily generated by horizontal tensile stresses and prevents them from propagating.

6. Nano clay

Amongst the potential nanoparticle, nano clay received much attention as a modifier due to the expectation to strengthen and enhance the physical and rheological properties of the asphalt binder. Nanoclay is a layered silicate that has a layer thickness between 1 and 100 nm and widely used in the binder modification to improve mechanical and thermal properties. In addition, nano clay is the most commonly used nanomaterials to modify bituminous material due to their low cost of production and abundance in nature. There are four main groups of clays which are kaolinite, montmorillonite, Illite and chloride. Generally, there are two main categories of nano clay used in asphalt binder, which is non-modified nano clay (NMN) and polymer modified nano clay (PMN). NMN is the most frequently used with 2-to-1 layered structure clay consisting of one octahedral alumina sheet sandwiched between two tetrahedral silica [26]. Based on previous studies, it has been proved that the utilisation of nano clay in bituminous material shows positive improvement in both rheological and mechanical properties of modified asphalt binder. However, no research was reported on the utilisation of NKC in terms of the physical and rheological properties of the asphalt binder as well as mechanical properties of asphalt mixture while this study was performed.

7. Conclusion

This review focuses on the performances and possible use of the nanoparticle as an asphalt modifier. Polymer modified asphalt binders have been immensely popular; the high cost and thermal instability thereof encouraged researchers to explore new materials to improve their performance. Many studies agreed that the usage of nanoparticle could enhance the physical, rheological and mechanical properties of asphalt binder and mixtures. The utilisation of nanoparticle in the asphalt binder has been introduced. The summary of the issues are presented below:

a) The performance application of nanomaterials for asphalt binder to improve the physical and chemical properties are inadequate, especially under durability context.
b) The existing research on the anti-aging performance of nanoparticle as the modified binder is inadequate.
c) The fundamental effect of nanoparticle with other modifiers to improve asphalt performance is needed.

Certainly, further investigation is needed to better understand the utilization of nanoparticle as bitumen modifier in the fundamental level along with its field performance.

8. References

[1] Airey G D 2002 *Constr. Build. Mater.* **16**(8) pp. 473-487.
[2] Yang X, You Z, Dai Q and Mills-Beale J 2014 *Constr. Build. Mater.* **51** pp. 424-431.
[3] Jeffry S N A, Ramadhansyah P J, Norhidayah A H, Haryati Y and Mohd Khairul Idham M S 2018 *Constr. Build. Mater.* **173** pp. 40-48.
[4] Yang J and Tighe S 2013 *Procedia Soc Behav Sci.* **96** pp. 1269-1276.
[5] Jeffry S N A, Jaya R P, Hassan N A, Mirza J and Yusak M I M 2019 *Proc. Ins. Civil Eng. Constr. Mater.* **172**(2) pp. 103-115.
[6] Jaya R P, Yusak M I M, Hainin M R, Mashros N, Warid M N M, Ali M I and Ibrahim M H W 2019 *AIP Conf. Proc.* **2151** pp. 1-8.
[7] Yusak M I M, Abdullah M E, Jaya R P, Hainin M R and Ibrahim M H W 2017 9 *IOP Conf. Ser.: Mater. Sci. Eng.* **226**(1) pp. 1-5.
[8] Hassan N A, Mahmud M Z H and Jaya R P 2014 *Adv. Mater. Res.* **911** pp. 443-448.
[9] Yao H, You Z, Li L, Goh S W, Lee C H and Yap Y K 2013 *Constr. Build. Mater.* **38** pp. 327-337.
[10] Md. Yusoff N I, Breem A A S, Alattug H N M, Hamim A and Ahmad J 2014 *Constr. Build. Mater.* **72** pp. 139-147.
[11] You Z, Mills-Beale J, Foley J M, Roy S, Odegrad G M, Dai Q and Goh S W 2011 *Constr. Build. Mater.* **25**(2) pp. 1072-1078.
[12] Masri K A, Arshad A K and Samsudin S 2015 *J. Teknologi.* **78** pp. 139-146.
[13] Iijima S 1991 *Nature.* **354** pp. 56-58.
[14] Saifuddin M, Raziah A Z and Junizah A R 2013 *J. Chem.* **213** pp. 1-18.
[15] Steyn W J, Bosman T E, Galle S and Heerden V J 2013 *Adv. Mater. Res.* **723** pp. 312-319.
[16] Rafique M M A and Iqbal J 2011 *J. Encap. Adsorption Sci.* **1**(2) pp. 29-34.
[17] Amin I, El-Badawy S M, Breakah T and Ibrahim M H Z 2016 *Constr. Build. Mater.* **121** pp. 361-372.
[18] Peng C M, Naveed A, Siddiqui A, Gad Marom B and Jang K K A 2010 *Compos.* **41**(A) pp. 1345-1367.
[19] Shirakawa T, Akio T and Noriyasu O 2012 *Int. J. Geomate.* **2**(1) pp. 161-165.
[20] Faramarzi M, Arabani M, Haghi A K and Mottaghitalab V 2105) *Int. J. Pavement Res. and Technol.* **8**(1) pp. 29-37.
[21] Santagata E, Baglieri O, Tsantilis L and Dalmazzo D 2012 *Procedia Soc Behav Sci.* **53** pp. 546-555.

[22] Ahmed A H, Ramadhansyah P J, Norhidayah A H, Haryati Y, Ghasan F H and Mohd Haziman W I 2017 *Constr. Build. Mater.* **156** pp. 496-505.
[23] Li R, Xiao F, Amirkhanian S, You Z and Huang J 2017 *Constr. Build. Mater.* **143** pp. 633-648.
[24] Shafabakhsh Gh, Mirabdolazimi S M and Sadeghnejad M 2014 *Constr. Build. Mater.* **54** pp. 566-571.
[25] Fang C, Yu R, Liu S and Li Y 2013 *J. Mater. Sci. Tech.* **29**(7) pp. 589-594.
[26] Ray S S and Okamoto M 2003 *Prog. Polym. Sci.* **28**(11) pp. 1539-1641.

Acknowledgements
The support provided by the Malaysian Ministry of Higher Education and Universiti Malaysia Pahang in the form of a research grant (RDU/UMP) vote number RDU1803160 for this study is highly appreciated.

The Effectiveness of Palm Oil Fuel Ash Cement Paste to Control Degradation in Ammonium Nitrate Solution

T Sadia[1], M E Rahman[2], A Raudhah[3] and S I Doh[4]

[1]Department of Civil Engineering, Curtin University Kent St, Bentley WA 6102, Australia
[2]Faculty of Engineering and Science, Curtin University Sarawak, CDT 250, 98009 Miri, Sarawak, Malaysia
[3]Department of Civil Engineering, University Malaysia Sarawak, 94300 Samarahan, Sarawak, Malaysia
[4]Faculty of Civil Engineering Technology, Universiti Malaysia Pahang, 26300 Gambang Kuantan, Pahang, Malaysia

Abstract. This paper presents the use of treated palm oil fuel ash (POFA) as a pozzolanic material in modified cement paste to control degradation in ammonium nitrate environment. The POFA used in this study had a mean diameter of 8.7μm. The treated POFA was used to replace cement in different percentage of 0%, 10%, 20% and 30% with a constant water to cement ratio of 0.4. 50mm cement cubes was prepared for mechanical, durability and micro-structural test. Mechanical test consisted of the compression strength test and durability test consisted of the Volume of Permeable Voids Test (VPV). Meanwhile, micro-structural test consisted of the Thermo-gravimetric analysis (TGA). After 28 days of full water curing, the 50mm cubes samples were immersed in 20% ammonium nitrate solution until 90 days. Based on the test results, it was observed that, cement cubes replaced with 20% POFA had good resistance against ammonium nitrate solution. However, in order to improve the current study, a number of recommendations had been suggested.

1. Introduction

Concrete is one of the oldest known man-made construction materials used for construction of various structures; such as bridges, dams, tunnels, buildings, sewerage systems, pavement, runways, roads and fertilizer plants. It is rare for concrete structures to fail since reinforced concrete structures are usually designed to a sufficiently high safety factor. However, over the last few decades, the lifespan of concrete structures has become a primary concern as many structures have collapsed, especially in buildings and transportation sectors. Failure of concrete structure happens due to various reasons. One particular reason is due to aggressive chemical solution attack. According to previous researchers, aggressive ammonium nitrate solution can cause failure of concrete structure by causing degradation to it **[1]**. A small amount of ammonium nitrate solution, approximately 0.5% by weight of cement, is sufficient to causes damage to concrete structures **[1]**.

1.1. Research Gap

According to literature review, previous researcher have used POFA of various sizes as a cement replacement to study properties such as compression, tension, flexural, setting time, workability, heat evolution, dry shrinkage, water absorption, porosity, and permeability in aggressive solution such as chloride, carbon dioxide and sulfate. However, studies are yet to be done using POFA against aggressive ammonium nitrate solutions to control deterioration. The aim of this paper is to investigate the effect of using palm oil fuel ash (POFA) as a cement replacement material to control degradation of cement paste in ammonium nitrate environment. POFA is a pozzolanic material, which has a high percentage of alumino-silicate content **[2]**.

2. Methods

2.1. Material and Properties

The material used in this are OPC, POFA and ammonium nitrate. Potable tap water was used for mixing and curing throughout the research. The OPC used in this research had fulfilled the requirements set under ASTM C-150 Type 1 Ordinary Portland Cement (OPC) as well as meeting the Malaysian standard MS 522: Part 1-2007, ISO 9001 and 14001 quality requirements. Apart from that, based on the physical test, it was found out that, the OPC used had a mean diameter of 24.67 μm, specific gravity of 3.1 and surface area of 10667.40 cm^2/g. The chemical composition of the OPC is stated as in Table 1. POFA used in this study was collected from a local palm oil mill. The raw POFA was later further treated by burning 600^0degrees Celsius for one hour and grinding. Based on the test result, the treated POFA had mean diameter of 8.7 μm, specific gravity of 2.50 and surface area of 53206.33cm^2/g. The chemical composition of the treated POFA is shown in Table 1. The POFA in this study is classified as a class N pozzolanic according to **ASTM [3]**. Ammonium nitrate was obtained from local market in pellet form and 20% ammonium nitrate was prepared by mixing with water, **[4]**.

2.2. Curing and Mix Proportion

The curing ages for the testing was 3, 7, 28, 56 and 90 days. The method of curing used in this research was water curing. Fresh cement paste was cast in 50mm moulds and demolded on the next day. The cement cubes were then immersed in water for 27 days for full water curing. After the desired cuing age was achieved, some of the samples were left at room temperature and some were placed in an ammonium nitrate solution for 90 days. Table 2 and 3 shows the curing condition and mix proportion of the samples.

2.3. Compressive Strength

Compressive strength test is a type of test which determines the maximum amount of load which can be applied to a material before fracturing. Compressive strength tests were carried out in accordance with **Australian Standard [5]**. Five samples were tested and an average result was calculated. The calibrated compression machine was used for a 50 mm cube test at a loading rate of 700 N/sec., which was within a loading range between 0.2 N/mm2 /sec and 0.4 N/mm2 /sec. Mechanical properties test consisted of the compressive strength test. The tests were carried out at 3, 7, 28, 56 and 90 days for 0%, 10%, 20% and 30% POFA samples. The compressive strength test was carried out in room temperature and in ammonium nitrate solution and comparison was made. 50mm samples were cast and cured in water until 28 days, after which samples were treated in room temperature and in ammonium nitrate solution until 90 days.

Table 1. Chemical Properties of Treated POFA and OPC

Type	SiO_2	AL_2O_3	Fe_2O_3	CaO	MgO	SO_3	K_2O	LOI
Treated POFA (%)	58.1	2.6	1.67	10.8	6.1	0.7	6.45	6.83
OPC (%)	19.28	4.67	3.3	65.9	1.7	0.3	0.5	n/a

Table 2. Curing Condition

Moulds	Waters	Ammonium Nitrates	Room Temperatures
1 day in mould	27 days curing in water	50% immersed till 90 days	50% left in room temperature till 90 days

Table 3. Mix Proportion

Mix. No	Symbol	OPC (gram)	POFA (gram)	W/C ratio	Water (gram)
1	0%POFA	200	0	0.4	80
2	10%POFA	180	20	0.4	80
3	20%POFA	160	40	0.4	80
4	30% POFA	140	60	0.4	80

2.4. Volume of Permeable Voids
Volume of permeable voids is a type of test which determines the amount of voids present in a sample. Volume of permeable void test was carried out according to ASTM C642 Standard. According to the standards, the VPV test was carried out in three steps. In the first step, samples were oven dried, in the second steps, the sample were immersed in water and in the third step, samples were immersed again in water to calculate the apparent mass. The VPV test was carried out for the 0%, 10%, 20% and 30 %POFA samples POFA samples treated in room temperature and in ammonium nitrate solution.

2.5. Thermo-gravimetric analysis (TGA)
Thermo-gravimetric analysis (TGA) test determines the mass gain or loss of a material due to oxidation, decomposition or losses of volatiles. The TGA test was carried out using STAR Software (METTLER TOLEDO) and test was carried out according to the procedure stated in the test machine manual. The micro-structure studies consisted of the Thermo-gravimetric Analysis (TGA). The test was carried out for 0%, 10%, 20% and 30% POFA samples treated in room temperature and in ammonium nitrate solution.

3. Result and discussion

3.1. Compressive strength of the modified cement paste
The compressive strength of the samples is stated as in Table 4 and in Figure 1 and 2. Based on Table 4 and Figure 1, for samples treated in room temperature, the compressive strength increased with an increase in the curing time. This is due to the fact that with an increase in the curing time, the cement matrix had been modified. There had been an addition formation of calcium silicate hydrate(C-S-H) and calcium hydroxide formed in the cement paste, as a result, the cement paste became denser and stronger.

Table 4. Comparison of compressive strength treated in room temperature and in ammonium nitrate solution

Mixes	Compressive strength (MPa)- Normalized compressive strength (%)				
	3 days	7 days	28days	56days	90days
Control	37.7 - 100	47.8 - 100	50.2 - 100	57.5 - 100	60.4 - 100
Control (AN)	37.7-100	47.8 - 100	50.2 - 100	42.7 - 74	39.2 - 65
10%POFA	26.2 - 100	41.9 - 100	52.7 - 100	57.9 - 100	63.7 - 100
10%POFA (AN)	26.2 - 100	41.9 - 100	52.7 - 100	47.4 - 82	37.4 - 59
20%POFA	20.9 - 100	34.4 - 100	52.0 - 100	60.7- 100	69.5 - 100
20%POFA (AN)	20.9 -100	34.4 - 100	52.0 - 100	44.2 - 73	42.4 - 61
30%POFA	20.0 - 100	24.4 - 100	41.8 - 100	45.7 - 100	57.0 - 100
30%POFA (AN)	20.0 - 100	24.4 - 100	41.8 - 100	36.1 - 79	31.8 - 56

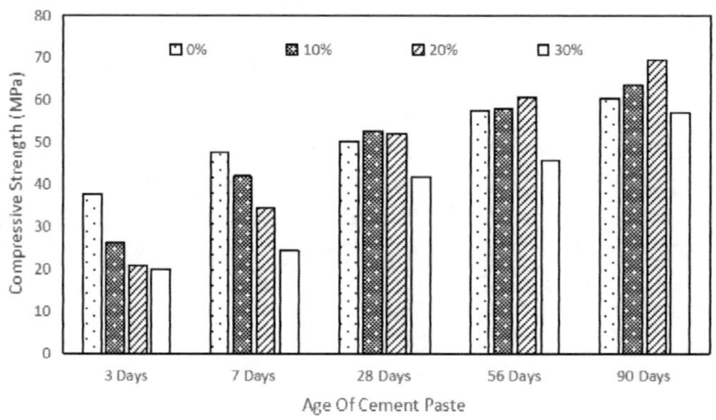

Figure 1. Compressive strength of cement paste in room temperature

For samples treated in room temperature, it was observed that, at 56 and 90 days, samples replaced with 20% POFA had the highest compressive strength. The compressive strength was 60.7MPa and 69.5MPa respectively for 20%POFA sample at 56 and 90 days. The highest compressive strength for 20%POFA samples is due to the pozzolanic reaction between the SiO_2 in the POFA with $Ca(OH)_2$ in the cement paste. However, based on Table 4 and Figure 2, for samples treated in ammonium nitrate solution, it was observed that the compressive strength of the samples shows a decrease after 28 days. This is due to the decalcification reaction, which had leached the $Ca(OH)_2$ out from the cement paste. Consequently, it was observed that, the percentage decrease in compressive strength at 56 days in an ammonium nitrate solution was almost 25% compared to samples at room temperature. The strength however continued to decrease to about 50% at 90 days in an ammonium nitrate solution. Findings, similar to this test results were also reported by other author **[4, 6, 7, 8, 9]**. It was therefore concluded that, samples replaced with 20% POFA had good performance in room temperature and against ammonium nitrate solution.

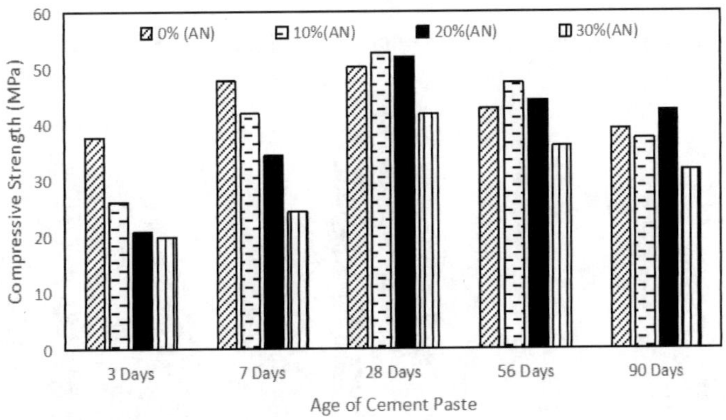

Figure 2. Compressive strength of cement paste in ammonium nitrate solution

3.2. Volume of Permeable Void (VPV) of the modified cement paste

The durability properties test consisted of volume of permeable voids (VPV) test. The VPV test was carried out to determine the amount of voids in a sample. The permeable void affects the transport properties and durability of concrete. The VPV test was carried out for the 0%, 10%, 20% and 30 %POFA samples POFA samples treated in room temperature and in ammonium nitrate solution. At 28 days, the VPV for 0%, 10% 20% and 30% POFA samples were 20.98%, 20.47%, 20% and 20.41% respectively. At 56 days, for sample treated in room temperature, the VPV for 0%, 10%, 20% and 30% were 20.77%, 19.63%, 18.01% and 20.35% respectively. At 56 days for samples treated in ammonium nitrate solution, the VPV for 0%, 10%, 20% and 30% POFA were 22.36%, 22.27&, 20.61% and 20.72%. Consequently, at 90 days for samples treated in room temperature, the VPV were 19.96%, 18.68%, 16.67% and 20.25% for 0%, 10%, 205 and 30% POFA samples. The VPV at 90 days for samples treated in ammonium nitrate solution were 23.5%, 25.84%, 23.85% and 23.72% for 0%, 10%, 20% and 30% POFA.

Based on the test results, for samples treated in room temperature, the void percentages in the cement paste decreased as the curing age increased. This is because more calcium silicate hydrate and calcium hydroxide is produced during the cement hydration process. Calcium silicate hydrate is responsible for decreasing the void percentages as the cement paste becomes denser. A lower void percentage is an indication the sample has a denser matrix and would increase durability. However, it was observed, that the VPV decreases as the curing age increases for samples treated in ammonium nitrate solution after 28 days. This is due to the dissolution of the calcium hydroxide. Calcium hydroxide is a hydrated product, which fills the pore system within the cement paste. Removal of the calcium hydroxide from the pore increases voids within the cement paste. As a result, the concrete becomes more permeable. According to previous research, similar test results were reported. The lower void percentage in samples with 10% and 20% POFA is due to the pozzolanic reaction, which produced additional calcium silicate hydrates. Previous author reported the VPV values decreased with an increase in the duration of the curing age **[4]**. However, the VPV values increased when immersed in an ammonium nitrate solution. According to literature review, the VPV value is influenced by the paste/aggregate ratio **[10]**. As the paste/aggregate ratio increases, the void ratio in the hardened concrete decreases. Previous researchers reported that the VPV value was lower for the 56 days samples compared to the 28 days samples **[11]**. In addition, the author also stated the VPV values decreased as the cement replacement percentage with pozzolanic material increased. Based on the test results, it is recommended to replace OPC with 20% of POFA if at room temperature and in an ammonium nitrate conditions.

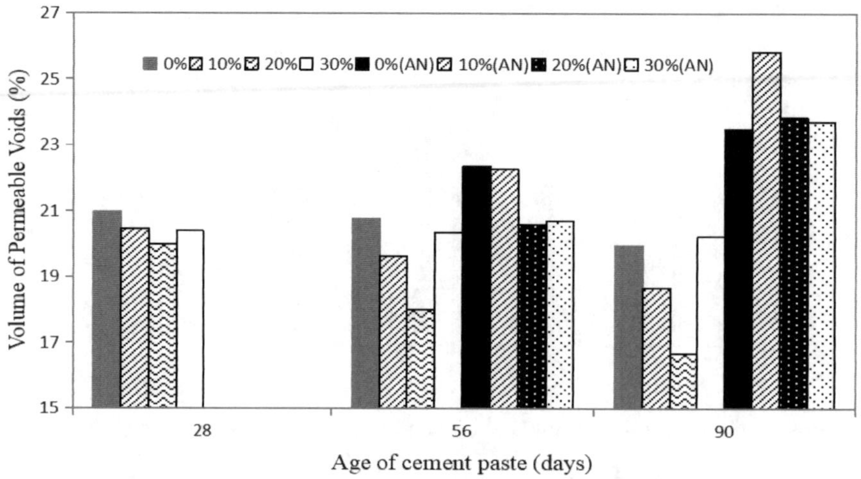

Figure 3. Volume of Permeable Voids (VPV) of the modified cement paste in room temperature and in ammonium nitrate solution

3.3. Thermo-gravimetric Analysis (TGA) of the modified cement paste

Thermo-gravimetric Analysis (TGA) is a widely accepted technique to assess $Ca(OH)_2$ content in hydrated cement paste. For this study, the TGA test was carried out to determine the total percentage of $Ca(OH)_2$ present in the 0%, 10%, 20% and 30% POFA samples at different curing ages and conditions. At 28 days, the $Ca(OH)_2$ content in 0%, 10%, 20% and 30% POFA was 21.07%, 18.92%, 16.82% and 19.37%. The $Ca(OH)_2$ content in 0%, 10%, 20% and 30% at 56 days for sample cured in room temperature was noted to be 23.61%, 21.05%, 17.82% and 16.81%. Consequently, the $Ca(OH)_2$ content in 0%, 10%, 20% and 30% at 56 days for sample treated in ammonium nitrate solution was noted to be 22.97%, 19.38%, 16.82% and 16.25%. The $Ca(OH)_2$ content at 90 days for sample cured in room temperature in 0%, 10%, 20% and 30% POFA was noted to be 27.99%, 22.25%, 18.40%, 15.78% respectively. At 90 days, the $Ca(OH)_2$ content in 0%, 10%, 20% and 30% POFA treated in ammonium nitrate solution was noted to be 23.86%, 19.79%, 18.06% and 15.21% respectively. According to Figure 4, it was observed the Ca(OH)2 content in the 0% POFA samples was the highest compared to samples containing POFA. This is attributed to the fact that there is a high amount of $Ca(OH)_2$ present in the paste due to the absence of the pozzolanic reaction. Based on the test results, it was observed that, the percentage of $Ca(OH)_2$ in 10%, 20% and 30% POFA was lower than sample containing 0%POFA. This is attributed to the fact that the $Ca(OH)_2$ content decreases as the replacement of POFA increases. Hence, the higher the amount of cement replaced with POFA, the higher the consumption of the $Ca(OH)_2$ due to the pozzolanic reaction. As a result, the $Ca(OH)_2$ content in the paste decreases with an increase in the percentage replacement of POFA. Apart from that, it was observed the $Ca(OH)_2$ content for samples in an ammonium nitrate solution were lower compared to samples at room temperature. This is attributed to the decalcification reaction, which caused some of the $Ca(OH)_2$ to reacted with the ammonium nitrate solution and hence its percentage is lower. Therefore, based on the test results, it is recommended to replace OPC with 20% POFA if at room temperature and in ammonium nitrate solution.

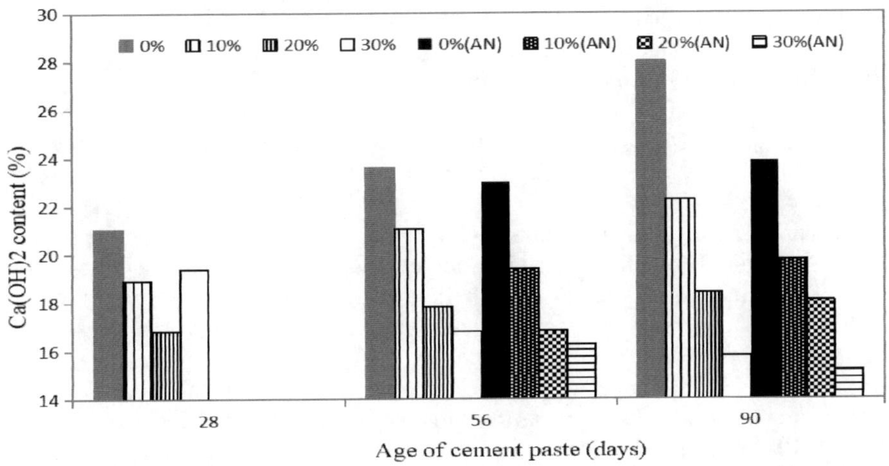

Figure 4. Calcium Hydroxide content of the cement paste in room temperature and in ammonium nitrate solution

4. Conclusion

Based on the test results, the following conclusions were drawn:

a. The compressive strength of the sample cured in room temperature increases with increase in the curing age. However, for sample immersed in ammonium nitrate solution, the compressive strength decreases after 28 days. Based on the test results, it is recommended to replace OPC with 20% POFA for good performance in room temperature and in ammonium nitrate solution.

b. The VPV value for sample cured in room temperature decreases with an increase in the curing time due to the modification of the cement matrix. The VPV of the samples treated in ammonium nitrate solution increases with curing time due to the decalcification reaction. It is suggested to replace OPC with 20% POFA for resistance against ammonium nitrate solution

c. According to the TGA test results, it was observed the calcium hydroxide content decreases as the percentage of POFA replacement increases in the cement paste. However, it was notice that, the calcium hydroxide content for sample immersed in ammonium nitrate solution was much lower than sample cured in room temperature.

5. References

[1] Hallaji M S M, Shekarchi M, Pagar Farhad P, Rahman S & Ghassemzadeh F 2012 Corrosion damage diagnosis of a 44 year old ammonium nitrate pill tower in a petrochemical complex. *Journal of Performance of Construction Facilities*. pp 532- 543.

[2] Oyeleke R B, Yusof M B, Salim M R and Ahmad K 2011. Physico-chemical properties of palm oil fuel ash as composite sorbent in kaolin landfill liner system. *International Journal of Renewable Energy Resources*: pp 1-8.

[3] ASTM C 618 2005. Standard specification for coal fly ash and raw or calcined.

[4] Wong L L, Rahman, M E, Asrah H & Mannan M A 2012 Effects of aggressive ammonium nitrate on durability properties of concrete using sandstone and granite aggregates, *International Journal of Civil, Environmental, Structural, Construction and Architectural Engineering* **Vol: 7, No:1. pp 49-53.**

[5] AS 1012.1(2014). Australian Standard. Methods of testing concrete.

[6] Abdullah K, Hussin M, Zakaria F, Muhammad R & Hamid Z 2006 POFA: A potential partial cement replacement material in aerated concrete. *Proceeding of the 6th Asia Pacific Structural Engineering and Construction Conference* (APSEC 2006), Kuala Lumpur, Malaysia. pp 132-140.

[7] Arafa M & Alqedra, M A 2016 Quantifying the effect of ammonium nitrate attack on the mechanical and physical properties of cement mortar. *Journal of Scientific Research & Reports.* pp 137-145.

[8] Tangchirapat W, Saeting T, Jaturapitakkul C, Kiattikomol, K & Siripanichgorn A 2007 Use of Waste Ash from Palm Oil Industry in Concrete. *Waste Management 27:* pp 81-88.

[9] Sata V, Jaturapitakkul C & Rattanashotinunt C 2010 Compressive strength and heat evolution of concretes containing palm oil fuel ash. *Journal of Material in Civil Engineering.* pp 1033-1038.

[10] Bashar S M, Muhd F N & Yogeswary 2016 Properties of high permeable concrete utilizing pozzolanic materials. *ARPN Journal of Engineering and Applied Sciences.* **Vol. 11, No. 3. ISSN 1819-6608. pp 1457-1466.**

[11] Jitendra B J & Saoji A C 2015 Comparative study of waste glass powder as the partial replacement of cement in concrete production-A Laboratory Investigation". *International Journal on Recent and Innovation Trends in Computing and Communication* **Volume: 3 Issue: 2. ISSN: 2321-8169. pp 63-66.**

Acknowledgement
This research was conducted mainly at Curtin University, through some experiment was also carried out at University Malaysia Sarawak and Universiti Malaysia Pahang. The research is partially funded by Universiti Malaysia Pahang (Grant No. RDU190151).

Numerical Modelling of Scale Effect on Mode of Failure and Strength of Offset Rock Joints

O Mughieda[1], M Alkhedher[2] and K Hazirbaba[3]

[1]Abu Dhabi University, Department of Civil Engineering, Abu Dhabi, UAE.
[2]Abu Dhabi University, Department of Mechanical Engineering, Abu Dhabi, UAE
[3]American University of RAK, Dept. of Civil and Infrast. Engineering, RAK, UAE

Abstract. The studies on the scale effect on the behavior of non-persistent rock joints are limited and the nature of scale dependency is still not well understood. In this study, the scale effect on failure mechanisms and compressive strength of rock blocks tested previously by the authors was investigated numerically. In the experimental study; two different block sizes, having dimensions of (63.5 x 28 x 20.3) cm and (30.5 x 15.24 x 10) cm, were tested. Samples of rock with non-persistent offset joints were subjected to uniaxial loading. The joint inclination angle was maintained at 22.5° in both cases. Also, degree of persistence was kept constant at 0.3 for all tested blocks. However, the offset angle which connects the inner tips of the joints was changed from 30°-90° with an increment of 15°. In the current study, finite element analyses were performed on the arrangements that were studied experimentally. ANSYS 19 Mechanical APDL was utilized to perform static structural analysis. A linear elastic material was assumed for these analyses, mainly because there were no signs of major material damage prior to fracture. Modulus of elasticity of approximately 10510 MPa and a Poisson's ratio of 0.25 were assumed. The boundary conditions applied to the finite element model were zero vertical displacement along the bottom edge, and a uniform distribution load on the top surface, the magnitude of which was approximately the measured coalescence load (in MPa) for the joint geometry analyzed. The stress distribution within the block and near the tips of joints in particular was analyzed in terms of maximum principal stresses, shear stresses and factor of safety resulted from maximum tensile stress theory. The FEM results were compared with the experimental results. The results of FEM emphasized the observed experimentally results regarding the decrease in strength as the size of block sample increased and no effect of block size on mode of failure.

1. Introduction

The strength of rock mass is complex due to the fact that many factors are involved in controlling the strength of rock mass under loading. One factor that needs more studies is the scale effect on strength of rock mass with non-persistent joints. The scale effect on rock mass with persistent rock joints has been studied previously by many researchers [1,2] and they indicated the existent of size effect on the strength of rock mass. While, the behavior of non-persistent rock joints at failure under uniaxial or biaxial loading condition is not well understood due to the interaction between the intact rock segments and the joint segments. Many researchers [3,4] have studied the effect of joints orientation and size on the strength of rock mass. The effect of joint orientation on mode of failure and propagation of crack has been studied also by many researchers [5-8] but still no common agreement on the modes of failure has been reached. The interaction between the intact rock and the joints depends on many parameters such as the geometry of the joint and bridge segments, the slope of the

IOP Conf. Series: Materials Science and Engineering **712** (2020) 012025 doi:10.1088/1757-899X/712/1/012025

joint segment, the stiffness of the rock among others, [9-15]. Generally, the process of failure begins with the initiation of cracks perpendicular to the joint segment which then becomes parallel to the principal loading axis. These cracks are called wing cracks. Wing cracks are followed by another type of cracks called secondary cracks that initiate from the inner joint segment tips and propagate toward each other and meet across the intact rock segment causing the coalescence of the joint segments leading eventually to failure of the rock. Propagation and coalescence of the secondary cracks are very quick compared to that of the wing cracks.

In the current work numerical modelling was used to study the effect of scale on the strength and more of failure of rock mass with offset nonpersistent joints. Numerical modelling is a powerful tool to understand the crack propagation and mode of failure in jointed rock mass and has been used intensively recently for understanding the behavior in rock mass and jointed rocks as well, [16-18]. Scale effect on the behavior of rock mass containing no-persistent rock joints has been acknowledged by many researchers and more studies are needed

2. Methods

2.1. Experimental work

This current work is a continuation to the previous work that has been done by the first author and his co-workers [19]. Mughieda et al. [19] have studied the effect of scale on the strength and mode of failure of offset non-persistent rock joints experimentally. In the experimental study; two different block sizes, having dimensions of (63.5 x 28 x 20.3) cm and (30.5 x 15.24 x 10) cm, were tested. Samples of rock with non-persistent offset joints were subjected to uniaxial loading. The joint inclination angle was maintained at 22.5° in both cases. Also, degree of persistence was kept constant at 0.3 for all tested blocks. However, the offset angle which connects the inner tips of the joints was changed from 30°-90° with an increment of 15°.

The procedure developed by Jamil [10] for preparing open non-persistent joints was used in this research with some modification. The sample preparation mold and set-up is shown in figure 1. A typical sample dimensions are shown in figure 2.

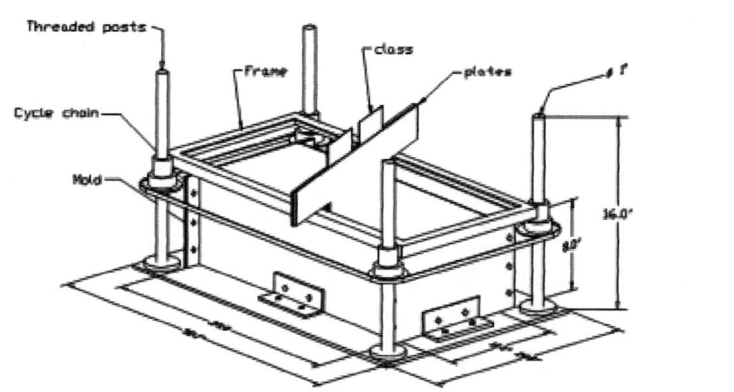

Figure 1. Sample preparation mold and set-up for making open joints [19]

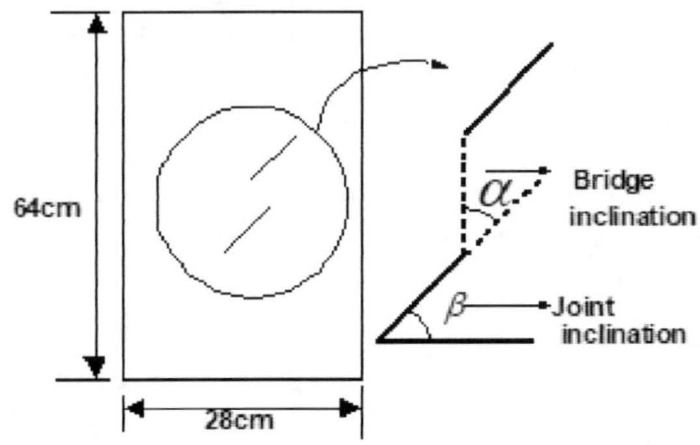

Figure 2. Geometry of the specimens and pre-existing cracks [19]

2.2. Description of finite element model

In order to determine the state of stress for uniaxial compressive loading on Rock blocks, finite element analyses were performed on the different arrangements that were studied experimentally. ANSYS 19 Mechanical APDL is utilized to perform static structural analysis. A linear elastic material was assumed for these analyses, mainly because there were no signs of major material damage prior to fracture. Modulus of elasticity of approximately 10510 MPa and a Poisson's ratio of 0.25 were assumed.

Quadratic eight-nodded quadrilateral plain strain element with two degree of freedom per node, and quadratic three-nodded constant strain triangular element with two degree of freedom per node were used in the present study. Refinement of mesh with order 3 is implemented around the joint surfaces for better results. The geometric model and the mesh used are shown in figure 3.

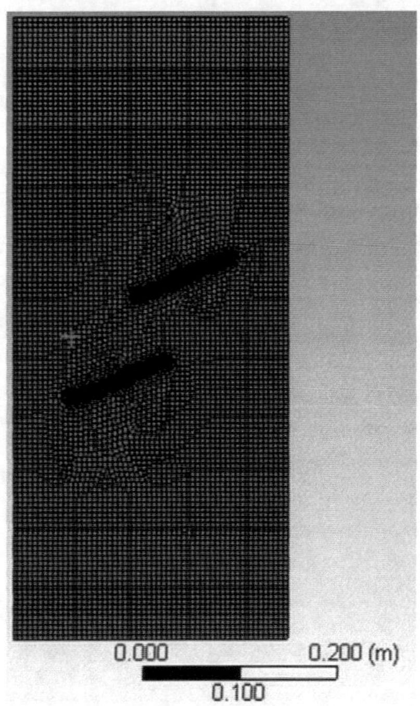

Figure 3. Geometric model and meshing

The boundary conditions applied to the finite element model are zero vertical displacement along the bottom edge, and a uniform distribution load on the top surface, the magnitude of which was approximately the measured coalescence load (in MPa) for the crack geometry analysed. The three-dimensional effect in the experiments is neglected since the model used in the present study was two-dimensional.

The two different block sizes, having dimensions of (63.5 x 28 x 20.3) cm and (30.5 x 15.24 x 10) cm, were modelled at joint inclination angle maintained at 22.5° in both cases. However, the offset angle is modelled for 30°,45°,60° and 90°

3. Results and discussion

3.1. Mode of failure

3.1.1. Slightly offset joints (β= 22.5°, α=30°)

The experimental study showed that failure was due to shear stresses by coalescence of secondary cracks at a point in the rock bridge. Crushed and pulverized materials have been noticed on the failure surface. No scale effect was noticed on mode failure. The failed specimen and sketch of the wing and secondary cracks are shown in figure 4. The shear stress, maximum principal stresses distribution in the tested blocks and the factor of safety based on tensile stress theory are shown in figure 5. The FE analysis also showed that shear and max principal stresses are high and concentrated at the inner tip of the joints causing the propagation of cracks till coalescence and failure for both sizes.

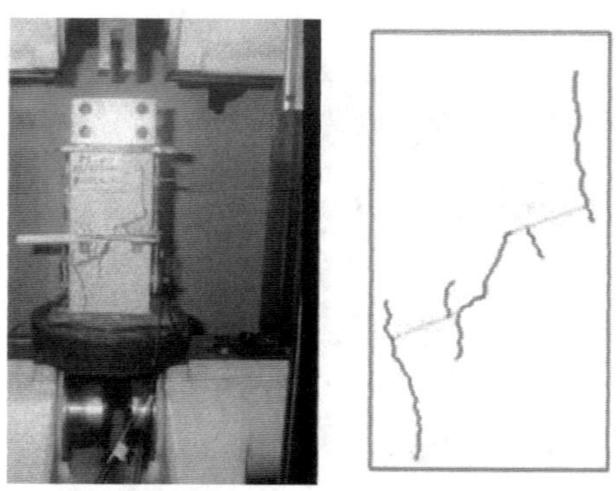

Figure 4. Crack growth in 22.5°-30° flaws [19]

30°-flaw	*Shear Stress*	*Safety Factor maximum tensile stress theory*	*Maximum Principal stress*
Large Block	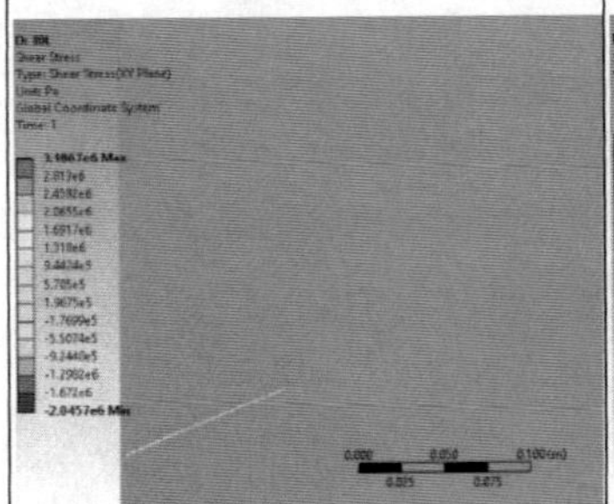	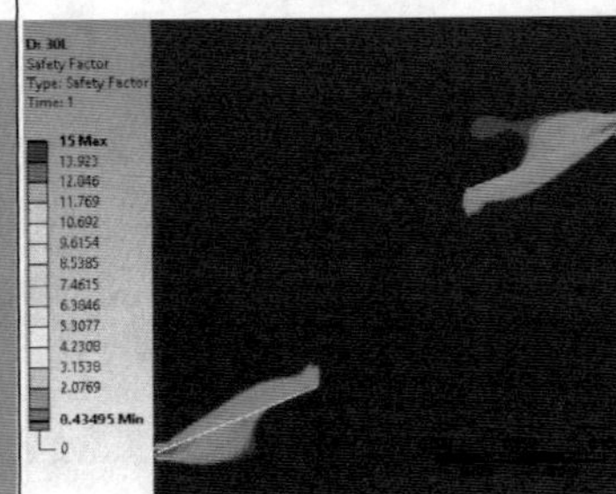	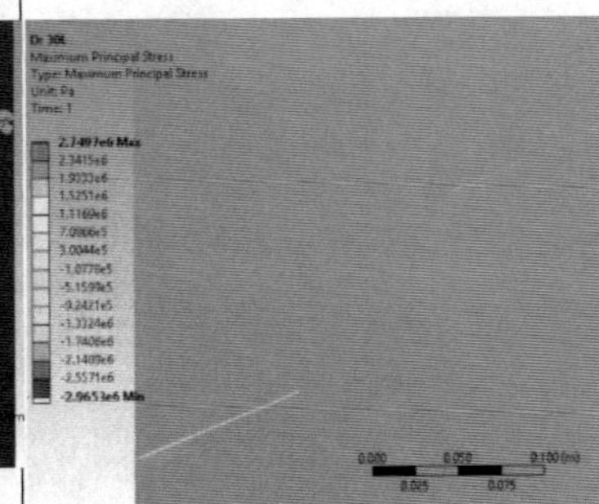
Small - Block	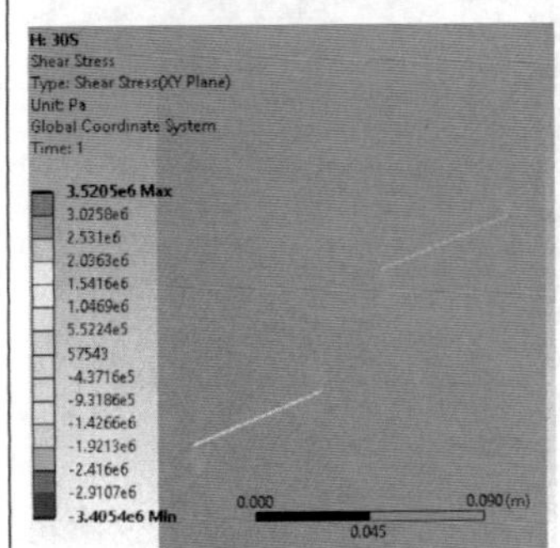	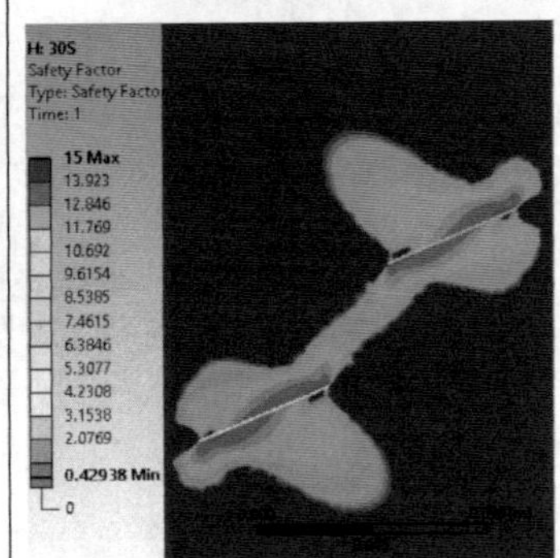	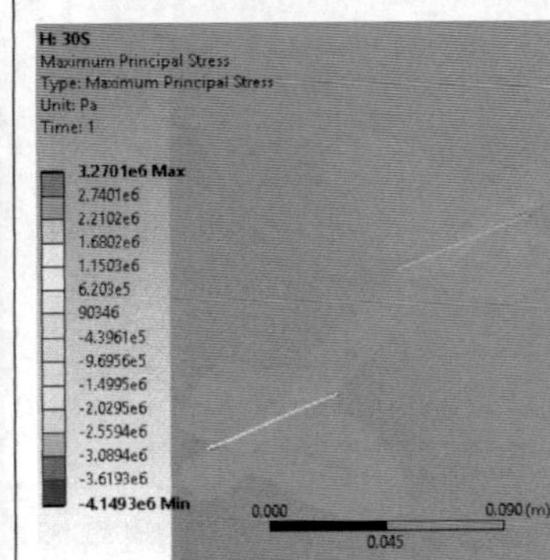

Figure 5. Shear stress, maximum principal stress and factor of safety based on max. tensile stress theory for 22.5°-30° flaws.

3.1.2. Offset joints (β= 22.5°, α=45°)

In these blocks coalescence occurred through the growth of secondary crack matching through the flaws. The FE analysis emphasized the experimental results and showed that both shear stress and max tensile stress are highly concentrated at the tip of the inner joints and were responsible for the failure of the block. The mode of failure in both sizes was identical. The failed specimen and sketch of the wing and secondary cracks are shown in figure 6. Shear stress distribution, max. principal stress and factor of safety based on max tensile tress theory results are shown in figure 7.

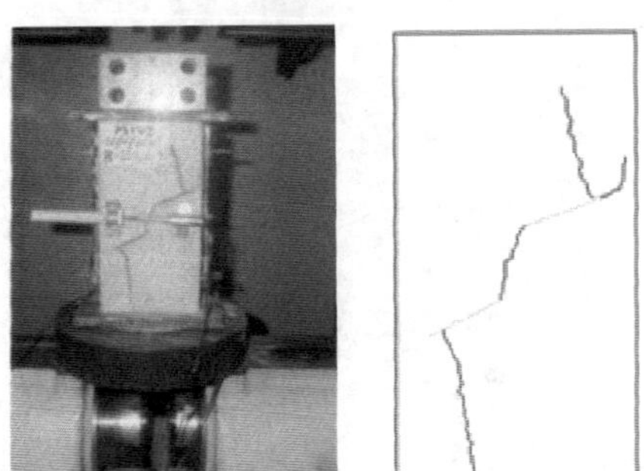

Figure 6. Crack growth in 22.5°-45° flaws [19]

IOP Conf. Series: Materials Science and Engineering **712** (2020) 012025 doi:10.1088/1757-899X/712/1/012025

45°-flaw	Shear Stress	Safety Factor-maximum tensile stress theory	Maximum Principal stress
Large-Block	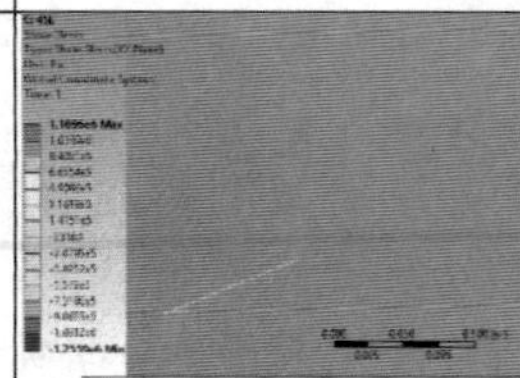	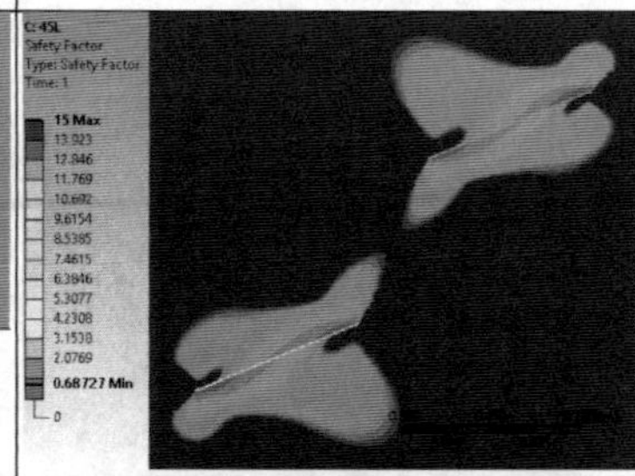	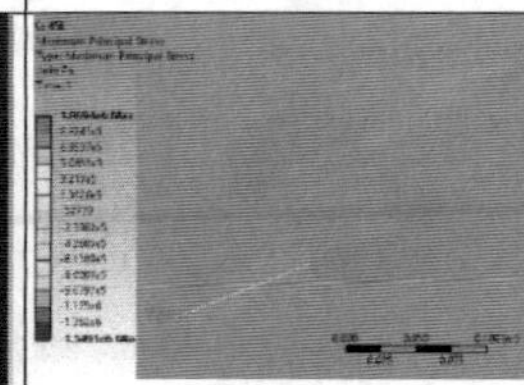
Small-Block	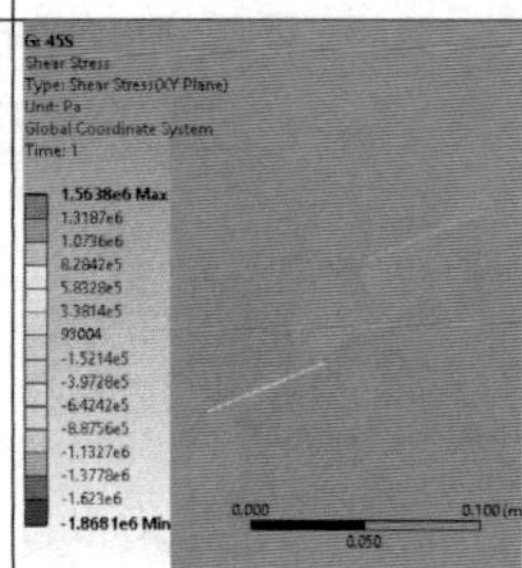	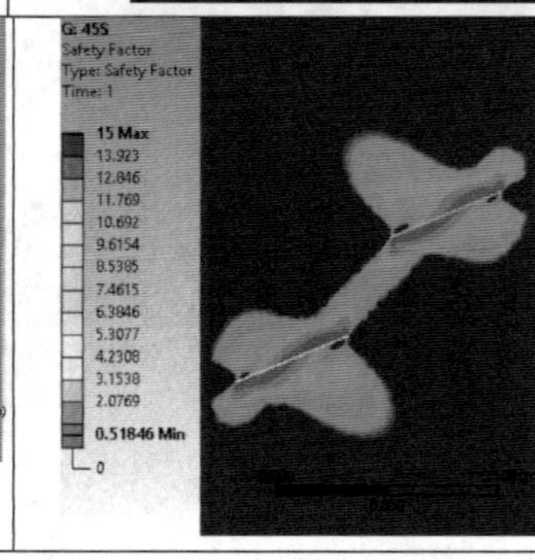	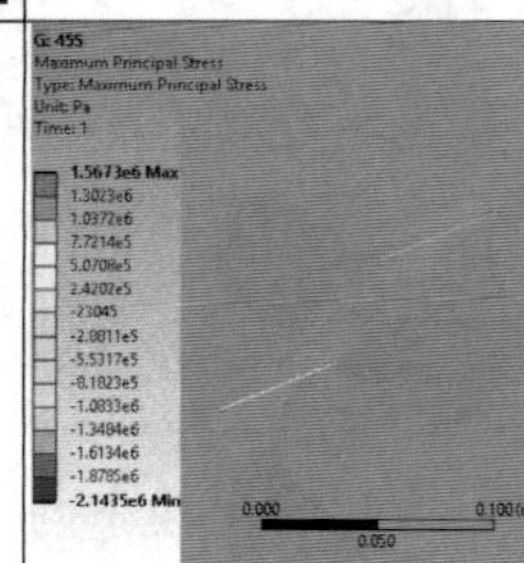

Figure 7. Shear stress, maximum principal stress and factor of safety based on max. tensile stress theory for 22.5°-45° flaws.

3.1.3. Offset joints (β= 22.5°, α=60°)

The FE analysis showed that shear and tensile stresses are both responsible for the failure of the blocks through the propagation of the secondary cracks along the rock bridge. This result explained the mode of failure obtained from the experimental study. In both block sizes the failure initiated by wing cracks propagated from the external flaw tips only, extended to a certain distance from the tips of the joints and stopped. Coalescence occurred through the growth of secondary crack matching through the flaws. The failed specimen and sketch of the wing and secondary cracks are shown in figure 8. Shear stress, maximum principal stress and factor of safety based on max. tensile stress theory distributions are shown in figure 9.

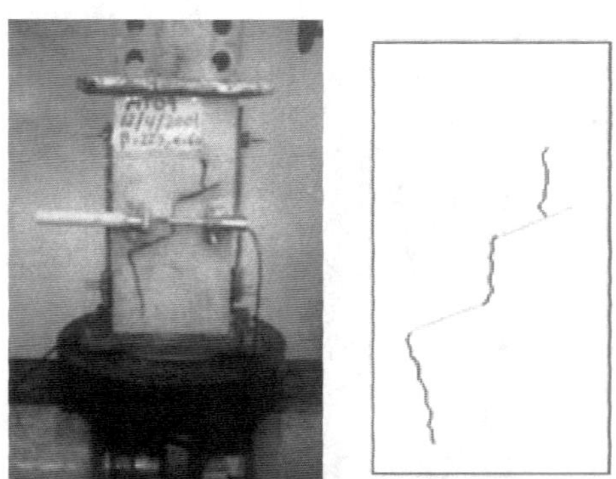

Figure 8. Crack growth in 22.5°-60° flaws [19]

IOP Conf. Series: Materials Science and Engineering **712** (2020) 012025 doi:10.1088/1757-899X/712/1/012025

flaw	Shear Stress	theory	stress
Large-Block	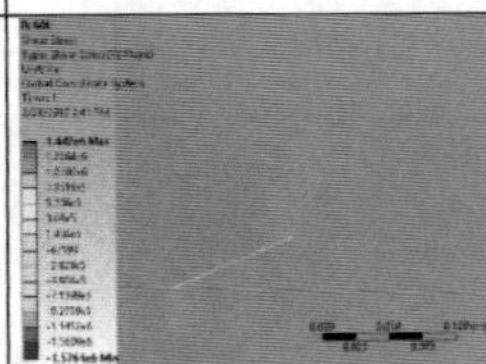	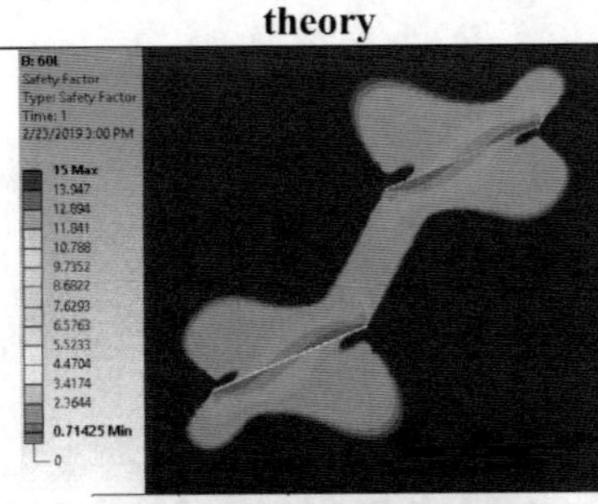	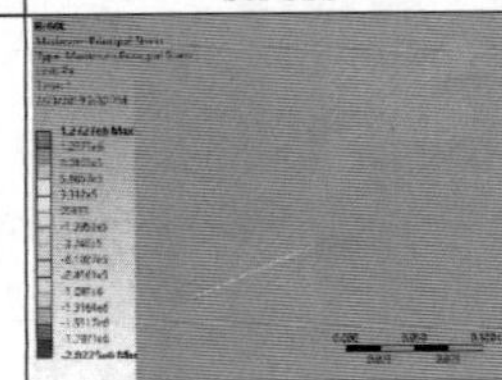
Small-Block	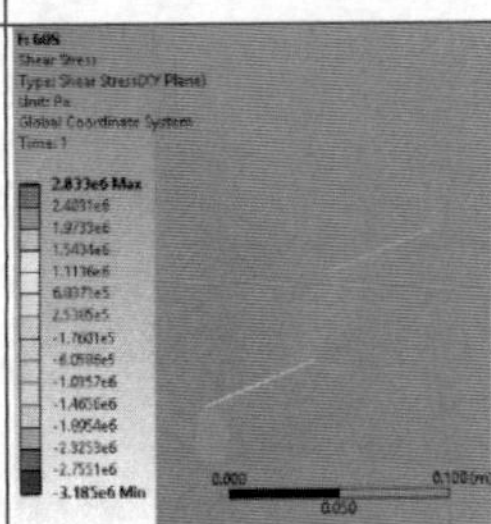	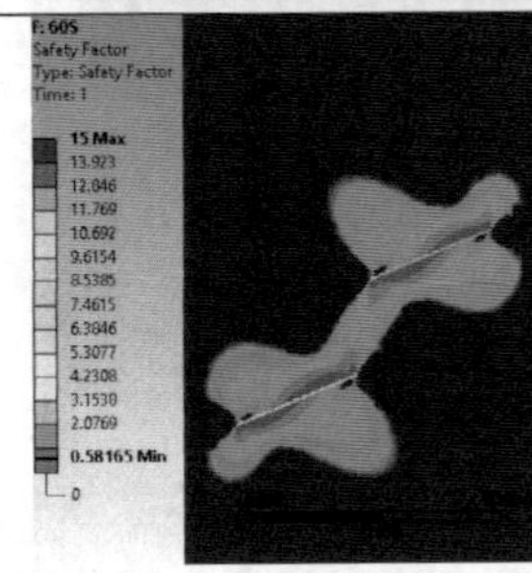	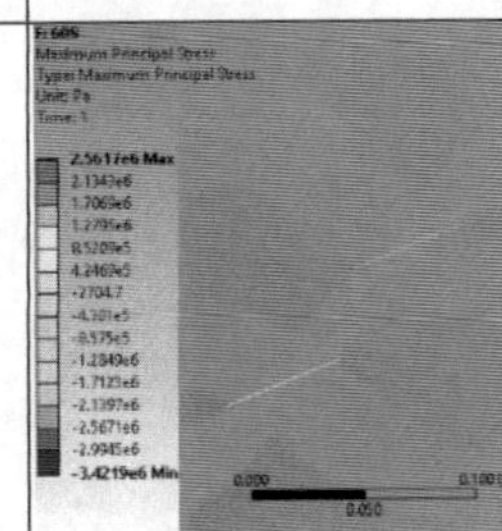

Figure 9. Shear stress, maximum principal stress and factor of safety based on max. tensile stress theory for 22.5°-60° flaws.

3.1.4. Offset joints (β= 22.5°, α=90°)

In these blocks failure occurred due to coalescence of wing cracks which are tension cracks. The FE analysis showed that that the area in between the inner tips of the joints were subjected to high value of tensile stresses which were responsible for the failure of the sample. The failed specimen and sketch of the wing and secondary cracks are shown in figure 10. Shear stress, maximum principal stress and factor of safety based on max. tensile stress theory are shown in figure 11.

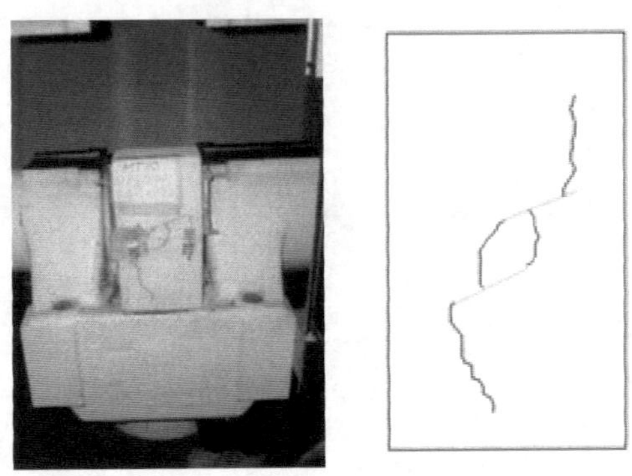

Figure 10. Crack growth in 22.5°-90° flaws [19]

IOP Conf. Series: Materials Science and Engineering **712** (2020) 012025 doi:10.1088/1757-899X/712/1/012025

90°-flaw	***Shear Stress***	***Safety Factor-maximum tensile stress theory***	***Maximum Principal stress***
Large-block	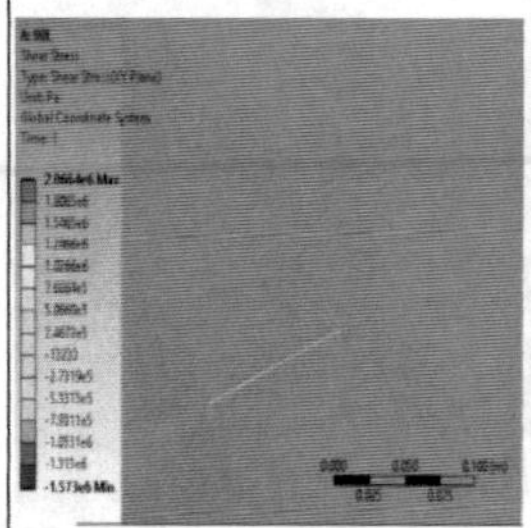	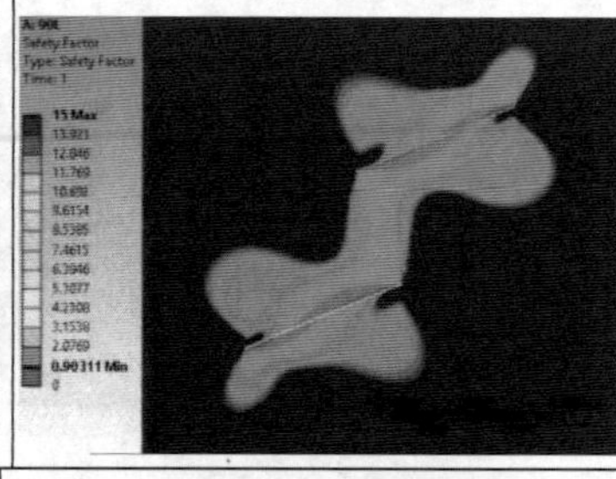	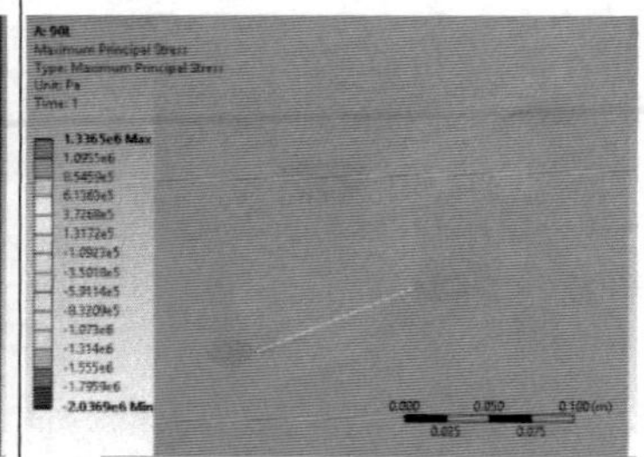
Small-Block	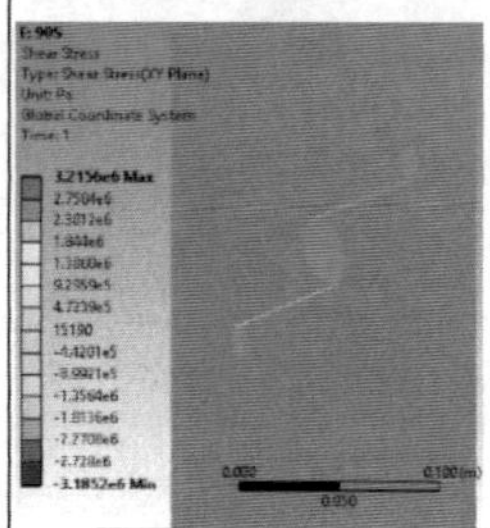	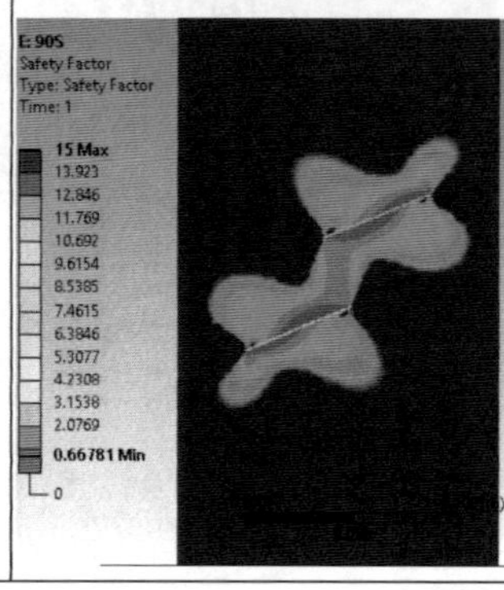	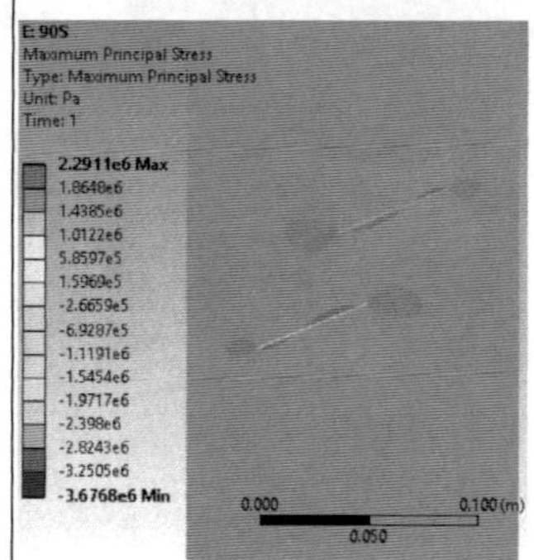

Figure 11. Shear stress, maximum principal stress and factor of safety based on max. tensile stress theory for 22.5°-90° flaws

The results of mode of failure of all tested samples are summarized in table 1.

Table 1. Mode of block failure with different bridge inclinations

	30°	**45°**	**60°**	**90°**
Large sample	Tensile + Shear failure	Tensile + Shear failure	Tensile failure	Tensile failure
Small sample	Tensile + Shear failure	Tensile + Shear failure	Tensile failure	Tensile failure

4. Conclusion

The FE results were compared with the experimental results. The results of FE emphasized the observed experimentally results as no effect of block size on mode of failure. Also, the FE analysis explained the experimentally observed modes of failures. Mode of failure of rock bridge varies from shear to tensile failure as the bridge inclination increases. For slightly offset joint ($\alpha = 30^{\circ}$ or 45°) the failure was due to combined shear and tension failure while it was purely tension for overlapped joints ($\alpha = 90^{\circ}$).

5. References

[1] Bandis C, Lumsden A and Barton N 1981 Experimental studies of scale effects on the shear behavior of rock joints *Int. J. of Rock Mech. Min. Sci. & Geomech. Abstr.* **18** 1-21
[2] Blejwas T and Hansen F 1990 Scale in the shear behavior of joints in welded tuff *Proc. Loen Conf. Rock Joints* Norway
[3] Yang Q, Liu R and Jing W 2013 Experimental investigation on fracture coalescence behavior of res sandstone containing two unparallel fissures under uniaxial compression *Int. J. Rock Mech. Min. Sci.* **63** 82-92
[4] Lu Y, Wang L and Elsworth D 2015 Uniaxial strength and failure in sandstone containing a pre-existing 3-D surface flaw *Int. J. Frcat.* **194(1)** 59-79.
[5] Mughieda O 1997 *Failure mechanisms and strength of non-persistent rock joints* Ph.D. thesis, University of Illinois at Urbana-Champaign, IL. U.S.A.
[6] Bobet A and Einstein H 1998 Fracture Coalescence in rock-type materials under uniaxial and biaxial compression *Int. J. Rock Mech Min Sci.* **35** 863-88
[7] Mughieda O and Alzo'ubi A 2004 Fracture mechanisms of offset rock joints- A laboratory investigations *Geotechnical and Geological Engineering Journal* **22** 545-562.
[8] Zhao Y, Zhang L, Wang W, Pu C, Wan W and Tang J 2016 Cracking and stress-strain behavior of rock-like material containing two flaws under uniaxial compression *Rock Mech. Rock Eng.* **49(7)** 2665-2687.
[9] Reyes O and Einstein H 1991 *Stochastic and centrifuge modeling of jointed rock Part I-fracturing of jointed rock* Final report submitted to the Air Force Office of Scientific Research and Air Force Engineering Services Center
[10] Jamil M 1992 *Strength of non-persistent rock joints* Ph.D. thesis, University of Illinois at Urbana-Champaign, IL, U.S.A.
[11] Boatang S 1993 *Mechanics of fractures and inverting bridges in hard rocks.* Doctoral Thesis, Royal Institute of Technology
[12] Wong C and Chau T 1998 Crack coalescence in a rock-like material containing two cracks *International Journal of Rock Mechanics and Mining Sciences* **35(2)** 147-164.
[13] Park H and Bobet A 2009 Crack coalescence in specimens with open and closed flaws: A comparison *International Journal of Rock Mechanics & Mining Sciences* **46(2)** 819-829.
[14] Xiao-Ping Z and Wong L N 2012 Cracking processes in rock-like material containing a single flaw under uniaxial compression: a numerical study based on parallel bonded-particle model approach *Rock Mechanics and Rock Engineering* **45(5)** 711-737.
[15] Jin J P, Cao P and Liu J 2015 The Influence of Bridge Length on Failure Patterns of Specimens Containing Two Coplanar Cracks under Uniaxial Compression Load *Applied Mechanics and Materials* **711** pp 410-413
[16] Aliha M R M, Pakzad R and Ayatollahi M R 2013 Numerical analysis of a cracked straight-through flattened Brazilain disk specimen under mixed-mode loading *J. of Eng. Mech.,* **140(1)** 219-224
[17] Zhang P and Wong Y 2012 Cracking processes in rock like material containing a single flaw under uniaxial compression: a numerical study based on parallel bonded-particle model approach *Rock Mech. Rock Eng.* **45(5)** 711-737
[18] Cao R H, Cao P, Fan X and Lin H 2016 An experimental and numerical study on mechanical behavior of ubiquitous-join brittle rock like specimens under uniaxial compression *Rock Mech. Rock Eng.* **49 (11)** 4139-4338
[19] Mughieda O, Hazirbaba K and Mohamed O 2017 Scale effect on mode of failure and strength of offset rock joints *MATEC Web of Conferences* **103** 07008

IOP Conf. Series: Materials Science and Engineering **712** (2020) 012026 doi:10.1088/1757-899X/712/1/012026

Mechanical Performance of Stone Mastic Asphalt Incorporating Steel Fiber

N E Jasni[1], K A Masri[1], P J Ramadhansyah[1], A K Arshad[2], E Shaffie[2], J Ahmad[3] and A H Norhidayah[4]

[1]Faculty of Civil Engineering Technology, Universiti Malaysia Pahang, 26300 Gambang, Pahang, Malaysia
[2]Institute for Infrastructure Engineering and Sustainable Management, Universiti Teknologi MARA, 40450, Shah Alam, Selangor, Malaysia.
[3]Faculty of Civil Engineering, Universiti Teknologi MARA, 40450, Shah Alam, Selangor, Malaysia
[4]Faculty of Engineering, School of Civil Engineering, Universiti Teknologi Malaysia, 81310 Johor Bahru, Malaysia

Abstract. Stone Mastic Asphalt (SMA) is a gap-graded hot mix asphalt that contains a large percentage of course aggregate and bitumen filler mastic. SMA is suffered from severe binder drain down due to the gap graded aggregates mixtures. Large difference between the sizes of aggregate tend to reduce tensile strength of the asphalt mixture. To overcome this, a type of fiber which is steel fiber is utilized to improve the tensile strength of SMA. Thus, the aim of this study is to increase the strength of SMA by utilizing steel fiber. To prepare SMA mixtures, specimens were compacted by applying 50 blows on each face using Marshall Compactor. Then, the modified specimens were tested to investigate the performance in terms of Los Angeles Abrasion Test, Marshall Stability Test, Resilient Modulus Test, and Dynamic Creep. From the results, it indicates that the addition of 0.3% fiber leads to better stability and stiffness while 0.5% fiber for resilient and enhanced modulus dynamic creep at 25°C and 0.4% fiber at 40°C. Thus, it can be concluded that the addition of steel fiber in the mixture significantly enhance the overall performance of SMA.

1. Introduction

Stone mastic asphalt is a gap graded hot mix asphalt with more than 65% of course aggregates, and according to [1], optimal asphalt binder for designing aggregate gradation of SMA mixtures are at 5.5 percent. Stone mastic asphalts the mixing of the aggregates with polymer-modified binder. The polymer modified binder in stone mastic asphalt mixing is in between the range of 5.5 to 7.5 percent. Stone mastic asphalt is good in skid resistance since it offers an improvements surface in texture depth where it is in between 0.7 to 1.0mm. The course aggregate composition is in contact from point to point, forming a skeleton structure tends to provide great internal friction between the wearing course and the tires of the vehicles. In general, the asphalt binder mixed with the aggregate at the desired temperature to fully coated aggregate and binder and prepare for suitable for paving. This reduced viscosity allows the aggregate to be fully coated at a lower temperature than what is traditionally required in HMA production Kilas et al. [2]. However, when the temperature is lower, the asphalt tends to lose its adhesive bond with the aggregate. These will contribute to the combined weakening of the mastic and weakening of the aggregate-mastic bond. According to Sengul et al. [3], stone mastic asphalt has its advantages as it has shown to be very durable surfacing, exhibit high resistance to rutting due to heavy axle loads since its stable aggregate skeleton structure and generates less tire noise inexperienced countries that have adopted stone mastic asphalt. Furthermore, it is reported that stone mastic asphalt has its ability to meet the United Kingdom surface texture requirements with a great choice of aggregate size. Unfortunately, SMA forming a stone skeleton structure by containing high course aggregate content that leads to having voids in the structural matrix where the voids are filled with high viscosity bituminous mastic with any stabilizing agent like bitumen, crushed sand or filler. The stabilizing agent is usually made up of fiber which prevents binder drainage in stone mastic asphalt mixing. Fibers reinforced bituminous mixture is now not rare in the industry. It is now widely used especially in SMA. A study by Arabani and Shabani [4] utilized the ceramic fiber to improve the properties of asphalt binder. In addition, Tanzadeh et al. [5] used glass fiber to improve the performance of open-graded friction course. The binder drain down happened when there is a high content of binder content, and the mix is at high temperature [6]. To prevent the binder drain down from getting worst, the addition of mineral fibers, steel fiber are well known to enhance the performance of the stone mastic asphalt. The production of steel fiber reinforced bituminous mixtures might be very costly as compared to other modified fibers, but at the end of the day, steel fiber may reduce and save the maintenance cost. This research study is done to identify the characteristics and the performance of stone mastic asphalt incorporating steel fiber. According to Bindu [7], the addition of fiber including steel fiber may play the significant roles in volumetric and mechanical properties as well as the binder drain down the problem of stone mastic asphalt mixture. This research might provide ways of optimizing fiber performance in asphalt pavement. It can be concluded that the addition of steel fiber in the mixes may enhance the performance of Stone Mastic Asphalt Mixtures.

2. Methods

2.1. Materials

This research was carried out by using 60/70 penetration-grade asphalt and Steel Fibre, as shown in Figure 1 as the additive of hot asphalt mix additive material. The amount of Steel Fibre contents was used in this study were 0%, 0.2%, 0.3%, 0.4%, 0.5%, and 0.6% by weight. The effects of Steel Fibre that mixed will be evaluated. It is typically mixed directly with the asphalt mixture at temperatures of 160°C–180°C. The crush aggregate granites have been dried and sieved into a selected size range with nominal size 20 mm of aggregates according to to a standard specification (SMA 20). Portland cement as the filler has been used.

Figure 1. Steel Fiber

2.2. Sample Preparation

The bituminous mixture was prepared by using SMA graded aggregates with 60/70 penetration grade bitumen with the mineral fiber used, Steel Fibre. The crush aggregate granites have been dried and sieved into a selected size range with nominal size 20 mm of aggregates. The dry blending method has been used in this research which the fiber was added to the mixture before the binder the optimum binder content for this researched was 6.2% by weight of the mix, where this percentage of optimum binder content comes from the past research. The fiber length in the mixture was preserved as a constant parameter that equal to mm with a maximum length of 6mm. All specimen was prepared

IOP Conf. Series: Materials Science and Engineering **712** (2020) 012026 doi:10.1088/1757-899X/712/1/012026

using Marshall Compactor machine to have the identity samples. The number of compaction for each of the samples are 50 blows at the top and the bottom side of the specimen. The temperature used for all specimens were in the range of 160°C to 180°C.

2.3. Performance Testings
Four tests are conducted to evaluate the performance of steel fiber-modified SMA which are Marshall Stability, Cantabro Loss, Resilient Modulus and Dynamic Creep. Marshall Stability was performed to evaluate the tensile strength capability of modified SMA [8]. The specimens were prepared using the optimum binder content (OBC) from Arshad et al. [9]. For Cantabro Loss Test, the test has been done to indicate aggregate toughness and abrasion characteristics to provide high-quality hot mastic asphalt. The specimens were kept at a temperature of 25°C for six hours before testing. The specimens were weighed after it had been kept for the specified time and placed into the Los Angeles machine without the steel balls. Then, the drum was switched on at a velocity between 188 and 208 rad/s and subjected to 300 revolutions without steel ball [10]. Resilient Modulus and dynamic creep were performed to evaluate the stiffness characteristics of modified SMA [9].

3. Results and discussion

3.1. Marshall Stability
Figure 2 presents the stability and Voids in Mix for different Steel fiber contents, respectively. Figure 2 shows that there was a slightly increase in stability from 0% to 0.2% fiber content. This is due to the small amount of steel fiber within the mix, which affect the foal of SMA by having contact points between the aggregates, therefore, lead to a good value of stability. The addition of steel fiber in figure 3 shows that the increase of steel fiber content in the mix results in a decrease in the voids in mixes. Mixes with higher fiber content experiencing higher compact ability and therefore, the air voids might be decreased.

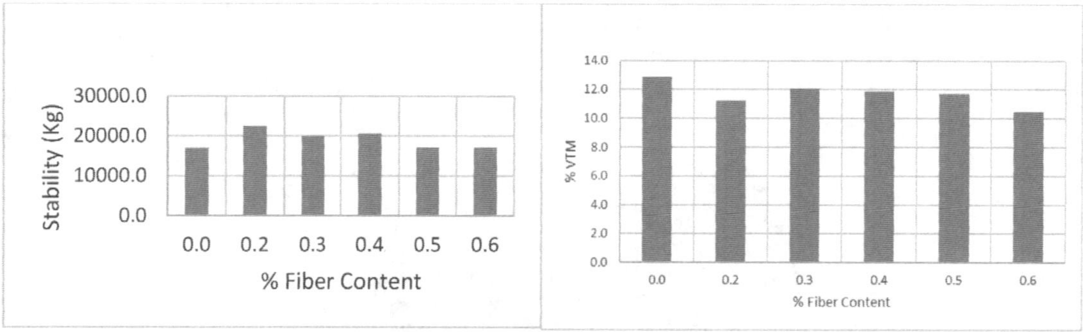

Figure 2. Stability and VTM

3.2. Cantabro Loss
Figure 3 shows that the addition of 0.4% of steel fiber reduced Los Angeles Abrasion loss after 300 turns. This proved that the addition of steel fiber had improved the cohesiveness and the abrasion of Stone Mastic Asphalt (SMA) mixture, making it more durable to accept higher forces due to traffic loads.

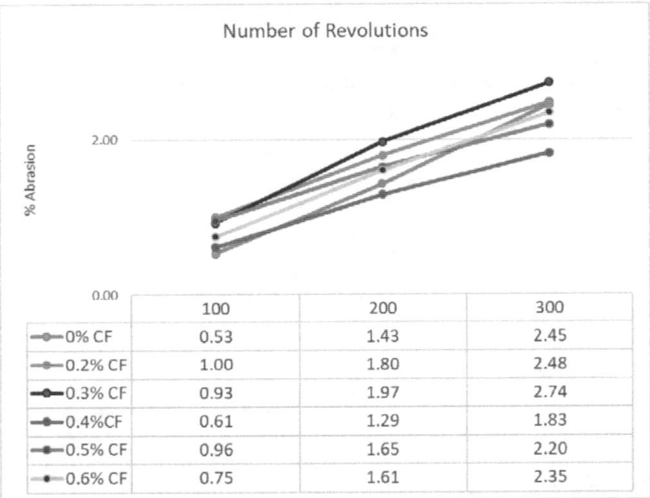

	100	200	300
0% CF	0.53	1.43	2.45
0.2% CF	1.00	1.80	2.48
0.3% CF	0.93	1.97	2.74
0.4%CF	0.61	1.29	1.83
0.5% CF	0.96	1.65	2.20
0.6% CF	0.75	1.61	2.35

Figure 3. Cantabro Loss Value

3.3. Resilient Modulus

The stiffness properties of the reinforced SMA mixes are at 25°C and 40°C are shown in Figure 4. The results show the trends where the specimens with fiber contents had slightly higher stiffness compared to the specimen without fiber content added. The result displayed that as the fiber content increases until the optimum fiber content reached, and then the result decreases back. The optimum value of fiber content for resilient modulus at the temperature of 25°C was at 0.5% while 0.4% at 40°C. Mixture with 0.5% fiber content exhibited higher stiffness modulus compared to other mixes at 25°C while 0.4% for 40°C. The increase in resilient modulus might be due to the higher modulus of elasticity and low ability of extension of the fiber and randomly oriented in a different direction. The result is consistent with a study by Arshad et al. [11], where the resilient modulus of SMA was increased due to modification process using alternative material. The declination in this test that was beyond certain value is most probably due to high inclusion of fiber thus higher surface area to be coated by the binder and therefore the specimens may be facing less stiff mix.

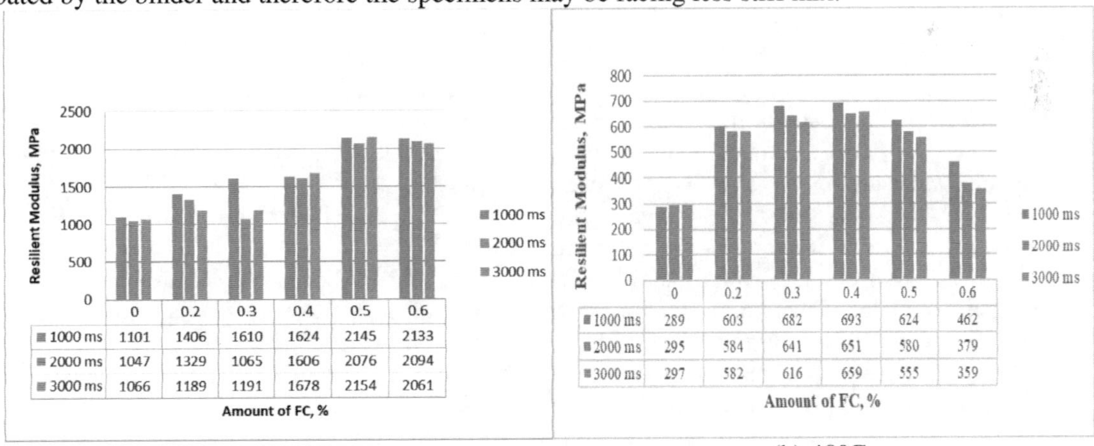

	0	0.2	0.3	0.4	0.5	0.6
1000 ms	1101	1406	1610	1624	2145	2133
2000 ms	1047	1329	1065	1606	2076	2094
3000 ms	1066	1189	1191	1678	2154	2061

	0	0.2	0.3	0.4	0.5	0.6
1000 ms	289	603	682	693	624	462
2000 ms	295	584	641	651	580	379
3000 ms	297	582	616	659	555	359

(a) 25°C (b) 40°C

Figure 4. Resilient Modulus

3.4. Dynamic Creep

Permanent strain deformation value of the reinforced SMA mix with steel fibers are illustrated in Figure 5 for both 25°C and 40°C. The results show the trends where the specimens with fiber contents had decreased the strain value. The result displayed that as the fiber content increases until the optimum fiber content reached, and then the result decreases back. The optimum value of fiber content for dynamic creep at both temperatures of at 0.4% with strain value of 1023.8 and 5440.4, respectively. Mixture with 0.5% fiber content exhibited higher permanent strain value compared to other mixes at 25°C while 0.4% for 40°C. The asphalt mixture is less prone to permanent deformation if the value for a permanent strain from the test is lower.

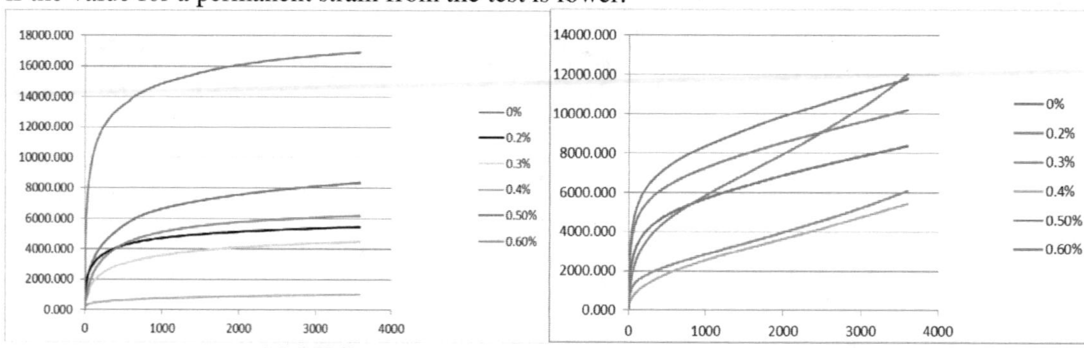

(a) 25°C (b) 40°C

Figure 5. Dynamic Creep

4. Conclusion

Based on the research work described:

- The addition of Steel Fibre significantly reducing the binder drain down a problem that might be caused by the voids of the aggregates structure.
- Addition of Steel fiber in the mixture also increases the stability, the resilient modulus, and the dynamic creep.
- The research objectives also have been achieved, which the steel fiber was proved in enhancing the performance of the stone mastic asphalt.
- Based on the mechanical performance tests, the optimum fiber content is 0.4%.

5. References

[1] Arshad A K, Masri KA, Ahmad J, Samsudin M S 2017 Dynamic Modulus of Nanosilica Modified Porous Asphalt, *IOP Conference Series: Materials Science and Engineering*, **271**, 012008.

[2] Kilas M, Vaitkus A, Paliukaite M 2010. Warm Mix Asphalts Research, Analysis, and Evaluation. *The 10th International Conference on Modern Building Materials, Structures and Techniques*. Vilnius Gediminas Technical University.

[3] Sengul C E, Oruc S, Iskender E, Aksoy A 2013 Evaluation of SBS Modified Stone Mastic Asphalt Pavement Performance. *Construction and Building Materials*, **41**, pp. 777-783.

[4] Arabani M and Shabani A 2019 Evaluation of The Ceramic Fiber Modified Asphalt Binder. *Construction and Building Materials*, **205,** pp. 377-386.

[5] Tanzadeh R, Tanzadeh J, Honarmand M, Tahami S A 2019 Experimental Study on The Effect of Basalt and Glass Fibers on Behavior of Open-Graded Friction Course Asphalt Modified with Nano-Silica. *Construction and Building Materials*, **212**, pp. 467-475.

[6] Masri K A, Arshad A K, Samsudin M S 2016 Mechanical Properties of Porous Asphalt with Nanosilica Modified Binder. *Jurnal Teknologi*, **78: 7-2**, pp. 139-146.

[7] Bindu C S 2012 Influence of Additives on the Characteristics of Stone Matrix Asphalt, *PhD Thesis*, Cochin University of Science and Technology.

[8] Masri KA, Awang H, Jaya R P, Ali M I, Ramli NI, Arshad A K 2019 Moisture Susceptibility of Porous Asphalt Mixture with Nano Silica Modified Asphalt Binder. *IOP Conference Series: Earth and Environmental Science*. **244**, 012028.

[9] Arshad A K, Masri K A, Ahmad J, Samsudin M S 2017 Investigation on Moisture Susceptibility and Rutting Resistance of Asphalt Mixtures incorporating Nanosilica Modified Binder. *Pertanika Journal of Science and Technology*, **25**, pp. 19-30.

[10] Arshad A K, Mansor S, Shaffie E, Hashim W 2016 Performance of Stone Mastic Asphalt Mix using Selected Fibers. *Jurnal Teknologi*, **78(7-2)**, pp. 99-103.
[11] Arshad A K, Shaffie E, Hashim W, Ismail F, Masri K A 2019 Evaluation of Nanosilica Modified Stone Mastic Asphalt. *International Journal of Civil Engineering and Technology*. **10**, pp. 1508-1516.

Acknowledgement

The authors would to acknowledge Universiti Malaysia Pahang for funding this research under the grant RDU 1803157.

IOP Conf. Series: Materials Science and Engineering **712** (2020) 012027 doi:10.1088/1757-899X/712/1/012027

Stability and Resilient Modulus of Porous Asphalt Incorporating Steel Fiber

N A M Radzi[1], K A Masri[1], P J Ramadhansyah[1], N E Jasni[1], A K Arshad[2], J Ahmad[3], N Mashros[4] and H Yaacob[4]

[1]Faculty of Civil Engineering Technology, Universiti Malaysia Pahang, 26300 Gambang, Pahang, Malaysia
[2]Institute for Infrastructure Engineering and Sustainable Management, Universiti Teknologi MARA, 40450, Shah Alam, Selangor, Malaysia.
[3]Faculty of Civil Engineering, Universiti Teknologi MARA, 40450, Shah Alam, Selangor, Malaysia.
[4]Faculty of Engineering, School of Civil Engineering, Universiti Teknologi Malaysia, 81310 Johor Bahru, Malaysia

Abstract. Porous Asphalt (PA) is known as a highly permeable asphalt surface that design to be permeable pavements for stormwater control and reduce the stormwater runoff. However, the structure is subjected to damage from cracking, rutting, stripping, and rapid aging under the effects of repeated vehicle loading, hot climates, and heavy rainfall. To overcome this, PA needs to modify using high tensile strength like steel fiber. Thus, this study aims to evaluate the performance of PA with the addition of steel fiber and overcome the issue that is related to PA. A mixture contains varying percentages of steel fiber were assessed to check which samples gives the best performance as per the requirement by using laboratory tests which are LA Abrasion, Resilient Modulus, Marshall Stability, and Density. The results show that the additions of 0.6% steel fiber give the lowest value of abrasion, while 0.5% fiber content contributes the highest value of Resilient Modulus and Marshall Stability respectively. PA mixtures modified with steel fiber produce the performance enhancement of PA as a road surfacing material. It is concluded that the asphalt mixtures containing steel fibers could increase the stability and strength of the mix.

1. Introduction

Porous asphalt pavements are known as highly permeable asphalt surface that also called as open-graded asphalt. The primary purpose of the design is to be water permeable pavements for stormwater control and reduce the storm water runoff. The benefits of open-graded asphalt are improving runoff quality, sustainability, long life through proper maintenance, and less costly. It also gives benefits in reduction of spray-on higher speed roads, reduces the noise of the tires, and prevents the pollution by eliminating surface runoff. However, the open structure of porous asphalt allows quick drainage of water from the road surface during wet weather. However, an inefficient drainage system can result in hydroplaning, which is dangerous to fast-moving traffic. This can reduce contact between tire and pavement, leading to the loss control of braking and steering during driving [1]. The problems caused by the high air void content consist of increased material deterioration, for example, raveling effects [2] and increased physical aging effects of bitumen due to water and air infiltration (for example causing moisture damages, stripping effects or embrittlement of the aggregate). Due to these factors, porous asphalt has a shorter structural service life than conventional asphalt pavements. In Malaysia, we are facing or dealing with high temperature and heavy rainfall throughout the day. An asphalts facing a lot of deficiency like rutting and stripping. Besides that, heavy traffic loads tend to affect the performance of the asphalt mixture in term of its resilient modulus. A common method to overcome this distressing problem is by modifying the asphalt by the properties. Hence, steel fiber tends to provide improvement of properties for porous asphalt pavements. Steel fiber can be defined as discrete, short length of steel having ration of its length to diameter in the range of 20 to 100 with any of that and that are sufficiently small to be easily and randomly dispersed in fresh Marshall mix using conventional mixing procedure throughout the porous asphalt [3]. Steel fibers are used to enhance their tensile strength and to increase the pavement toughness and help to control cracking. Normal asphalt pavement is brittle with low tensile strength and strain capacity. Fibers are generally utilized in the pavement or mix to manage the problem. The fibers are bonded to the material and allow to withstand considerable stresses. As a result, steel fibers additions can be used in the binder course of flexible pavements because of its positive stability impact. In terms of efficiency, fiber-asphalt mixture shows a slight increase in the optimum asphalt binder content compared with the pure asphalt mixture. In this way, adding fibers to asphalt is very similar to the addition of very fine aggregates. Thus, fibers can stabilize asphalt to prevent leakage. This is due to the adsorption of asphalt on fibers [4,5]. The reported results indicate that fiber–asphalt mixtures have good moisture resistance, creep compliance, and rutting resistance, low-temperature anti-cracking properties, and durability [6]. Thus, the objectives of this study are to evaluate the performance of PA incorporated with steel fiber. The addition of fiber into the asphalt mixture may improve certain properties of PA, such as reducing drain down, enhancing the moisture sensitivity and compressive strength in PA. Therefore, this research shows how optimum fiber content perform in asphalt pavement.

2. Methods

2.1. Materials

For this study, the asphalt binder used was 60/70 PEN, and porous asphalt mixtures were modified with steel fiber to enhance the performance of porous asphalt. A total of 42 samples were prepared for porous asphalt mixtures at different fibre content ranging from 0%, 0.2%, 0.3%,0.4%, 0.5% and 0.6% and the 0% as a reference between modified and unmodified asphalt binder. The design binder content adopted from a study by Masri et al. [7]. The specification used for this research is by the Malaysian Public Work Department for road works [8].

2.2. Material Testing

Abrasion Among the test of asphalt binders, the physical test is including penetration test at 25oC, softening point test and ductility test were carried out on asphalt binders. From a penetration test, we can know the hardness and consistency of bitumen before it may be applied on the road. To determine the temperature at which given bitumen reaches a certain degree of softness and get the result for flow and consistency of asphalt binder can be known from the softening point test. Last but not least, the ductility test is to characterize the ductility of asphalt binders.

2.3. Mixture Testing

After all of the samples of porous asphalt mixtures modified with steel, fiber are done by using the Marshall mix design method; the performance test porous asphalt is performed. Among the tests involve evaluating the performance of fiber modified porous asphalt was LA Abrasion, Marshall Stability, and Flow test, and Resilient Modulus. Abrasion loss value for performance test was evaluated using the Los Angeles Abrasion machine without steel ball. Stability and density values, specific gravity and void analysis, which is to determine the percentage of air voids in mineral aggregate, and air voids in the compacted mixture can be known from the Marshall Stability test.

3. Results and discussion

3.1. Loss Angeles Abrasion

As shown in Figure 1 (a), the abrasion loss value for control samples is much higher than modified samples of porous asphalt which is 8.60%, 9.22% and 14.76% for 100 to 300 revolution respectively. Then, the abrasion loss value for the modified sample is lessened than the control sample, as seen in Figure 1 (b). From the graph above, we can conclude that the fiber content of 0.6% is much lower from others. Therefore 0.6% fiber content is the best fiber content for LA abrasion test. Besides that, the average loss of mass is less than 15%, which is acceptable according to the JKR specification.

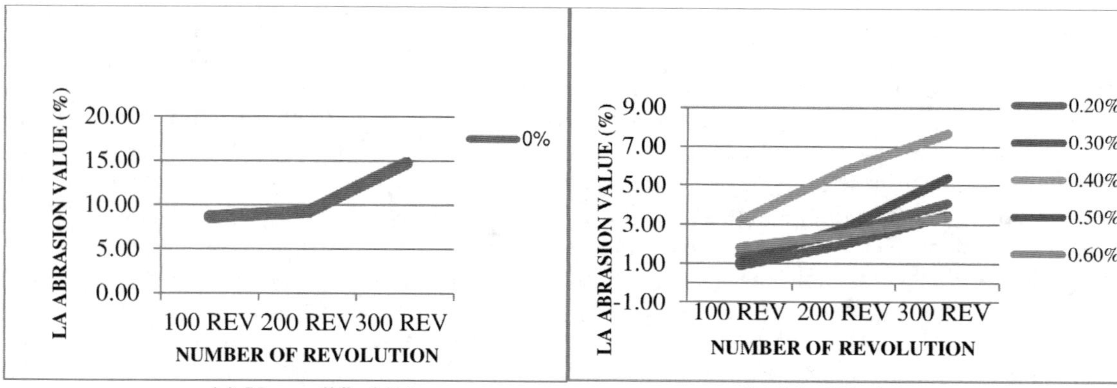

(a) Unmodified PA (b) Modified PA

Figure 1. Abrasion Loss

3.2. Marshall Stability and Density

The Marshall method is to analyze two features of pavement mixes, which is stability (flow analysis) and density (voids analysis). The stability and flow values are directly recorded during the test. These will show the maximum load which the mix can take before deforming to failure. From Figure 2 (a), the result is taken from the Marshall Stability and Flow test. The higher stability is from fiber content 0.5% while the lowest value of stability is 0.2% of fiber content. As shown in Figure 2 (b), the most efficient fiber content can be determined with the highest number of density value, which is 0.2% of fiber content.

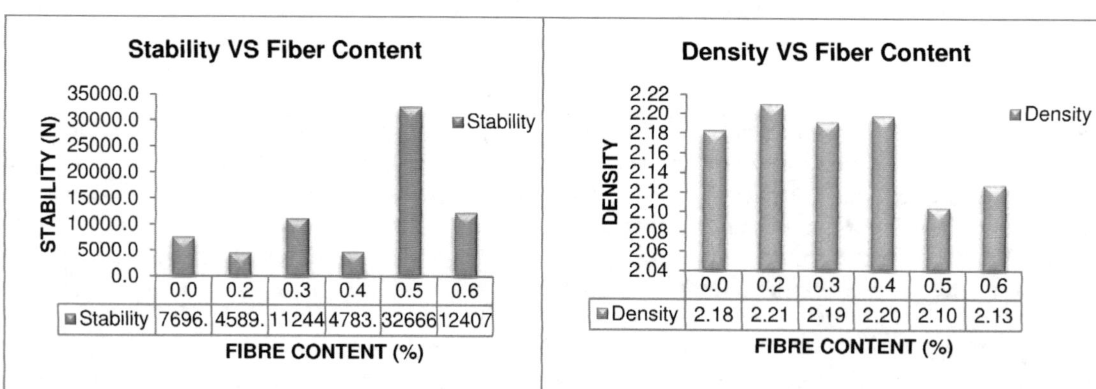

(a) Stability (b) Density

Figure 2. Marshall Properties

3.3. Resilient Modulus

The resilient modulus is the elastic modulus based on the strain under repeated loads. The testing is carried out by using 1000, 2000, and 3000 pulse repetition with two different temperatures, which are 25°C and 40°C. Based on Figure 3, the number percentage of fiber content with highest resilient modulus is 0.5% at temperature 25°C while 0.6% for temperature 40°C. This result is consistent with a study by Masri et al. [7] and Arshad et al. [9], where the modification of mixture improves the resilient modulus value significantly [10,11].

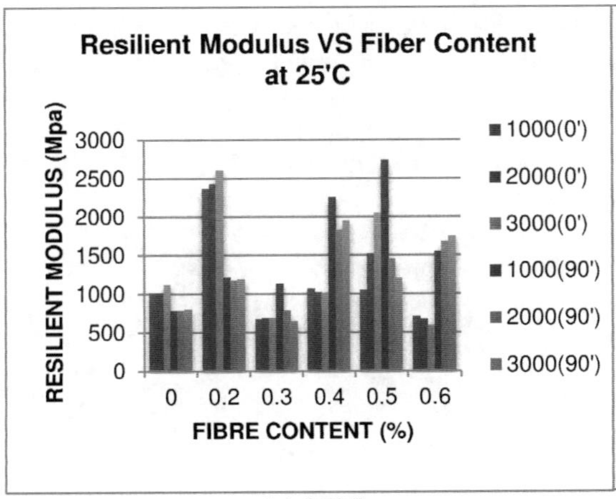

(a) 25°C

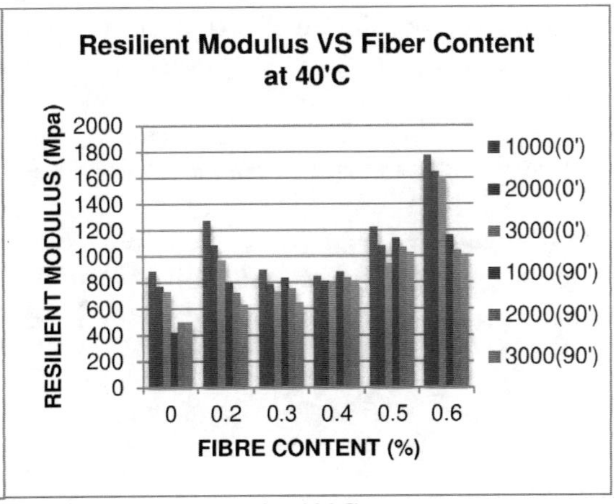

(b) 40°C

Figure 3. Resilient Modulus

4. Conclusion

a) Steel fiber possessed a greater effect on porous asphalt absorption and stabilization than other fibre. This indicated that fiber could effectively improve the abrasion resistance of asphalt.
b) With the addition of steel fiber in porous asphalt mix, the Marshall stability increases up to higher value.
c) With the addition of steel fiber, the average resilient modulus increases up to 15% at 25°C, whereas the increases in resilient modulus are up to 10% at 40°C.
d) It can be said that the steel fiber can be used to alter the phase composition and improve the engineering properties of porous asphalt.

5. References

[1] Ajam H K, Meijide B J, Artamendi I, Garcia A 2018. Mechanical and Healing Properties of Asphalt Mixes Reinforced with Different Types of Waste and Commercial Metal Particles. *Journal of Cleaner Production*, **192**, pp. 138-150.
[2] Aman M Y and Hamzah M O 2014. Effects of Anti-Stripping Additives on Moisture Sensitivity of Warm Porous Asphalt Mixtures. *International Journal of Construction Technology and Management,* **1(1)**, pp.10–16.
[3] Arshad A K, Masri K A, Ahmad J, Samsudin M S 2017. Dynamic Modulus of Nanosilica Modified Porous Asphalt, *IOP Conference Series: Materials Science and Engineering*, **271**, 012008.
[4] Arshad, A K, Ahmad J, Masri K A 2019. Rutting Resistance of Nanosilica Modified Porous Asphalt, *International Journal of Civil Engineering and Technology*, **10(1)**, pp. 2274-2284.
[5] Masri K A, Awang H, Jaya R P, Ali M I, Ramli N I, Arshad A K 2019. Moisture Susceptibility of Porous Asphalt Mixture with Nano Silica Modified Asphalt Binder. *IOP Conference Series: Earth and Environmental Science*. **244,** 012028.
[6] Tapkin S, Tuncan A, Tuncan M 2010. Repeated Creep Behaviour of Polypropylene Fiber-Reinforced Bituminous Mixtures. *Journal of Transportation Engineering*, **135(4)**, pp. 240-249.
[7] Masri K A, Arshad A K, Samsudin M S, 2016. Mechanical Properties of Porous Asphalt with Nanosilica Modified Binder. *Jurnal Teknologi*, **78: 7-2**, pp. 139-146.
[8] Jabatan Kerja Raya Malaysia (JKR). 2008. Standard Specification for Road Works, Section 4: Flexible Pavement. No. JKR/SPJ/2008-S4, pp. S4-58-S4-69
[9] Arshad A K, Masri K A, Ahmad J, Samsudin M S 2017. Investigation on Moisture Susceptibility and Rutting Resistance of Asphalt Mixtures Incorporating Nanosilica Modified Binder, *Pertanika Journal of Science and Technology*, **25,** pp. 19-30.

[10] Wu S, Roger R, Loke Y, Pfeiffer D 2014. Interaction between Entomopathogenic Nematodes and Entomopathogenic Fungi Applied to Third Instar Southern Masked Chafer White Grubs, Cyclocephala lurida (Coleoptera: Scarabaeidae), under Laboratory and Greenhouse Conditions. *Biological Control*, **76**.
[11] Zhang Y, Martin V D V, Molenaar A, Wu S 2016. Preventive Maintenance of Porous Asphalt Concrete using Surface Treatment Technology. *Materials & Design*, **99**, pp. 262-272.

Acknowledgement
The authors would to acknowledge Universiti Malaysia Pahang for funding this research under the grant RDU 1803157.

IOP Conf. Series: Materials Science and Engineering **712** (2020) 012028 doi:10.1088/1757-899X/712/1/012028

Volumetric Properties and Resilient Modulus of Stone Mastic Asphalt incorporating Cellulose Fiber

N A N Md Fauzi[1], K A Masri[1], P J Ramadhansyah[1], M S Samsudin[1], A Ismail[1], A K Arshad[2], E Shaffie[2], A H Norhidayah[3] and M R Hainin[3]

[1]Faculty of Civil Engineering Technology, Universiti Malaysia Pahang, 26300 Gambang, Pahang, Malaysia
[2]Institute for Infrastructure Engineering and Sustainable Management, Universiti Teknologi MARA, 40450, Shah Alam, Selangor, Malaysia
[3]Faculty of Engineering, School of Civil Engineering, Universiti Teknologi Malaysia, 81310 Johor Bahru, Malaysia

Abstract. Stone mastic asphalt (SMA) is well known as a high coarse aggregate content that interlocks to form a stone skeleton that resist permanent deformation. However, it facing a lot of problems such as rutting and stripping because of the high temperature and repeated axial load. It also suffers with creep issue. The cellulose fiber is a type of synthetic fiber that can enhance the properties of asphalt mixture. Thus, the aim of this study is to utilize the cellulose fiber that has high tensile strength to overcome the problem of SMA. Among the tests involve are Resilient Modulus, Marshall Stability and Cantabro Loss. From the results, it shows that the addition of cellulose fiber improved the recovery ability of asphalt binder. From each test, the addition of 0.2% cellulose fiber contributes to lowest value of abrasion, while 0.3% producing highest value of resilient modulus. Thus, the addition of cellulose fiber is capable in enhancing the properties of SMA.

1. Introduction

Stone Mastic Asphalt (SMA), which has been utilized in Europe for around 40 years ago to supply resistance to scraped spot by studded tires [1]. Within the 1970s, studded tires were prohibited in Germany, and the utilize of SMA blends declined since of the higher material and construction costs and there now not showed up to be a basic require for these mixtures. SMA has a high coarse aggregate content that interlocks to form a stone skeleton that resist permanent deformation [2]. Hence, the asphalt concrete facing a lot of problems such as rutting and dynamic creep. Subsequently, the mixture's resistance to moisture damage and rutting specifically will effects the life span of the paving mixture. Sufficiently, high binder contents cannot be achieved using unmodified or unstabilised bitumen [3]. The drainage of bitumen would occur during transport and laying. Therefore, most stone mastic asphalt mixture uses a fiber stabilizer mixed with asphalt binder in the bitumen to keep it homogenous [4]. Cellulose fibers has been selected as the modified binder to improve the pavement performance of asphalt mixture. The addition of cellulose fibers into the control asphalt improved the recovery ability of asphalt binder. The natural cellulose fibers are fibers that are still recognized as being from a part of the original plant because they are only processed as much as needed to clean the fibers for use [5]. This studies were done to enhance the performance of asphalt mix with cellulose fiber. One type of binder modification is the cellulose fiber which promising result in improving binder properties. This study focus on the cellulose fiber act as addictive for the asphalt binder. The aim of this study to enhance the properties of SMA in terms of resilient modulus, dynamic creep, Marshall Stability and abrasion with the existence of fibre content.

2. Methods

2.1. Materials

In this study, Penetration Grade 60/70 (PEN60/70) types of binder was mixed with cellulose fibers with amount of 0%, 0.2%, 0.3%, 0.4%, 0.5% and 0.6% by weight of mixture. Cellulose fiber acts as the modifier for SMA. The size of the cellulose fiber is about 100 mm. Table 1 and Table 2 show the properties of aggregate and asphalt binder. Both materials passed the requirement stated in JKR specification [6].

Table 1. Aggregate Properties

Aggregate Testing	Observed Test Result	Limit
AIV test	15%	<25%
ACV test	15.09%	<35%
LA Abrasion	21%	<25%
Flakiness	20%	<25%
Elongation	20%	<25%

Table 2. Asphalt Binder Properties

Asphalt binder Testing	Observed Test Result	Limit
Penetration Test in 25°C	60 mm/10	60-70 mm/10
Softening Point Test	50°C	49-56°C
Ductility at 25°C	102.5 cm/s	.>100 cm/s

2.2. Marshall Mix Design

To prepare SMA Mixture, Marshall Mix Design process was carried out in accordance to the Malaysian Public Works Department [6]. Two specimen were prepared for each binder content with the range of 0.2% - 0.6% of cellulose fiber for SMA with the increment of 0.1%. The bulk specific gravity, the stiffness, the stability and flow value, the percentage air voids in mineral aggregate (VMA) and the percentage of air voids in the compacted mix (VIM) was obtained and plotted separately against the fiber content and a smooth curve was drawn through the plotted values to determine the volumetric properties of modified specimen. The optimum binder content (OBC) use in this study is adopted from a study by Arshad et al. [7].

2.3. Mixture Testing

Among the tests involve to evaluate the performance of modified specimen are Marshall Stability, Resilient Modulus and Cantabro Loss. The resilient modulus is determined by the elastic modulus based on the recoverable strain under repeated loads [8,9]. The data provides the total resilient axial deformation response of a specimen subjecting with the cylindrical specimen to loads using Universal Testing Machine (UTM) Machine. For this test, two specimens were prepared for each cellulose fiber content (0%,0.2%,0.3%,0.4%, 0.5% and 0.6%). All the samples were tested at fixed temperature of 25°C and 40°C with two positions and three different pulse repetitions (1000ms, 2000ms and 3000ms). The Cantabro Loss test is performed to determine the abrasion loss of compacted SMA specimens. This test measures the breakdown of compacted specimens due to repeated rotation in Los Angeles drum mixer equipment. Then, the percentage of weight loss (Cantabro loss) is an indication of PFC durability and relates to the quantity and quality of the asphalt binder [10].

3. Results and discussion

3.1. Loss Angeles Abrasion

From Figure 1, at 300 revolutions, 0.3% CF has the highest value of abrasion and 0.2% has the lowest value of abrasion loss. This may cause by the increased tensile strength of the SMA due to the existence of cellulose fiber. The cellulose fiber helps to increase the tensile strength of SMA, thus improving the performance of the SMA. From the result, 0.2% CF has the lowest abrasion loss value.

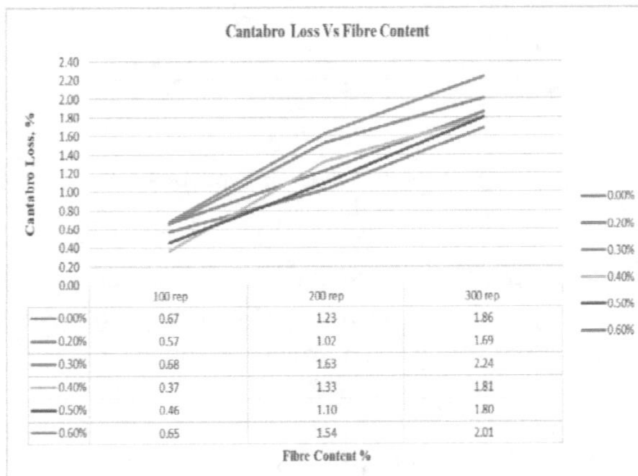

Figure 1. Abrasion Loss

3.2. Volumetric Properties

From Figure 1, 0.4% CF produces the highest density, while 0.5% CF produces highest stiffness and 0.2% CF produces the highest stability for SMA. It is found that the stability value is lower at less binder content, then increases with increment of content. The stability of the mix is primarily controlled by the cohesion and internal friction of the matrix which supports the coarse aggregates. The volumetric properties result verified the suitability of cellulose fiber to be used for SMA.

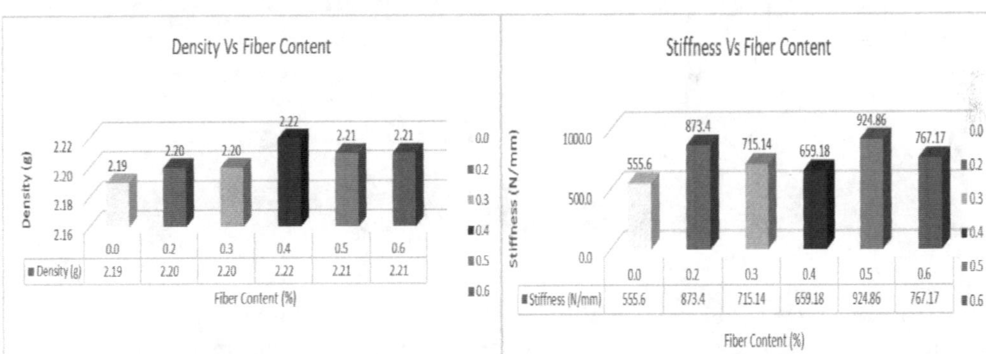

(a) Density (b) Stiffness

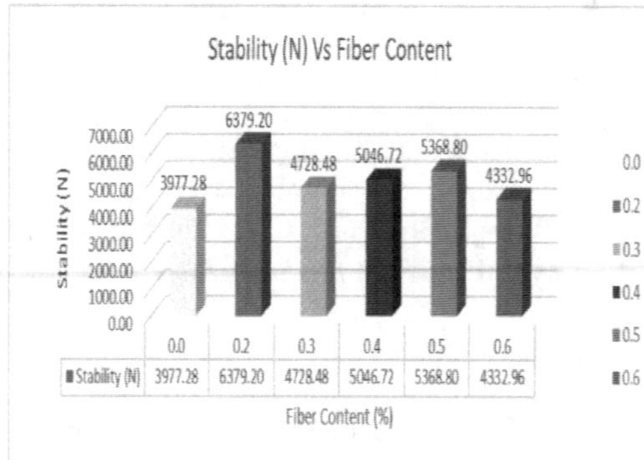

(c) Stability
Figure 2. Marshall Properties

3.3. Resilient Modulus
From Figure 3, 0.3% CF provides the highest resilient modulus value (Mr) at 25°C while 0.3% for 40°C. For SMA20, the increment in the amount of cellulose fiber will result in a higher value of Mr. At 25°C and at pulse 1000ms period, Mr for control specimen is lowest which is 1050 MPa. The highest Mr value obtained at 0.3% CF-SMA20 is 2269 MPa. However, the result for Mr values at different pulse period did not show any trend in uniformity. In addition, at 3000ms pulse repetition period, the Mr for 0.3% CF-SMA20 is 2011 MPa. In general, higher pulse repetitive period gives lower resilient modulus value. The result is consistent with a study by Arshad et al. [7]] and Masri et al. [11].

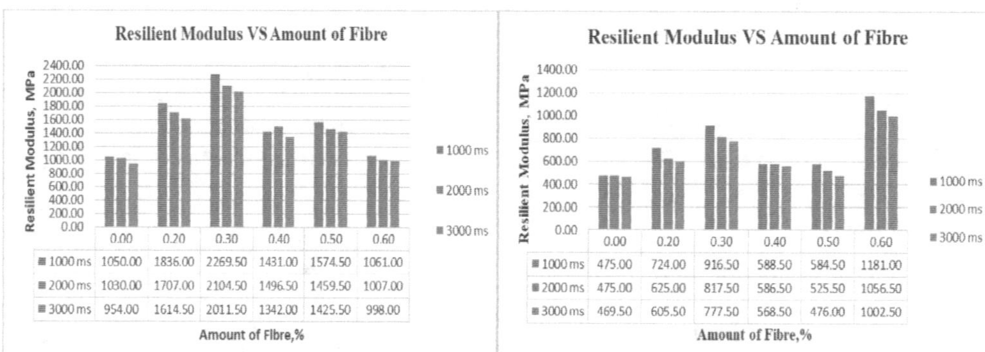

(a) 25°C (b) 40°C
Figure 3. Resilient Modulus

4. Conclusion
From the laboratory works and data analysis, the optimum cellulose fibre content is 0.2% and considered as the most effective amount for the performance enhancement of SMA mixtures. The addition of cellulose fiber also increased the overall performance of SMA in terms of volumetric properties, abrasion resistance as well as resilient modulus.

5. References
[1] Woodward D, Millar P, Lantieri C, Vignali V, 2016. The Wear of Stone Mastic Asphalt due to Slow Speed High Stress Simulated Laboratory Trafficking, *Construction and Building Materials*, **110**, pp. 270-277.
[2] Arshad A K, Masri K A, Ahmad J, Samsudin M S 2017. Dynamic Modulus of Nanosilica Modified Porous Asphalt, *IOP Conference Series: Materials Science and Engineering*, **271**, 012008.

[3] Arshad A K, Ahmad J, Masri K A 2019. Rutting Resistance of Nanosilica Modified Porous Asphalt, *International Journal of Civil Engineering and Technology,* **10(1)**, pp. 2274-2284.

[4] Masri K A, Arshad A K, Samsudin M S, 2016. Mechanical Properties of Porous Asphalt with Nanosilica Modified Binder**.** *Jurnal Teknologi,* **78: 7-2,** pp. 139-146.

[5] Hainin M R, Reshi W F, Niroumand H, 2012. The Importance of Stone Mastic Asphalt in Construction. *Journal of Geotechnical Engineering* **17:49**, pp. 56-74.

[6] Jabatan Kerja Raya Malaysia (JKR). 2008. Standard Specification for Road Works, Section 4: Flexible Pavement. No. JKR/SPJ/2008-S4, pp. S4-58-S4-69

[7] Arshad A K, Masri K A, Ahmad J, Samsudin M S 2017. Investigation on Moisture Susceptibility and Rutting Resistance of Asphalt Mixtures Incorporating Nanosilica Modified Binder, *Pertanika Journal of Science and Technology*, **25**, pp. 19-30

[8] Irfan M, Ali Y, Ahmed S, Iqbal S, Wang H, 2018. Rutting and Fatigue Properties of Cellulose Fiber-Added Stone Mastic Asphalt Concrete Mixtures. *Advances in Materials Science and Engineering*, 5604197.

[9] Muniandy R and Aburkaba E E 2018. Effect of Filler Type and Particle Size on Moisture Susceptibility of Stone Mastic Asphalt Mixtures. *Australian Journal of Basic and Applied Sciences.* **4(11)**, pp. 5522–5532.

[10] Zulhaidi M J, Mohd Hafzi M I, Rohayu S, Wong S V 2018. An Exploration of Weather Threats to Road Safety in Tropical Country. Road Engineering & Environment Research Centre Malaysian Institute of Road Safety Research (MIROS), Malaysia.

[11] Masri K A, Awang H, Jaya R P, Ali M I, Ramli N I, Arshad A K 2019. Moisture Susceptibility of Porous Asphalt Mixture with Nano Silica Modified Asphalt Binder. *IOP Conference Series: Earth and Environmental Science*. **244**, 012028.

Acknowledgement

The authors would to acknowledge Universiti Malaysia Pahang for funding this research under the grant RDU 1803157.

IOP Conf. Series: Materials Science and Engineering **712** (2020) 012029 doi:10.1088/1757-899X/712/1/012029

Construction readiness parameters for highway projects

A R Radzi[1], R A Rahman[1,2], S I Doh[1] and M Esa[3]

[1]Faculty of Civil Engineering Technology, Universiti Malaysia Pahang, 26300 Gambang, Pahang, Malaysia
[2]Earth Resources & Sustainability Center, Universiti Malaysia Pahang, 26300 Gambang, Pahang, Malaysia
[3]School of Housing Building and Planning, Universiti Sains Malaysia, 11800, USM, Penang, Malaysia

Abstract. Highways greatly contribute to a country's economic development and growth. Therefore, it is vital for highway construction projects to complete on time and budget. However, when a project starts construction prematurely, projects frequently experience interruption that eventually resulted in delays, causing a multitude of adverse effects on all project stakeholders. In other words, premature starts are one of the factors of construction delays. However, prior studies have little understanding of information that can help practitioners determine whether a highway project is premature or ready for construction. This study aims to identify the parameters used in practice to differentiate whether a highway construction project is ready or not to begin construction. To achieve this objective, interviews with sixteen industry practitioners working on highway construction projects are being conducted and analyzed. The significant findings include: (1) construction readiness can be assessed even as early as during the project start-up phase; and (2) not fulfilling the construction readiness parameters may cause a work stoppage, inefficient work, rework, and shortages in labor, equipment or materials. This study contributes to the current body of knowledge by identifying the parameters that indicate whether a highway project is construction-ready or construction-not-ready. The lessons from this study could help the industry to avoid premature starts in highway construction projects.

1. Introduction

A highway can be defined as a public road, especially an important road that joins cities or towns together. Highway construction is vital for every nation because of its contribution to social and economic development. As an essential part of modern transportation systems, highways represent an essential infrastructure for the rapid delivery of people and goods to meet the demands of regional economic development. Empirical evidence suggests that there are significant and positive correlations between highway transportation infrastructure and economic activity [1]. While governments are pushing for successful highway projects because of its importance, highway projects are negatively affecting Malaysia's economic growth from delays and cost overruns [2]. Therefore, identifying approaches to improve the chance of having a successful highway construction project is crucial to nations economic and social development.

A construction project is commonly acknowledged as successful when it is completed on time, within budget, and to stakeholders' satisfaction. Since at least one stakeholder on virtually every construction project benefit from an early start to construction, project teams nearly always feel pressure to begin construction, whether or not they are in fact ready. Premature starts to construction happened when a decision, by at least one party, to start construction with at least one risk that exceeds an acceptable tolerance to a party and which can result in an interruption to construction. Cost overruns, schedule slippage, overtime, and poor productivity are among the most common and significant impacts projects face when they are not ready to start [3]. Therefore, the industry must prevent premature starts in a construction project that will ultimately lead to various unwanted impacts. Construction readiness is defined as the series of activities and procedures that should be completed or substantially completed before construction so that construction can productively start and sustain operations. Numerous projects begin construction before they are ready to do so (premature start), leading to reduced productivity, and unsatisfactory project performance. Construction-ready projects have on average 22% schedule reduction, 29% productivity improvement, 20% cost savings, 7% less rework, and 21% less change, relative to construction-not-ready projects [4]. Moreover, adequate construction readiness can contribute to avoiding many of the factors causing out of sequence work, and ultimately result in improved project performance [4]. Thus, it is crucial for industry practitioners to assess the readiness of the project before starting the construction. This research sets out to investigate the parameters used in practice to differentiate whether a highway project is ready or not to begin construction. To achieve the objective, this paper addresses research question related to: What are the parameters that can be used to differentiate whether a highway construction project is ready or not ready to begin construction? The author answers the question by conducting interviews with sixteen industry practitioners working on highway construction projects in Malaysia. Then, the data acquired from the interviews were analyzed using thematic analysis. Finally, a set of construction readiness parameter for highway construction is determined. The lessons from this study could help the industry to prevent premature starts in highway construction projects by assessing the construction readiness of a project.

2. Background

2.1. Construction Readiness

Readiness in the construction industry is one of the topics that have been studied by researchers. A readiness assessment identifies the potential challenges that might arise when implementing new procedures, structures, and processes within a current organizational context. The purpose of a readiness assessment is to determine if there are potential barriers to success and provide the individuals or organizations the ability to overcome such barriers before beginning the project. Assessing construction readiness of a project helps project team prevent premature starts by quantitatively determining if a project is "construction ready" resulting to increase in construction cost performance, construction schedule performance, and construction productivity [4]. In order to prevent premature starts to construction, prior research has identified parameters to evaluate construction readiness in general. However, different types of construction projects have different characteristics. For example, highway construction is a horizontal construction, involves more land area and geographic scale is larger compared to building construction. Moreover, it consists of working around a broad and challenging range of existing conditions such as traffic, wetlands, underground utilities, or above-ground infrastructure. Therefore, it is worthwhile to identify the construction readiness parameters specific to highway projects.

2.2. Highway Construction in Malaysia

Due to the importance of highway construction for the development of a nation's economic and social, numerous studies have been conducted by researchers and industry practitioners in Malaysia. Effective communication system, good project governance, responsibility and commitment of public and private sectors, competitive procurement process and ability to delegate authority have been identified as the top five critical success factors for the Butterworth Outer Ring Road BORR expressway in Malaysia [5]. Meanwhile, the top five factors that contribute to the delay in the highway construction projects is the improper planning, weather, poor site management, inadequate site investigation and underground utilities [6]. On the other hand, five aspects of green highway terminologies namely; conservation and ecosystem management, life cycle energy and emission reduction, recycle, reuse, and recycle, watershed driven stormwater management, and overall societal benefits were identified as common understandings among the stakeholders [7]. Various studies have been investigating both the common issues in highway projects and the different aspects of highway projects in Malaysia because it is an important topic. However, there is still a lack of research on how to determine whether the highway project is ready or not ready to begin construction. Therefore, this study will fill the gap by investigating on the parameters to assess the readiness of highway construction projects.

3. Methods
The data collection involves acquiring data from interviews with industry practitioners working on highway construction projects. Qualitative approaches are used to analyze the collected data. The following subsections discuss the methods of collecting and analyzing the construction readiness parameters. Figure 1 summarizes this study's methodology in investigating the parameters used in practice to differentiate whether a highway project is ready or not to begin construction.

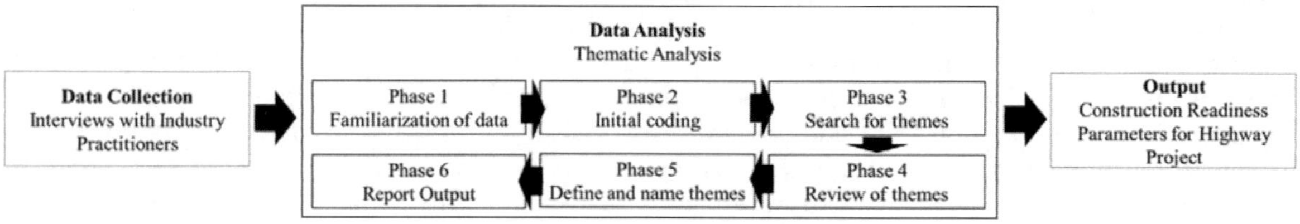

Figure 1. Research methodology

3.1. Data Collection
This study collects information through interviews with industry practitioners. Individual interview has been chosen as the approach for data collection because it helps the interviewer to explain, better understand, and explore research subjects' opinions and experiences. Industry practitioners were consulted to share their unique perspectives, practical knowledge and experiences. The personnel involved in the interviews are all from companies that hold a CIDB's grade G7 license. A contractor with Grade G7 license approved by the Construction Industry Development Board (CIDB) Malaysia is permitted to undertake Civil Engineering Construction and Building Construction projects for an unlimited amount. Therefore, this research purposely selected contractors with grade G7 license because these contractors are well-established organizations in the Malaysian construction industry with strong financial capability. The authors use open-ended questions because these encourage participants to contribute as much detailed information as desired [8]. This study's data collection involves interviewing sixteen valid respondents. The interview started with an introduction that explained why the author wanted to talk with them and what the author wanted to discuss. The question for the interview is: What parameters can be used to differentiate whether a highway construction project is ready or not ready to begin construction? The question was followed by additional questions depending on the participants' response. The follow-up questions aimed to obtain a deeper understanding of the information that they gave and to check whether their statements were understood correctly. If the participant was unable to respond or elaborate on the questions asked, the interviewer tried to rephrase the interview question in another way and gave time for a response. The interviewer encouraged the participants to continue if they had started on an answer without finishing their explanation. After each interview, a summary of the interview was made and sent to the participants for validation purposes.

3.2. Data Analysis
The qualitative analysis involves performing thematic analysis to identify the construction readiness parameters of highway construction because the approach can assist in making sense of qualitative data [9]. This method has been employed by Rahman and Ayer [10] and Radzi et al. [11] to analyze qualitative data associated with construction management topics. The thematic analysis was conducted based on the six phases described in Braun and Clarke [9]. The first phase is becoming familiarized with the data. The authors transcribed the interview data, read, reread and noted the initial ideas. The second phase is to generate the initial codes. The authors coded for as many potential themes and

patterns as possible from the data. The authors then reviewed, discussed, and agreed on any additions and/or changes to the coding. The third phase is to search for themes based on the initial codes. During the process of creating the themes, the authors frequently revisited the codes from the second phase and the original data from the first phase. The fourth phase is to review the themes. To ensure saturation of the data, the authors continually reviewed the subthemes, defined and refined them, checking if themes work in relation to the coded extracts and the entire data set, and reviewing data to search for additional themes. The fifth phase is to define and name the themes. The authors continually went back and forth between the themes, codes, and transcription of the interview to ensure that the themes were true to the independently coded responses. The final phase (sixth phase) is to report the output of the analysis.

4. Results and Discussions

Figure 2 summarizes the themes and subthemes of the construction readiness parameters for highway construction projects that were identified through analyzing this study's interviews with sixteen industry practitioners from Malaysia. A total of 28 parameters have been identified in this study. The parameters can be categorized into five categories: "Approval", "General Requirement", "Drawing Requirement", "On-site" and "Material". These five categories can be grouped into two themes: "Project Start-up" and "Execution." The details of each parameter are discussed in the subsequent subsections.

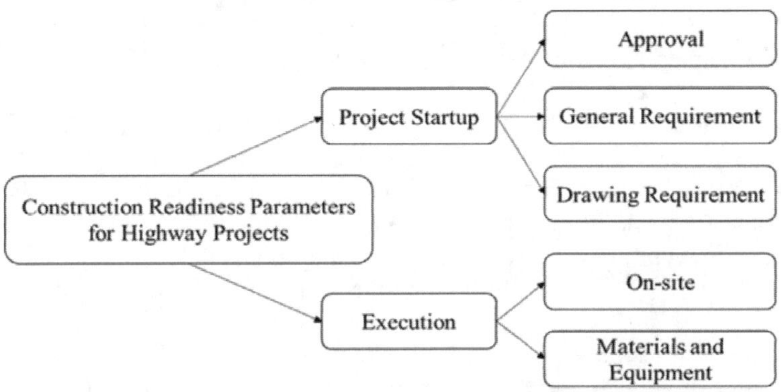

Figure 2. Overview of the construction readiness parameters of highway projects

4.1. Parameters related to Project Start-up

4.1.1. Approval. According to the interview results, one of the parameters used to determine whether the construction is ready or not is making sure that the project received approvals from different legal agencies in Malaysia such as, local authorities, utility companies, Department of Occupational and Safety (DOSH), and Construction Industry Developmental Board (CIDB). Approval in construction project refers to obtaining permissions from the relevant authorities to ensure the development follows the standards laid down in the building regulations and certain stages of the construction are inspected by officers from the local authority. Sometimes during construction, authority officers will perform inspections at the construction site to ensure the works progress as intended, both in terms of quality and compliance. Failed to show notice of approvals to the inspection officers might cause them to stop the construction work until decision have been made leading to loss of productivity and delay.

4.1.2. General Requirement: (a) Received letter of award from client. Letter of award is sent by the client to the contractor as written confirmation that the contractor has been successful and will be awarded the project. Without a letter of award, various problems might occur because there is no written and signed contract stated that the owner had awarded the project to the contractor. Without a contract, there's no agreement, no protection for the contractor and no guarantee that the contractor will be paid for the work they do. Furthermore, failure to agree on the contract terms and conditions before work starting means that when the project manager finally reviews them, they have often incurred costs on the project and are not in the position to persuade the client to alter the terms and conditions to those that are more appropriate or acceptable. When a contractor incurred too much cost, it might affect their cash flow. Consequently, workers might refuse to work due to contractors unable to pay their salary on time resulting from halting in construction work. Without workers, construction work could not be done causing loss of productivity.

(b) Finished land acquisition. Land acquisition is the process of acquiring the required or compulsory land for the project. It is one of the critical elements in the land development process. The process of land acquisition provides a legal procedure for the public sector in acquiring any land for development projects that benefits the country. Highway development may require the acquisition of land from local people for the infrastructure and economic development of the country. In many cases, acquiring land from the public is not easy for many reasons, primarily related to compensation or the price of land consequently prolongs the duration of the project. If project have begun, construction work might need to stop until the land acquisition is completed leading to loss of productivity.

(c) Sufficient funding. Funding is vital for a construction project. Project managers should ensure that the project has enough money throughout the construction. Inadequate funding might cause workers and suppliers refuse to work and supply material to the site affecting construction work resulting to loss of productivity. Also, lack of funds may affect the project's cash flow and lead to delay in site possession, which consequently causes delays in the project as a whole [12].

(d) Obtained project insurance. Construction insurance is a practice of exchanging a contingent claim for a fixed payment to protect the interests of parties involved in a construction project. It is a major method of managing risks in the construction industry. Construction is a dangerous and usually unpredictable industry to work in and presents a considerable risk of personal injury, damage to existing property and financial loss. Lack of insurance may make workers refused to work because if accidents happened, they might not have enough money to pay for medical expenses.

(e) Verified official commencement date and construction duration. The commencement date is the date at which the time begins to run, based upon the project schedule. Conversely, construction duration is defined as the time frame given by the owner for the contractor to complete the project under normal work conditions. It is a necessary step for project managers to verify the commencement date and construction duration to plan construction works accordingly. Poor project planning can result to project delay [13]. Poor planning means that the schedule that the workers are supposed to follow is not set out. The workers will have no clear picture of what is expected of them as they work on the projects. There will be no deadlines to meet hence creating a lazy atmosphere among the workers decreasing productivity. This means that the project will not be completed on time resulting to project delay.

(f) Verified nearest authority. Construction site theft can cause various problems to the project. Therefore, working with local authorities is necessary to prevent crime at the construction site. Project managers should verify nearest local authorities and inform them about the construction site. Besides, project managers can ask the authority to monitor the site on weekends or holidays when the site is unattended to avoid theft. In addition to the direct costs of replacing the stolen goods, there is also rental costs to replace stolen equipment and lost productivity. Interruptions to the project schedule due to a shortage of equipment and materials can also lead to delay.

(g) Meeting with consultant and client. The client is the owner of the project and consultants are professionals such as architect, structural engineers, and others, appointed by the client to perform expert tasks on a project. During this meeting, stakeholders can discuss and answers outstanding questions about the project. Productivity might decrease due to outstanding questions about the

IOP Conf. Series: Materials Science and Engineering **712** (2020) 012029 doi:10.1088/1757-899X/712/1/012029

project. Also, some decisions on the project could not be made resulting in interruption to project schedule.

(h) Verified construction work plan. A work plan is a sequence of step that must be taken or activities that must be performed well for a project to succeed. Furthermore, a work plan shows all the tasks involved in a project, who is responsible for each task, and when the tasks will be completed. It can help the contractor to stay organized while working on projects and helps them to complete the project in time and within the budget. Starting construction without a work plan could cause various problems such as reduced productivity and schedule slippage.

(i) Client and consultant approved construction work plan. Client and consultant should know the work plan that has been developed by the contractor so that there is no dispute between stakeholders. Lack of understanding of the project work plan might cause conflict between stakeholders. If disputes are not adequately managed, they may cause project delays, undetermined team spirit, increase project costs and above all, damage continuing business relationships [14]. Also, work stoppage might happen because of a decision could not be made due to a dispute between stakeholders causing loss of productivity and delay. Therefore, client and consultant should approve the work plan before starting a project.

4.1.3. Drawing Requirement. (a) Drawings approved by consultant. It is important that the consultant can review and approve the construction drawing before the project manager can start the project. Consultant's approval can ensure conformity to construction drawings, specifications, and standards. Clients always change the design during the construction period and at the same time, it will affect the whole project duration. The contractor cannot carry out the construction work until the latest drawing issue by the architect. Consequently, the work cannot be done within the contract period resulting from delaying and reduced productivity. Therefore, to prevent work stoppage, drawings must be approved by the consultant before construction.

(b) Drawings approved by authorities. Approval of drawings by authorities is a process undertaken by authorities to ensure new construction safe for human and as per planning regulations. The approval process checks whether the building is safe and structurally sound and whether the building materials proposed are safe for human habitation. Starting construction with a drawing that had yet been approved by an authority can risk project schedule. If authority officers inspect at the site and acknowledge that the drawing has not been approved, they might stop the construction work resulting to delay and lost productivity.

(c) Complete drawings are all supplied. Complete drawings are vital to ensure the project can be completed within a time frame. Errors and omission can cause the drawings to be incomplete leading to project delay. During construction, due to drawing is unclear and important details are not shown in the drawing, workers could not continue the work. As a result, the works have to stop immediately until details of that particular drawing can be obtained leading to loss of productivity and delay.

(d) Verified discrepancies between tender drawings and construction drawings. Discrepancies between tender drawings and construction drawings can cause project delay. Tender drawings are drawing given to the bidders(contractors) at the tender time. Construction drawings are given by the consultant for construction purposes. Sometimes, there are discrepancies between these drawings. These discrepancies result in rework, changes in quantities, and delays and defects in construction. Furthermore, reworks lead to lower productivity [15].

4.2. Parameters related to Project Execution

4.2.1. On-site. (a) Site office. Office facilities are necessary at the construction site to provide accommodation for project managers, provide space for meetings, and to provide storage for site documentation. Besides, the site office is vital to make sure that all project's documentation is kept safe. When the construction project starts before site office have been built, documentation such as projects files and drawings can be all over the place. Missing project documentation such as contract files or drawings can cause interruption to project schedule. For example, project managers need to find the missing drawings before the work can resume resulting in lost productivity and delay.

(b) Site condition as per the contract. Project managers should make sure that the site condition is the same as stated in the contract and there is no differing site condition. A differing site condition occurs when the project manager experiences a condition on the site that differs materially from the conditions indicated in the contract with the owner of the project or from what is to be normally expected in the site area. Differing site condition will cause time delay to the project due to work stoppage at the site until relevant decisions are made by stakeholders [16].

(c) Verified utility cables and relocate if necessary. Utility relocation is the adjustment, replacement, or relocation of utility facilities as required by a highway construction project, such as: removing and reinstalling the utility, acquiring necessary right-of-way, moving or rearranging existing facilities, changing the type of utility and applying any necessary safety and protective measures [17]. Damage to underground utilities due to construction work can result in weeks of delays. For example, accidentally hitting a water pipe can lead to a temporary halt of the project. The water would spill out and flood the area, which makes the site unworkable. Also, there will be community complaints and legal issues that will further the delay.

(d) Verified traffic diversion and control system. Traffic diversion system relates to making sure the traffic around the construction site will not be interrupted by the project and if necessary, traffic diversion is made to minimize interruption. On the other hand, the traffic control system relates to a system of managing traffic in and around construction sites. The above parameters can be related in terms of preventing an accident at a construction site. Vehicles move in and around the workplace, reversing, loading and unloading are often the leading cause of death and injuries to workers and members of the public. Inadequate planning and control are the root cause of workers and construction vehicle accidents. Accident involving construction vehicle can lead to equipment shortage due to the accident eventually result to time overrun. Also, an accident involving construction workers can lead to a labor shortage, ultimately resulting in a construction delay.

(e) Placed safety signboard. Safety signboards are crucial in any work environment. The primary importance of displaying safety signboard is to prevent accidents and ensure workers are well aware of the possible dangers and hazards ahead in certain situations and environments. Without signboards, many workers would lack the necessary direction in times of crisis, and employers might find themselves in significant legal difficulties if any accidents were to arise as a result. Accidents involving construction workers due to inadequate site safety can reduce the work rate and efficiency, which will lead to work disruption and ultimately to project schedule delay [18]. Besides, accidents can affect the number of workers leading to interruption to the project schedule. In some cases, finding a replacement for the injured workers takes more time than expected.

(f) Installed closed-circuit television (CCTV) and verified parking space for construction machinery. CCTV is a system that sends television signals to a limited number of screens and is often used in shops and public places to prevent crime. Conversely, verified parking space for construction machinery relates to designing parking space for construction machinery. Installing CCTV and verifying parking space for construction machinery can be linked in terms of preventing crime at the construction site. Nowadays, theft and vandalism on construction sites in the commercial construction

industry is a problem that can affect productivity and drain profits. One of the techniques used to prevent crime at the construction site is by installing security cameras [19]. Besides, construction machinery should be parked in the designated parking area at the site to prevent crime. Machinery with no parking space at the site and left unattended in remote locations can be easily targeted by thieves. Companies need to pay to replace the loss machinery or rent it in the interim to keep projects on schedule. Besides, there are other costs involved as well, not the least of which is the resulting downtime caused by the lack of availability of machinery. Hence, it is essential to make sure there is parking space designated for construction machinery before construction and to ensure that CCTV had been installed before starting the construction to avoid a shortage of equipment.

(g) Verified temporary utilities on site. Before work can begin on a construction site, several services such as electricity, water, lighting, internet connection, and others must be temporarily set up. Construction site without electricity can disrupt office works leading to delay in a construction project. Also, internet connection and telephones are necessary at the construction site for project managers to contact the client about the project. Lack of readily available utilities on site is one of the causes of delays in the Ethiopian construction industry [20].

(h) Adequate workforce. Construction workforce means the workers who are engaged in the construction industry. Shortage of workers causes interruptions to project schedule. For example, inadequate equipment operators at the site will disrupt construction work because there are not enough workers to operate construction equipment. Moreover, although materials are ready there is an inadequate worker to carry the work can affect project duration. Lack of foreign and local workers is listed as one of the most frequent factors that lead to delay in the Malaysian construction industry [21].

4.2.2. Materials and Equipment. (a) Verified nearest material supplier and quarry. Material supplier means any entity that supplies materials, services, or equipment to be used in conjunction with the performance of work on a construction contract. On the other hand, a quarry is a place where rocks, gravel or sand, dimension stone, slate construction aggregate and others are excavated from the ground. The geographical location of material supplier and quarry might affect the delivery lead-time, transport and logistics costs. For example, selecting a supplier or a quarry that is near to the site can ensure the material arrives at the right time. Dealing with distant suppliers or quarry might mean longer delivery times and extra freight costs. Supplier failure to deliver material on time can disrupt operations and delay the completion of a project. Unavailability of materials on site is one of the factors that influences productivity.

(b) Equipment availability. Highway construction requires a variety of specialized equipment. Lack of equipment affects labor productivity since labor relies on equipment to aid the construction process. For example, the excavator is used to excavating and move a large object. Without it, construction material could not be move resulting to project disruption and work stoppage. Also, equipment availability is one the cause of delay in the Malaysian construction industry [22].

5. Conclusion

This study identified the construction readiness parameters for highway projects by analyzing interviews with sixteen industry practitioners in Malaysia. In conclusion, readiness parameters related to either the project start-up or execution phase have been identified. In other words, construction readiness can be assessed as early during the start-up phase of a highway construction project. Also, starting construction without adequately satisfying the parameters can result to a work stoppage, inefficient work, rework, and shortages in labor, equipment, or materials. These findings highlight the need for industry practitioners to assess the readiness of a project before starting construction. While some of the parameters identified in this study can be associated with other types of construction projects, there are several parameters that are specific to highway construction. Specifically, parameters such as finished land acquisition, verified traffic around the construction site, and traffic control system are not part of the readiness factors for general construction [4]. These findings suggest the land acquisition process is vital in highway construction because highway projects are usually long and involve lots of lands. Meanwhile, because of construction workers face the danger of working in the close vicinity of passing traffic, verification of traffic and traffic control system are essential before construction. In other words, this study contributes to the existing body of knowledge on highway construction by providing additional parameters that are specific to evaluating construction readiness of highway projects. The lesson of this study, when properly used, would help the industry to prevent premature starts in highway construction projects by assessing the project construction readiness.

6. References

[1] Chandra A and Thompson E 2000 Does public infrastructure affect economic activity?: Evidence from the rural interstate highway system *Regional Science and Urban Economics* **30** (4) 457-490
[2] Bank Negara Malaysia 2018 *Bank Negara Malaysia Quarterly Bulletin Q4 2018* (Malaysia)
[3] Griego R and Leite F 2016 Premature construction start interruption: How awareness could prevent disputes and litigations *Journal of Legal Affairs and Dispute Resolution in Engineering and Construction* **9** (2) 04516016
[4] Ibrahim M W 2018 *Improving Project Performance by Mitigating Out-of-Sequence Work and Assessing Construction Readiness* The University of Wisconsin-Madison
[5] Ghazali F M, Rashid S A and Sadullah A M 2017 The critical success factors for public-private partnership highway construction project in Malaysia *Journal of Engineering and Technology (JET)* **8** (1)
[6] Karunakaran S, Ramli M Z, Malek M A, Musir A A, Imran N F, Fuad N F, Zawawi M H and Zainal M Z 2018 Causes of delay on highway construction project in Klang valley *AIP Conference Proceedings* vol 2030 p 020242
[7] Nusa F N M, Nasir S and Endut I R 2018 Awareness of Green Highway Concept and Terminology: A Perspective of On-Site Personnel in Malaysian Highway Construction Industry *Advances in Transportation and Logistics Research* **1** (1) 475-487
[8] Turner III D W 2010 Qualitative interview design: A practical guide for novice investigators *The qualitative report* **15** (3) 754-760
[9] Braun V and Clarke V 2006 Using thematic analysis in psychology *Qualitative research in psychology* **3** (2) 77-101
[10] Rahman R A and Ayer S K 2017 Prevalent issues in BIM-based construction projects *In Proceedings of Joint Conference on Computing in Construction* vol 1 pp 645-652
[11] Radzi A R, Bokhari H R, Rahman R A and Ayer S K 2019 Key attributes of change agents for successful technology adoptions in construction companies: A thematic analysis *Computing in Civil Engineering 2019: Data, Sensing, and Analytics* (Reston, VA: American Society of Civil Engineers) pp 430-437
[12] Abdul-Rahman H, Berawi M A, Berawi A R, Mohamed O, Othman M and Yahya I A 2006 Delay mitigation in the Malaysian construction industry *Journal of construction engineering and management* **132** (2) 125-133
[13] Marzouk M M and El-Rasas T I 2014 Analyzing delay causes in Egyptian construction projects *Journal of advanced research* **5** (1) 49-55
[14] Cheung SO and Suen H C 2002 A multi-attribute utility model for dispute resolution strategy selection *Construction Management & Economics* **20** (7) 557-568
[15] Jarkas A M and Bitar C G 2011 Factors affecting construction labor productivity in Kuwait *Journal of construction engineering and management* **138** (7) 811-820
[16] Amarasekara W D, Perera B A and Rodrigo M N N 2018 Impact of differing site conditions on construction projects *Journal of Legal Affairs and Dispute Resolution in Engineering and Construction* **10** (3) 04518006
[17] Jung Y J 2012 Evaluation of subsurface utility engineering for highway projects: Benefit–cost analysis *Tunnelling and underground space technology* **27** (1) 111-122
[18] Durdyev S, Omarov M and Ismail S 2017 Causes of delay in residential construction projects in Cambodia *Cogent Engineering* **4** (1) 1291117

[19] Berg R and Hinze J 2005 Theft and vandalism on construction sites *Journal of Construction Engineering and Management* **131** (7) 826-833
[20] Gebrehiwet T and Luo H 2017 *Procedia engineering* vol 196 p 366-374
[21] Abdul Kadir M R, Lee W P, Jaafar M S, Sapuan S M and Ali A A A 2005 Factors affecting construction labour productivity for Malaysian residential projects *Structural survey* **23** (1) 42-54
[22] Sambasivan M and Soon Y W 2007 Causes and effects of delays in Malaysian construction industry *International Journal of project management* **25** (5) 517-526

Acknowledgments
The authors would like to thank Universiti Malaysia Pahang for supporting this study through financial grant RDU190340 as well as the industry practitioners that agreed to participate in this work.

IOP Conf. Series: Materials Science and Engineering **712** (2020) 012030 doi:10.1088/1757-899X/712/1/012030

Factors affecting the success of highway construction projects: the case of Malaysia

R A Rahman[1,2], A R Radzi[1], M S H Saad[1] and S I Doh[1]

[1]Faculty of Civil Engineering Technology, Universiti Malaysia Pahang, 26300 Gambang, Pahang, Malaysia
[2]Earth Resources & Sustainability Center, Universiti Malaysia Pahang, 26300 Gambang, Pahang, Malaysia

Abstract. Highways are often high-risk projects due to its importance to a nation's economic, societal, and political development. Therefore, highway projects need to be successful. Numerous researchers have identified the key factors affecting their local highway projects to identify opportunities for improvements. However, the current body of literature lacks that type of information in regards to Malaysia's construction industry. This study's objective is to identify factors that are affecting Malaysia's highway projects from industry practitioners' perspective. To achieve this objective, interview data with highway project managers are analyzed. Eleven factors affecting the success of highway projects in Malaysia were identified. The factors are: workflow for the design process, competencies of project managers, availability of workforce, availability of materials, availability of equipment, weather, site location, involvement from upper management, support from clients, public acceptance, and efficiency of authorities. This study contributes to the current body of knowledge by providing researchers and practitioners with a set of alternative factors that are affecting highway project success. The lesson of this study would help the industry to promote project success of highway projects in Malaysia.

1. Introduction

Highways are public roads that connect cities or towns. Highways play a significant role in a country's development as these infrastructures are suggested to have positive correlations towards economic activities by allowing the rapid delivery of people and goods to meet regional demands [1, 2]. Furthermore, compared to other construction projects, highway projects are often inevitably recognized as high-risk projects due to its importance to a nation's economic, societal, and political development [3]. While being highly visible to the public, these projects are only established as successful if and only completed within the allocated time and budget, meets predetermined requirements and objectives, and cause minimal disruptions to the environment [4]. While governments are pushing for successful highway projects because of its importance, highway projects are negatively affecting Malaysia's economic growth from delays and cost overruns [5]. Therefore, highway projects need to meet these success criteria to ensure project success.

While ensuring project success is vital, various factors can hinder the progress of highway projects including unsettled project funding, lack of project planning, and errors or omissions in construction work [6]. Until now, researchers from numerous countries are identifying the key factors affecting their local highway projects to have a thorough understanding of their nation's construction industry including in Egypt, Cambodia, and Kenya [7, 8, 9]. However, the current body of literature lacks that type of information in regards to Malaysia's construction industry. Understanding the factors affecting local highway projects can assist researchers and industry practitioners in identifying opportunities for improvement. Especially, overcoming the key factors may help highway projects avoid unnecessary complications and unwanted misfortunes. Therefore, it is necessary to explore the factors that are affecting Malaysia's highway projects.

The objective of this paper is to identify factors that are affecting Malaysia's highway projects from industry practitioners' perspective. The authors address this objective by analyzing individual interviews with project managers that are managing highway projects throughout Malaysia. In addition to providing additional insights into the existing highway construction body of knowledge, the findings of this study also offer references for assisting researchers and practitioners in promoting highway construction project success.

2. Background

2.1. Factors affecting Highway Projects

Topics related to highway projects have been investigated by researchers and practitioners from both the construction project management and highway transportation researchers field. Studies have determined the key factors that are recurring in their local highway projects [7, 8, 9]. For example, in Kenya, the top factors affecting highway projects include late payment by clients, slow decision making and bureaucracy in the client organization, inadequate planning and scheduling, and rain [9]. While in Egypt, the top factors are political situations, segmentation of the west bank causing limited movements between areas, owner awarding projects to the lowest bidder, delays in progress payment by owners, and shortage of equipment [7]. The top factors for highway projects in Cambodia are rain, flood, land acquisition, owner awarding projects to the lowest bidder, and equipment breakdowns [8]. In other words, although these studies provide insights into the factors affecting highway projects, the results also illustrate that different factors are affecting the local construction sector. Therefore, exploring the factors that are affecting local highway projects is essential.

2.2. Highway Projects in Malaysia

In Malaysia, researchers have explored various topics related to highway projects including critical success factors for private-public partnership, assessments for privatization, processes for managing knowledge, and green behavior of project team members [10, 11, 12, 13, 14]. A prior study has identified the causes of delays in highway projects of Malaysia through a questionnaire survey including improper planning, weather, poor site management, inadequate site investigation, and inappropriate underground utilities [15]. However, the study may not capture additional factors that might affect the industry nor acquire in-depth insights from the industry practitioners because questionnaire surveys are standardized and inflexible to changes throughout the data collection. On the contrary, although findings from individual interviews have difficulties in generalization, the approach provides opportunities to acquire unique perspectives, practical knowledge, and experiences from research participants. Therefore, to provide additional insights to this topic that might not be captured from the questionnaire survey, this study identifies factors that are affecting Malaysia's highway projects through individual interviews with industry practitioners.

3. Methods

The data collection involves acquiring qualitative data from individual interviews with highway project managers. The qualitative data are analyzed using the thematic analysis method. The subsequent subsections describe this study's data collection, and data analysis approaches.

3.1. Data Collection

This study collects data on factors affecting highway projects through open-ended interviews with project managers. Contrary to questionnaire surveys, individual interviews allow investigators to explain, better understand, and explore respondents' opinions and experiences. Furthermore, open-ended questions encourage participants to contribute as much detailed information as they desire [16].

Therefore, project managers are purposefully interviewed using these approaches to acquire their unique perspectives, practical knowledge, and experiences on the topic.

The target population of this study is project managers from highway construction companies that hold a Construction Industry Development Board (CIDB) grade G7 license. In Malaysia, organizations in other license categories are not entitled to undertake any construction projects which exceeds the predetermined value of the construction works in the registration grade. For example, the next highest rank, G6 licensed companies are only allowed to undertake projects not exceeding MYR 10 million (around USD 2.5 million). However, companies with grade G7 license are permitted to take any construction projects without restrictions. Therefore, this research purposely selects project managers from grade G7 highway construction companies because most major highway projects exceed that MYR 10 million threshold.

The individual interviews start with an introduction to the topic and open-ended question. The question introduced to the participant is: What challenges do you usually face in your highway projects? In addition to that question, additional open-ended questions are provided to the participants based on responses received. The follow-up questions aimed to obtain a deeper understanding of the information collected and to clarify whether the participants' statements were understood correctly. If the participant was unable to respond or elaborate on the questions asked, the interviewer tried to rephrase the interview question in another way and gave time for a response. The interviewer encouraged the participants to continue if they had started on an answer without finishing their explanation. After each interview, the discussions are summarized and sent to the participants for validation. This study's data collection involves interviewing sixteen valid respondents.

3.2. Data Analysis

The data analysis consists of performing the thematic analysis to identify patterns from the interview data because the approach can assist in making sense of qualitative data [17]. This method has been employed by Rahman and Ayer [18] and Radzi et al. [19] to analyze qualitative data associated with construction management topics.

The thematic analysis was conducted based on the six phases described in Braun and Clarke [17]. The first phase is becoming familiarized with the data. The authors transcribed the interview data, read, reread, and noted the initial ideas. The second phase is to generate the initial codes. The authors coded for as many potential themes and patterns as possible from the data. The authors then reviewed, discussed, and agreed on any additions and/or changes to the coding. The third phase is to search for themes based on the initial codes. During the process of creating the themes, the authors frequently revisited the codes from the second phase and the original data from the first phase. The fourth phase is to review the themes. To ensure saturation of the data, the authors continually reviewed the subthemes, defined and refined them, checking if themes work to the coded extracts and the entire data set, and reviewing data to search for additional themes. The fifth phase is to define and name the themes. The authors continually went back and forth between the themes, codes, and transcription of the interview to ensure that the themes were true to the independently coded responses. The final phase (sixth phase) is to report the output of the analysis.

4. Results and Discussion

Figure 1 summarizes the factors affecting the success of highway projects in Malaysia that were identified by analyzing individual interview data with sixteen project managers of highway projects in Malaysia. In this study, the factors are themed into internal and external factors from a project manager's perspective. In other words, the internal factors include variables that are usually actionable by project managers (e.g., shortage of equipment that can be improved through resource management processes). Conversely, the external factors are those that often uncontrollable by project managers (e.g., delays in progress payment by clients). The internal factors include variables related to process,

IOP Conf. Series: Materials Science and Engineering **712** (2020) 012030 doi:10.1088/1757-899X/712/1/012030

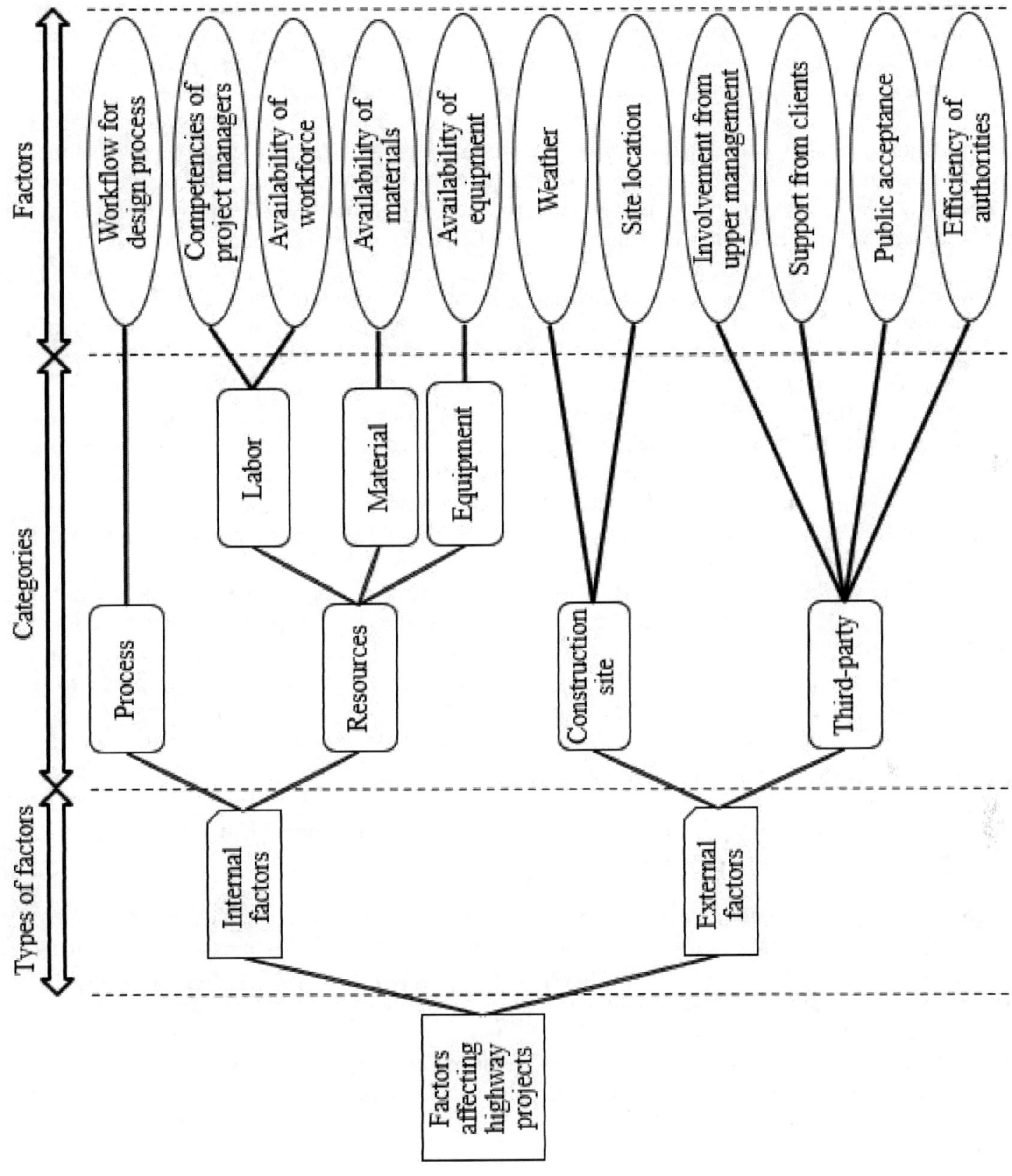

Figure 1. Overview of the factors affecting the success of highway projects in Malaysia

IOP Conf. Series: Materials Science and Engineering **712** (2020) 012030 doi:10.1088/1757-899X/712/1/012030

labor, material, and equipment, while the external factors consist of variables associated with the construction site and third-parties. These variables are further described in the subsequent subsections.

4.1 Internal Factors affecting Highway Projects

4.1.1. Process-related internal factors. Factors in this category include workflows for the design process. Several participants suggested that projects without an effective workflow for the design process usually have drawings that are incomplete, inaccurate, or not finalized that can result in incomplete construction documents for project execution. In addition to incomplete documents, the process of reviewing the documents and issues associated with project planning that are caused by these documents can also cause work idling by contractors and their project team members. Participants also suggest that the effective workflow is specifically crucial to ensure minimal disruptions since designs in highway projects keep changing throughout the project. Also, it is suggested that with an effective workflow, human errors in the design process can be reduced.

4.1.2. Labor-related internal factors. Human-related internal factors in this study involve competencies of project managers and the availability of workforce. Participants suggest that the competencies of project managers play a significant role in ensuring the success of highway projects. Specifically, project managers are suggested to possess skills to solve issues all sorts of issues and problems at construction sites, manage numerous people and various types of individuals, and plan and schedule projects. While most of these skills can be associated with soft skills or people skills such as analytical and problem-solving, communication, teamwork, and planning and organizational [20], participants also suggest that project managers should possess knowledge on highway construction to be effective. Therefore, these results illustrate that the competencies of project managers are affecting the success of highway projects.

Also, having an adequate workforce is suggested to play a crucial role in ensuring project success. Generally, highway projects in Malaysia are reported facing a shortage in labor because the local citizens are not interested in joining the construction industry since the work is dirty, dangerous, and difficult. While employing foreign workers is a plausible option, acquiring working permits from the authority is somewhat challenging since there are specific requirements that need to be fulfilled. Therefore, having the necessary amount of the workforce is a challenge for highway projects, especially in large-scale projects that require a higher number of individuals. In addition to the labor shortage, participants also suggest that highway projects are having difficulties in acquiring individuals with adequate skills, knowledge, and abilities for the workforce. This situation increases the challenge of having a proper workforce in highway projects.

4.1.3. Material-related internal factors. Factors for material-related internal factors involves the availability of materials during construction. While theoretically materials should be that instant during construction, highway projects are suggested to face difficulties in having those materials right on time due to several reasons including absence of the required materials in local markets, suppliers unable to provide the materials due to material shortages, unreasonable pricing for monopolizing practices, and insufficient ground transportation to deliver materials to the site. While projects can opt to obtain materials from international markets, this approach is suggested to have a higher risk due to fluctuation in currencies.

4.1.4. Equipment-related internal factors. Similar to materials, equipment-related internal factors involve the availability of the resource (in this case, equipment and machinery) during construction. While some may suggest that machinery breakdowns are unavoidable, project managers can avoid equipment obsolescence by ensuring proper operation, performing preventive maintenance adequately, and continually monitoring equipment [21]. In addition to equipment failures, several participants suggested that there are also project with inadequate equipment or machinery. Therefore, project managers should ensure to have a proper equipment management system to promote success in highway projects.

IOP Conf. Series: Materials Science and Engineering **712** (2020) 012030 doi:10.1088/1757-899X/712/1/012030

4.2. External Factors affecting Highway Projects

4.2.1. Construction-site related external factors. From the analysis of individual interviews with project managers, factors of site situation that affects the success of highway projects in Malaysia include the weather and site location. First, in addition to slowing down and delaying the workforce from working, the rain has a stronger influence on highway projects compared to building projects because highway projects are more sensitive towards rainwater. Therefore, weather plays a significant role in the success of highway projects, which is also suggested by prior studies including the study on the causes of delay of highway projects in Malaysia [15, 9].

Next, respondents suggest that site location can heavily influence the work progress of highway projects. For example, highway projects that involve areas that are challenging for construction, such as terrains that are hilly or rocky are more demanding compared to those with typical terrains during earthmoving and avoiding drainage issues. Also, terrains that are forestry or swampy areas are less accessible. Furthermore, highways that cross rivers or lakes are challenging because they involve constructing bridges. While these challenges can be associated with highway projects in rural areas, construction sites in urban areas also have their challenges including difficulties in acquiring permits from authorities, managing traffic, relocating underground utilities and services, and setting up storage areas. Also, regardless of a project being in urban or rural areas, the distance between sites and the nearest quarry can affect success due to issues associated with accessibility and cost. In other words, in addition to challenges that emerge from different site conditions caused by poor site investigation [15], site location can affect highway projects in various ways.

4.2.2. Third party-related external factors. Third parties that are suggested to affect the success of highway projects include the upper management, clients, the public, and authorities. Top management can influence success by ensuring projects have front monies for the project before funds from progress payments are available (i.e., financial support), and visiting the sites to provide feedback and motivation to project team members (i.e., emotional support). Also, management can promote project success by ensuring that projects have an adequate number of labor with the appropriate skills. One approach to achieve that includes having responsible parties to identify gaps in the competencies of individuals that will be involved in the project.

Conversely, another third-party that is indirectly involved in the projects – clients can support projects by paying progress payments on time and providing constructive feedback to avoid unnecessary stress among project team members. Conversely, third-parties that are indirectly involved in highway projects can also affect these projects such as the public by causing issues that are emerged from different political views, random amount of compensation and disputes during the land acquisition process, and complaints due to traffic congestion to the authorities. These problems may result in idling in projects until the issues are resolved. While there are complaints from the public, several participants suggest that highway projects do involve constructing alternative roads and working during the night due to traffic concerns, safety issues, and public perception. Furthermore, authorities, the other third-party that are indirectly involved with the highway projects, are suggested to influence the success of highway projects due to authorities' processing times of submittals. Unpredictable and slower than usual processing times might cause issues to highway projects such as idling and unnecessary rescheduling. In other words, public acceptance and efficient authorities are also factors that are affecting highway project success.

5. Conclusion

To promote the success of highway projects in Malaysia, this study identifies factors that are affecting highway project success by analyzing interview data with sixteen highway project managers. Through thematic analysis, the results identified 11 factors affecting the success of highway projects in Malaysia. The factors are: workflow for the design process, competencies of project managers, availability of workforce, availability of materials, availability of equipment, weather, site location, involvement from upper management, support from clients, public acceptance, and efficiency of authorities. These factors can be grouped into six categories: process, labor, material, equipment, construction site, and third-party. Then, these categories can be divided into two groups: internal factors and external factors. These findings highlight the need to identify opportunities in avoiding factors that can negatively affect project success, especially the internal factors that are more actionable by project managers. Thus, the lesson of this study would help the industry to promote project success of highway projects in Malaysia. The key theoretical contribution of this research is by providing researchers and practitioners with a set of alternative factors that are affecting highway project success.

6. References

[1] Högselius P, Kaijser A and Van Der Vleuten E 2015 *Europe's infrastructure transition: Economy, war, nature* (Basingstoke UK)

[2] Rietveld P and Bruinsma F 2012 *Is transport infrastructure effective?: transport infrastructure and accessibility: impacts on the space economy* Springer Science & Business Media.

[3] Donaldson D 2018 Railroads of the Raj: Estimating the impact of transportation infrastructure *American Economic Review* **108** (4-5) 899-934

[4] Wibowo A and Alfen H W 2015 Government-led critical success factors in PPP infrastructure development *Built Environment Project and Asset Management* **5** (1) 121-134.

[5] Bank Negara Malaysia 2018 *Bank Negara Malaysia Quarterly Bulletin Q4 2018* (Malaysia)

[6] Larsen J K, Shen G Q, Lindhard S M and Brunoe T D 2015 Factors affecting schedule delay, cost overrun, and quality level in public construction projects *Journal of Management in Engineering* **32** (1) 04015032

[7] Aziz R F and Abdel-Hakam A A 2016 Exploring delay causes of road construction projects in Egypt *Alexandria Engineering Journal* **55** (2) 1515-1539

[8] Santoso D S and Soeng S 2016 Analyzing delays of road construction projects in Cambodia: Causes and effects *Journal of Management in Engineering* **32** (6) 05016020

[9] Seboru M A 2015 An investigation into factors causing delays in road construction projects in Kenya *American Journal of Civil Engineering* **3** (3) 51-63

[10] Alashwal A M, Abdul-Rahman H and Radzi J 2016 Knowledge utilization process in highway construction projects *Journal of Management in Engineering* **32** (4) 05016006

[11] Ghazali F M, Rashid S A and Sadullah A M 2017 The critical success factors for public-private partnership highway construction project in malaysia *Journal of Engineering and Technology* **8** (1)

[12] Mohamed A, Yusof Z M, Mohamed S F, Misnan M S and Islam R 2017 A framework and evaluation technique for project's viability of privatization highway projects in Malaysia. *International Journal of Engineering and Technology* **9** (6)

[13] Nusa F N M, Endut I R., Takim R and Ishak S Z 2015 Green highway for Malaysia: a literature review *Journal of Civil Engineering and Architecture* **9** (1) 64-71

[14] Nusa F N M, Nasir S and Endut I R 2018 Awareness of green highway concept and terminology: a perspective of on-site personnel in Malaysian highway construction industry *Advances in Transportation and Logistics Research* **1** (1) 475-487

[15] Karunakaran S, Ramli M Z, Malek M A, Musir A A, Imran N F, Fuad N F, Zawawi M H and Zainal M Z 2018 Causes of delay on highway construction project in Klang valley *AIP Conference Proceedings* vol 2030 p 020242

[16] Turner III D W 2010 Qualitative interview design: A practical guide for novice investigators *The qualitative report* **15** (3) 754-760

[17] Braun V and Clarke V 2006 Using thematic analysis in psychology *Qualitative research in psychology* **3** (2) 77-101

[18] Rahman R A and Ayer S K 2017 Prevalent issues in BIM-based construction projects *Proceedings of Joint Conference on Computing in Construction* vol 1 pp 645-652

[19] Radzi A R, Bokhari H R, Rahman R A and Ayer S K 2019 Key attributes of change agents for successful technology adoptions in construction companies: a thematic analysis *Computing in Civil Engineering 2019: Data, Sensing, and Analytics* (Reston, VA: American Society of Civil Engineers) pp 430-437

[20] Rahman R A and Ayer S K 2019 Enhancing the non-technological skills required for effective building information modeling through problem-based learning *Journal of Information Technology in Construction* **24** (9) 154-166
[21] Schaufelberger J E and Migliaccio G C 2019 *Construction equipment management* Routledge

Acknowledgments
The authors would like to thank Universiti Malaysia Pahang for supporting this study through financial grant RDU190340 as well as the project managers that agreed to participate in this work.

Incorporation of Risk Index for Risk Response and Risk Mitigation Strategies of Public-Private Partnership (PPP) Housing Construction Project in Malaysia

P Z Razi, M I Ali and N I Ramli

Faculty of Civil Engineering Technology, University Malaysia Pahang, 26300, Gambang, Kuantan, Malaysia

Abstract. The present work attempt to derive a risk index that causing performance failures among Public-private partnership (PPP) housing construction project, and to propose a risk response and monitoring strategy based on the risk index obtained. A total of thirty-three (33) respondent involved in the PPP housing construction assessing the risk elements employing the Analytical Hierarchy Process (AHP). Findings reveal that among the risks that were captured for high to extreme risks ($0.100 < RI \leq 0.150$) are geologic hazard risk (0.125), and inflation and interest rate risk (0.116) whereas strategy proposed for both of the extreme risks are avoiding the risk and it best to be assigned the risks to the private sector. On contrary, results for moderate risks captured with unreliable value for money (0.066), fluctuation in currency exchange rate (0.058), absence of transparency and accountability during procurement process (0.084), absence of robust and clear agreement (0.077), unforeseen ground condition (0.058) and shortage of technical expertise (0.054) the strategy advocated for the moderate risk is to transfer the risks and shared within both parties i.e. public and private. Eventually, low risks occupied within the risk index of ($0.000 < RI \leq 0.050$) are financial incapacity if private partners (0.034), weak state intervention (0.039), persistent land acquisition (0.039), insufficient capacity in procurement and negotiation (0.035), frequent design change (0.039), design over specification (0.023), poor quality workmanship (0.029), absence of specific PPP framework (0.046) and inadequate PPP skills and knowledge leading to poor planning (0.043) all the risks best to accept and retained within the public sectors. Taken together, the development of risk response and risk mitigation plan that emerged from the risk index offered significant contribution which has gain a new understanding that risks with severe or low exposure can be reduced or avoided taking into account its strategic and effective response and mitigation approaches.

1. Introduction

Public-private partnerships (PPP) since its establishment is often associated with infrastructure development [1], however in line with current developments, PPP has extended to other sectors such as housing, tourism and hospitals [2]. The principle of construction of PPP housing project is that the private sector is expected to carry out the project whereas the government provided the land and other incentives that may benefit both parties [1]. PPP in housing project bonded under the turnkey contract, whereas the government will motivate the private sector by providing an attractive payment method [3]. Malaysia government have tremendously introduced many PPP housing programmes in order to provide adequate housing for the targeted groups to name some, including One Malaysia housing programme (PR1MA), 1Malaysia Civil Servants Housing (PPA1M) and others. In a similar vein, to ensure an adequate supply of low-cost houses, government, as the public sector sets that for any mixed development projects undertaken by private developers are required to allocate a minimum of 30% to low-cost housing [4].

The adoption of Public-Private Partnership (PPP) in a housing facility in Malaysia is intended to increase urban housing standard and address housing affordability and convenience. However, there is still a lack of or limited consensus on the key risk factors that adversely impact on PPP housing success performances in terms of the schedule delay, significant increase of cost and the poor quality of the house built. Thus, this study incorporates the risks involved in the construction of PPP housing through risk management processes comprising risk identification, risk analysis, risk response, risk mitigation and the construction performance of the housing construction itself through several risk elements that presented in the m section.

2. Literature Review

Although literature review underscores the tremendous development of PPP projects in the infrastructure project, however this is not happening for the construction of PPP housing. Even though the developed countries have achieved a successful PPP implementation, however there are evident in the accounts of many failed PPP projects around the world [5], [6]. While in developing countries, PPP is still at the beginning of the learning process [7]. Despite the fact that few previous studies reports on the PPP housing including [3], [8]–[11], however there is still a lack of or limited consensus on the key risk factors that adversely impact on PPP housing success performances. Such failure of this PPP demonstrates fallacies on the risk management approach on which, absence of convincing risk assessment models [12] and arguments on the risk transference for which parties best to bear the risk [13]. Apart from that, the relevancy is called upon for a dearth of publications on the actual implementation of housing public-private partnerships (PPPs), especially in developing countries [10], [14] couple with the lack of success reporting on the PPP housing approach [9]. Aggravating the current situation is much depended to the lack of knowledge on the mechanism in PPP procurement, criticisms on the early stage of development are becoming more and more significant [15] couple with absence of integration between public and private in handling the housing issues [16] as the level of understanding of the implications of risk and constraints differs in the public and private sectors, since public sector seems to underestimate the extent of risk that needs seriously addressed before considering implementing a reliable PPP project [17]. Risk Management is a proactive decision-making process used to minimise and manage the risks most efficiently and appropriately. Risk management is indeed a dynamic tool which must be continuous throughout the project life cycle, and it is based on intuition and experience for a high level of judgment [18]. Risk management is primarily a decision to be made, rather than a predetermined outcome [19].

3. Research Methodology

This study employing questionnaire survey as its tools, designed with a pair-wise comparison of scale 1 to 9 (Table 1) following the AHP method, to determine the prioritisation of risks that caused poor performance of PPP housing construction project. The weightage obtained for each risk elements deemed as the probability (P). In addition, as AHP is a multi-criteria decision making tools, the methodological of AHP requires an expert to be the subject respondent. Experts are needed to determine the relative importance of each element and sub-elements using pair-wise comparison and their constructive insight thus provides a weightage for each REs. A quantitative approach was also used to determine the impact of occurrence (I) adopting the mean rank analysis using SPSS version.

Table 1. AHP Rating Scale [20], [21]

Intensity	Definition	Explanation
1	Equal importance	Two activities contribute equally to the objective
3	Moderate importance	Experience and judgement slightly favors one activity over another
5	Essential or strong importance	Experience and judgement strongly favor one activity over another
7	Very strongly importance	An activity is strongly favored and its dominance demonstrated in practice
9	Extremely importance	Evidence favoring one over another of

		highest possible order of affirmation
2,4,6,8	Intermediate values	When compromise is needed

4. Results and Discussion

4.1. Demographic Data

A total of sixty-four (64) construction project team that inclusive of the project developer, civil & structural engineer, architect, main contractor, and mechanical and electrical engineer in accordance to the listing provided in PR1MA corporation official website further contacted requesting for their willingness to participate in answering the survey. Hence, sixty-four (64) questionnaire distributed and, thirty three (33) responded giving the response rate of the survey instrument at 52%. The response rate is considered sufficient as recommended by [22], a response rate of 50% for a questionnaire survey is considered adequate. Table 2 depicts the expert respondent profession. Civil Engineer monopolised the highest number of respondents with (N = 11), followed by Quantity Surveyor (N=7) and Architects (N=6). While the remaining is Main Contractor (N=3), “Others” as their profession categories that consist of (land surveyor, property manager and project manager). Similarly, Subcontractors (N=2) while Mechanical and Electrical Engineers (N=1). As this study relying on the expert judgement, the expertise of each of the respondent undeniable since most of them have more than ten years of experience and the detail tabulated in the table below.

Table 2. Expert’s respondent profiles

Profession (N)	Experience (years)	
	10 to 14 years	Over 15 years
Civil Engineer (11)	8	3
Architect (6)	5	1
Mechanical Engineer (1)	1	-
Quantity Surveyor (7)	4	3
Main Contractor (3)	1	2
Sub-Contractors (2)	1	1
Others (3)	2	1
Total (33)	22	11

4.2. Risk probability and impact

Analysis for risk probability employing the AHP method whereas it gave the prioritisation of risk elements and sub-elements (REs) in terms of their weighted importance. On the other hand, analysis of the risk impact presented in descriptive statistics using the mean score. The scale value of risk impact presented in Table 3 was proposed and guided from rating impacts for the risk from the project management body of knowledge (PMBOK® guide) [23]. The description shown in the table below was adjusted to suit within the context of this study.

Table 3. Scale of impact and description

Scale Value	Scale	Description of Scale
0.1	Little impact	The impact is insignificant; minimal impact or nor apparent impact at all
0.3	Minor impact	Minor impact to the project progress performance
0.5	Moderate impact	Significant impact on the project performance caused to high cost, time and quality of work

0.7	Major impact	Serious or major impact on the project performance caused to stop work of the project
0.9	Catastrophic impact	Catastrophic impact to the project performance including project termination

Table 4 depicts the results of the risk probability and impact analysis. It is worth noting that the risk element was divided into two core element namely the external (country level) and internal (project level). From an analysis of impact, it is apparent from the mean values of the external risks for the risk impact of sub-elements risk ranged from the highest mean at (0.682) and the lowest mean at (0.536), which shows that, the impact upon performance of PPP housing construction ranging from moderate impact to major impact for the external risks. On the contrary, for the internal risks it is apparent that the mean values ranging from the highest (0.664) to the lowest (0.367). The impact upon the performance of PPP housing construction ranging from moderate impact to minor impact for the internal risks.

The ranking for the risk sub-elements captured the highest for geologic hazards risks (earthquakes, tsunami, landslide and floods) at 0.184, followed by inflation and interest rate risk at 0.179, unreliable value for money (VfM) at 0.104, fluctuation in currency exchange rate at 0.092, weak state intervention at 0.064 followed by persistent land acquisition issues at 0.058. On the other hand, absence of transparency and accountability during procurement process were rated the highest risk sub-elements at 0.126, followed by absence of robust and clear agreement at 0.120, unforeseen ground conditions at 0.091, need for a specific PPP framework at 0.077 and lastly inadequate PPP skills and knowledge leading to poor planning at 0.070. The findings of ranking for the risk sub-elements is quite revealing that most of the prioritised risk monopolised by the economic and financial risk for external risk, meanwhile procurement and contractual risks dominate in the internal risks.

Table 4. Result of risk probability and impact

Core Element	Risk Element	Risk Sub-Elements (REs)	Probability (P)	Impact (I)
External	EF Economic & Financial Risk	EF1 (Inflation & interest rate risk)	0.179	0.645
		EF2 (Fluctuation in the currency exchange rate)	0.092	0.633
		EF3 (Financial incapacity of private partners)	0.055	0.615
		EF4 (Unstable macro-economic condition & absence of sound economic policy)	0.051	0.567
		EF5 (Unreliable value for money project)	0.104	0.633
	SP Socio & Politic Risk	SP1 (Absence of good & favourable governance and political support)	0.049	0.063
		SP2 (Absence of uniform policy on PPP housing provision)	0.025	0.585
		SP3 (Weak state intervention)	0.064	0.603
		SP4 (Persistent land acquisition issue)	0.058	0.676
	MR Market Risk	MR1 (Government interference in PPP construction market)	0.056	0.597
		MR2 (Unanticipated housing demand risk)	0.019	0.567
		MR3 (Inadequate PPP housing ownership)	0.020	0.536
	NH Natural Hazard Risk	NH1 (Geologic hazards – earthquakes, tsunami, landslide, floods)	0.184	0.682
		NH2 (Atmospheric hazards – tropical cyclone, droughts, severe thunderstorms)	0.044	0.633
	PC	PC1 (Inappropriate indemnity provision)	0.035	0.561

Internal	Procurement & Contractual Risk	PC2 (Absence of transparency & accountability during the procurement process)	0.126	0.664
		PC3 (Absence of robust and clear agreement)	0.120	0.645
		PC4 (Insufficient capacity in procurement & negotiations)	0.059	0.591
	CR Construction Resources Risk	CR1 (Shortage of workers)	0.021	0.579
		CR2 (Poor quality workmanship)	0.046	0.633
		CR3 (Suppliers inability to supply material)	0.018	0.585
		CR4 (Shortage of technical expertise)	0.085	0.633
Internal	TF Technical Faulty Risk	TF1(Insufficient drawings & specifications)	0.028	0.597
		TF2 (Frequent design change)	0.062	0.627
		TF3 (Unforeseen ground conditions)	0.091	0.639
		TF4(Misinterpretation of technical specifications)	0.037	0.615
		TF5 (Design over specification)	0.038	0.615
	PO Project Organization Risk	PO1 (Inadequate PPP skills & knowledge leading to poor planning)	0.079	0.609
		PO2 (Lack of cordial relationship among construction parties)	0.027	0.518
		PO3 (Need for a specific PPP framework)	0.077	0.597
		PO4(Disputes & conflicts between parties)	0.021	0.579
	TA, Technological Advancement Risk	TA1 (Insufficient technology investment)	0.035	0.397
		TA2 (Unavailability of technology advancement)	0.019	0.367

4.3. Risk Index

Generally, the risk index (RI) is derived from the probability (P) and impact (I) analysis as illustrated by the following equation.

Risk Index (RI) = Risk Probability (P) x Risk Impact (I) Equation (1)

This risk index may serve as a construction indicator from a PPP housing construction perspective. Figure 1 below illustrates the risk index for external and internal risk sub-elements respectively. The evidence from this study captured risk index for external risk dominates by geologic hazards (earthquakes, tsunami, landslides, floods) - NH1 (0.125), inflation and interest rate risk - EF1 (0.116), unreliable value for money (vfm) project - EF5 (0.066), fluctuation in currency exchange rate - EF2 (0.058), persistent land acquisition issue - SP4 (0.039), weak state intervention - SP3 (0.039) and financial incapacity of private partners - EF3 (0.034). On the other hand, risk index of internal risk demonstrates the highest risk index monopolized by absence of transparency and accountability during procurement process - PC2 (0.084), absence of robust and clear agreement - PC3 (0.077), unforeseen ground conditions - TF3 (0.058), shortage of technical expertise - CR4 (0.054), absence of a specific PPP framework - PO3 (0.046), inadequate PPP skills and knowledge leading to poor planning - PO1 (0.043), frequent design change - TF2 (0.039), Insufficient capacity in procurement and negotiations - PC4 (0.035), poor quality workmanship - CR2 (0.029) and design over specification - TF5 (0.023).

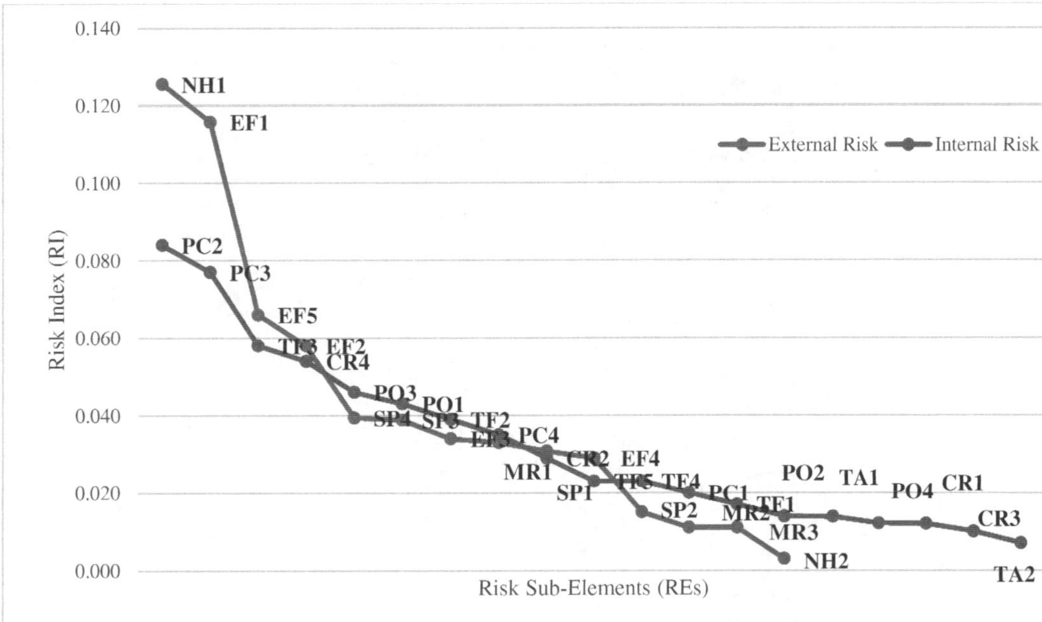

Figure 1. Risk index of external and internal risk

4.4. Risk Response

The risk response mechanism for this study is drawing largely on "risk allocation preferences" developed by [24] whom also borrowing similar study during earlier analyses conducted by [25] focusing on PPP infrastructure project. For the context of this study, to suit the risk index that was developed in previous section, the risk allocation preferences were develop based on the equalities and inequalities function form as described below:

$0.100 < RI \leq 0.150$	Extreme/High risk (Threshold 1)
$0.050 < RI \leq 0.100$	Medium risk (Threshold 2)
$0 < RI \leq 0.050$	Low risk (Threshold 3)

Whereas, for the risks falls within threshold 1, all the risks should borne by the private sectors, while, in threshold 2 risk shared within the public and private and else threshold 3 all the risk will borne by the public sector. Therefore, figure 2 illustrates the risks that divided according to their respective risk index and the preferences on the risk allocation strategies.

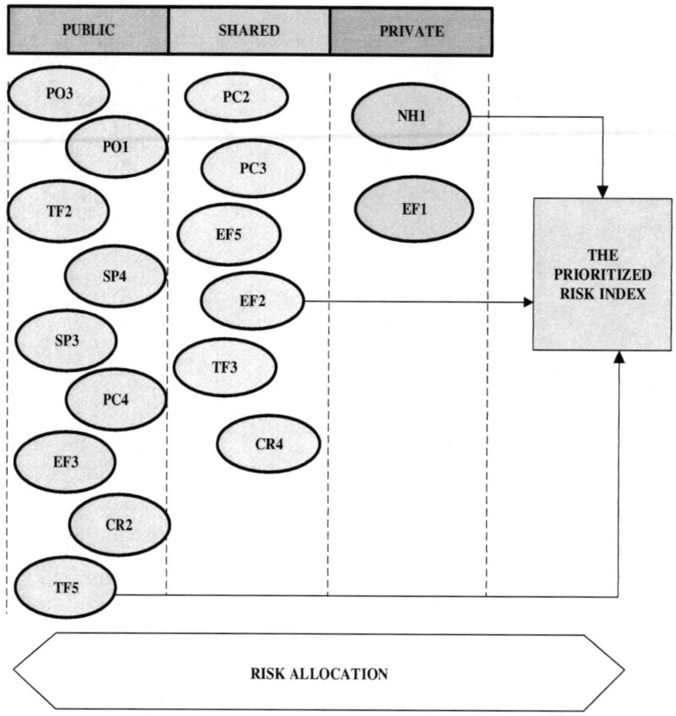

Figure 2. Integration Model of Risk Allocation for PPP Housing Construction

4.5. Risk Mitigation

In ensuring the success of PPP projects, it is important for all partners to manage the risks from a project life cycle perspective, in which risks are identified and assessed in the earliest possible project stage and allocated to the parties who are in the best position to control them. Furthermore, it is also important to continuous monitor the risks and develops proactive risk respond strategies throughout the project life cycle [26]. Within the context of risk mitigation, for threshold 1 shows that within the risk index falls in the category will classified as extreme/high thus the mitigation requires immediate government decision for intervening in the issue with involvement of state and highest project management, threshold 2 will be served as medium risk that requires higher senior management decision and manages by specific monitoring or response procedures, if the resources allow, this kind of risk requires to develop more detailed actions, while threshold 3 all the risk regards as low risk, thus the mitigation merely requires a project routine procedure.

5. Conclusion

By integrating the performance measurement (time, cost and quality) in the forms of the risk impact towards the PPP housing construction and risk management pillars process, it strengthens the main goals of this study. The incorporation of risk index based upon the weighting rates by its importance for the Malaysian context, the weightings may need for restructuring before the tools employed in other countries outside of Malaysia. Notwithstanding the limitation, this studies main contributions relevant to academia and industry practice: it open insight of the construction industry practitioners in particular the PPP housing construction needs and the importance of risk management in PPP housing construction projects. Also, the risk analysis process, besides giving weighted to prioritisation of each risk, it can be used as a key performance indicator to anticipate the risks to be encountered, the response to those risks and the effective strategies to mitigate those risks.

6. References

[1] Abdul Aziz AR 2010 Housing private public partnerships: perspectives from the government agencies in *4th NAPREC Conf.* 2010 September 1–24.

[2] Hossain M, Guest R and Smith C 2019 Performance indicators of public private partnership in Bangladesh *Int. J. Product. Perform. Manag.* **68** 1 pp 46–68
[3] Trangkanont S and Charoenngam C 2014 Private partner's risk response in PPP low-cost housing projects *Prop. Manag.* **32** 1 pp 67–94
[4] Bakhtyar B, Zaharim A, Sopian K, Saadatian O and Abdulateef J 2012 Affordable quality housing for urban low income earners in Malaysia *SSRN Electron. J.* December
[5] Loosemore M and Cheung E 2015 Implementing systems thinking to manage risk in public private partnership projects *Int. J. Proj. Manag.* **33** 6 pp 1325–34
[6] El-Gohary NM, Osman H and El-Diraby TE 2006 Stakeholder management for public private partnerships *Int. J. Proj. Manag.* **24** 7 pp 595–604.
[7] Sastoque LM, Arboleda CA and Ponz JL 2016 A proposal for risk allocation in social infrastructure projects applying PPP in Colombia *Procedia Eng.* **145** pp 1354–61
[8] Kavishe N 2016 Risks associated with public private partnership (PPP) housing projects delivery in Tanzania *Int. J. Constr. Eng. Manag.* 7 2 pp 81–7.
[9] Kwofie TE, Afram S and Botchway E 2016 A critical success model for PPP public housing delivery in Ghana *Built Environ. Proj. Asset Manag.* **6** 1 pp 58–73
[10] Abdul-Aziz AR 2012 Control mechanisms exercised in Malaysian housing public-private partnerships *Constr. Manag. Econ.* **30** 1 pp 37–55
[11] Liu T, Chan A and Wang S 2014 PPP framework for public rental housing projects in China *Int.Conf. on Construction and Real Estate Management (ICCREM)* pp 573–581
[12] Tang L, Shen Q and Cheng EWL 2010 A review of studies on Public–Private Partnership projects in the construction industry *Int. J. Proj. Manag.* **28** 7 pp 683–94
[13] Reeves E 2013 The not so good, the bad and the ugly: over twelve years of PPP in Ireland *Local Gov. Stud.* **39** 3 pp 375–95
[14] Sengupta U 2006 Government intervention and public-private partnerships in housing delivery in Kolkata *Habitat Int.* **30** 3 pp 448–61
[15] Wan Zahari WY and Xuen JWM 2013 Private Finance Initiative (PFI)/ Public Private Partnership (PPP) in Malaysia context *Int. J. Inf. Technol. Bus. Manag.* **17** 1 pp 29–42
[16] Rahardjo H, Suryani F and Trikariastoto ST 2014 Key success factors for public private partnership in urban renewal in Jakarta *Int. J. Eng. Technol.* **6** 3 pp 217–19
[17] Akerele D and Gidado K 3-5 September 2003 The risks and constraints in the implementation of PFI/PPP in Nigeria Greenwood DJ (Ed.) *19th Ann.Assoc.of Researchers in Construction Management ARCOM Conf.* University of Brighton **1** pp 379-91
[18] Abdul-Rahman H, Wang C and Sheik Mohamad F 2015 Implementation of risk management in Malaysian construction industry: case studies *J. Constr. Eng.* pp 1–6
[19] Kang BG, Fazlie MA, Goh BH, Song MK and Zhang C 2015 Current practice of risk management in the Malaysia construction industry–the process and tools/techniques *Int. J. Struct. Civ. Eng. Res.* **4** 4 pp 371–77
[20] Saaty TL 1988 What is the Analytic Hierarchy Process? *Mathematical Models for Decision Support* **48** Berlin, Heidelberg: Springer Berlin Heidelberg pp 109–21.
[21] Saaty TL and Vargas LG 2012 The possibility of group choice: Pairwise comparisons and merging functions *Soc. Choice Welfare* **38** 3 pp. 481–496
[22] Creswell JW 2014 *Research design* Sage Publications Inc California 4th ed.
[23] Project Management Institute (PMI) 2013 A guide to the project management body of knowledge Pennsylvania USA
[24] Ke Y, Wang SQ, Chan APC and Lam PTI 2010 Preferred risk allocation in China's public-private partnership (PPP) projects *Int. J. Proj. Manag.* **28** 5 pp 482–92
[25] Bing L, Akintoye A, Edwards P and Hardcastle C 2005 The allocation of risk in PPP/PFI construction projects in the UK *Int. J. Proj. Manag.* **23** 1 pp 25–35
[26] Zou PXW, Wang S and Fang D 2008 A life- cycle risk management framework for PPP infrastructure projects *J. Financ. Manag. Prop. Constr.* **13** 2 pp 123–42

Acknowledgement
Authors would like to thank all relevant respondents that willingly participate in the survey.

IOP Conf. Series: Materials Science and Engineering **712** (2020) 012032 doi:10.1088/1757-899X/712/1/012032

Methodological Study to Classification of Damage State Immediately Subsequent to the Banjarnegara Indonesia Earthquake On 2018

A S Bawono[1,2], M I Ali[1], S Kusumadewi[3] and N I Ramli[1]

[1]Faculty of Civil Engineering Technology, Universiti Malaysia Pahang, 26300 Gambang, Pahang, Malaysia
[2]Department of Civil Engineering, Faculty of Science and Technology, Universitas Teknologi Yogyakarta, Jl. Siliwangi Ringroad Utara, Jombor, Mlati, Kab. Sleman, DIY 55285, Indonesia
[3]Department of Informatic, Faculty of Industrial Engineering, Universitas Islam Indonesia Jl. Kaliurang Km.14,4, Kab. Sleman, DIY 55584, Indonesia

Abstract. This paper presents a preliminary methodological study of a seismic risk assessment in a residential area to classification of damage state immediately subsequent after earthquake happen. This paper compared with difference research of a seismic risk assessment in the same area, i.e. Kertosari, Banjarnegara Indonesia. This methodological study to identified compatibility of the building vulnerability assessment with Naïve Bayes Classifier (NBC) method for determine the probability of damage that has 5 different damage statuses (slight, moderate, extensive, complete and none), where the criteria damage states in vulnerable building has additional damage parameters based on the latest building damage regulations issued in 2013, which have totally 104 data difference criteria. The results from determine the probability of damage processing using the NBC methodology in Kertosari Banjarnegara that most type of damage states is extensive damage. There is a difference with 2018 research data, where the most damage state is the moderate damage, and there is also a difference with the Banjarnegara Municipal Disaster Management Authority (BPBD Banjarnegara) data, which has the most damage state is the complete damage. The difference is caused by the criteria in determining the damage conditions are still too general and the uncertainty value is high. Therefore, it is necessary to improve and detailed the building type model and damaged condition criteria for Indonesian buildings, based on the assessment and suggestion of experts who have good and long experience in earthquake risk management and understand the state of damage to Indonesian building regulations, such as: (1) researchers/scientists; (2) small and medium-sized contractors; (3) construction supervisors; (4) government officials; (5) business leaders; (6) academics/educators; (7) non-governmental organizations, in order to obtain an assessment of damage to residential houses quickly and accurately.

1. Introduction

The seismic risk assessment process can be divided into two main groups namely seismic hazard and vulnerability. Seismic vulnerability depends upon model building type, and damage state effected from the hazard and seismic hazard is the event capable of causing damage while seismic vulnerability represents the degree of loss of an element resulting from hazard. It can depict in the equation R = H x V / C, where R is a risk, H is a hazard, V is a vulnerability, and C is capacity [1,2]. It can be understood that the failure of buildings and many victims when the earthquake came one of them due to vulnerable buildings [3–5]. In seismic risk assessment, many researchers need experts to judgment an evaluation and decision-making process. The expert could have totally different opinions about the potential earthquake vulnerability of a building if compared to the opinions of another expert, due to the differences in their knowledge and experience [6,7]. Research on the vulnerability of buildings in a developing country, is still low, including Indonesia. Contrary, information from the building vulnerability assessment can be helpful for risk mitigation and emergency response planning and important as a based guideline to build any new building for preventing of losses [5], [8–11]. In line that, this research aim to develop a simple guide to resolution building type model and the criteria of damage state for compatible Indonesian building using NBC methodology in determine the probability of damage that has 5 different damage state (slight, moderate, extensive, complete, and none).

2. The general background of the study

2.1 Banjarnegara Earthquake
The Meteorological, Climatological, and Geophysical Agency (BMKG) obtain information data as to Banjarnegara Earthquake in April 2018 [12]. Time of the incident on Tuesday, 18 April 2018 at 13:28:35 WIB, scale: 4,4 SR, Location in 7,21°LS and 109,65°BT with Depth 4 Km, earthquake shocks felt strong enough in Kertosari Villages, Banjarnegara District and surrounding areas. There were various damage states of building i.e. slight by 88 houses damage, moderate by 31 houses damage, extensive and complete by 82 houses damage [13]

2.2 Predicting and estimating accuracy damage state using Naïve Bayes Classifier (NBC)
Classification is one of the most important techniques in datamining for data analysis. In datamining, different classification techniques are available to predict the outcome for a given dataset. There are many classification methods for predicting and estimating accuracy; one such famous method is Naïve Bayes Classifier. Naïve Bayes is very popular as it is easy to build, however to the assumption of conditional independence among predictor's results in loss of accuracy damage states. One of the main reasons for the better performance of Naïve Bayes Classifier is the assumption of independence among predictors [6,7].

3. Methods

3.1. Naïve Bayes Classifier (NBC)
Classification is a process of grouping data based on certain characteristics into predetermined classes. The Naïve Bayes Classifier is a datamining classification method which takes probabilities of attributes belonging to class for prediction. NBC is a supervised classification approach which can be used effectively to model a predictive problem probabilistically [14]. Naïve Bayes classifier is based on Bayes' Theorem where predictors are treated as Independent. In Naïve Bayes method the overall probabilities of attributes belonging to a class are calculated by resuming that the likelihood of an attribute on a given class value is not dependent on other attributes. This presumption leads NBC to better results and is called conditional independence [14–16]. Is known $X = \{x_{1,}x_2, \ldots, x_k\}$ is the attribute influence the C class. Naïve Bayes theorem is described as follows:

$$P(C|X) = P(X|C) * \frac{P(C)}{P(X)} \quad (1)$$

Where:
$P(X)$ = predictor prior probability
$P(C)$ = the relative frequency of the sample C class
$P(X|C)$ =likelihood
$P(C|X)$ = posterior Probability
The conditional Independence is explained in this scenario as, the predictor (x) value on class (c) has no effect on the other predictor's values.

3.2 Discrete Probability
There are given free attributes (independence), the value of probability can be given as follows:

$$P(x_1 \ldots\ldots, x_k|C) = P(X_1|C)^* \ldots {}^* P(X_k|C) \quad (2)$$

If the attribute is to discretion, then $P(X_i|C)$ estimated as frequentation relative of the sample which has a value of xi, as attribute in class C. The conditional Independence is explained in this scenario as; the predictor (x) value on class (c) has no effect on the other predictor's values.

4. Results and Discussion

The methods of NBC are to calculate the discrete probability of damage states, there are none, slight, moderate, extensive and complete. Where, every house will get impact of all damage states, but the highest rank/the most type of damage states can be difference among them. Based on research before [8] and data additional criteria of damage states from Ministry of Public Works (PU) [17], there are 7 objects of assessment for classified of damage states [18]: 1.Building Condition, 2.Damage of column and beam, 3.Damage of wall, 4.Damage of roof, 5.Damage of plafond, 6.Damage of electrical installation and 7.Damage of door/window. The 7 criteria divide on to 104 damage states (slight, moderate, extensive, complete and none), as sample as in Table 1.

Table 1. Sample Object of Assessment and Damage State

No.	Object of Assessment							Damage States
	House Condition (X1)	Column and Beam (X2)	Wall (X3)	Roof (X4)	Plafond (X5)	Electrical Installation (X6)	Door/window (X7)	
1	Building still stands	Most columns, beams and/or roofs are damaged	A small part of wall is broken	Roof collapsed	Most of walls and ceiling are collapsed	Electrical installation is totally damaged	Door/window is totally damaged	Extensive (E)
2	Building still stands	Column beam partially is broken	Cracks on the plastering walls	Loose roof/tile cover	Some of ceiling coverings are broken	Some electrical installations are damaged	Door/window is partially damaged	Slight (S)
3	Building collapsed completely	Most columns, beams and/or roofs are damaged	Most of walls and ceiling are collapsed	Some of roof trusses are broken	Some ceiling coverings are damaged	Some electrical installations are damaged	Door/window is totally damaged	Complete (C)
4	Building still standing	Structure collapsed mostly	Wall partially collapsed	Roof/tile cover cannot be separated	Broken ceiling	Electrical installation is not damaged	Door/window is not broken	Moderate (M)
5	Building still stands	Structure/beam column is not damaged	Wall does not crack	Roof/tile cover cannot be separated	Ceiling is not damaged	Electrical installation is not damaged	Door/window is not broken	None (N)

And then an object of assessment calculated depends on probability which has difference damage states use equation 1, as in Table 2.

IOP Conf. Series: Materials Science and Engineering **712** (2020) 012032 doi:10.1088/1757-899X/712/1/012032

Table 2. Probability of Damage States

Slight (S)	Moderate (M)	Extensive (E)	Complete (C)	None (N)
0.17308	0.41346	0.36538	0.03846	0.009615385

And then criteria of damage states, are calculated depends on damage states using Equation 1, as sample as in Table 3.

Table 3. Sample of scoring Damage Criteria from Building Conditions (X1)

X1	Building Conditions				
	Slight (S)	Moderate (M)	Extensive (E)	Complete (C)	None (N)
P (Building is remains standing) =	1	1	0.3421	0	1
P (Building is totally colapse) =	0	0	0.0263	1	0
P (Building is incline) =	0	0	0.6316	0	0

To calculate the highest rank/the most type of damage states of damage probability of decision damage state use Equation 2, as in Table 4.

Table 4. Discrete Probability of Decision Damage State (NBC) in Kertosari Village

Number	Photo	Coordinate	Discrete Probability of Decision Damage State (NBC)	
			Highest Rank	Categories
1		7°13'21.068"S109°40'12.551"E	1.0000	Extensive Damage (E)
2		7°13'899"S109°40'026"E	1.0000	Extensive Damage (E)
3		7°12'758"S109°40'673"E	1.0000	Extensive Damage (E)

IOP Conf. Series: Materials Science and Engineering **712** (2020) 012032 doi:10.1088/1757-899X/712/1/012032

4		7°12'006"S109°40'263"E	0.9777	Slight Damage (S)

In Table 4 the most damage states in Kertosari village is Extensive Damage, which have a difference result to a preliminary survey [8] and BPBD Banjarnegara data [13] in year 2018 as Table 5.

Table 5. Damage data based on a preliminary survey and BPBD Banjarnegara in Kertosari village

Number	Photo	Building	Latitude	Longitude	Damage State
1		House 1	-7.21415	109.67872	Complete Damage (C)
2		House 2	-7.22300	109.67012	Complete Damage (C)
3		House 3	-7.22333	109.66958	Complete Damage (C)
4		House 4	-7.22232	109.66930	Complete Damage (C)

In Table 5 the most damage state in Kertosari village is complete damage, and the values of cumulative damage probabilities summarized is moderate damage [8].

5. Conclusion

The results from determine the probability of damage processing using the NBC methodology in Kertosari Banjarnegara that most of damage state is extensive damage. There is a difference with 2018 research data, where the most damage states is the moderate damage, and there is also a difference with the BPBD Banjarnegara data, which has the most damage states is the complete damage. The difference is caused by the criteria in determining the damage conditions are still too general and the uncertainty value is high. Therefore, it is necessary to improve and detailed the building type model and damaged condition criteria for Indonesian buildings, based on the assessment and suggestion of experts who have good and long experience in earthquake risk management and understand the state of damage to Indonesian building regulations, such as: (1) researchers/scientists; (2) small and medium-sized contractors; (3) construction supervisors; (4) government officials; (5) business leaders; (6) academics/educators; (7) non-governmental organizations, in order to obtain an assessment of damage to residential houses quickly and accurately.

6. References

[1] A. Heijmans and L. P. Victoria, *Citizenry-Based & Development-Oriented Disaster Response*. Center for Disaster Preparadness, 2001.

[2] D. van Niekerk, *Introduction To Disaster Risk Reduction*, no. August. 2011, p. 59.

[3] A. S. Elnashai and L. di Sarno, *Fundamentals of earthquake engineering from source to fragility*, no. 1. John Willey and Sons, Ltd, 2015.

[4] A. H. M. Muntasir Billah and M. Shahria Alam, "Seismic fragility assessment of highway bridges: a state-of-the-art review," *Structure and Infrastructure Engineering*, vol. 11, no. 6. Taylor & Francis, pp. 804–832, 2015.

[5] V. Rahimi, M. I. Khodakarami, and R. Vahdani, "Determination of Structural Fragility Curves of buildings with Hazus methodology for Seismic risk Assessment in the city of Semnan , Iran Determination of Structural Fragility Curves of buildings with," in *10th International Cingress on Civil Engineering 5-7 May 2015*, 2015, no. May.

[6] K. Netti and Y. Radhika, "Minimizing Loss of Accuracy for Seismic Hazard Prediction using Naive Bayes Classifier," pp. 75–77, 2016.

[7] O. Otobo, "RISK MANAGEMENT SYSTEM TO GUIDE BUILDING CONSTRUCTION PROJECTS ' IN DEVELOPING COUNTRIES : A CASE STUDY OF NIGERIA Odimabo Otobo , Onengiyeofori February RISK MANAGEMENT SYSTEM TO GUIDE BUILDING CONSTRUCTION PROJECTS ' IN DEVELOPING COUNTRIES : A CASE STU," no. February, 2016.

[8] N. Bawono, Adi Setiabudi;Ali,MI,Ramli, "A Preliminary Study of Seismic Risk Assessment Shortly After the Banjarnegara Indonesia Earthquake on 2018." 2018.

[9] A. S. Bawono, "Studi Kerentanan Bangunan Akibat Gempa (Studi Kasus Perumahan di Bantul)," Universitas Islam Indonesia, 2009.

[10] M. Khalfan, "FRAGILITY CURVES FOR RESIDENTIAL BUILDINGS IN DEVELOPING COUNTRIES: A CASE STUDY ON NON-ENGINEERED UNREINFORCED MASONRY HOMES IN BANTUL, INDONESIA," McMaster, 2013.

[11] G. G. Penelis and G. G. Penelis, *Concrete Buildings in Seismic Regions*. Boca Raton: CRC PRESS, 2014.

[12] BMKG, "Press Release No.UM.505/10/D3/IV/2018," Jakarta, 2018.

[13] BPBD Banjarnegara, "REKAPITULASI DATA KERUSAKAN INFRASTRUKTUR BANGUNA RUMAH BENCANA GEMPA BUMI KECAMATAN KALIBENING KABUPATEN BANJARNEGARA TAHUN 2018," BANJARNEGARA, 2018.

[14] J. P. Jiawei Han, Micheline Kamber, *Data Mining Concepts and Techniques Third Edition*, Third edit. USA: Elsevier Inc., 2012.

[15] X. Wang, Y. He, S. Member, D. D. Wang, and S. Member, "Non-Naive Bayesian Classifiers for Classification Problems With Continuous Attributes," vol. 44, no. 1, pp. 21–39, 2014.

[16] S. Kusumadewi, "Ketidakpastian dalam Sistem Pakar," 2019.

[17] P. KEMENPU, BALITBANG, "Panduan Praktis Pemeriksaan Kerusakan Bangunan akibat Gempa Bumi," 2013.

[18] D. Departemen Pekerjaan Umum, "Laporan penanganan bencana gempa bumi yogyakarta dan jawa tengah STATUS, MINGGU 4 JUNI 2006," 2006.

Acknowledgement

The support provided in the preliminary survey for getting information/data about the preliminary study of seismic risk assessment in Banjanegara Earthquake 18th April 2018 on 21st April 2018, between Universitas Teknologi Yogyakarta and Magister Teknik Sipil Universitas Islam Indonesia for this study is highly appreciated.

IOP Conf. Series: Materials Science and Engineering **712** (2020) 012033 doi:10.1088/1757-899X/712/1/012033

Effect of government policy towards the abandoned housing project in Malaysia

S Saidah An'nisaa, A Nur Farhayu, R Nur Asyekin, A Mohamad Idris and N I Ramli

Faculty of Civil Engineering Technology, Universiti Malaysia Pahang, Lebuhraya Tun Razak, Gambang, 26300, Pahang, Malaysia

ABSTRACT

The issue of abandoned housing project happens across the globe, and it brings nothing positive to the stakeholders as well as to the economic and environment. Even though there are numerous policies and the Malaysian government has promulgated laws, yet the problem of abandoned housing projects is still occurring and not yet been fully addressed. The objectives are to investigate the factor of government policy that contribute to the abandoned housing project and to suggest the suitable solution in mitigating the problem. After the extensive literature review and multi-choice questionnaire survey method been done, the author found that the requirement of 30% low cost house for each development and unstandardized policy together with strict regulation from authorities is one of the main factor that contribute to the abandoned housing project. This paper suggested that the requirement of 30% low cost house for each development should be revised and the requirements for statutory approval should be standardized and short. This study will be contributing an information to the stakeholder involved in housing development.

1. Introduction

Housing is a basic human need that not only serves as a shelter, but also is a simultaneous source of luxury, investment, privacy, and comfort. Since 1957 until 11th Malaysian Plan, Malaysian government policy has put the objective of meeting housing needs to the public [1]. Despite the emphasis on improving the provision of housing by the government in Malaysia, housing is still a controversial issue [2]. One of the continuous housing problematic issues was the abandoned housing project [3]. The Ministry of Housing and Local Government (MHLG) was considers that the project can been abandoned if 1) there has been no substantial activity on site for six consecutive months, or 2) it is involved in a winding-up petition registered at the High Court under Section 218 of the Companies Act, or 3) it is under receivership, or 4) the developer has informed 2 the Housing Controller in writing of his inability to complete the project, or 5) the project has been certified to be abandoned by the Minister under Section 11 (1) I of Housing Development (Control and Licensing) Act 1966 (i.e. Act 118). Abandoned housing projects are not a new issue since it considered as a major problem in Malaysian construction industry as well as in many other countries including United State, Spain, Russia, Dubai and Kuwait [4]. Previous study found that there are several factor that contribute to the abandoned housing project and one of the factor is unfavourable government policies [3]–[5]. This study aims to investigate the sub factors that include in unfavourable government policies toward

IOP Conf. Series: Materials Science and Engineering **712** (2020) 012033 doi:10.1088/1757-899X/712/1/012033

abandoned housing project. Furthermore, the suggestion in mitigate the problem due to unfavourable government policies will be suggested.

2. Government programs and policy in housing development

There are two categories for housing developer in Malaysia, which is public and private developer. In Malaysia, the federal government is responsible to formulate the policy, laws and regulations governing housing especially for the private sector. The local government generally did not involve directly in housing production except the low cost house development and approving the planning permission and building plans for housing construction by both public and private developers [6]. In achieving the agenda of providing suitable and affordable housing for society in Malaysia, the government has set up various housing programs and policies. Table 1 illustrate the housing program that been set up by the government in order to provide adequate housing to the targeted group. The most recent statute relating to housing development that just be introduced in Malaysia is the 1Malaysia Housing Programme (PR1MA) [7].

Table 1. Housing programmes development in Tenth Malaysia Plan (Eleventh Malaysia Plan, 2016)

Programmes	Monthly Household Income (RM)	Ministries / Agencies	Number of Housing Units		
			Complete d	Under Construction	Total
Program Bantuan Rumah (PBR)	Below Poverty Line Income1	Ministry of Rural and Regional Development	56,668	8,298	64,966
Program Perumahan Rakyat (PPR)	2,500 and below	Ministry if Urban Wellbeing, Housing and Local Government	12,025	27,087	39,112
Rumah Mesra Rakyat 1Malaysia (RMR1M)	750 to 3,000	Syarikat Perumahan Negara Berhad (SPNB)	32,948	2,803	35,751
Perumahan Rakyat 1Malaysia (PR1MA)	2,500 to 10,000	Perbadanan PR1MA Malaysia	560	18,400	18,960
1Malaysia Civil Servants Housing (PPA1M)	2,500 to 10,000	Prime Minister's Department	-	13,539	13,539
Rumah Wilayah Persekutuan (RUMAWIP)	6,000 and below	Ministry of Federal Territories	-	9,309	9,309

Besides, the Malaysian government has formulated a set of policies to strengthen the involvement of the private sector in housing production and delivery [8]. The housing policy is definitive, and is articulated adequately in all of the five-year Malaysia Plans as demonstrate in

Table 2 [9].

IOP Conf. Series: Materials Science and Engineering **712** (2020) 012033 doi:10.1088/1757-899X/712/1/012033

Table 2. The summary of National Housing Policy based on the Five Year Malaysian Plan (The Economic Planning Unit & National Institute of public Administration)

No of planning	Main agenda
1st Malaya Plan (1966-1970)	Providing low cost housing units to lower income groups
2nd Malaysian Plan (1971-1975)	Providing national housing programmes to all citizens that categorized under low income groups
3rd Malaysian Plan (1976-1980)	To develop a national unity via housing programmes To improve life quality for rural area
4th Malaysian Plan(1981-1985)	To ensure all citizens especially for low income groups have a fair chances to have their own house in town and rural area
5th Malaysian Plan (1986-1990)	Private sector has given more chances to provide housing and the government responsibility has been reduced
6th Malaysian Plan (1991-1995)	The National housing policy has been focus on providing shelter that can be owned by all citizens and to encourage national integration
7th Malaysian Plan (1996-2000)	Private sector has been given more responsibility to provide 70% of housing target during particular period
8th Malaysian Plan (2001-2005)	Emphasized has been given to provide high quality of housing scheme at prime location
9th Malaysian Plan (2006-2010)	To provide enough housing schemes that can be owned by all citizens via government agency
10th Malaysian Plan (2011-2015	Ensuring access to quality and affordable sustainable housing
11th Malaysian Plan (2016-2020)	Providing adequate and quality affordable housing to poor, low and middle-income households

In year 2007, the government is engaging the private housing developer to cooperate in their new housing programmes. The government give a responsible to the private developer in providing good quality and affordable prices for "Low-Medium Cost" houses [10]. The private developer required by Malaysian government to build 30% of low cost house from their total development [11]. The incompetence private developer left the low cost house abandoned due to low demand and found unprofitable to be completed [5]. The requirement is not flexible and cause a burden to the smaller project [4]. Another unfavourable government policy is the housing delivery system. Malaysia has two system Sell-Then-Build (STB) and Build-Then-Sell (BTS) system. The STB system, in an effort to provide more housing to Malaysians and it has contributed to a certain extent the problem of abandoned housing when the problematic developer left the project uncompleted [12]. [13] found that the funding risk for STB system is lesser since the agreement is only between banker and the housing buyer. In 2006, the Malaysian government already introducing the Build Then Sell (BTS) system, which give more advantages to the house buyer since the buyer only pay 10% after signing the sell and purchase(S&P) agreement and 90% after the project has received the certificates of completion [14]. However most of the housing developer reluctant to implementing the new system and refuse to face the risk of that system [15]. One of the risk in BTS system is the developer need to provide a big fund for the whole project until its completed and only the strong financially developer are able to implement the system [16].

Moreover, the delay and unstandardized policy for a statutory approval may cause cost overrun and project abandonment [4], [5]. The statutory approval refers to the obtaining of permissions from the relevant authorities to initiate and construct a housing and upon its completion to occupy and use the completed housing[17]. The planning system in Malaysia adopts a 'top-down' approach, starting at the federal level, then at the state level and finally at the local authority level **Error! Reference source not found.**]. In order to smoothen the process of approval matters the MHLG introducing the One-Spot-Centre (OSC). However, due to lack of workers and expertise to do task such as the fee calculation, pre-consultation and technical discussion at the OSC department the delay for approval matter still exist [16]. [9] Also, found that the weakness of OSC is happen due to lack of communication skill staff, unclear procedures and the administration problems. Although, the MHLG has implemented a lot of policy in order to improved and reduce the numbers of abandoned housing project, the study on factor contribute to the unfavourable government policy should be done to know the root causes. Together, the suggestion for solution in managing the problem will be discuss in this paper.

3. Methodology

In order to achieve the listed objective, the quantitative research methodology that involves the measurement of variables and their interrelationships conducted. The data obtained using the quantitative method is quantified data which measurable using scientific techniques and instruments. Data from the quantitative method are normally analysed by using statistical procedures [18]. Analyses of the data yield quantified results and conclusions derived from evaluation of the results in the light of the theory and literature [19]. For the purpose of this study, the research methodology can be briefly divided into four stages, namely: 1) Literature Review, 2) Data Collection, 3) Results and Analysis and 4) Conclusions.

3.1 Data Collection

Data collection techniques would be split into several stages to get information that is orderly. In early stage, a structure work shaped to get picture on study journey. This structure will divide to various parts namely data observation and questionnaire survey. An extensive review of the literature was conducted. The researcher study on problem occurs and the causes of the abandoned housing project in Malaysia. From the literature review, the factor, causes, impacts and the stakeholder that leads to the problem also can be identified. This research also studies the solution has been taken by government of Malaysia to preventing those problems. Afterward, the questionnaire survey will be designed using all of the information that been collected.

In order to ensure the reliability of the survey, thirty (30) sets of the questionnaire interview face to face distributed to all respondents from expert in housing development background around Peninsular Malaysia. The questionnaire consists of three (3) sections, section A; respondent personal background. Section B; to determine the perception of respondent towards the relationship between all sub-factors for unfavourable government policy towards the abandoned housing project. Section C is an open-ended question asking for suggestion from respondent in order to mitigating the unfavourable government policy that causes the abandoned housing project in Malaysia. Likert's Scale is used in the multiple choices question in questionnaire which is five ordinal measure of agreement for each statement from 1 to 5.

Ordinary scale from 1 to 5 in ascending order
→
1 2 3 4 5

Increasing degree of disagreement

Figure 1. Likert's scale

Each scale represents the following rating:

IOP Conf. Series: Materials Science and Engineering **712** (2020) 012033 doi:10.1088/1757-899X/712/1/012033

1 = Totally Agree
2 = Agree
3 = Moderate
4 = Disagree
5= Totally Disagree

3.2 Data Analysis
Relative Index Inequality (RII) is adopted to evaluate the ranking of different factors by the respondents. The higher index value donates higher unnecessary statement with maximum index value factor. The classification of the rating scale is following below:

Table 3. Relative Importance Index

Rating Scale	Relative Index	Category
1	0.00 ≤ Relative Index < 0.20	Totally Agree
2	0.20 ≤ Relative Index < 0.40	Agree
3	0.40 ≤ Relative Index < 0.60	Neither agree or disagree
4	0.60 ≤ Relative Index < 0.80	Disagree
5	0.80 ≤ Relative Index < 1.00	Totally Disagree

The formula for Relative Importance Index is as follows:

$$\text{RII Value} = \frac{\Sigma w}{A \cdot N} \; (0 \leq \text{RII} \leq 1) \qquad \text{(Eq. 1)}$$

RII : relative importance index
W : the weight given to each factor by the respondents and ranges from 1 to 5, (where '1' is "strongly agree" and "5" is "strongly disagree")
A : the highest weight (i.e. 5 in this case)
N : the total number of respondents

4. Result and discussion

4.1 Respondent background
Figure 2 represent the numbers of respondent from housing development expertise such as developers, consultants, contractors and government authority been collected and analysed. The total of 30 numbers respondent has been select for this research. Most of the respondent are experience in abandoned housing project and some of them involved in housing development industry more than 10 years.

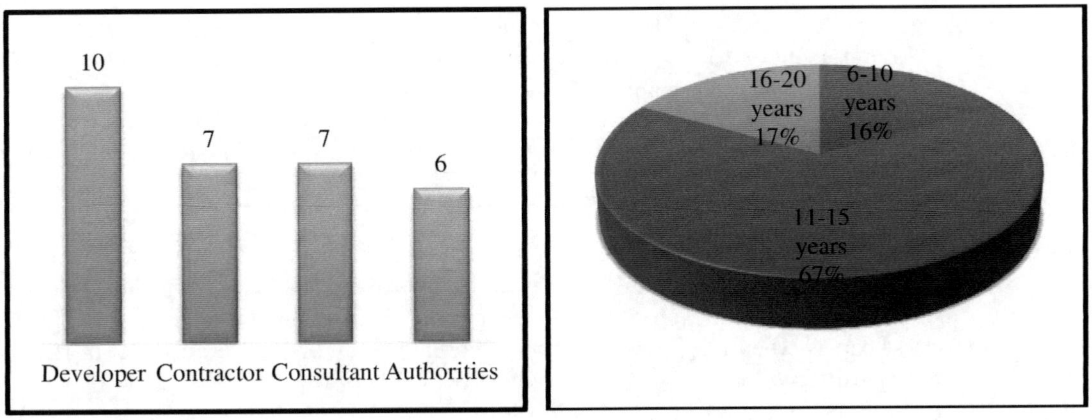

Figure 2. Numbers of Respondent and Year of Experience in Housing Industry

Government has been implement many policy regarding the housing development in Malaysia and some of the policy is considered unfavourable to be implemented as per discuss in literature review. There were many other factors contribute to abandoned project, but in study the focus will be on the government policy factors only. The purpose of this study is to identify how much the unfavourable policy could influence the abandoned housing project in Malaysia.

Based on Figure 3, 52% of the respondent agree that the unfavourable government policy is one of the factor contribute to abandoned project. Followed by 43% strongly agree and 5% more are moderately think that abandoned project is cause by unfavourable government policy. The unfavourable government policy is including the housing delivery system and it is supported by [3], [12], [20], [21]. Secondly is the requirement to every developer to provide 30% of low cost housing for every new development found by [4], [5], [22]. Followed by the unstandardized regulation and policy for every state for a statutory approval application [4], [5], [23], [24] and another factor is the limitation of the tribunal for homebuyer claims found by [25].

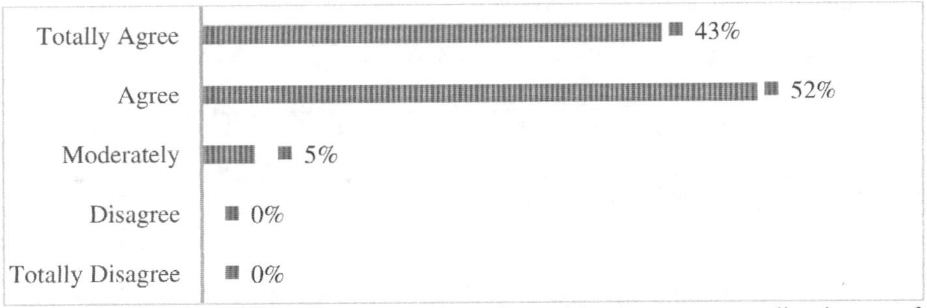

Figure 3. Percentage of respondent agreement on unfavorable government policy is one of the factor that cause abandoned housing project

4.2 Unfavorable government policy that contribute to the abandoned project

Result in **Error! Reference source not found.** show the ranking for unfavorable government policy in Malaysia housing development. As can see the rank 1 with RII value 0.90 is the requirement of providing 30% of low cost housing and unstandardized policy and strict regulation for statutory approval can contribute to abandoned housing project. The housing developer reluctant to construct the low cost housing due to low profit return[5], [26]. [4]Also found that the inconsistence and non-standardized practices in statutory will make the project abandoned.

Table 4. Factor of unfavourable government policy

Rank	Unfavorable Government Policy	RII
1	Requirement of 30% low cost house	0.90
	Unstandardized policy and strict regulation from authorities	0.90
2	STB Housing delivery system	0.85
3	Limitation of tribunal for homebuyer claims	0.75

While rank 2 with RII 0.85 is the STB housing delivery system, which is found bias to the homebuyer. The STB system give more advantages to the developer to improve cash flow from the progress payment by the homebuyer and the developer can use it as an investment to other project [27]. Nevertheless, the worst part is when the project is incomplete and abandoned; the only victim is the homebuyer because they still need to pay the housing loan to the banker [28].

The rank 3 for factor of unfavorable government policy with RII 0.75 is the limitation of the tribunal for homebuyer claims when the developer abandoned their incomplete house. [26] Found that the limitation of the jurisdiction of the tribunal for homebuyer claims is one of the causes of abandoned housing projects. The limitation is the homebuyer need to file the claim before the ordinary court and the total claim not exceeding fifty thousand ringgit Malaysia, which is only fair for low cost house [29].

In order to improve the government policy, the solution for every factor should be study. This study also investigates the proper solution to mitigate the problem of unfavorable government policy in the next chapter.

4.3 The solution for unfavorable government policy.

The suggestion solution by the respondent in improving the government policy listed in

Table 5. The first solution suggested by the respondent is by revised the requirement of 30% low cost house for every new development. The government should consider the location, standard of living and land price for the development in order to preventing the abandonment of low cost house [5], [30].

Table 5. Suggestion solution

Solution
Review the requirement of 30% low cost house to meet the demand
Standardize the requirement and short approval from authorities
BTS system as a priority housing delivery system in Malaysia
Offering incentive to the developer using BTS system in term of cash flow support
Monitoring the housing price using BTS system
Enact legislation to provide a win-win situation for all parties, especially developers and buyers

Secondly, the government should standardize and revise the unnecessary requirement for statutory approval and improve the OSC department in order to reduce the time consuming for application of approval [4], [5]. The need of expertise in OSC department and clear procedure is crucial and continuous communication training crucially needed to prevent misunderstanding [9], [31]. Other solution is by making the BTS system as a priority for Malaysian housing development. By implementing this regulation only capable and financially strong private developer are able to do the development and the abandoned housing project may not occur [32]. The homebuyer feeling secure in term of money and quality for BTS housing [33]. The author also added that the banker is easily release the housing loan to the qualify homebuyer. At the same time, the government need to monitor the housing price for BTS system to meet the current housing price [16]. To make the BTS more convenient and do able, the cooperation from financial institution to give a lenient project financing for the low financially developer until the project is completed [14]. Finally, is by enact legislation that giving a protection to the buyer and strict action taken to the defaulting private developer. The defaulting developer who are causing the abandoned housing project can easily run away from their culpability due to weakness of enforcement by the government [3].

5. Conclusion

The study proof that the unfavourable government policy also contributes to the abandoned housing project in Malaysia. The analysis found that, the most influence sub-factor for unfavourable government policy is the requirement of housing provision and the inconsistence of statutory approval for housing development. The revising and implementing new relevant policy crucially needed in order to improve the housing development in Malaysia. This study also listed some of the suggested solution to reducing the problem. It can be conclude that by revising the government policy is not the only way to prevent the abandoned housing project but surely the numbers can be reduce.

6. References

[1] N. Soffian, A. Ahmad, and N. Rahman, "Housing Development in Malaysia," *Int. J. Acad. Res. Bus. Soc. Sci.*, vol. 8, no. 2, pp. 852–860, 2018.
[2] T. Tan, "Base lending rate and Housing Price: : Their Impacts on Residential Housing Activities in Malaysia," pp. 1–10, 2010.
[3] N. H. M. Dahlan, "The Need To Define ' Abandoned Housing Project' In Peninsular Malaysia: An Analysis," *IIUM Law J.*, vol. 16, no. 1, pp. 85–107, 2008.
[4] E. H. Yap, "Causes of Abandoned Construction Project in Malaysia (Master Thesis)," University Tunku Abdul Rahman, 2013.
[5] M. S. Khalid, "Abandoned Housing Development: The Malaysian Experience," Heriot Watt University Edinburgh, 2010.
[6] S. Shuid, "The housing provision system in Malaysia," *Habitat Int.*, vol. 54, pp. 210–223, 2016.
[7] A. Sufian, "Housing Development and PR1MA : Myth or Reality ?," in *1st International Conference on Innovation and Technology for Sustainable Built Environment 2012 (ICITSBE 2012)*, 2012.
[8] A. Othman, "The effect of the planning system on housing development: a study of developers behaviour in Kuala Lumpur and Johor Bahru, Malaysia," University of Aberdeen, 1999.
[9] A. A. Abdullah, Z. Harun, and H. Abdul Rahman, "Planning Process of Development Project in the Malaysian Context : A Crucial Brief Overview," *Int. J. Appl. Sci. Technol.*, vol. 1, no. 2, pp. 74–81, 2011.
[10] A. C. Ezeanya, "Malaysian housing policy: Prospects and obstacles of National Vision 2020," *Int. Conf. Adequate Afford. Hous. All*, p. 12, 2004.
[11] N. Idrus and C. S. Ho, "Affordable and quality housing through the low cost housing provision in," 2008.
[12] S. Fauzi, N. Yusof, and N. Abidin, "Evaluation of Housing Defects in Build-Then-Sell Houses: A Study of Six Residential Areas," *Int. J. Econ. Manag. Eng.*, vol. 5, no. 10, pp. 554–560, 2011.
[13] F. A. Ibrahim, M. Wira, M. Shafiei, and H. Jusoh, "Sell-then-build and Build-then-sell :," no. June, 2015.
[14] N. A. Almi and K. Husin, "Success Factors of Housing Delivery System Among Malaysian Developers," *Int. J. Real Estate Stud.*, vol. 11, no. 5, p. 2017, 2017.
[15] E. Mustafa Kamal, S. Hany Haron, N. Md Ulang, and F. Baharum, "The Critical Review on the Malaysian Construction Industry," *Issn*, vol. 3, no. 13, pp. 2222–1700, 2012.
[16] N. Yusof, I. Y. Abu Jarad, and M. H. Badree, "The Effectiveness of Government Incentives to Facilitate an Innovative Housing Delivery System : The Perspective of Housing Developers," 2012.
[17] A. Mustafa and M. Ghazali, "Issues in housing delivery systems and customer satisfaction," *Mark. Manag.*, vol. 48, pp. 9395–9399, 2012.
[18] J. W. Creswell, *Data Analysis Workshop*, vol. 1, no. 9. SAGE Publications, Inc, 2019.
[19] R. Fellows and A. Liu, "Research Methods for Construction," in *Research Methods for*

Construction, 4th Edition, vol. 4, 2015, p. 302.
[20] A. N. Abdul Hadi, N. H. Mohd Salleh, and G. M. Tan, "Abandoned Housing Project:Issues and Challenge," *Word J. Int. Linguist. Assoc.*, pp. 1–22, 2014.
[21] W. Daud and W. Norhayate, "What Causes Abandoned Residential Housing Projects in Malaysia ? An Insight from Economic and Management Perspectives," vol. 6, no. 4, pp. 57–73, 2017.
[22] T. H. Tan, "Sustainability and Housing Provision in Malaysia," *J. Strateg. Innov. Sustain.*, vol. 7, no. 1, pp. 62–71, 2011.
[23] S. B. Raheem, "Causes , Effects and Possible Remedies to Delay in Completion of Building Projects," no. December, pp. 11–14, 2015.
[24] S. V. Doraisamy, Z. A. Akasha, and R. Yunus, "A Review on Abandoned Construction Projects: Causes & Effects," *Appl. Mech. Mater.*, vol. 773–774, pp. 979–983, 2015.
[25] F. Ibrahim, "Faktor-faktor kritikal bagi pemulihan projek perumahan terbengkalai," Universiti Teknologi Malaysia, 2006.
[26] S. Shuid, "Low Medium Cost Housing in Malaysia: Issues and Challenges," *Int. Islam. Univ. Malaysia. cost-housing*, pp. 1–13, 2003.
[27] L. F. Ng, "Determinant Factors of Implementing Build Then Sell in," University Sians Malaysia, 2007.
[28] S. N. Phang and T. H. Tan, "Challenges of Implementing Build-Then-Sell Housing Delivery System to Address the Abandoned Housing Problem in Malaysia," *Malaysian J. Econ. Stud.*, vol. 53, no. 1, pp. 135–151, 2016.
[29] A. Sufian, "Housing tribunals: Comparative analysis of the practices in peninsular Malaysia, Sabah and Sarawak," *Int. J. Bus. Soc.*, vol. 13, no. 2, pp. 151–162, 2012.
[30] H. Yakob, F. Yusof, and H. Hamdan, "Land use Regulations Towards a Sustainable Urban Housing: Klang Valley Conurbation," *Procedia - Soc. Behav. Sci.*, vol. 68, pp. 578–589, 2013.
[31] I. Mohd, F. Ahmad, and E. Arbi, "One Stop Centre as a Boon to Property Development Approval Process. A Case Study: City Hall of Kuala Lumpur," *J. Des. Built Environ.*, vol. 8, pp. 1–15, 2011.
[32] M. Zairul, M.N, and R. Ibrahim, "Abandoned Housing Projects in Malaysia : Proposal for Promoting the Bts (Build Then Sell) Concept for Malaysian Housing Development," *Conf. 2nd Int. Conf. Built Environ. Dev. Countries, USM*, no. 2003, pp. 1–18, 2004.
[33] F. A. Ibrahim, M. S. Mohd Wira, and J. Hanafi, "Sell-then-build and Build-then-sell : House Buyers Preferences," *Int. J. Appl. Eng. Res.*, vol. 9, no. 15, pp. 2923–2936, 2014.

Acknowledgment
The authors are grateful to the Ministry of Higher Education, Malaysia (MOHE) and Research Management Centre (RMC), University Malaysia Pahang (UMP) for financial support under grant RDU170341.

IOP Conf. Series: Materials Science and Engineering **712** (2020) 012034 doi:10.1088/1757-899X/712/1/012034

Effect of Unground Palm Oil Fuel Ash as Partial Sand Replacement on Compressive Strength of Oil Palm Shell Lightweight Concrete

S Wan Ahmad[1], K Muthusamy[1,2], M H Hashim[1], A M Albshir Budiea[3] and N F Ariffin[1]

[1]Faculty of Civil Engineering Technology, Universiti Malaysia Pahang, Lebuhraya Tun Razak, 26300 Gambang, Kuantan, Pahang, Malaysia
[2]Earth Resources and Sustainability Centre, Universiti Malaysia Pahang, Lebuhraya Tun Razak, 26300 Gambang, Kuantan, Pahang, Malaysia
[3]Faculty of Civil and Environmental Engineering, Universiti Tun Hussein Onn Malaysia, 86400 Parit Raja, Batu Pahat, Johor, Malaysia.

Abstract. Concern towards degradation of environment due to increasing use of natural sand in construction industry and dumping of solid wastes from palm oil industry namely palm oil fuel ash and oil palm shell has lead towards the development of environmental friendly concrete. The present study investigates the effect of unground palm oil fuel ash as partial sand replacement towards workability and compressive strength of oil palm shell lightweight aggregate concrete. Two types of mixes were used. Control specimen was prepared using 100% natural sand. Another type of mix were prepared by integrating 5%, 10% and 15% unground palm oil fuel ash by weight of sand. The concrete mixes workability were investigated by conducting slump test in accordance to standard. All specimens were made in form of cubes (150mm x150mm) and water cured up to 28 days. The compressive strength test was carried out in accordance to BS EN12390: 3 at 1, 4, 7 and 28 days. The finding shows that integration of 10% unground palm oil fuel ash contributes to the enhancement of oil palm shell lightweight aggregate concrete properties. Success in using unground POFA as partial sand replacement in concrete production would reduce quantity of ash disposed as waste and save the consumption of natural river sand.

1. Introduction

Malaysia is one of the largest palm oil producers in the world. Throughout the year, as the demand for palm oil continues to increase, the palm oil industry also flourishes in the country. Total exports of oil palm products rose 2.9% to 23.97 million tonnes in 2017 from 23.29 million tonnes exported in 2016 [1]. At the same time, the palm oil processing mills are generating many types of wastes namely palm oil fuel ash (POFA), clinker, oil palm shell (OPS), empty fruit bunch and others. As the palm oil industry continues to grow, more waste are generated and need to managed. Annually, the estimated total production of POFA was 2.6 million tonne in Malaysia [2]. According to Teo et al. [3], approximately 4 MT of OPS were generated yearly by Malaysian palm oil mills. Both oil palm kernel shell (OPKS) and palm oil fuel ash (POFA) were thrown as waste, consuming larger area for dumping purposes and causes environmental pollution. Other researchers Mohammad et al. [4] and Kim et al. [5] have reported the pollution problem caused by palm oil wastes dumping. Thus, it is seen that, converting this waste material into construction products would reduce amount of by-product disposed as waste and contributes to the environment cleanliness.

At the same time, the development of construction industry has resulted in the increase of natural sand use in construction. Based on the report by Department of Minerals & Geosciences of Malaysia [6], the sand mining activity in Malaysia has increased from 24.471 million tonne in 2008 to 34.341 million tonne in 2015. Excessive sand mining activity causes destruction of natural habitats, soil erosion and reduced river water quality. Previous researcher Ashraf et al. [7] has pointed out that environmental problems occur when the rate of extraction of sand, gravel and other materials exceeds the rate at which natural processes generate these materials. Discovery of other alternative material to reduce the use of natural sand would be able to preserve environment and contribute to sustainable construction. Realization on the environmental problems caused by the above-mentioned industrial activities, the present research investigates the potential of using palm oil fuel ash as partial sand replacement in oil palm shell lightweight aggregate concrete production. Many researchers, Mohammad et al. [4], Kim et al. [5], Teo et al. [8], Shafigh et al. [9], Jumaat et al. [10] and Ming et al. [11] have investigated the properties and potential of oil palm shell lightweight aggregate concrete. However, research on the use of palm oil fuel ash as partial sand replacement in this type of concrete are not found in the literature so far. Thus, the present research aims to investigate the effect of integrating unground palm oil fuel ash as partial sand replacement in oil palm shell lightweight concrete.

2. Methods

The experimental works were divided into three stages. Firstly, materials needed were obtained and processed ready for use. Then, in the second stage, the concrete specimens were prepared for testing. At the third stage, the concrete specimens were tested in accordance to the standard.

2.1. Materials

The material used in this experiment are cement, sand, water, water reducing admixture, oil palm shell and palm oil fuel ash from the palm oil mill. Ordinary Portland cement (OPC) was used for preparing all the concrete specimens involved in the experiment. The type of sand used in natural river sand with specific gravity of 2.77 obtained from the nearby supplier. Tap water at the laboratory was used for concrete mixing and curing work. Both solid wastes oil palm shell (OPS) and palm oil fuel ash (POFA) were supplied by nearby palm oil mills in the state of Pahang, East Coast Malaysia. Oil palm shells were packed in gunny as shown in Figure 1 before brought to the laboratory for concrete preparation work. The specific gravity of palm oil fuel ash used is 2.21. Figure 2 illustrates the image of natural sand and palm oil fuel ash used in this experimental work.

Figure 1. OPS collection from the palm oil mill.

IOP Conf. Series: Materials Science and Engineering **712** (2020) 012034 doi:10.1088/1757-899X/712/1/012034

Figure 2. Sand (left) and palm oil fuel ash (right).

2.2. Concrete Specimen Preparation
Two types of mixes were used in this experimental work. The control specimen of Grade 30 produced using 100% natural sand was used as reference specimen. Another type of mix were prepared using various percentage of unground palm oil fuel ash (UPOFA). The weight of sand was substituted with 5%, 10% and 15% of UPOFA. All the mixing ingredient content is kept constant except for the unground palm oil fuel ash (UPOFA). Table 1 tabulates the mix proportion of plain concrete. The preparations of the concrete specimens were done using a clean drum mixing machine. Firstly, oil palm shell (OPS), unground palm oil fuel ash (UPOFA) and sand was blended in the mixer. Next, the water was gradually added with small amount of superplasticizer onto the mix. Then, the cement was added and the balance of water was poured to the mixer until the concrete are ready and homogeneous. The fresh mix are placed in the cube mould, compacted before covered with wet gunny and left overnight. The next day, the demoulded samples were placed in water tank for curing purpose.

Table 1. Details of 1m³ control mixes.

Materials	Quantity (kg)
Cement	500
Oil Palm Shell	300
Water	225
Sand	700

2.3. Testing
In order to determine the concrete workability, the freshly prepared mixed were immediately subjected to slump test. The slump test were conducted following the procedure in BS EN12350-2 [12].The effect of unground palm oil fuel ash content on the concrete strength is determined through compressive strength test. The concrete samples were tested at various curing ages up to 28 days. The compressive strength test were carried in accordance to BS EN 12390-3 [13]. Scanning Electron Microscopy analysis were conducted on UPOFA and sand used in this research to justify the effect of the particles on the results obtained.

IOP Conf. Series: Materials Science and Engineering **712** (2020) 012034 doi:10.1088/1757-899X/712/1/012034

3. Results and discussion

3.1. Workability

Figure 3 illustrates the result of slump test conducted on mixes containing various percentage of unground POFA. Evidently, inclusion of POFA as partial sand replacement affects the workability of concrete. As higher percentage of unground POFA replaces the sand in the mix, the concrete mixture becomes stiffer thus resulting in lower slump value. Mixture containing 15% POFA replacement is categorized as very low workability compare with the control specimen 0% POFA that have high workability. The workable mix of control specimen as shown in Figure 4 changed to be stiffer mix with the inclusion of 15% of ash as can be observed in Figure 5. One of the contributing factors is the finer size of POFA particles with specific gravity of 2.21 having larger surface area than sand with specific gravity of 2.77, which increases the mixing water requirement resulting in the lower workability. Aggregate characteristic is one of the factors that influence concrete workability [14]. A similar trend result is observed when the waste material that is used as partial sand replacement in concrete is smaller size than sand [15].

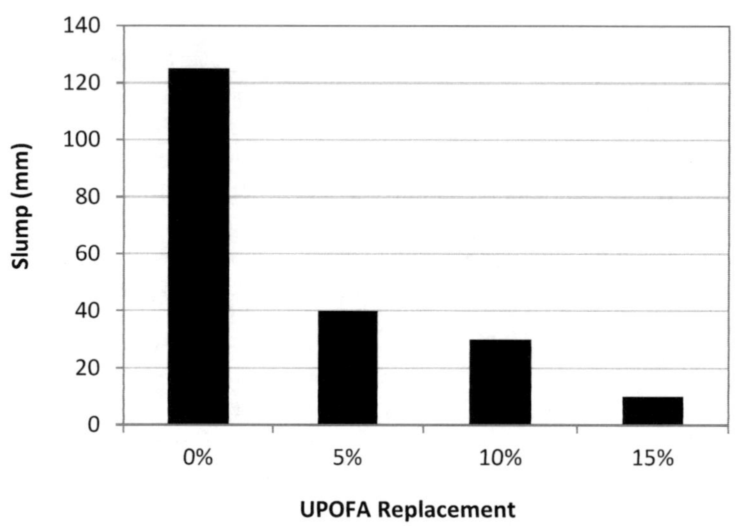

Figure 3. Slump test result

Figure 4. Slump of plain concrete mix

Figure 5. Slump of mix with 15% UPOFA

3.2. Compressive Strength

Figure 6 illustrates the compressive strength result of oil palm shell lightweight aggregate concrete containing unground palm oil fuel ash up to 28 days. Generally, all mixes exhibit continuous strength increment throughout the curing age. The amount of UPOFA added in the concrete mix affect the strength performance of concrete. The result also indicates that inclusion of unground POFA content at 10% able to produce concrete with enhanced strength than control specimen. The use of UPOFA which is finer as illustrated in Figure 7 than natural sand shown in Figure 8, enables the ash contribute towards concrete internal structure densification by filling in the existing void in concrete. Utilization of 15% UPOFA causes the concrete strength to drop. This probably attributed to the characteristic of fine UPOFA possessing higher specific area than sand thus requires more water to coat the UPOFA particles. As a result, the concrete mix becomes drier, more difficult to be mixed and to be compacted when quantity of UPOFA used is increased. Finally, the concrete exhibit lowers strength due to lack of bonding between particle and existence of more voids in the hardened concrete. Similar pattern in increase and drop in concrete strength depending on quantity of waste materials used as partial sand replacement were reported by Singh and Siddique [16].

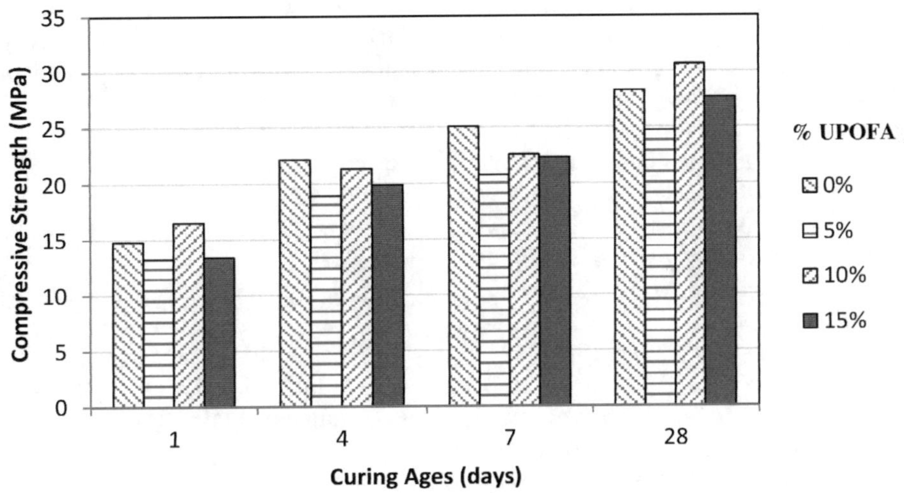

Figure 6. Compressive strength test result

Figure 7. Unground POFA

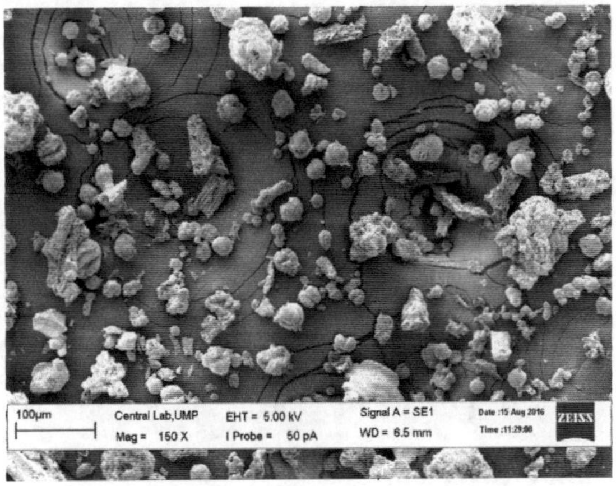

Figure 8. Natural river sand

4. Conclusion

The finding shows that use of 10% unground palm oil fuel ash able to produce concrete with strength higher than control specimen. The recent discovery on the potential of unground POFA as partial sand replacement in concrete is anticipated to reduce amount of POFA disposed as waste and reduces the dependency of concrete industry on natural river sand supply.

5. References

[1] MPOB 2018 Export trend 2018 http://bepi.mpob.gov.my/index.php/en/statistics/export/192-export-2018/868-export-trend-2018.html
[2] Hassan J U, Noh M Z and Ahmad Z A 2014 Effects of palm oil fuel ash composition on the properties and morphology of porcelain palm oil fuel ash composite *Jurnal Teknologi* vol 70(5) pp 5-10
[3] Teo D C L, Mannan M A and Kurian V J 2010 Durability of lightweight OPS concrete under different curing conditions. *Material Structures* vol 43 pp 1–3
[4] Mohammad M U I, Kim H M, Alengaram U J and Jumaat M Z 2016 Mechanical and fresh properties of sustainable oil palm shell lightweight concrete incorporating palm oil fuel ash *Journal of Cleaner Production* vol 115 pp 307–514
[5] Kim H M, Alengaram U J, Jumaat M Z 2016 Assessing some durability properties of sustainable lightweight oil palm shell concrete incorporating slag and manufactured sand *Journal of Cleaner Production* vol 112 pp 763–770
[6] Department of Mineral and Geoscience, Malaysia. 2016 http://malaysianminerals.com/index.php?option=com_content&task=view&id=21&Itemid=45
[7] Ashraf M A, Maah M J, Yusof I., Wajid A and Mahmood K 2011 Sand mining effects, causes and concerns: A case study from Bestari Jaya, Selangor, Peninsular Malaysia *Scientific Research and Essays* vol 6(6) pp 1216-1231
[8] Teo D C L, Mannan M A, Kurian V J and Ganapathy C 2007 Lightweight concrete made from oil palm shell (OPS): Structural bond and durability properties *Building and Environment* vol 42(7) pp 2614-2621
[9] Shafigh P, Jumaat M Z, Mahmud H B and Alengaram U J 2013 Oil palm shell lightweight concrete containing high volume ground granulated blast furnace slag *Construction and Building Materials* vol 40 pp 231–238
[10] Jumaat M Z, Alengaram U J, Ahmmad R, Bahri S and Saiful Islam A B M 2015 Characteristics of palm oil clinker as replacement for oil palm shell in lightweight concrete subjected to elevated temperature *Construction and Building Materials* vol 101(1) pp 949–951
[11] Ming K Y, Mahmud H B, Bee C A, Ming C Y 2015 Influence of different types of polypropolyne fibre on the mechanical properties of high strength oil palm shell lightweight concrete *Construction and Building Materials* vol 90 pp 36–43
[12] B.S. EN 12350-2 2009 Testing fresh concrete: Slump-Test., Ed. London Br. Stand. Inst.
[13] B.S. EN 12390-3 2009 Testing hardened concrete. Compressive Strength of Test Specimens, Br. Stand. Inst. (n.d.)
[14] Neville A M 2011 Properties of Concrete (Longman London)
[15] Saffuan W A, Muthusamy K, Mohd Salleh N A and Nordin N 2017 Properties of concrete containing ground palm oil fuel ash as fine aggregate replacement. *IOP Conference. Series: Materials Science and Engineering* vol 264 p 012008
[16] Singh M R and Siddique R 2014 Strength properties and micro-structural properties of concrete containing coal bottom ash as partial replacement of fine aggregate *Construction and Building Materials* vol 50 pp 246–256.

Acknowledgements

The authors would like to acknowledge the financial support received from Universiti Malaysia Pahang through internal grant RDU190342.

IOP Conf. Series: Materials Science and Engineering **712** (2020) 012035 doi:10.1088/1757-899X/712/1/012035

Rock mass classification for rock slope stability assessment in Malaysia: a review

A F Salmanfarsi, H Awang and M I Ali

Faculty of Civil Engineering Technology, Universiti Malaysia Pahang, 26300 Gambang, Pahang, Malaysia

Abstract. Rock mass classification systems are commonly used in the design and construction of rock engineering, and have seen widespread modifications and validations by various researchers over the last few decades. The rock mass classification, in particular the slope mass rating (SMR), continues to be the preferred preliminary method in small-scale assessment of rock slope stability. In Malaysia, parameters related to rock slope stability assessment have been modified to consider the condition of the rock mass such as the effect of heterogeneous rock units and weathering of rocks. The application of rock mass classifications however have been shown to contain some discrepancies, and the acknowledgement of the limitations of the system is important for an optimum use in the design stage. This paper reviews several development of rock mass classifications in Malaysia, as well as looking at potential direction of further development of the rock mass classification system in the context of local slope stability analysis.

1. Introduction

Development involving rock slope are usually related to highway construction, where large rock surface are excavated. Often times, these slopes are prone to instability problems, due to rock mass conditions and environmental external factors. Internally, factors such as the rock materials, slope height, slope face angle, and discontinuities affect the slope's stability. Over the years, various techniques and methods have been developed by researchers working with rock slopes, from the fields of tunnelling, mining, or conventional slope cutting. Malaysia have been subjected to several major landslides over the years, with several involving cut rock slopes on highway [1]. Geological condition have been reported to account only a portion of the contributing factor to landslide in Malaysia, accounting for a total of 8% [2]. However, due to the safety and economic factor involved in rock slope stability, the input of engineering geology to the process of excavation and treatment of cut slopes is still of great importance [3]. To mitigate potential slope failures in cut rock slopes, a proper understanding of the lithology and discontinuities in rock mass is thus necessary.

2. Assessment methods for rock slope stability

It has been noted by [4] that eight modelling methods for the purpose of rock engineering has been categorized, which includes pre-existing standard methods, analytical methods, basic and extended numerical methods, precedent type analysis, empirical classification, the basic system approach and the combined system approach. In this review the kinematic analysis and rock mass classification is discussed:

IOP Conf. Series: Materials Science and Engineering **712** (2020) 012035 doi:10.1088/1757-899X/712/1/012035

2.1. Kinematic analysis

Kinematic analysis represents one of the conventional methods of slope stability analysis, and is a purely geometric method which examines potential modes of failures in jointed rock mass through the usage of stereographic projection technique. The common method, originally proposed by [5], is later redefined by [6-7]. In the test, the great circle of the slope face and circle of friction angle, φ, is plotted on stereographic projection. The zone between the great circle and the friction circle (sliding envelope) represent the condition for failure, where the plunge value of the joint is less than the slope angle and greater than the friction angle of the joint. A review for case studies of engineering geology in Malaysia by [3] have cited the usage of kinematic analysis as the standard for local geologists and engineer geologists working on rock slopes. It could be observed that numerous local slope stability analyses employed kinematic analysis, either in conjunction or independent of rock mass classification systems. The method is however limited to the case of structurally controlled cut slopes, and have been noted to ignore the strength parameters of the discontinuities and of the rock mass, as well as acting forces on the slope the quantifiable slope stability condition is not given, as only the potential for slope failure is given [8]. Kinematic analysis still remains essential for evaluation of structurally controlled rock slopes, and has been recommended as the first step before proceeding to other analytical techniques of slope stability [9].

2.2. Rock mass classification

Rock mass classifications are one of the most widely known empirical classification for rock engineering. They represent the means for evaluating the performance of rock cut slopes based on important parameters, describing the rock mass condition quantitatively [10]. Due to their simplicity and reliability, the system enjoys wide usage among practitioners, having been time-tested for more than three decades [11]. Summary and discussion of existing rock mass classification systems can be found in the work of [10] and [12]. Rock Mass Rating (RMR) perhaps one of the most widely used empirical method for rock mass classification. Originally designed by [13] to evaluate the quality of rock mass while working in underground projects, the system contain five parameters representing different conditions of rock and discontinuities: strength of intact rock material (uniaxial compressive test or point load strength) (R_δ), rock quality designation (R_{RQD}), spacing between discontinuities (R_{SD}), condition of discontinuities (R_{CD}), and groundwater condition (R_{CG}) (1):

$$RMR = R_\delta + R_{RQD} + R_{SD} + R_{CD} + R_{CG} \tag{1}$$

The rock mass could be sorted into five classes: very good (RMR 100–81), good (80–61), fair (60–41), poor (40–21), and very poor (<20). The list of parameters and assigned value in the rating is given in table 1. From the classes, [15] provided guidelines for supports to tunnel excavated through conventional drilling and blasting. The Slope Mass Rating (SMR), devised by [16] modifies the RMR of [13]. The SMR system aims to remove ambiguities in RMR for the purpose of classifying rock slope. The SMR index adds four adjustment factors, with parameters that reflect joint-slope relationship (F1–F3), as well as method of excavation (F4) (2):

$$SMR = RMR + F1\ F2\ F3 + F4 \tag{2}$$

As with RMR, slope stability is divided five classes: completely stable (SMR 100–81), stable (80–61), partially stable (60–41), unstable (40–21), and completely unstable (20-0). SMR is one of the most widely used classification system used for the purpose of slope stability analysis, with subsequent modifications by workers further modifying the parameters from the system. [14] recommends the usage of [16] SMR for rock slope stability analysis.

Table 1. Rock rating system [14].

Parameter			Range of values				
1	Strength of intact	Point-load	>10	4–10	2–4	1–2	For the low range, uniaxial compression

rock mineral	strength index (MPa)					test is preferred		
	UCS (MPa)	>250	100–250	50–100	25–50	5–25	1–5	<1
Rating		15	12	7	4	2	1	0
2 Drill core RQD (%)		90–100	75–90	50–75	25–50	<25		
Rating		20	17	13	8	3		
3 Spacing of discontinuities		>2 m	0.6–2 m	200-600 mm	60-200 mm	<60 mm		
Rating		20	15	10	8	5		
4 Condition of discontinuities		Very rough surfaces Not conti-nuous No separa-tion Unwea-thered wall rock	Slightly rough surfaces Separa-tion <1 mm Slightly weather-ed walls	Slightly rough surfaces Separa-tion <1 mm Highly weather-ed walls	Slicken-sided surfaces, or Gouge < 5 mm thick, or Separa-tion 1–5 mm (Conti-nuous)	Soft gouge >5 mm thick, or Separa-tion > 5 mm (Continuous)		
Rating		30	25	20	10	0		
5 Ground-water	Inflow per 10 m tunnel length (L/min	None	<10	10–25	25–125	>125		
	Ratio of joint water pressure to major princi-pal stress	0	<0.1	0.1–0.2	0.2–0.5	>0.5		
	General condi-tion	Complete-ly dry	Damp	Wet	Dripping	Flowing		
Rating		15	10	7	4	0		

Other notable rock mass classification systems include the Rock Tunneling Quality Index (Q) ([17], and the Geological Strength Index (GSI) [18-19]. The former is used for tunnel support work, while the latter deals with heterogenous and poor-quality rock mass.

3. Modification of rock mass classification

3.1. Modification in Malaysian context

As Malaysia is particularly vulnerable to landslides occurrence [1-2], several empirical methods for slope assessment and management have been developed over the years for in large-scale and medium-scale assessment of slope. [20] reviewed the accuracy of five slope assessment systems (SAS) developed by the Malaysian Public Work Department (PWD). The study found that none of the systems were satisfactory in predicting landslides in rock cut slopes, although one of the system (Slope Management and Risk Tracking System, SMART) seems satisfactory in predicting failures for

IOP Conf. Series: Materials Science and Engineering **712** (2020) 012035 doi:10.1088/1757-899X/712/1/012035

metasedimentary rocks (table 2–3). The various reasons for the unsatisfactory prediction of landslide were cited to be the result of usage of hazard score developed from other country, insufficient database, the use of an oversimplified approach, and the use of database derived from a different rock/soil formation.

Table 2. Accuracy of the SAS in predicting landslides: granitic formation cut slopes [20].

Prediction	SMS	SPRS	SIMS	SMART	LHRA
(1) Number of slopes assessed	139	139	139	139	139
(2) Number of recent landslides or failed slopes	44	44	44	44	44
(3) Number of slopes classified as High and Very High Hazard that actually failed	17	23	1	27	1
(4) Percentage of (3) compared with (2)	39%	52%	2%	61%	2%

Table 3. Accuracy of the SAS in predicting landslides: meta-sediment formation cut slopes [20].

Prediction	SMS	SPRS	SIMS	SMART	LHRA
(1) Number of slopes assessed	47	47	47	47	47
(2) Number of recent landslides or failed slopes	29	29	29	29	29
(3) Number of slopes classified as High and Very High Hazard that actually failed	13	17	5	26	0
(4) Percentage of (3) compared with (2)	45%	59%	17%	90%	0%

Out of the rock mass classification systems, the SMR method has proven to be widely accepted for local practitioners working on rock slopes. As an example, recent case study on slope stability analysis of limestone cliff at Gunung Kandu, Gopeng by [21] highlight how the usage of SMR is significant for its quantification of rock slope stability in a practical method for large area of rock slope assessment. RMR and SMR have been noted as being useful for preliminary assessment of slope stability, incorporating geological, geometric, and engineering parameters to arrive at a quantitative value of rock mass quality [22].

Only few works have been found to modify the rock mass system in the context of local conditions. The most notable example is the Modified Slope Mass Rating (M-SMR) by [23–25] based on the works on the Crocker Formation in Kota Kinabalu. The system modifies RMR of [14] and SMR of [16] to consider the effect of alternating lithologies in heterogeneous rock formation, introducing the concept of 'lithological unit thickness' in lieu of assigning a single value for strength of intact rock material (UCS) for the whole rock unit (figure 1). The system is divided into six classes: very good (M-SMR 100–81), good (80–61), moderate (60–41), poor (40–21), very poor (20–1) and extremely poor (<1). Slope stabilization and protection measures are proposed for each classes (figure 2).

Another notable modification is the development of systematic cut slope stability evaluation by [26–27]. Here, the RMR and SMR values were compared with dip angle of the discontinuity (βi) and the peak friction angle, αp of discontinuity surfaces from laboratory tests for slope stability. The evaluation is based on the derived polynomial equations by [28] that correlates the αp of discontinuity planes from schist bedrocks with Joint Roughness Coefficient (JRC), which in effect include the parameter of discontinuity surface roughness for cut slope stability evaluation. The systematic approach propose four classifications for potential for failure: very high failure potential, intermediate failure potential, low failure potential, and stable (figure 3).

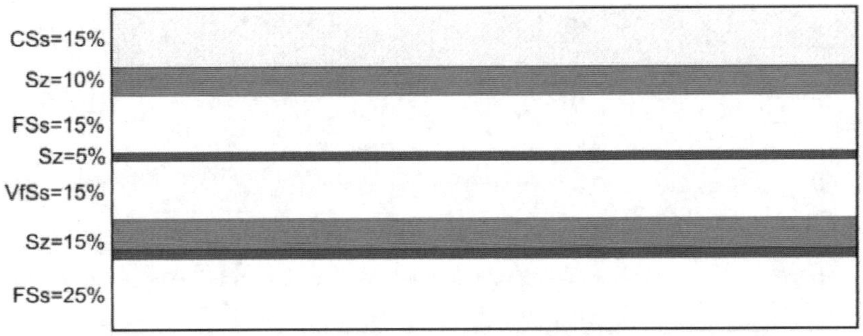

CSc	15%
FSs	40%
VfSs	15%
Sz	30%

Lithology	Thickness (m)	Thickness %	Strength (UCS) MPa	Strength by lithology (MPa)	Strength of the SFRM (Mpa)
MSs	k	15.00	100.00	15.00	107.85
FSs	l	40.00	165.00	66.00	
VfSs	m	15.00	105.00	15.75	
Sz	n	30.00	37.00	11.10	
	Total	100.00			Very strong

Figure 1. Lithological unit thickness model. The final intact rock strength of slope forming rock material (SFRM) is represented by the sum of strength of all lithological in any particular slope [23].

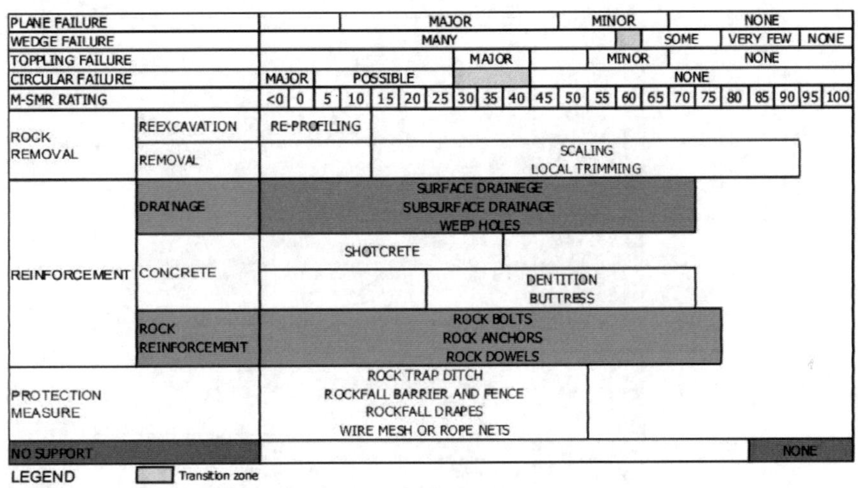

Figure 2. Slope stabilization and protection measures from M-SMR values [25].

Due to the tropical condition of Malaysia, weathering in rock mass is extremely common. Extensive studies on the weathering of bedrock in Malaysia have notably been carried out [29-31] with clear indication of the uppermost zone of weathering profile consisting of completely weathered bedrock material with unclear relict original bedrock texture. It is clear that any significant assessment of rock mass has to factor in the influence of weathering. [32] developed a typical mass weathering profile of tropically weathered granite (figure 4). The profile includes geological and structural parameters (joint characteristics, corestone occurrence, rock/soil ratio, mass homogeneity, colour of rock, and discoloration at joint' surfaces). The classification provides useful parameters for the preliminary stage of any civil engineering design, which potentially will save cost and time during site investigations for development of engineering work design parameters.

IOP Conf. Series: Materials Science and Engineering **712** (2020) 012035 doi:10.1088/1757-899X/712/1/012035

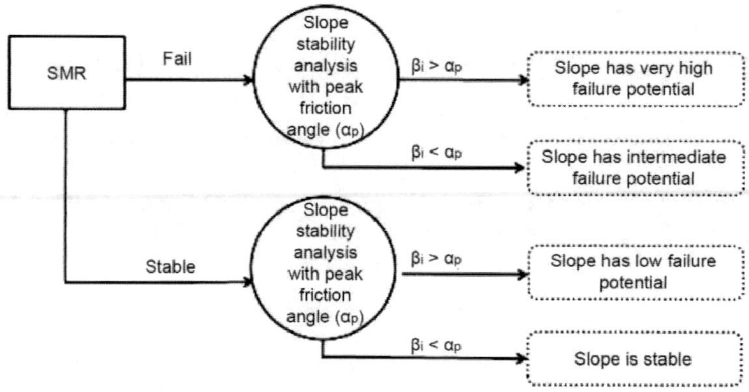

Figure 3. Diagram for systematic cut slope stability evaluation [26].

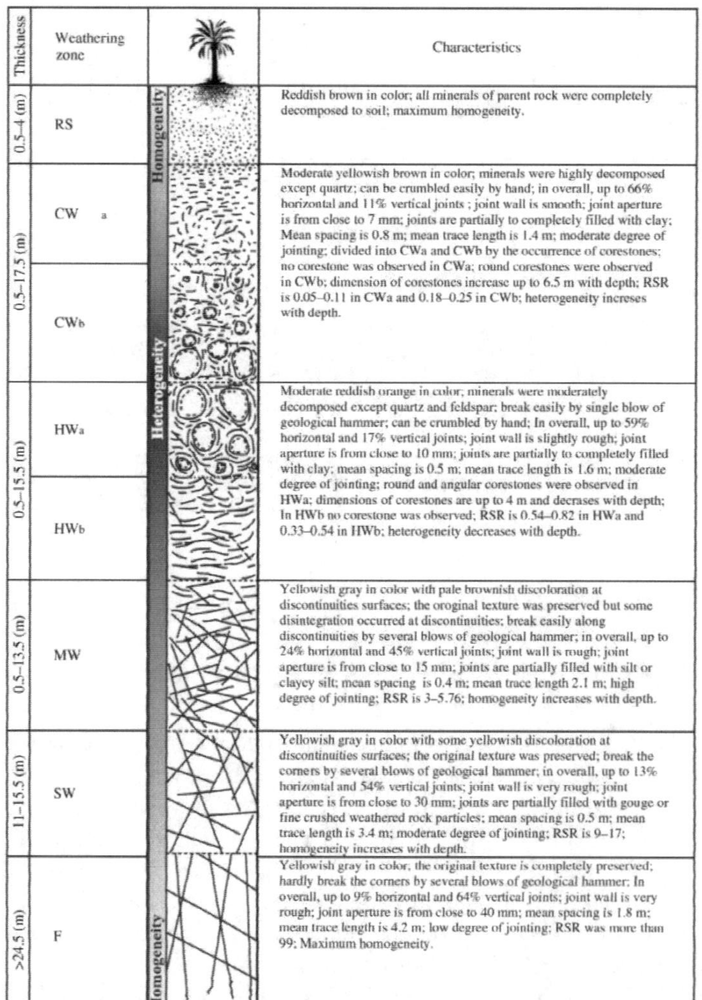

Figure 4. Typical mass weathering profile of tropically weathered granite [32].

3.2. Evaluation of rock mass classification

From the literatures mentioned in previous sections, it has been shown that rock mass classifications have been subjected to wide usage over a long period of time – leading to the identification of some inherent weakness and deficiency of the classification systems in reflecting the actual condition of rock mass. Some of the more inherent issue include discussions on the validity of rock quality designation (RQD) as parameters in rock mass classifications [33-35] or the correction factors applied to SMR (2) [12]. Perhaps more pressing in the context of Malaysia is that in most rock classification systems, the role of water movement has not been given significant proportion in the parameters [10]. This is especially significant for local climate, with water movement being the largest contribution factors for landslide, making up to 58% of landslide cases [2]. [10] suggests quantifying the hazard for failure in rock slopes (figure 5), where the use of factors related to precipitation and temperature characteristic of a study area allow adaptation of rock mass classification system to local climatic conditions.

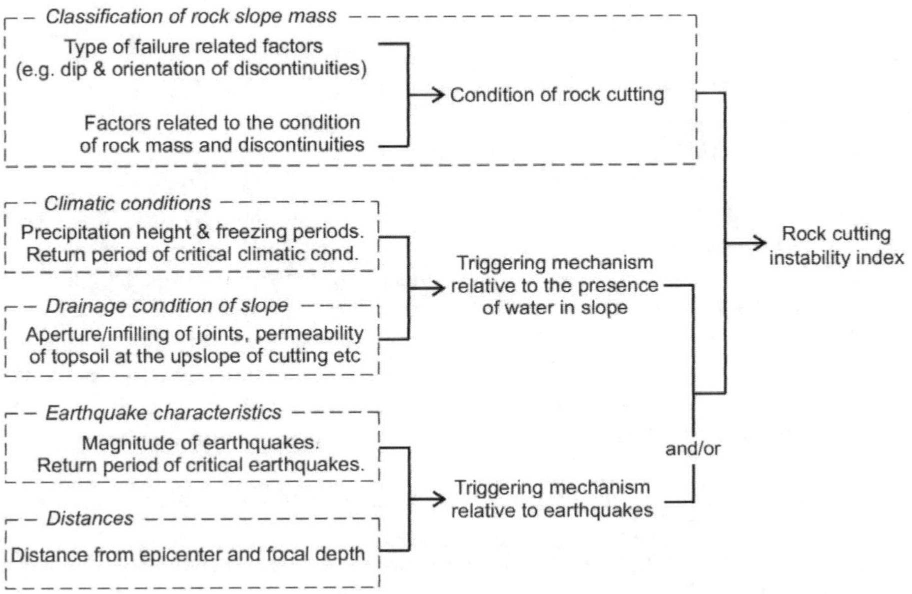

Figure 5. Proposed flow chart for quantification of the hazard for failure of rock cuttings [10].

[36] in their review of SMR have acknowledged some of the reported common issues found in the system, which includes: 1) rather conservative value of the classification in general; 2) extreme values of correction factor F3 proposed by Bieniawski is difficult to cope with in actual stability analysis of slope; 3) the failure modes derived from SMR occurs in reality; 4) excavation method is highly influential for slope's stability, and is necessary to include in the system; 5) practical difficulties for classification of slopes with berms; and 6) system does not consider the effect of slope height. It has also been noted that the rock mass classification systems are not applicable to complex cases involving variable slope geometric, coupled problems, and/or complex conditions of discontinuities [8]. The system however remains widely used, as it has been proven to be a powerful system in the initial stage of slope stability analysis, and continues to act as a common language for both engineering geologist and geotechnical engineers. Development of remote sensing technology, in particular the Light Detection and Ranging (LiDAR), have been incorporated into slope stability assessment and post-failure slope investigation. In their review of the development of SMR, [36] have pointed the usage of LiDAR that allows the generation of precise 3D point clouds from slopes which can be utilized to obtain parameters that are relevant for SMR or other rock mass classification. Usage of LiDAR for characterizing the parameters of rock slope (i.e. discontinuities) in local context have seen limited usage, with recent notable case involving the stability assessment of limestone rock cave [37]. The usage of LiDAR has been noted for its possibility of characterizing complex landslides along the transportation route in mountainous region [38], and offer the possibility of usage alongside conventional field data gathering due to the ability to cover large surface area in relatively short time.

IOP Conf. Series: Materials Science and Engineering **712** (2020) 012035 doi:10.1088/1757-899X/712/1/012035

4. Conclusion

Both kinematic analysis and rock mass classification have been established as valid and reliable methods for assessment of rock slope stability over the years, and have continued to be widely used in Malaysia, being widely accepted by both fields of engineering geology and geotechnical engineering. Although some flaws and limitations to the classification system have been discussed over the years, the simple nature of the system makes it desirable for practitioners to modify the parameters to better fit the context of local rock mass conditions. Any subsequent modifications to the system in the context of local conditions should consider the role of weathering and water movement in rock mass, as they have not been given much emphasis in current scope of available systems. The usage of laser scanning in slope stability assessment is an unexplored potential, and appears to be the next step forward in rock mass classification for slope.

5. References

[1] Abdul Rahman H and Mapjabil J 2017 Landslides Disaster in Malaysia: an Overview *Health & the Environment Journal* **8** 58–71

[2] Kazmi D, Qasim S, Harahap I S H, Baharom S, Imran M and Moin S 2016 A Study on the Contributing Factors of Major Landslides in Malaysia *Civ. Eng. J.* **2** 669-78

[3] Tan B K 2017 Engineering geology in Malaysia – some case studies *Bulletin of the Geological Society of Malaysia* **64** 65–79

[4] Feng X-T and Hudson J A 2011 *Rock Engineering Design* (Boca Raton: CRC Press)

[5] Markland J T 1972 *A useful technique for estimating the stability of rock slopes when the rigid wedge slide type of failure is expected* (London: Imperial College Rock Mechanics Research)

[6] Hocking G 1976 A method for distinguishing between single and double plane sliding of tetrahedral wedges *Int. J. Rock Mech. Min.* **13** 225–6

[7] Hoek E and Bray J W 1981 *Rock Slope Engineering* (London: Institute of Mining and Metallurgy)

[8] Alzo'ubi A K 2016 Rock slopes processes and recommended methods for analysis *International Journal of GEOMATE* **11** 2520–7

[9] Raghuvanshi T K 2017 Plane failure in rock slopes – A review on stability analysis techniques. *J. King Saud Univ.* Sci **31** 101–9

[10] Pantelidis L 2009 Rock slope stability assessment through rock mass classification systems *Int. J. Rock Mech. Min.* Volume **46** 315–25

[11] Singh B and Goel R K 2011 *Engineering Rock Mass* Classification (Amsterdam:Elsevier Inc. Publication)

[12] Zheng J Zhao Y Lü Q Deng J Pan X and Li Y 2016 A discussion on the adjustment parameters of the Slope Mass Rating (SMR) system for rock slopes *Eng. Geol.* **206** 42–9

[13] Bieniawski Z T 1979 The geomechanics classification in rock engineering applications *Proc. of the 4th International Cong. on Rock Mechanics* (Rotterdam: Balkema and Swiss Society for Soil and Rock Mechanics) pp 41–8

[14] Bieniawski Z T 1989 Engineering rock mass classifications: a complete manual for engineers *and geologists in mining, civil, and petroleum engineering* (New York: John Wiley & Sons)

[15] Bieniawski Z T 1984 The design process in rock engineering *Rock Mech. Rock Eng.* **17** 183–90

[16] Romana M 1985 New adjustment ratings for application of Bieniawski classification to slopes *Proc. of int. symp. on the role of rock mechanism* (Zacatecas: International Society of Rock Mechanics) pp 49–53

[17] Barton N R, Lien R and Lundge J 1974 Engineering classification of rock masses for the design of tunnel support *Rock Mech. Rock Eng.* **6** 189–239

[18] Hoek E, Kaiser P K and Bawden W F 1995 *Support of Underground Excavations in Hard Rock* (Rotterdam: Balkema)

[19] Hoek E, Marinos P and Benissi M 1998 Applicability of the geological strength index (GSI) classification for very weak and sheared rock masses: the case of the Athens schist formation *B. Eng. Geol. Environ.* **57** 151–60

[20] Singh H Huat B B K and Jamaludin S 2008 Slope Assessment Systems: A Review and Evaluation of Current Techniques Used for Cut Slopes in the Mountainous Terrain of West Malaysia *Electron. J. Geotech. Eng.* **13** 1–24

[21] Mohd Razib A M, Goh T L, Mazlan N A, Abdul Ghani M F, Tuan Mohamed T R, Rafek A G, Serasa A S, Chen Y and Zhang M 2018 A Systematic Approach of Rock Slope Stability Assessment: A Case Study at Gunung Kandu, Gopeng, Perak, Malaysia *Sains Malaysiana* **47** 1413–21

[22] Nkpadobi J I, Raj J K and Ng T F 2016 Classification of cut slopes in weathered meta-sedimentary bedrocks *Earth Sci. Res. J.* **20** J1–9

[23] Abd Rahim I, Hj Tahir S, Musta B and Omang S A K 2009 Lithological Unit Thickness Approach For Determining Intact Rock Strength (IRS) of Slope Forming Rock Material Of Crocker Formation *Borneo Science* **25** 23–32

[24] Abd Rahim I 2011 *Rock mass classification system of the Crocker formation in Kota Kinabalu for rock slope engineering purpose, Sabah, Malaysia* (Unpublished doctoral dissertation) (Malaysia: Universiti Malaysia Sabah)

[25] Abd Rahim I 2015 Geomechanical classification scheme for heterogeneous Crocker Formation in Kota Kinabalu, Sabah, Malaysia *Borneo Science* **36** 12–20

[26] Rafek A G and Goh T L 2015 *Engineering Geology for Society and Territory – Volume 2,* ed G Lollino et al (Switzerland: Springer International Publishing) pp 787–90

[27] Rafek A G, Mohd Jamin N H, Goh T L, Simon N and Hussin A 2016 Systematic Approach to Sustainable Rock Slope Stability Evaluation *Procedia Chem.* **19** 981–5

[28] Rafek A G and Goh T L 2012 Correlation of Joint Roughness Coefficient (JRC) and Peak Friction Angles of Discontinuities of Malaysian Schists *Earth Science Research* **1** 57–63

[29] Raj J K 1985 Characterisation of the weathering profile developed over a porphyritic biotite granite bedrock in Peninsular Malaysia *Bulletin of the International Association of Engineering Geology* **32** 121–8

[30] Raj J K 1998 The failure of a slope cut into the weathering profile developed over a porphyritic biotite granite *J. Asian Earth Sci.* **16** 419–27

[31] Raj J K 2010 Soil-moisture retention characteristics of earth materials in the weathering profile over a porphyritic biotite granite *Current Research in Geoscience* **1** 12–20.

[32] Alavi Nezhad Khalil Abad S V, Tugrul A, Cokceogluc C and Jahed Armaghanic D 2016 Characteristics of weathering zones of granitic rocks in Malaysia for geotechnical design *Eng. Geol.* **200** 94–103

[33] Palmström A 2005 Measurements of and correlations between block size and rock quality designation (RQD) *Tunn. Undergr. Sp. Tech.* **20** 362–77

[34] Pells P J, Bieniawski Z T, Hencher S R and Pells S E 2017 Rock quality designation (RQD): time to rest in peace *Can. Geotech. J.* **54** 825–34

[35] Chen Q and Yin T 2018. Should the Use of Rock Quality Designation Be Discontinued in the Rock Mass Rating System? *Rock Mech. Rock Eng.* **52** 1075–94

[36] Romana M, Tomás R and Serón J B 2015 Slope Mass Rating (SMR) geomechanics classification: thirty years review *13th ISRM Int. Cong. of Rock Mechanics (Montreal, 10-13 May)*

[37] Idrees M O and Pradhan B 2018 Geostructural stability assessment of cave using rock surface discontinuity extracted from terrestrial laser scanning point cloud *Journal of Rock Mechanics and Geotechnical Engineering* **10** 534–44

[38] Razak K A, Hasan R C, Aitin A, Sheng L C, Mohamed Z, Qalam-A'Zad and Abu Bakar R2014 Hyperspatial resolution of ground based remote sensing for natural limestone characterization: A critical input for rock slope hazard assessment in the tropics *35th Asian Conf. on Remote Sensing 2014 (ACRS 2014)* (New York: Curran Associates, Inc) pp 90–5.

Acknowledgments
The authors would like to acknowledge the research support provided by Universiti Malaysia Pahang under Geran Universiti Malaysia Pahang, grant number RDU190345.

IOP Conf. Series: Materials Science and Engineering **712** (2020) 012036 doi:10.1088/1757-899X/712/1/012036

Short Term Aging Effect of Asphaltic Concrete Incorporating Charcoal Ash from Coconut Shell

P J Ramadhansyah[1], K A Masri[1], H Awang[1], M K I M Satar[2], M R Hainin[2], A H Norhidayah[2], M N Mohd Warid[2], H Yaacob[2] and A Juraidah[3]

[1]Faculty of Civil Engineering Technology, Universiti Malaysia Pahang, 26300 Gambang, Pahang, Malaysia
[2]Faculty of Engineering, School of Civil Engineering, Universiti Teknologi Malaysia, 81310 Johor Bahru, Malaysia
[3]Faculty of Civil Engineering, Universiti Teknologi MARA, 40450 Shah Alam, Selangor, Malaysia

Abstract. Numerous studies have been conducted to improve the performance of asphalt mixture properties in comparison with conventional mixes. One of the alternatives is utilizing waste agriculture product as additive or replacement into bitumen mixture. Coconut shell is one of the agricultural wastes that can be utilized. This paper aims to study the effect of asphaltic concrete incorporating charcoal ash from a coconut shell under short term aging condition. Charcoal coconut shell ash (CCSA) was added into a bitumen with 0%, 2%, 4%, 6%, and 8% by weight. The original CCSA was initially ground until getting the particles sizes less than 0.075mm. Laboratory simulation of aging was applied in this investigation. Marshall Stability, volumetric properties, resilient modulus, and dynamic creep test were performed to analyses asphalt mixture containing CCSA. It was observed that the addition of CCSA in asphaltic concrete gives the significant development in stability, modulus of resilient and creep stiffness. This study suggests that 4% and 6% of CBA as a supplementary binder could improve the mechanical properties of asphaltic concrete even under aging conditions.

1. Introduction

Coconut shell (CS) being hard and not easily degraded the material and bring the contribution to environmental pollution [1]. However, researchers in developing countries have found an alternative solution to reduce pollution by making sustainable use of waste. One of the common use of CS in the road industry is as aggregate replacement or filler in the binder [2]. The advantages of coconut shell are strong, rigid and lightweight material [3]. Also, it is very economical as large amounts are available as agricultural waste material. Furthermore, it is environmentally friendly because it is a natural material that is biodegradable and releases less of carbon dioxide when burned [4]. The utilisation of various agricultural waste materials in road construction has been carried for many years [5,6]. According to Slebi-Acevedo et al. [7], the use of various waste materials in the hot mix asphalt industry was due to concerns with the engineering, environment and economic factors. Agricultural wastes are lead to environmental problems. One solution is that these materials be recycled into useful materials; for example, by transforming them into modifying binder. Of all the agricultural wastes, coconut shell is seen as the material with the most potential to be produced as modify binder because of its strength and good quality in various construction materials [8]. Coconut shell has become a promising material that can be transformed into modifier due to its performance, environmental and economic advantages. Therefore, the potential of coconut shell for use as bitumen modifiers must be investigated in order to ensure the compatibility between the materials. Coconut consists of various components such as flesh, shell and fiber which can be recycled after the coconut water and flesh are taken [9]. Of the coconut components, coconut shell is mostly used both in the production of coconut-based products and in research areas. Based on past studies, coconut shell consists of three main compositions, which are hemicellulose, cellulose and lignin [10]. Study on coconut shell as bitumen modifier is very limited. For example, Buhari et al. [11] conducted research using coconut shell powder (CSP). The CSP was added at 2, 4, 6, 8 and 10% by weight of bitumen PEN 80/100. The results of the physical properties of the modified bitumen revealed that CSP improved the performance of the bitumen by decreasing the penetration and increasing the softening point, respectively. Another research that used a coconut shell as bitumen modifier was performed by Abdullah et al. [12]. Charcoal ash from coconut shell (CSC) at three different sizes was used to replace bitumen PEN 60/70 at 0, 10, 15 and 20%. The results showed that the penetration of the modified bitumen decreased while the softening point and viscosity increased up to 20% replacement. The results implied that the CSC improved the temperature susceptibility of the bitumen. Based on the past researches, it can be seen that coconut shell was most often used as filler in the construction materials. However, in bitumen modification is limited especially under aging session. Hence, the goal of this study is to determine the performance of asphaltic concrete incorporating charcoal coconut shell ash under short term aging.

2. Materials and Method

2.1. Binder

The binder used for this study is bitumen with the penetration grade 60/70. This bitumen was heated at 100°C for a period of 1 hour before being added to the aggregate mixes. The physical properties of 60/70 bitumen were 68 dmm penetrations (at 25°C) with a softening point approximately 52°C. In addition, the relative density of bitumen is 1.03 while the viscosity at a temperature of 135°C was 600cp.

2.2. Aggregate and gradation

In this investigation, crushed granite aggregates were used to prepare the asphalt mixture. While the asphaltic concrete grade 14 (AC14) was chosen for the mixture design based on the JKR specification [13]. The mix design incorporated a range of aggregate particle sizes, with each aggregate particle size in a given design mix contributing to the enhanced strength and durability characteristics of the asphalt structure. The combined gradation for each aggregate particle size (coarse aggregate, fine aggregate and mineral filler) for AC14 is presented in Table 1.

Table 1: Gradation limits of the combined aggregates for ACW14

Sieve size (mm)	Gradation (JKR, 2008)
20	100 – 100
14	100 – 90
10	86 – 76
5	62 – 50
3.35	54 – 40
1.18	34 – 18
0.425	24 – 12
0.15	14 – 6
0.075	8 – 4
Pan	-

2.3. Charcoal Coconut Shell Ash

At the laboratory, coconut shells were sun-dried to ease the removal of the coconut fibre. The shells were ripped out and washed out from any unwanted substances. Coconut shells were subjected under the heating process at 400°C to 450°C for five minutes using the furnace to change its properties to charcoal. The charcoal coconut shell was then left to cool down in room temperature before being crushed manually. Small sizes of crushed charcoal coconut shells were then were ground using Los Angeles Abrasion Value test (LAAV) with 1200 cycles until became ash in physically. CCSA was then sieved with the requirement of passing BS sieve 0.075mm. Charcoal coconut shell ash was then weighted in 2%, 4%, 6%, and 8% by the weight of bitumen for every each of testing conducted.

2.4. Marshall Stability Test
Marshall Stability test was used to determine the strength of an asphalt mixture that was compacted to a standard laboratory compaction effort. The Marshall Test procedure involved applying a compressive load to a cylindrical specimen with 101 mm diameter and 63.5 mm high. The temperature of the specimen was 60°C and the load applied was at a rate of 50 mm per minute. The test was carried out in compliance with ASTM D5581-07 [14].

2.5. Resilient Modulus Test
The 5-Pulse resilient modulus test was conducted in accordance with ASTM D7369-11 [15]. The specimens for this study were divided into two sets. Each set was conditioned at 25°C and 40°C correspondingly for 4 hours prior to testing. The resilient modulus value was calculated using Equation 1.

$$M_R = \frac{P}{Ht}(0.27 + \mu) \qquad (1)$$

Where **MR** is resilient modulus (MPa), **P** is applied force (N), **t** is thickness (m), **H** is horizontal displacement (m), and **μ** is Poisson ratio.

2.6. Dynamic Creep Test
Dynamic creep test was used to determine the rutting potential of asphaltic concrete. This test was carried out in accordance with BS EN 12697-25 [16]. Based on the standard, the specimen was conditioned at 40°C for at least 4 hours. The dynamic creep modulus was determined using Equation 2.

$$E = \frac{\sigma}{10 \times \varepsilon n} \qquad (2)$$

Where **E** is dynamic creep modulus (MPa); **Σ** is applied stress (kPa), and **ε** is measured vertical strain (mm).

2.7. Aging Test
Aging test was used to determine the deterioration of specimens. In this study, the aging procedure adopted was based on AASHTO R30-02 [17]. At the laboratory, the specimen was prepared under the same parameters as controlled mix up to the step of mixing. Then, these loose mixes were evenly placed in trays and arranged in a force draft oven at a temperature of 135±3°C for 4 hours and namely non-aging (NA). The force draft oven is used to simulate the hardening that binder would experience during seven to ten years of service. In short term aging (STA), the aging process ended with the removal of mixes from the oven after this period; then the loose mixes were compacted.

3. Results and Discussion

3.1. Stability
The stability of asphaltic concrete under NA and STA increases until a maximum at 4% CCSA content then reduces as the CCSA content increases as illustrated in Figure 1. In all cases, the stability of the asphaltic concrete increases with aging, the increase in stability partially due to the hardening of bitumen. Previously, Romastarika et al. [18] reported that the bitumen became harder after a certain aging condition. The results also indicate that the stability increases with increasing CCSA content up to a peak level and then decreases with further additions. For example, at the unaged condition, the stability value increases from 11,276 N to 13,532 N when the CCSA content increases from 0% to 4%.

IOP Conf. Series: Materials Science and Engineering **712** (2020) 012036 doi:10.1088/1757-899X/712/1/012036

Then start to decreases between 12,572 N to 12,170 N when the CCSA content increases from 6% to 8%. This means the addition of 2% CCSA to the asphaltic concrete mix can increase stability by up to 16.25%. Furthermore, the stability of specimens increased up to 14,818 N, and then relatively decreased between 14,591 N and 13,934 N after short term aging. The results show that the maximum stability at unaged is 13,532 N while under STA is 14,818 N, respectively. It can be seen that the addition of CCSA content is the main factor contributing mixes with higher stability.

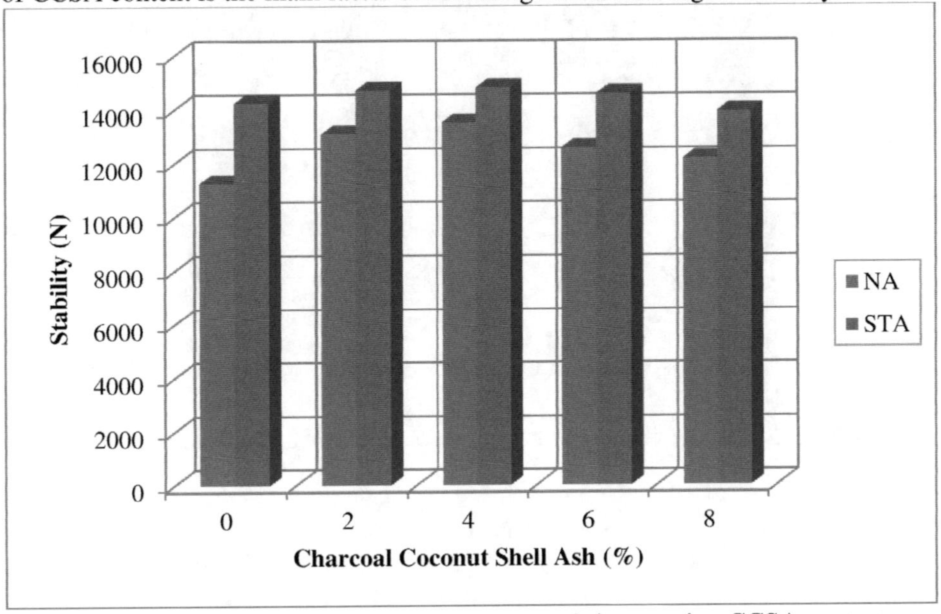

Figure 1. Stability of asphaltic concrete containing varying CCSA content

3.2. Stiffness

An overview of the stiffness results of CCSA asphalt mixture under NA and STA is graphically illustrated in Figure 2. In general, the stiffness value of specimens under STA is higher than those of NA due to hardening of the binder in CCSA mixture. For instance, at 0% CCSA, the stiffness of asphalt mixture increases by almost 4.3% after STA. The results also indicate that an increase in CCSA content causes the stiffness of the all mix types to increase until a maximum value is reached, beyond which the stiffness value decreases as the CCSA content continue to increase. For example, at the unaged condition, the stiffness of specimens with CCSA content increase by 8.7% after added 2% CCSA and, by 5.2% after STA. This showed that the aging process causes oxidation and increases the hardening rate of bitumen, thus resulting in increased stiffness. However, the different pattern is shown at 6% and 8% CCSA content where the stiffness of specimens decreases by 4.4% and 5.75, respectively.

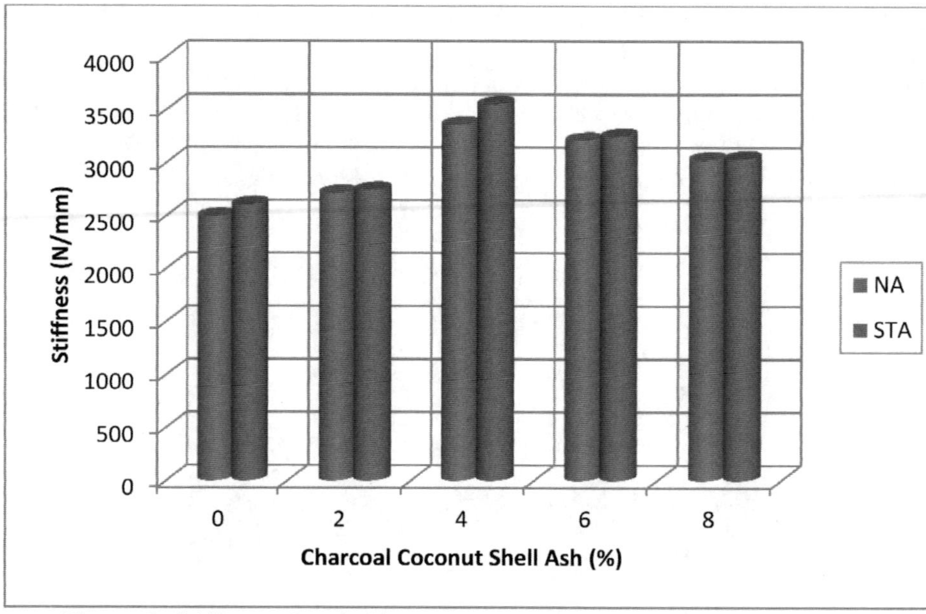

Figure 2. Stiffness of asphaltic concrete containing varying CCSA content

3.3. Flow

Curve illustrated in Figure 3 represent the relationship between the flow of asphalt mixed and CCSA contents under varying aging. In general, there were not many differences in flow observed from specimens subjected to NA and STA conditions. However, the flow values of all mix types were met all the requirements as specified by JKR standard for wearing course [13]. Based on Figure 3, the flow of asphaltic concrete initially increased during the CCSA content increases between 0 to 2% and then started to decrease until the specimens reached 8% CCSA. Asphaltic concrete incorporating 2% CCSA content exhibits a higher flow value compared to other mixture under both aging conditions. Previously, Wang et al. [19] reported that a modified mixture with waste material achieved the lowest flow when compared to the control mixture. The flow value represents the deformation rate potential in the asphalt mixture. High flow indicated that the high tendency in pavement mixture easily exposed the deformation.

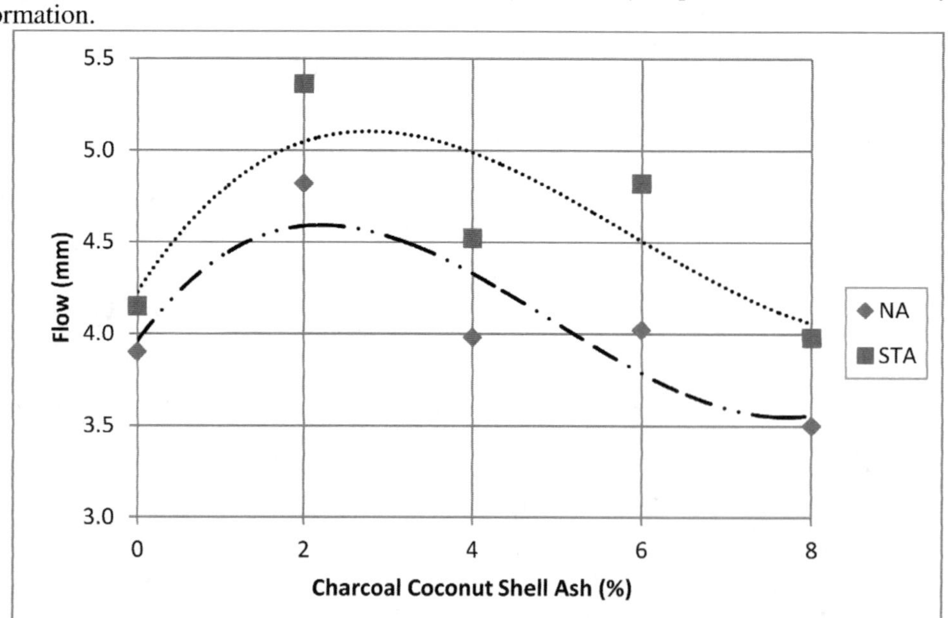

Figure 3. Flow of asphaltic concrete containing varying CCSA content

3.4. Voids in Total Mix

The voids in total mix versus CCSA content relationships of asphaltic concrete under different aging condition are presented in Figure 4. The results show that at the unaged condition, the VTM of 2% CCSA specimens are lower than the VTM of other mixture where the VTM of the controlled specimen under short term aging is less than that of other mixes. Lower air void contents minimize the aging of the asphalt cement films within the aggregate mass. Lower the air-voids as results less permeable of the mixture. However, higher air void content provides passageways through the mix for the entrance of damaging air and water. Test results also indicate that the VTM of the specimen under unaged increases between 2.5% to 3.7% when the CCSA content increases from 0% to 8% while the corresponding VTM under STA situation increases from 3.2% to 4.2%. This phenomena due to allowing more air voids in the specimens under NA compare to STA. On the other hand, poor air void contents minimize the possibility of moisture penetrating the thin asphalt cement film and strip the asphalt cement off the aggregates.

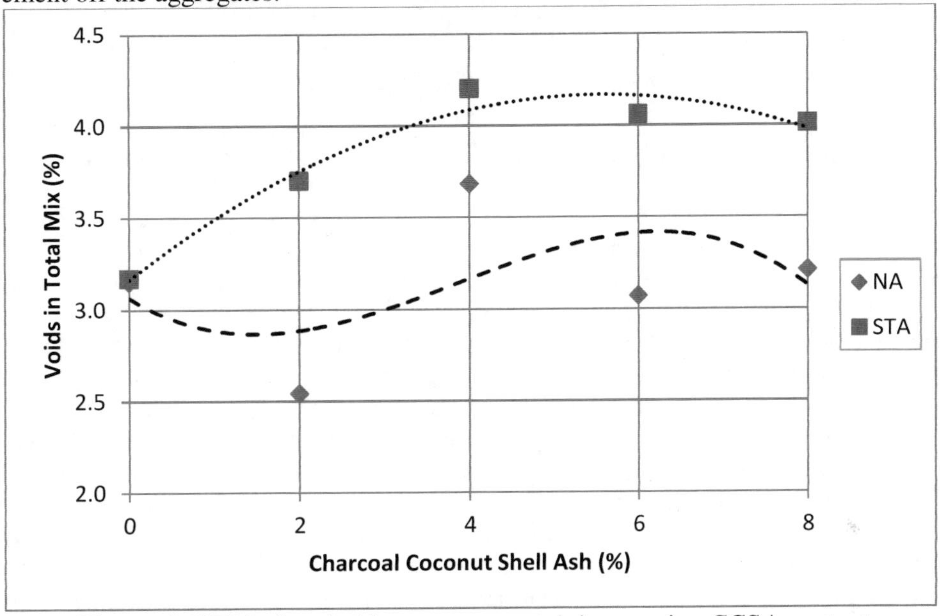

Figure 4. VTM of asphaltic concrete containing varying CCSA content

3.5. Voids Filled with Bitumen

VFB represents the number of voids which are filled with a binder in an asphalt mixture. As shown in Figure 5, the VFB of asphaltic concrete specimens increases from 39% to 96% and 36% to 95% when the CCSA content increases from 0% to 8%, respectively for NA and STA. The VFB of specimens peaks at the CCSA content of 8%. At all aging session, the minimum VFB value of mixes is 36% (0% CCSA) content while the maximum VFB value is 96% (8% CCSA). According to JKR specification [13] for dense mixes, the VFB should range from 75% to 85%. However, the specification does not specify limitation values for dense asphalt mixture, especially for incorporated with waste agriculture product.

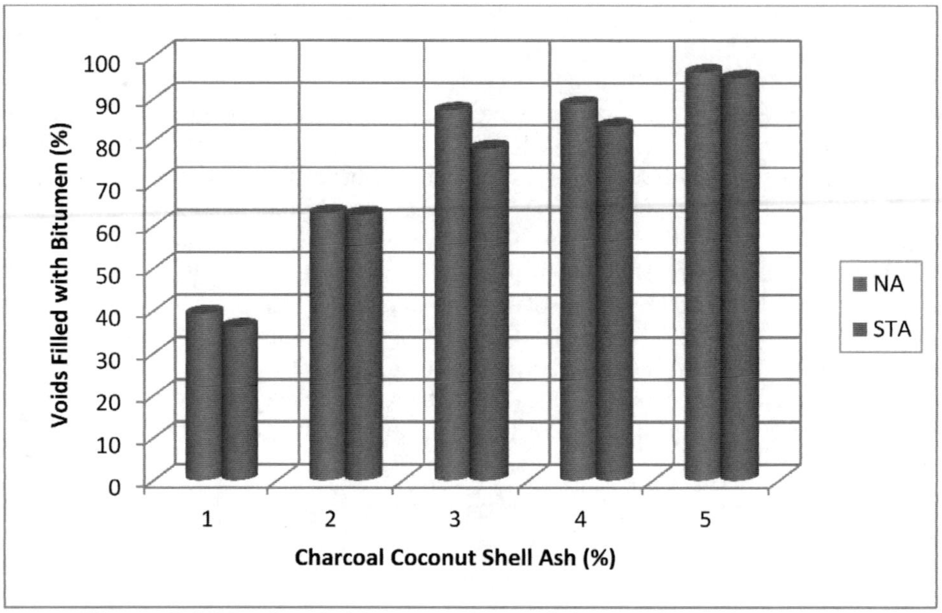

Figure 5. VFB of asphaltic concrete containing varying CCSA content

3.6. Resilient Modulus

The results of the resilient modulus test for aged specimens at varying CCSA content and temperature are graphically illustrated in Figure 6. Generally, the results indicate that an increase in CCSA content causes the resilient modulus of the mix to increase until a maximum value is reached, beyond which the resilient modulus decreases as the CCSA content continue to increase. At all conditions, the resilient modulus increases with increasing CCSA content up to a 6%, then start to decreases with further additions of CCSA. In addition, the resilient modulus of specimens increases after every aging session. For example, at 2% CCSA, the resilient modulus at the temperature of 25°C increases by 10.8% after STA and 85% when exposed under 40°C. It can be seen that the aging process causes oxidation and increases the hardening rate of the bitumen, thus resulting in the increasing resilient modulus. This phenomenon is usually termed as age hardening [20]. Furthermore, the resilient modulus of aged specimens also reduces when the temperature increases. The resilient modulus of unaged at 6% CCSA is higher than the resilient modulus of 0%, 2%, 4% and 8% CCSA at every test temperature. However, at STA session, 4% CCSA is higher than other mixtures.

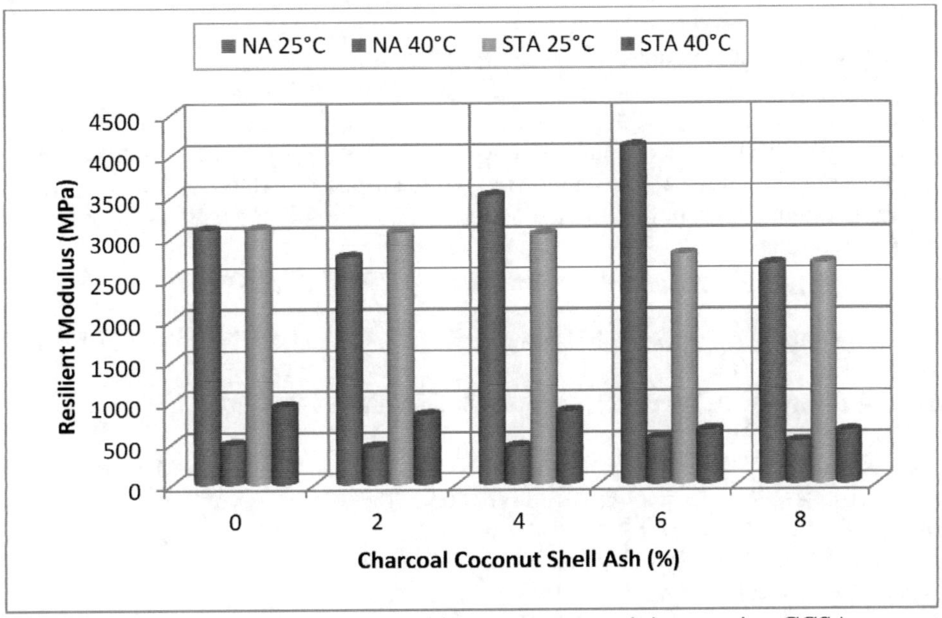

Figure 6. Resilient Modulus of asphaltic concrete containing varying CCSA content

3.7. Dynamic creep

Creep modulus from the dynamic creep test on aged specimens was used to predict the ability of mixes containing CCSA to resist permanent deformation after 10 to 15 years' service period. In this study, the dynamic creep test was determined after 3600 load cycles, and the test results are illustrated in Figure 7. The test results indicate that the creep modulus of specimens at unaged condition increases up to a distinct peak beyond which the creep modulus reduces as the CCSA content increases. The creep modulus under STA exhibits similar trend peaks at different CCSA content. The creep modulus of aged CCSA asphalt mixes is generally higher than unaged mixes. The test results also show that aging causes creep modulus of CCSA asphalt mixes to exhibit higher resistant against rutting. Age hardening of the binder causes an increment of approximately 10% in creep modulus after short term aging.

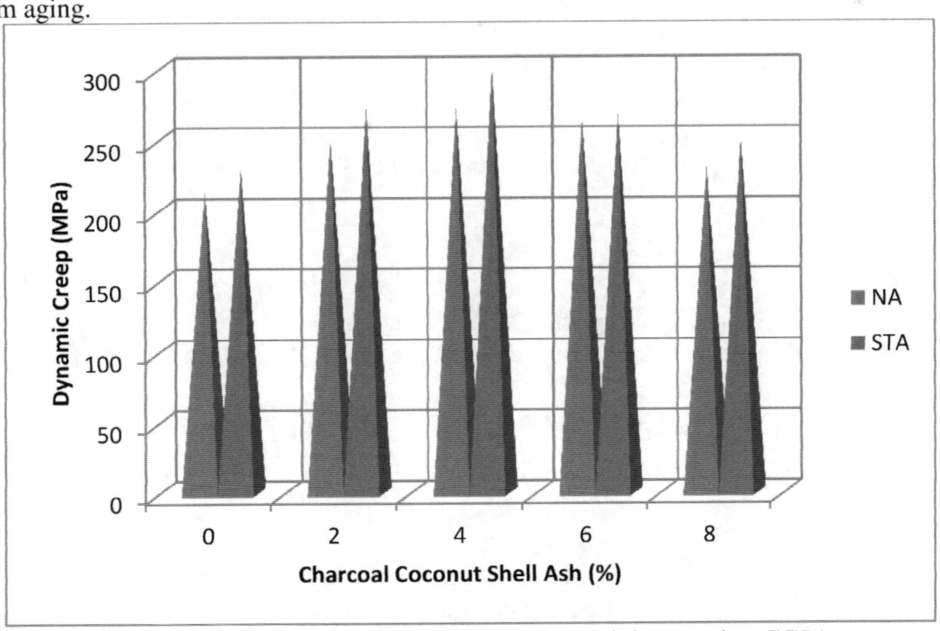

Figure 7. Creep modulus of asphaltic concrete containing varying CCSA content

4. Conclusions

Following conclusions could be drawn from the experimental study:

a) This study indicated the potentiality of charcoal coconut shell ash (CCSA) as modify binder in hot mix asphalt.
b) The addition of charcoal coconut shell ash as binder replacement improved the stability, stiffness, resilient modulus and dynamic creep of the asphalt mixture.
c) Asphalt mixture incorporating charcoal coconut shell ash exhibits higher performance after short term aging session.

5. References

[1] Lay Ting T, Putra Jaya R, Abdul Hassan N, Yaacob H and Sri Jayanti D 2015 *J. Teknologi.* **76**(14) pp. 121-125
[2] Jeffry S N A, Jaya R P, Manap N, Miron N A and Hassan N A 2016 *Key Eng. Mater.* **700** pp. 227-237.
[3] Ramadhansyah P J, Nurfatin Aqeela M, Siti Nur Amiera J, Norhafizah M, Norhidayah A H and Dewi S J 2016 *ARPN J. Eng. Appl. Sci.* **11**(12) pp. 7457-7462.
[4] Chen H and Xu Q 2010 *Fuel.* **89**(7) pp. 1616-1622.
[5] Yin J M and Wu W 2018 *Waste Manage.* **78** pp. 948-954.
[6] Liu X and Wu S 2011 *Constr. Build. Mater.* **25** pp. 1807-1811.
[7] Carlos J. S-A, Pedro L-G, Pablo P-M and Daniel C-F 2019 *Constr. Build. Mater.* **200** pp. 756-769.
[8] Norhafizah M, Ramadhansyah P J, Siti Nur Amiera J, Nurfatin Aqeela M, Norhidayah A H, Hainin M R and Che Norazman C W 2016 *J. Teknologi.* **78**(7-2) pp. 127-132.
[9] Abiola O S, Kupolati W K, Sadiku E R and Ndambuki J M 2014 *Constr. Build. Mater.* **54** pp. 305-312.
[10] Jeffry S N A, Jaya R P, Abdul Hassan N, Yaacob H and Satar M K I M 2018 *Constr. Build. Mater.* **173** pp. 40-48.
[11] Buhari R, Nur Fareesya Z, Mohd Ezree A, Siti Khatijah A B and Nurul Hidayah M K 2016 *J. Teknologi* **78**(7–3) pp. 17–22.
[12] Abdullah M E, Rosni N N M, Jaya R P, Yaacob H, Hassan N A and Agussabti 2017 *Key Eng. Mater.* **744** pp. 121-125
[13] Jabatan Kerja Raya Malaysia 2008 *Standard Specification for Road Works – Section 4: Flexible Pavement.* Kuala Lumpur, Malaysia.
[14] ASTM D5581 2007 *Standard Test Method for Resistance to Plastic Flow of Bituminous Mixtures Using Marshall Apparatus.* ASTM International, West Conshohocken, PA, USA.
[15] ASTM D7369 2011 *Standard Test Method For Determining The Resilient Modulus Of Bituminous Mixtures By Indirect Tension Test.* ASTM International, West Conshohocken, PA, USA.
[16] BS EN 12697-25 2016 *Bituminous mixtures. Test methods. Cyclic compression test.* British Standards Institution, UK.
[17] AASHTO R30 2002 *Mixture Conditioning of Hot Mix Asphalt (HMA).* American Association of State Highway and Transportation Officials, Washington D.C, USA.
[18] Romastarika R, Jaya R P, Yaacob H, Nazri F M, Agussabti, Ichwana and Jayanti D S 2017 *AIP Conf. Proc.* **1875** pp. 1–8.
[19] Wang H, Shihao Z, Guojun L and Junwei X 2019 *Constr. Build. Mater.* **226** pp. 442-448.
[20] Ting T L, Ramadhansyah P J, Norhidayah A H, Yaacob H, Hainin M R, Wan Ibrahim M H, Jayanti D S and Abdullahi A M 2018 *IOP Conf. Ser.: Mater. Sci. Eng.* **140**(1) pp. 1-8.

Acknowledgement

The support provided by the Malaysian Ministry of Higher Education, Universiti Malaysia Pahang in the form of a research grant (RDU/UMP) vote number RDU1803160 and Universiti Teknologi Malaysia with research grant number Q.J130000.2651.16J30 (GUP Tier 2) for this study is highly appreciated.

Strength and Porosity of Porous Concrete Pavement Containing Nano Black Rice Husk Ash

P J Ramadhansyah[1], K A Masri[1], S A Mangi[2], M I Mohd Yusak[3], N Mashros[4], M N Mohd Warid[4], M K I M Satar[4] and W I Mohd Haziman[5]

[1]Faculty of Civil Engineering Technology, Universiti Malaysia Pahang, 26300 Gambang, Pahang, Malaysia
[2]Department of Civil Engineering, Mehran University of Engineering & Technology, SZAB Campus Khairpur Mir's, 76060 Sindh, Pakistan
[3]SPC Industries Sdn. Bhd, 21 Miles, Jalan Pontian, 81150 Ulu Choh, Johor Bahru, Malaysia
[4]Faculty of Engineering, School of Civil Engineering, Universiti Teknologi Malaysia, 81310 Johor Bahru, Malaysia
[5]Faculty of Civil and Environmental Engineering, Universiti Tun Hussein Onn Malaysia, 86400 Batu Pahat, Johor Bahru, Malaysia

Abstract. Rice husk ash (RHA) as replacement material in the conventional concrete mixture has been widely studying around the world. However, there is a lack of study on nanoparticle produced from black rice husk ash (BRHA) used as a replacement material in porous concrete pavement mixture. Therefore, this study aims to evaluate the performance of porous concrete pavement containing nano black rice husk ash in terms of compressive strength, porosity and its correlation. A nano BRHA dosage of 10, 20, 30 and 40% by weight of binder was used throughout the experiments. Four different grindings of BRHA were examined, and it was found that BRHA ground had a nanoparticle between 64 nm to 85 nm. There appears to be an optimum replacement of approximately 10% nano BRHA, during which time the compressive strength and porosity increase significantly.

1. Introduction

Porous concrete is also referred to as pervious concrete, no-fines concrete and Permeable concrete. Porous concrete is a special type of cementitious material composed of gap-graded aggregates, coated with a thin layer of cement paste and bonded by the cement paste layers partially being in contact [1]. Porous concrete is concrete with continuous voids, which are purposely incorporated into concrete. The porous concrete has a large number of voids; it differs from conventional concrete, which has small or microvoids. The finished surface is not tight, and uniform, but is open and varied, to admit large quantities of stormwater (Specifier's Guide for Pervious Concrete Pavement Design). The ability of the porous concrete pavement to allow water to percolate into the underlying strata depends on its porosity, which is one of its most important pore structure features [2]. This type of concrete is a completely different category from conventional concrete, and therefore its physical characteristics differ greatly from those of normal concrete [3]. In recent years, the use of nanomaterials has received particularly interested in various fields of applications to improve existing technologies and also produce materials with new functionalities [4]. Research and developments have established that the application of nanotechnology can improve the performance of conventional construction materials, such as concrete and steel [5]. Nano-materials are used as a replacement for cement parts. The nanoscale size of the materials can result in significantly enhanced properties from conventional grain-size materials of the same chemical composition. Previous research studies on mixing nanomaterials in

IOP Conf. Series: Materials Science and Engineering **712** (2020) 012037 doi:10.1088/1757-899X/712/1/012037

cement indicate that the inclusion of nanomaterials modifies the fresh and hardened properties [6]. Nanoparticles have a high surface area to volume ratio and also provide high chemical reactivity [7]. Previous researchers agree that the finer pozzolanic ash has a better reaction compared to coarser pozzolanic ash [8,9]. The fineness of the rice husk ash is important because it influences the rate of reaction and the rate of gain in concrete strength [10,11]. In producing fineness or nano-size of materials, Pacheco-Torgal et al. [7] reported that the production of nanomaterials could be obtained using two methods, namely through high milling energy or by chemical synthesis. The methods consist of ball milling, flame spray pyrolysis, chemical vapour deposition, micro-emulsion, sol-gel and many more [12]. Various grinding methods have been applied by previous researchers to produce fine rice husk ash. Some researchers pulverised the rice husk ash by using a vibrating ball mill [13]. Fernandes et al. [14] grinding the passing rice husk ash using an eccentric ball mill. Salas et al. [15] the RHA was ground by ball milling using ceramic balls and Habeeb and Fayyadh [16] was ground the RHA using a Los Angeles mill. The lack of research conducted on porous concrete containing nanoparticle produced by BRHA is the main reason for this study. Porous concrete has been used for many years, but there are still many outstanding issues related to its structural performance [17]. The fundamental information, i.e. the influence of the water-cement (W/C) ratio, void ratio, cement paste characteristic, the volume ratio of coarse aggregate and etc. has been studied. However, the relationship between strength and porosity of porous concrete pavement is still not yet established [18]. Therefore, this research was conducted.

2. Materials and methods

2.1. OPC and aggregate
The ordinary Portland cement type I with the requirements of BS EN 197-1 [19] was supplied by the Tasek Corporation of Malaysia Berhad and was used as the major binder material in the production of strength concrete pavement grade 20 MPa. On the other hand, crushed granite of size 20 mm was used as the coarse aggregate with a specific gravity of 2.66 and water absorption of 0.86%, respectively.

2.2. Production Nano BRHA through the grinding process
The black rice husk ash (BRHA) samples were collected from the rice mill factory. The rice husk was burned at the factory without controlling the specific temperature. At the laboratory, the raw BRHA was sieved using a sieve size of 150μm to ensure the homogeneity of the BRHA size before the grinding process. Then, the BRHA passing the 150μm sieve was taken and ground for four different grinding times to obtain the nano size (NS). The original BRHA was generally designated as NS_0; further explanation exhibited in Table 1.

Table 1. Grinding designation

Nano BRHA	Time of grinding (hrs)
NS_0	0
NS_1	33
NS_2	48
NS_3	63
NS_4	81

2.3. Mix proportions and curing
The OPC was partially replaced with nano BRHA at dosages of 10%, 20%, 30% and 40% by weight of the cementitious material. A mass water-to-binder ratio of 0.34 was used. A total cement content

used in this investigation was 450 kg/m^3, coarse aggregate (1115 kg/m^3) and water (153 kg/m^3). Further, the curing condition is one of the important factors that need to be considered. The standard curing no-fines test cubes were followed according to established British standard BS EN 12390-2 [20]. Immediately after making the cube specimens, the specimens were stored in a place free from vibration and under a condition that prevented moisture loss. Upon demoulding after a period of 16 to 28 hours, the specimens were immersed it in water until air bubbles cease to rise. The specimens were drained and immediately placed in a polyethylene bag. Then, the polyethylene bags were sealed and stored. Tests were performed when the samples were 7, 14, 28, 56 and 90 days old. The results obtained at each age were reported as an average of the three tested samples.

2.4. Compressive strength test

The compressive strength test of the porous concrete pavement mixes was performed on 100 x 100 x 100 mm cubes. The specimens were compressed using a compression machine with a loading rate of 3.5kN/s. The reported compressive strength was the average of the three specimens tested. The test was conducted according to the British standard test method BS EN 12390-3 [21].

2.5. Porosity test

The porosity test was conducted by ASTM D7063/D7063M [22]. At laboratory, the dry specimens were weighed and designated as mass A. The specimens were then vacuum sealed using the Corelok® vacuum sealer, inside a bag having a mass, B, and then submerged underwater and weighed both sealed, E, and unsealed, C. The value of the porosity was calculated using Eq. (1), (2) and (3).

$$\text{Porosity} = \frac{SG2-SG1}{SG2}\ x\ 100 \quad (1)$$

$$\text{Bulk specific gravity} = SG1 = \frac{A}{B-E-\frac{B-A}{F_T}} \quad (2)$$

$$\text{Apparent specific gravity} = SG2 = \frac{A}{B-C-\frac{B-A}{F_{T1}}} \quad (3)$$

Where F_T and F_{T1} are the apparent specific gravity of the plastic sealing material when sealed and unsealed, respectively.

3. Results and discussion

3.1. Chemical properties

The chemical test of the OPC and nano BRHA was conducted using the X-ray Fluorescence spectrometer (XRF). It was found that the calcium oxide had the highest percentage for OPC (62.27%) followed by silica (22.68%). The other elements had a small percentage for example alumina (4.72%), iron (3.5%), oxide (1.89%), magnesium oxide (0.31%) and potassium oxide (4.29%). It can be seen that the chemical composition of OPC fulfilled the standard requirements, as stated in the American Society for Testing and Materials [23]. Furthermore, SiO_2 (91%) was identified as the main component of the BRHA. The other elements in BRHA are alumina (0.07%), iron oxide (0.07%), magnesium oxide (0.45%), sodium oxide (0.28%), potassium oxide (0.01%) and sulphur trioxide (2.64%). Compared with OPC, the BRHA has a small percentage of calcium oxide and a high percentage of silica. According to ASTM C618 [24], the BRHA used in this study satisfied the Class N requirement for natural pozzolan, which indicates that the total amount of silicon dioxide, aluminium oxide and iron oxide must be equal or greater than 70%.

3.2. Physical properties

The specific gravity of the OPC used in this study was 2.23, where the specific gravity of nano BRHA at the different grinding time are summarized in Table 2. It was found that the specific gravity of nano BRHA was less than that of the OPC. This is due to the porous particles of the materials becoming fine particles after the crushing process. Also, the mean particle size of the OPC is 1353 nm, while the mean particle size of the nano BRHA is presented in Table 2. It clearly shows that an increase in the grinding time, decrease the mean particle size of nano BRHA.

IOP Conf. Series: Materials Science and Engineering **712** (2020) 012037 doi:10.1088/1757-899X/712/1/012037

Table 2. Grinding designation

Nano BRHA	Specific gravity	Particle size (m)
NS_1	1.62	85
NS_2	1.64	76
NS_3	1.72	66
NS_4	1.76	64

3.3. Compressive strength

Figure 1 shows the result of compressive strength for the various nano BRHA replacements. Generally, the compressive strength for all specimens increases with the increase in curing age. From the figure, it can be seen that the highest compressive strength for each age is 10% for the replacement specimen, which is 20.88 MPa, 24.45 MPa, 27.80 MPa, 34.20 MPa and 35.23 MPa for 7, 14, 28, 56 and 90 days, respectively. The 40% replacement of nano BRHA is the lowest compressive strength, which is 14.64 MPa, 16.28 MPa, 22.92 MPa, 27.53 MPa and 29.30 MPa for 7, 14, 28, 56 and 90 days, respectively. The compressive strength for each day of testing shows the same trend where the compressive strength starts to increase from 0 % replacement to 10 % replacement of the nano BRHA. Then, the compressive strength starts to decrease from 10 % to 20% replacement of the nano BRHA. The same applies to 20 % to 30 % and 30 % to 40 % replacement. This finding similar to Jaya et al. [11] where the compressive strength starts to increase with 10% replacement, and it starts to decrease when the replacement is beyond 10%. This phenomenon probably occurred due to workability and the lack of required water in the mixture. The use of NS in porous concrete had greater compressive strength than the control specimens [25]. Nevertheless, Wang et al. [26] found in their study that too high or too low a content of NS is not beneficial to upgrading the cement strength. This is probably due to the amount of NS in the mix being higher than the amount required combining with the liberated lime during the process of hydration. From this figure, the 10 % replacement of nano BRHA has the best property for the compressive strength evaluation.

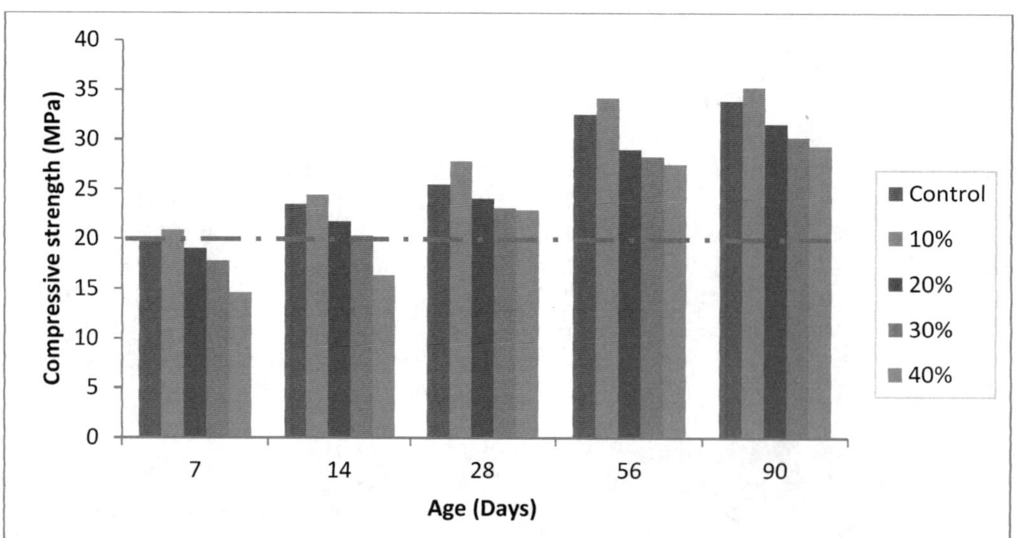

Figure 1. Compressive strength at different nano BRHA replacement

3.4. Porosity

Figure 2 illustrated the porosity of porous concrete pavement at different level of nano BRHA. It can be seen that the porosity of the porous concrete pavement increases with the increase in the nano BRHA replacement. The porosity starts to increase from 0% to 40% when nano BRHA was replaced with cement. The increased in porosity was about 14.75%, 14.85%, 17.40%, 17.99% and 23.94% respectively. There is no significant increment of porosity when nano BRHA was replaced between 0% and 10%. The same applies to 20% to 30% nano BRHA replacement; there is no significant increment in porosity. The difference between 0% to 10% and 20% to 30% is only 0.1% and 0.59%, respectively. The increase in porosity with the increasing percentage of the nano BRHA replacement is due to the nano BRHA in the mixtures absorbing the water. When the amount of nano BRHA in the mixture increases, it will absorb more water. The results also found that there is a strong correlation between the percentage of porosity and the percentage of nano BRHA replacement. This indicating that when the percentage of the nano BRHA replacement increases, the porosity will also increase.

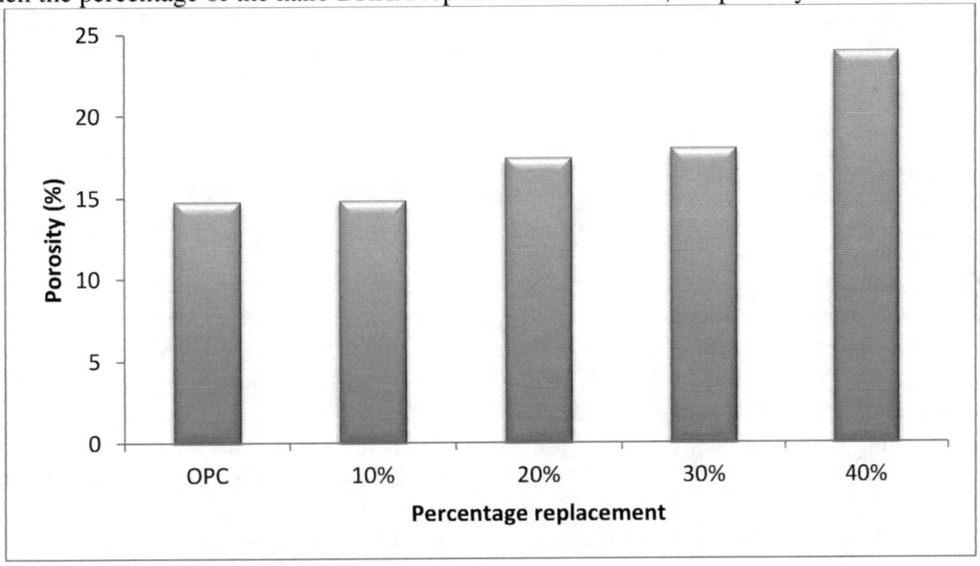

Figure 2. The porosity of porous concrete pavement

3.5. Correlation between strength and porosity

Figure 3 shows the relationship between compressive strength and porosity for the porous concrete pavement. The relationship was determined using a value of 14.75, 14.85, 17.40, 17.99 and 23.94 % for porosity and 25.48, 27.80, 24.07, 23.06 and 22.92 MPa for compressive strength, respectively. The R^2 value for the relationship is 0.8083, and it shows a negative correlation. It was found that increasing porosity will decrease the compressive strength. It is due to the interconnected voids decreasing the strength of the specimens.

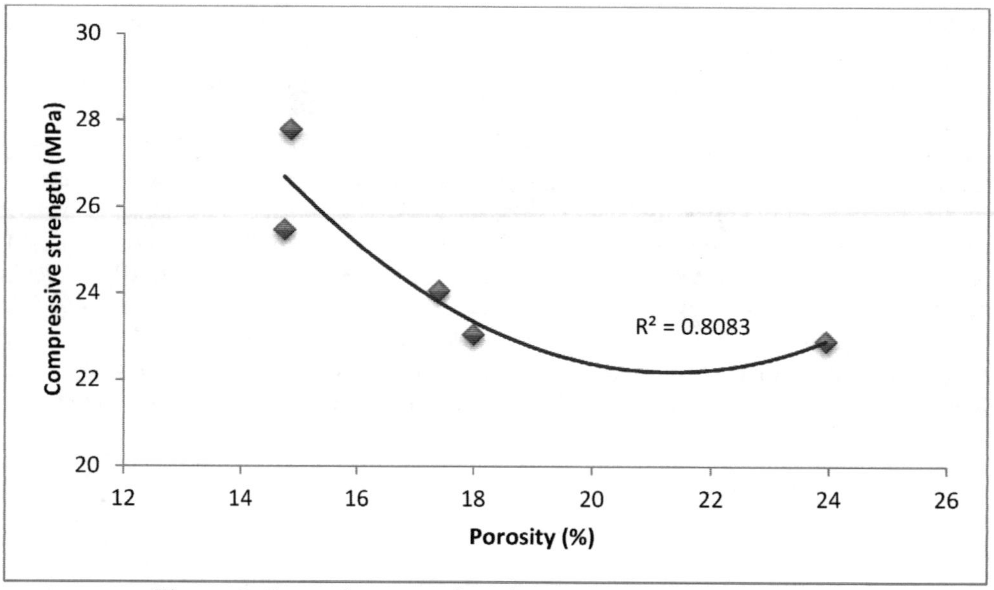

Figure 3. Strength vs porosity of porous concrete pavement

4. Conclusions

a. It was found that the nano BRHA and OPC had the highest percentage of calcium oxide and followed by other elements. The results also show that the nano BRHA used was satisfied with the ASTM requirement for natural pozzolan.

b. The specific gravity of nano BRHA increased with increased grinding time where nanoparticle BRHA decreased as grinding time increased.

c. The compressive strength of the porous concrete pavement incorporating nano BRHA investigated initially increased up to 10% replacement and then began to decrease until 40% level.

d. There appears to be an optimum nano BRHA level of approximately 10%, for which compressive strength and porosity values increased significantly.

5. References

[1] Agar Ozbek A S, Weerheijm J, Schlangen E and Van Breugel K 2013 *Cement and Concrete Comp*. **39** pp. 1-11.

[2] Deo O and Neithalath N 2011 *Constr. Build. Mater.* **25** pp. 4181-4189.

[3] Chindaprasirt P, Hatanaka S, Chareerat T, Mishima N and Yuasa Y 2008 *Constr. Build. Mater.* **22**(5) pp. 894-901.

[4] Senff L, Labrincha J A, Ferreira V M, Hotza D and Repette W L 2009 *Constr. Build. Mater.* **23**(7) pp. 2487-2491.

[5] Hanus M J and Harris A T 2013 *Prog. Mater. Sci.* **58** pp. 1056-1102.

[6] Salemi N and Behfarnia K 2013 *Constr. Build. Mater.* **48** pp. 934-941.

[7] Pacheco-Torgal F, Miraldo S, Ding Y and Labrincha J A 2013 *Constr. Build. Mater*, 38, 365-370.

[8] Hainin M R, Ramadhansyah P J, Chan T H, Norhidayah A H, Nazri F M, Ichwana 2017 *Mater. Sci. Forum.* **889** pp. 265-269.

[9] Abu Bakar B H, Ramadhansyah P J and Megat Azmi M J 2011 *Mag. Concrete Res.* **63**(5) pp. 313-320.

[10] Zain M F M, Islam M N, Mahmud F and Jamil M 2011 *Constr. Build. Mater.* **25** pp. 798-805.

[11] Jaya R P, Abu Bakar B H, Megat Johari M A, Wan Ibrahim M H, Hainin M R and Jayanti D S 2014 *Adv. Cement Res.* **26**(3) pp. 145-154.

[12] Singh L P, Bhattacharyya S K, Kumar R, Mishra G, Sharma U, Singh G and Ahalawat S 2014 *Adv. Colloid Interface Sci.* **214** pp. 17-37.
[13] Bui D D, Hu J and Stroeven P 2005 *Cement Concrete Comp.* **27** pp. 357-366.
[14] Fernandes I J, Calheiro D, Kieling A G, Moraes C A M, Rocha T L A C, Brehm F A and Modolo R C E 2016 *Fuel.* **165** pp. 351-359.
[15] Salas A, Delvasto S, de Gutierrez R M and Lange D 2009 *Cement and Concrete Res.* **39** pp. 773-778.
[16] Habeeb G A and Fayyadh M M 2009 *Aust. J. Basic & Appl. Sci.* **3**(3) pp. 1616-1622.
[17] Gupta R 2014 *Case Stud. Constr. Mater.* **1** pp. 1-9.
[18] Jaya R P, Yusak M I M, Hainin M R, Mashros N, Warid M N M, Ali M I and Ibrahim M H W 2019 *AIP Conf. Proc.* **2151** pp. 1-8,
[19] BS EN 197-1 2011 *Cement. Composition, specifications and conformity criteria for common cements*. British Standards Institution, United Kingdom.
[20] BS EN 12390-2 2019 *Testing hardened concrete. Making and curing specimens for strength tests*. British Standards Institution, United Kingdom.
[21] BS EN 12390-6 2009 *Testing hardened concrete – Part 6: Tensile splitting strength of test specimens*. British Standards Institution, United Kingdom
[22] ASTM D7063/D7063M 2011 *Standard Test Method for Effective Porosity and Effective Air Voids of Compacted Bituminous Paving Mixture Samples*. Annual Book of ASTM Standards, American Society for Testing and Materials, West Conshohocken, PA, USA.
[23] ASTM C150/C150M 2011 *Standard Specification for Portland Cement. Annual Book of ASTM Standards*. American Society for Testing and Materials, West Conshohocken, PA, USA.
[24] ASTM C618 2012 *Standard Specification for Coal Fly Ash and Raw or Calcined Natural Pozzolan for Use in Concrete*. Annual Book of ASTM Standards, American Society for Testing and Materials, West Conshohocken, PA, USA.
[25] López-Carrasquillo V and Hwang S 2017 *Constr. Build. Mater.* **139** pp. 148-158.
[26] Wang L, Zheng D, Zhang S, Cui H and Li D 2016 *Nanomaterials.* **6**(12) pp. 241.

Acknowledgements
The support provided by Malaysian Ministry of Higher Education and Universiti Malaysia Pahang in the form of a research grant (RDU/UMP) vote number RDU190339 for this study is highly appreciated.

IOP Conf. Series: Materials Science and Engineering **712** (2020) 012038 doi:10.1088/1757-899X/712/1/012038

Strength Properties of Porous Concrete Pavement Blended with Nano Black Rice Husk Ash

P J Ramadhansyah[1], K A Masri[1], S A Mangi[2], M I Mohd Yusak[3], M R Hainin[4], M N Mohd Warid[4], N Mashros[4], H Yaacob[4] and W I Mohd Haziman[5]

[1]Faculty of Civil Engineering Technology, Universiti Malaysia Pahang, 26300 Gambang, Pahang, Malaysia
[2]Department of Civil Engineering, Mehran University of Engineering & Technology, SZAB Campus Khairpur Mir's, 76060 Sindh, Pakistan
[3]SPC Industries Sdn. Bhd, 21 Miles, Jalan Pontian, 81150 Ulu Choh, Johor Bahru, Malaysia
[4]Faculty of Engineering, School of Civil Engineering, Universiti Teknologi Malaysia, 81310 Johor Bahru, Malaysia
[5]Faculty of Civil and Environmental Engineering, Universiti Tun Hussein Onn Malaysia, 86400 Batu Pahat, Johor Bahru, Malaysia

Abstract. In order to improve the porous concrete pavement strength, one of the elements in the porous concrete structure that needs to be considered is the cement paste binder. This study aimed to evaluate the strength properties of porous concrete pavement blended with nanoparticle from black rice husk ash (BRHA). The performance was evaluated based on compressive strength, density, age of curing and strength activity index. Four nano BRHA replacement levels were considered in the study, i.e. 10%, 20%, 30%, and 40% by weight of cement. Test results showed that porous concrete pavement containing 10% nano BRHA replacement levels exhibits excellent compressive strength and strength activity index. The results also indicate that there is no significant correlation between density of the porous concrete pavement and compressive strength.

1. Introduction

Porous concrete has become increasingly used in a variety of infrastructures such as pavements and overlays subjected to the heavy traffic load [1]. Due to these extended applications, superior strength and durability constitute the main concerns associated with porous concrete pavement [2]. Porous concrete pavement is concrete with limited or no fine aggregates [3]. The concept of porous concrete tends to increase the strength and reduce surrounding noise [4]. Different from conventional concrete pavement, the porous concrete pavement has a high percentage of voids (between 15 % to 25 %) [5]. One of the approaches to strengthen the concrete strength is to minimise the number of voids [6]. Compared with conventional concrete pavement, porous concrete pavement is designed to have a high percentage of voids in its structure. At the same time, the required strength must be achieved to cater vehicle loads [7]. In order to increase the porous concrete pavement strength, one of the elements that need to be considered is the cement paste binder [8]. Besides aggregates, the strength of the porous concrete pavement depends on the cement paste binder [9]. A various study has been conducted in order to explore the advantages of using nano-size in their products. For example, Heikal et al. [10] found that by using nanoparticle size, the strength of the concrete was improved significantly. Several sources can be used to produce nanoparticle size, for instance, black rice husk ash (BRHA), palm oil fuel ash (POFA), fly ash (FA), Kaolin clay and etc. In this investigation, the nanosize produced from

IOP Conf. Series: Materials Science and Engineering **712** (2020) 012038 doi:10.1088/1757-899X/712/1/012038

black rice husk ash was selected as a main replacement component for the porous concrete pavement mixture.

2. Materials and methods

2.1. OPC

The ordinary Portland cement (OPC) was supplied by the Cement Industries of Malaysia satisfying the requirements of British European Standard specification for ordinary and rapid hardening Portland cement [11]. The chemical compositions of the OPC are listed in Table 1. The chemical analysis was carried out using XRF apparatus in accordance with the procedure given in BS EN 197-1 [11].

Table 1: Chemical compositions of PC

Oxide compounds	Chemical composition (%)
Silicon dioxide (SiO_2)	17.0
Aluminum oxide (Al_2O_3)	3.90
Iron oxide (Fe_2O_3)	3.20
Calcium oxide (CaO)	70.0
Magnesium oxide (MgO)	1.50
Sodium oxide (Na_2O)	0.02
Potassium oxide (K_2O)	0.53
Sulfur trioxide (SO_3)	3.60
Loss on ignition (LOI)	0.25

2.2. Aggregates

In order to obtain the final porous concrete pavement proportioning, the aggregates were washed, dried, and then sieved into their respective size ranges. The size required would depend on the kind of gradation used. The gradation of coarse aggregates used in this study is tabulated in Table 2. Based on the table, a sieve analysis carried out on a representative sample of the coarse aggregates showed that they complied with the requirement of ASTM C33/C33M [12].

Table 2: Coarse aggregate used

Aggregate size (mm)	Upper and lower limits (%)	Mid-gradation (%)
19.00	100	100
12.50	90-100	95
9.50	40-70	55
4.75	0-15	0
2.36	0-5	0

2.3. Water

Water is an important constituent in the manufacture of porous concrete pavement. In this investigation, tap water was used, which is safe for drinking. The general chemical analyses of tap water are shown in Table 3.

Table 3: Chemical analysis of the tap water

Element	Dissolved solid	Calcium	Chloride	Magnesium	Sulphate	Bicarbonate	Potassium
mg/L	80	23	2	2.6	4	81	3.8

pH	6.9

2.4. Nano from BRHA

At the laboratory, the raw BRHA was sieved using a sieve size of 150μm to ensure the homogeneity before the grinding process. Then, the BRHA passing the 150μm was taken and ground by a grinding machine for different grinding times. From the test that has been conducted, it was found that the optimum grinding time of BRHA to achieve nanosize was 63 hours with a mean particle size of 66 nm.

2.5. Mix design and compressive strength test

At the laboratory, a total cement content of 450 kg/m^3, 1115 kg/m^3 of coarse aggregate and 153 kg/m^3 of water were found and used as the control mix proportion for this study. The water to cement ratio (w/c) used was 0.34. Then, the compressive strength test of the porous concrete pavement mixes was performed on 100 × 100 × 100 mm cubes. The specimens were compressed by a compression machine with maximum capacity of 3000 kN with a loading rate of 150 kN/min. The reported compressive strength was the average of the three samples. The test was conducted according to the British standard test method BS EN 12390-3 [13].

3. Results and discussion

3.1. Compressive strength

The compressive strength of the porous concrete pavement containing OPC and nano BRHA at different replacements is illustrated in Figure 1. The test results indicate that the strength of porous concrete pavement incorporating nano BRHA was higher than the controlled mixes even at 90 days. Clearly, strength properties of porous concrete have increased significantly by the aged process. The amorphous silica and the nanoparticle size of BRHA are the principal reasons for the excellent pozzolanic activity and increase in compressive strength. On the other hand, the compressive strength of the porous concrete pavement replacement with 40% nano BRHA is lower than that of the others mixture at all ages as shown in Figure 1. Based on the results, it appears that there is an optimum percentage replacement of nano BRHA in porous concrete pavement approximately 10% beyond which the compressive strength increased significantly.

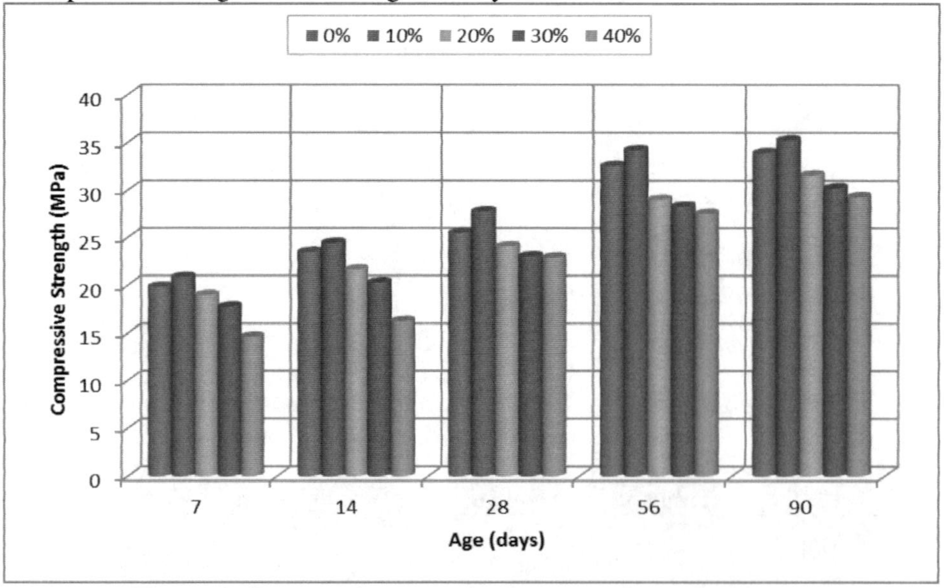

Figure 1: Compressive strength of porous concrete pavement

3.2. Density

The density of the porous concrete pavement specimens was measured right before the specimens were tested. The results are shown in Figure 2. The density of the porous concrete pavement is in the range of 1900 kg/m^3 to 2190 kg/m^3. As for the control specimens, the mean value was calculated as 2106 kg/m^3. While for the specimens with nano BRHA replacement, the mean value was calculated as 2012 kg/m^3. The mean density for specimens with nano BRHA replacement reduced by 4.5% from the control specimens. This is probably due to the effect of the nano BRHA replacement in the porous concrete specimens.

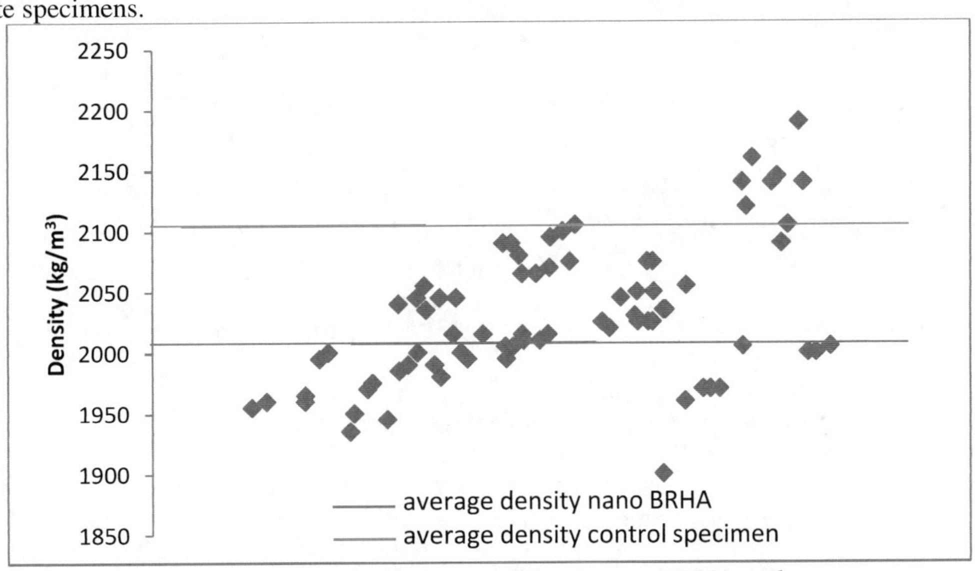

Figure 2: Density of specimens at different nano BRHA replacement

3.3. Compressive strength vs density

The relationship between the density and compressive strength of porous concrete pavement containing different nano BRHA percentage replacements is shown in Figure 3. From the figure, it can be seen that the density of the porous concrete pavement specimens has no significant relationship with the compressive strength. It can be seen that the R^2 of the relationship is 0.2626. Both high and low-density values have high strength. This phenomenon contrasts the typical theory which stipulates that when the density of concrete increases, the compressive strength of the concrete also increases and vice versa. This is probably due to the reaction of the nano BRHA replacement in the porous concrete mixture. A different percentage of replacement gives a different reaction. Although the porous concrete pavement specimens have low density, if the amount of nano BRHA contained in the porous concrete mixture is correct and enough to have a good reaction, it will result in good compressive strength.

IOP Conf. Series: Materials Science and Engineering **712** (2020) 012038 doi:10.1088/1757-899X/712/1/012038

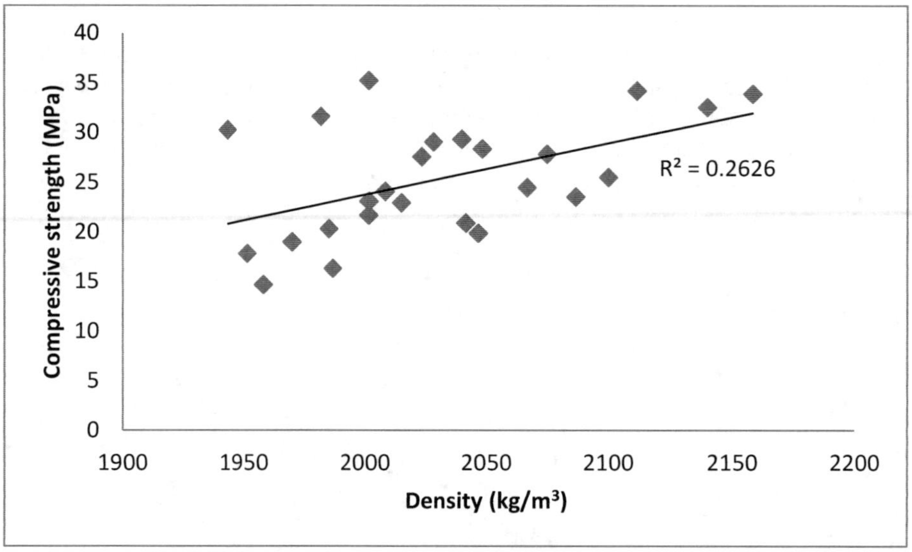

Figure 3: Density vs compressive strength of the porous concrete pavement

3.4. Compressive strength vs age of curing

Figure 4 shows the trend lines between the compressive strength and the age of the specimens for different percentages of nano BRHA replacements. While Table 4 shows the coefficient of determination (R^2) value for the relationship between the compressive strength and the age of the specimens for different percentages of nano BRHA replacements. The R^2 values are 0.90, 0.88, 0.95, 0.93 and 0.87 for 0%, 10%, 20%, 30% and 40%, respectively. All of the specimens show a good relationship between compressive strength and the ages of the specimens. The compressive strength of the specimens for each of the nano BRHA percentage replacements increases with the increase in the specimens' age. The continued hydration develops the strength of cement-based materials [14]. The increase in age gives more time for the hydration process in the specimens. Indirectly, the specimens become more hardened with respect to age. This is because the figure shows that the strength of the specimens increases with the increase in age.

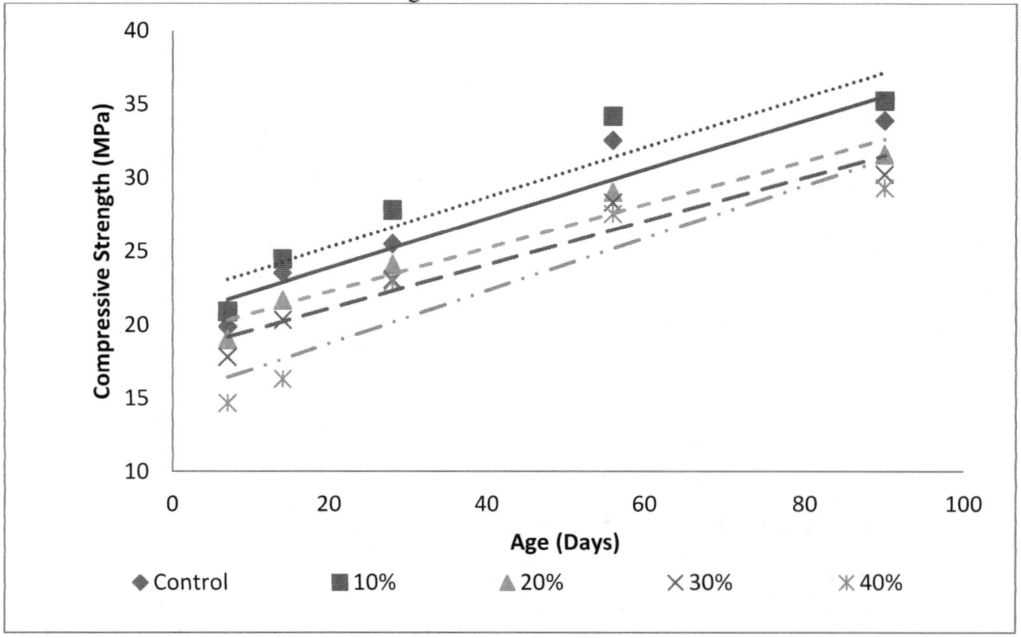

Figure 4: Compressive strength vs age of specimens

IOP Conf. Series: Materials Science and Engineering **712** (2020) 012038 doi:10.1088/1757-899X/712/1/012038

Table 4: Correlation value of strength vs curing age

Nano BRHA	0%	10%	20%	30%	40%
R^2 value	0.90	0.88	0.95	0.93	0.87

3.5. Strength activity index

As prescribed by ASTM C311/C311M [15], the strength activity index is used to determine whether natural pozzolan results in an acceptable level of strength development when used with hydraulic cement in concrete. This calculation was adopted here in order to determine the different strength values when a different percentage of replacement is used. The strength activity index was calculated following Equation 1.

$$\text{Strength activity index} = \frac{A}{B} X\ 100 \qquad (1)$$

Where A is compressive strength of the test mixture cube (MPa), and B is compressive strength of the control mix cube (MPa).

The strength activity index of the compressive strength results for all of the nano BRHA replacement percentages is shown in Figure 5. At any specimen age, 10% of nano BRHA replacement shows an increase in strength activity index. The strength activity index increased by 5.14%, 4.01%, 9.08%, 5.10% and 3.94% for 7 days, 14 days, 28 days, 56 days and 90 days, respectively. For 20%, 30% and 40% of nano BRHA replacement, a decrease in the strength activity index is observed at all ages. The lowest strength activity index at all ages belongs to specimens with 40% nano BRHA replacement. The strength activity index decreased to 73.72%, 69.26%, 89.93%, 84.60% and 86.45% for 7 days, 14 days, 28 days, 56 days and 90 days, respectively. This may be due to the quantity of nanoparticle present in the mixes being higher than the amount required to combine with the liberated lime during the hydration process, consequently leading to excess silica leaching out causing a deficiency in strength as it replaces part of the cementitious material [16]. Overall, for the strength activity index, all the specimens have a strength activity index of more than 75% for the 28 days age.

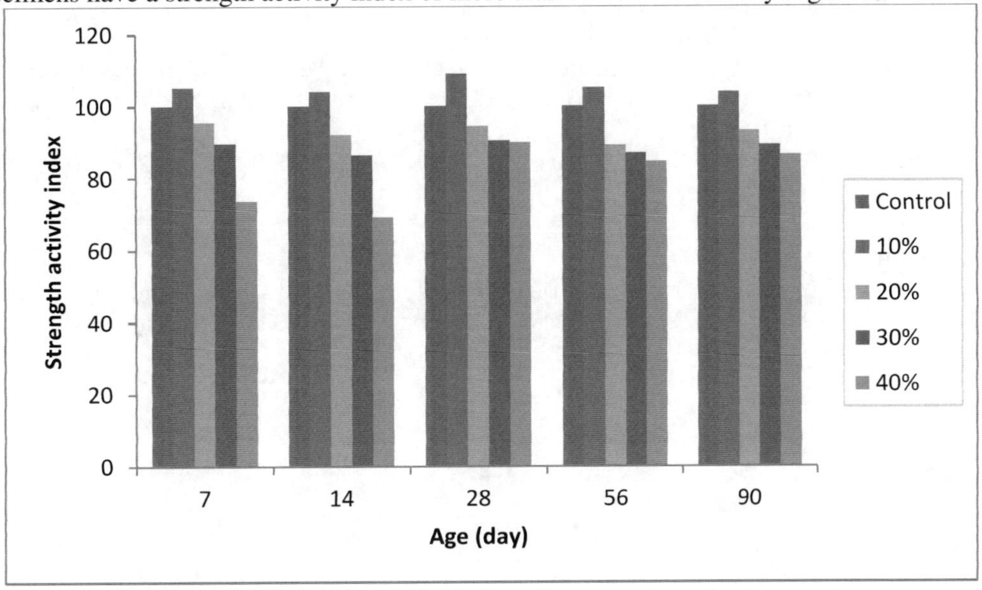

Figure 5: Strength activity index of porous concrete pavement

4. Conclusions

a) From the tests, it was found that the strength of the porous concrete pavement increases with age, the increase in strength partially due to the pozzolanic reaction. Furthermore, the compressive strength increases with increasing nano BRHA content up to a peak level and then decreases with further replacement.

b) The strength activity index of the nano BRHA in porous concrete pavement investigated increased with increased replacement level.

c) The use of nano BRHA at 10% replacement with cement produces the porous concrete pavement with good compressive strength and strength indexed.

5. References

[1] Wuguang L, Dae-Geun P, Sung W R, Byeong-Tae L and Yoon-Ho C 2016 *Constr. Build. Mater.* 118 pp. 20-26.
[2] Ning X, Michelle A and Xianming S 2019 *J. Clean. Prod.* 210 pp. 1605-1621.
[3] Hainin M R, Ibrahim M Y M, Khairul Idham M S M, Nordiana M, Ramadhansyah P J, Idris A M, Ibrahim M H W and Fadzli M N 2019 *IOP. Conf. Ser.: Earth. Environ. Sci.* 244(1) pp. 1-7.
[4] Jiusu L, Yi Z, Guanlan L and Xinghai P 2017 *Constr. Build. Mater.* 138 pp. 479-485.
[5] Xiaodi H, Kang D and Pan Pan 2019 *J. Clean. Prod.* 209 pp. 1484-1493.
[6] Jaya R P, Mohamad Idris A, Mohd Ibrahim M Y, Hainin M R, Yaacob H, Hassan N A, Mohd Haziman W I and Ekarizan S 2019 *IOP. Conf. Ser.: Mater. Sci. Eng.* 527(1) pp. 1-6.
[7] Guoyang L, Lukas R, Tianshuai L, Dawei W, Hui L and Markus O 2019 *Constr. Build. Mater.* 202 pp. 838-850.
[8] Mohd Ibrahim M Y, Ramadhansyah P J, Mohd Rosli H and Wan Ibrahim M H 2018 *IOP. Conf. Ser.: Mater. Sci. Eng.* 290(1) pp. 1-8.
[9] Xiaodan C, Hao W, Husam N, Giri V and John Hencken 2019 *J. Clean. Prod.* 237 pp. 1-11.
[10] Heikal M, Abd El Aleem S and Morsi W M 2013 HBRC J. 9(3) pp. 243-255.
[11] BS EN 197-1 2011 Cement. Composition, specifications and conformity criteria for common cements. British Standards Institution, London, UK.
[12] ASTM C33 / C33M 2018 *Standard Specification for Concrete Aggregates*. ASTM International. West Conshohocken, PA, USA.
[13] BS EN 12390-3 2019 Testing hardened concrete. Compressive strength of test specimens. British Standards Institution, London, UK.
[14] Metwally A A A E 2014 *HBRC J.* 10(2) pp. 145-155.
[15] ASTM C311 / C311M 2018 Standard Test Methods for Sampling and Testing Fly Ash or Natural Pozzolans for Use in Portland-Cement Concrete. ASTM International. West Conshohocken, PA, USA.
[16] Naji Givi A, Suraya A R, Farah N A A and Mohamad A M S 2010 *Constr. Build. Mater.* 24 pp. 2145-2150.

Acknowledgements

The support provided by Malaysian Ministry of Higher Education and Universiti Malaysia Pahang in the form of a research grant (RDU/UMP) vote number RDU190339 for this study is highly appreciated.

Investigating the Relationship between Site Supervision and IBS Based Construction Projects

M S Syamsyul Hairi, W R Wan Salleh, R A Rahman and S I Doh

Faculty of Civil Engineering Technology, Universiti Malaysia Pahang, 26300 Gambang, Pahang, Malaysia

Abstract. Successful delivery of construction of Industrialized Building System (IBS) projects needs an effective supervisor's that have a combination of strong technical, interpersonal, communication, and managerial skills. Unfortunately, a consistent finding is that there are many factors that affecting site supervision and most construction works are not delivered on time and consequently have a problem during installation of IBS. Therefore, the main aim of the study is to identify the factors affecting site supervision in improving IBS productivity. The study adopts a questionnaire survey approach to elicit information from the respondents, and out of the seventy questionnaires administered fifty-five were retrieved. The information and the data from the questionnaires were analyzed using the average index method. From the finding analysis, the main factors affecting site supervision is on-site construction process. While the main problem during installation of IBS is required planning on time delivery. It is recommended that the quality of IBS product or components should be enhanced to promote better site supervision

1. Introduction

Industrialized Building System (IBS) is one of technologies which can be categorized as an old in a developed country but yet considered as an emerging technology in developing countries such as the Asian region. Representatives of several prominent construction industries recommended the other construction sectors to extend the use of modern construction methods and information technology. Specifically, the use of mechanization in construction that leads to the production of IBS may help to ease the pressures of labour requirements while boosting quality and productivity [1]. IBS can be defined as a construction system which components are manufactured in a factory, on or off-site, positioned and assembled into structures with minimal additional site work [2]. In other words, IBS is a system that involves constructing buildings using pre-fabricated components. The manufacturing of the components is systematically done using machines, formwork, and other forms of mechanical equipment [3]. The components are manufactured offsite and once completed will be delivered to construction sites for assembly and erection. Since the first project of IBS in the year 1964 until today, IBS in Malaysia is not well accepted by the construction parties because of failure to adequately deal with risks. Failure to keep the cost within the budget is still common in Malaysia, and it is one of the reasons that limit the development of IBS [4]. Successful site management system for IBS construction had been categorized under five factors which include management (success integration, installed integration, caused delay, improvement), and workplace (site preparation, space, components keeping) [5]. IBS is admitted to be comparatively not new construction technique in Malaysia. Nevertheless, to ensure work on site is accomplished smoothly and without any risk, site supervision is required. In the IBS construction, improper site supervision will lead to problems on site, including incorrect assembly, leakage, and crack in the future. Moreover, it is essential for site supervisors of

IBS construction to understand the provisions that have been published to supervise sites adequately. Competent supervisors need a combination of strong technical, interpersonal, communication, and managerial skills. Successful delivery of construction of projects within the measurable tenets is driven by effective site supervision [6]. In many organizations, the efficiency of supervision is low even though the latest technology and trained workforce are made available. Also, it indicated that there is a lack of consideration in site supervision toward the improvement of IBS construction [7]. During the installation of IBS components, it is essential to have a good working collaboration, effective communication channel, and improvement in planning and scheduling of the project, risk management [4]. During IBS site management, delays are caused by late-arriving of IBS components at the construction site, erecting problems always occur without proper handling by the supervisor and installers, unskilled and inexperienced labour, insufficient plant and equipment, lack of teamwork and coordination, and weather problems [5]. Contractors are currently facing difficulties during installation at the site to comply with the design and manufacturing requirement [8]. This study aims to provide a general perspective of the factors that are affecting site supervision in improving IBS construction productivity. The specific objectives of this study are as follows:-

a) To study the factors affecting site supervision in IBS construction.
b) To identify the problems during installation of IBS.
c) To analyse and suggest ways to enhance site supervision in improving IBS productivity.

2. Methods

This section focuses on the method of study to be carried out to achieve the objective of this paper. Therefore, every measure is appropriate and relevant to the study's topic. Three approaches have been throughout this study to gather reliable and relevant data. The approaches are:

a) Disseminating a questionnaire survey
b) Computing the relative average index
c) Reviewing the effective strategic formulation method

2. 1 Data collection

This study's data collection involves acquiring information through a questionnaire survey because surveys involve a set of pre-formulated and written questions that the researchers would like to ask the respondents and record their answers. Surveys can be an efficient data collection tool when researchers know precisely the information that is needed and how to measure the variables of interest. Thus, all questions should be understandable, clear, and obtain no ambiguity. Before disseminating the questionnaire survey, a literature review was performed to outline an overview of the research topic. Moreover, the necessary information and additional references needed for research were gained from the literature review. The materials for the literature review include books, articles, magazines, internet, journals, newspaper, and others. Thus, the literature review guides the preparation of the survey questionnaire. Data validation occurs after the questionnaires had been collected. In the process of validation, the answer obtained from questionnaires will be checked for accuracy and suitability for this study purpose.

2.2 Sampling data

Handing out questionnaires is an approach to determine the current factors affecting site supervision in improving IBS construction productivity among IBS players in peninsular Malaysia. The target respondents for this study are contractors, suppliers/manufacturers, government officials, consultant firms, developers, and architectural firms for IBS construction. About 100 copies of the questionnaire were distributed to the targeted respondents. The method of distribution and collection of the questionnaire survey encompass the following:-

a) By hand distributions for selected respondents
b) By conforming through telephone calls and dispatching the questionnaire
c) By mail and returned via mail through a stamped self-addressed envelope

2.3 Data analysis

After the process of collecting data, the information and data gathered through questionnaire were carried out using the relative average index (AI) method in relation to the objectives and scope of the study. The AI of each factor was computed using the following formula:

$$\textbf{\textit{Average Index (AI)}} = \frac{\Sigma xiai}{\Sigma xi}$$

Where a = is a constant, weighting factor for i, x is the frequency of the respondent, and i is 1, 2, 3…….n. Then, the relative importance of each factor was interpreted based on the scale of five (5) categories as shown in Table 1.

Table 1. Interpretation for the AI values

AI range value	Interpretation
$1.00 \leq AI < 1.50$	Least Important
$1.50 \leq AI < 2.50$	Less Important
$2.50 \leq AI < 3.50$	Moderate Important
$3.50 \leq AI < 4.50$	Important
$4.50 \leq AI < 5.00$	Most Important

3. Result and discussion

3.1 Demographic of respondents

In total, data from fifty-five respondents were collected in this study. Figure 1 and 2 show the characteristics of the respondents. Figure 1 shows the distribution of the respondents' current profession. Figure 2 shows the distribution of the number of projects undertaken or engaged by the respondents. 29% of the respondents had undertaken 1-5 projects, 11% had undertaken 6-10 projects, 42% had undertaken 11-15 projects, while 18% had undertaken above 16 projects. The results show that the respondents have experience in IBS projects. The subsequent subsections present the respondents' views on-site supervision and IBS.

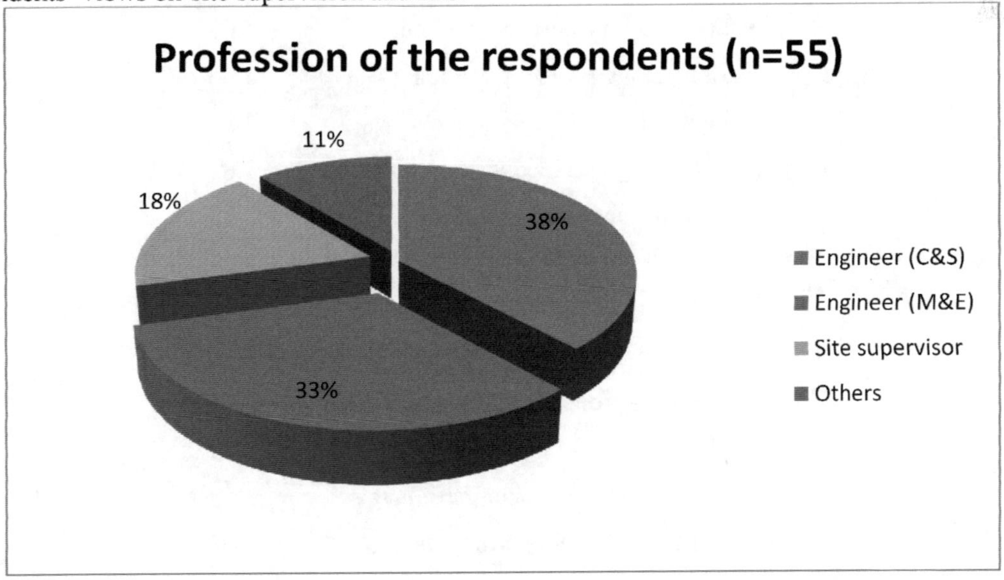

Figure 1. Profession of the respondents

IOP Conf. Series: Materials Science and Engineering **712** (2020) 012039 doi:10.1088/1757-899X/712/1/012039

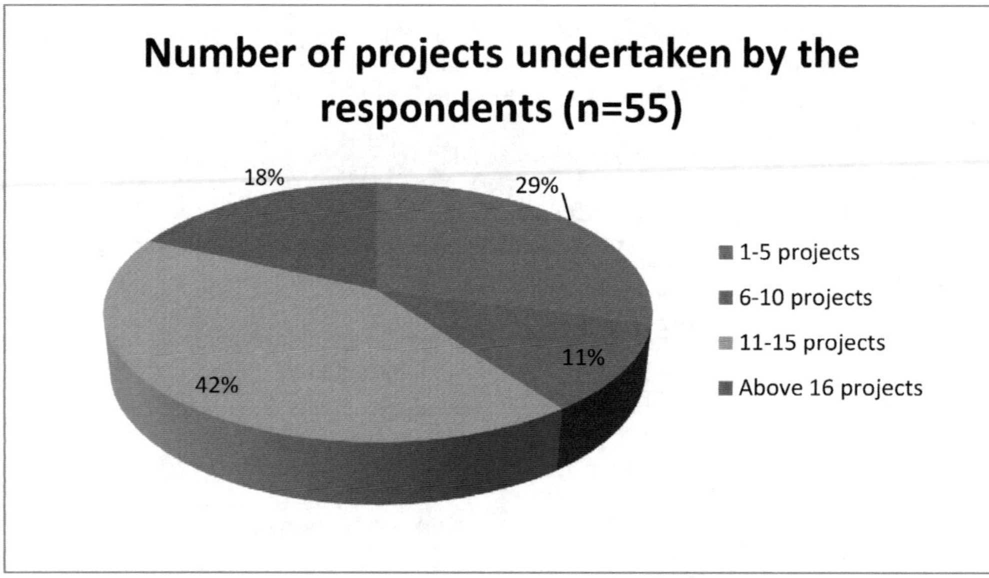

Figure 2. Number of IBS projects undertaken by the respondents.

3.2 Factors affecting site supervision of IBS projects

Table 2 shows the AI value of each factor based on the collected data. Based on the importance scale that considers the effectiveness of the result, this study considers factors with AI values above 3.5 as important. Conversely, factors with AI values below 3.5 are considered as less important. Out of the fifteen factors, three factors have AI values below the 3.5 thresholds. Based on the list of factors, the researcher come out four (4) the most influential factors to discuss.

Table 2. Factors affecting site supervision in IBS projects

Factors	Average index
Improper site layout	4.30
To meet design and manufacturer requirements	4.25
Poor quality of IBS product/components	4.09
Coordination among project participants	3.96
Interface problem between systems	3.85
Improper monitoring and feedback systems	3.84
Loss of management control	3.78
Insufficient training on-site levels	3.73
Climate conditions	3.71
Lack of knowledge in IBS construction	3.69
Conflict among project participants	3.62
Poor quality control	3.53
Lack of commitment to all project participants	3.18
Indecisiveness of project participants	3.07
Negative attitude among project participants	3.03

3.2.1 Site layout

Table 2 shows that improper site layout is the most factors that affect site supervision, which the main factors of that factor are on-site construction process. Since IBS is a construction system which components are manufactured in a factory, it may be a different problem on the site during site supervision. Thus, the installation of the components needs to be monitoring by the experienced and skilled person. Site layout refers to the arrangement of the site facilities, the tools and supporting utilities for optimum product flow [9, 10]. Hence, poor arrangement of the site can obstruct workflow, hinder proper supervision, increase multiple handling, and ultimately reduce productivity that results in project delays and failure. Unfortunately, site planning is often neglected, and the attitude of engineers has been that it will correct itself as the work progresses. Strategic arrangement of facilities, including workers accommodation, materials all has an influence on-site supervision [5, 9, 10] and in effect, affect construction site supervision. Consequently, respondents were recognized of these facts and ranked this factor as significant on factors affecting site supervision.

3.2.2 Design and manufacturer requirements

The other factor is to meet design and manufacturing requirements. Difficulties during the installation of IBS at the site are to comply with that factor. This classification is intended to meet a spatial necessity of the client's that is the spaces required for different capacities in the structure just as the particular compositional plan. In this occasion, the client's needs are vital that sometimes different to manufacturing requirements. Thus, it will be a problem on the site supervision to deal with both of that.

3.2.3. IBS product and components

On the other hand, poor quality of IBS product and components also would affect site supervision. This problem occurred due to a low exposure of IBS knowledge among the installers during the installation work of IBS components where the employees employ workers that do not possess sufficient knowledge of the installation process of IBS components. From the reading of the researcher, lack of experienced supervisors during the component production process may affect the quality of IBS product.

3.2.4. Coordination between project participants

Lastly, the other factors that point out in this research are on working collaboration, which is coordination among project participants. The coordination among the project participants and the dimension of responsibility are fundamental for any venture to be effective. Coordination and responsibility are relied upon to improve execution in various ways. Timely communication can produce a quick response to minimizing delays. The researcher identified that based on reading when selecting IBS as an option, it is necessary to understand the extensive coordination required before construction operations, for instances coordination of design, transportation, tracking, and installation to ensure successful site supervision. Moreover, with the expanded of coordination needed for the construction operations, the needs of effective communication becoming vital for the distribution of information respecting decisions, designs, schedules, and transportation requirements.

3.2.5. Least impacting factors

While prior studies suggest that these factors are affecting the success of site supervising in IBS projects, based on the result, most of the respondents agreed that these factors have the least amount of influence. The factors are: negative attitudes among project participants, the indecisiveness of project participants, and lack of commitment of all project participants' factors least affect the site supervision in IBS construction among the IBS players based on their experiences.

4. Conclusion

The main objective of this research is to list out the main factors of site supervision in IBS construction to improve the productivity of IBS. The success of a project invariably depends on the

IOP Conf. Series: Materials Science and Engineering **712** (2020) 012039 doi:10.1088/1757-899X/712/1/012039

level of supervision of the project or managerial skills. Project supervision somewhat can make or unmake a project. From the research, the successful delivery of projects within cost, time, and quality is hinged on the proper supervision of works. Thus, this makes IBS project supervision imperative to crucial construction activities. This study has identified key factors that affect site supervision in the IBS construction Industry, which is the challenge to IBS players is to identify the ways to reduce the occurrence of such factors on IBS projects. However, several factors count against the successful supervision of construction works. Lastly, the commitment to enhance site supervision in improving IBS productivity is required for all the IBS parties involved.

5. References

[1] Nasrollahzadeh T, Marsono A K and Masine M 2017 Critical Success Factor for Industrialized Building System Process Protocol Map by Lean Construction management', *Journal of Civil & Environmental Engineering*, 06(05).

[2] Mydin M A O, Sani N and Taib M 2014 Industrialised Building System in Malaysia : A Review *MATEC Web of Conferences*, 01002, pp. 1–9.

[3] Abdul Rahim A, Latif Qureshi S 2018 A Review of IBS Implementation in Malaysia and Singapore, *Journal of the Malaysian Institute of Planners,* Vol 16(2) pp. 323-333.

[4] Ismail F, Mohamad Yusuwan N, Baharuddin H E A 2012 Management Factors for Successful IBS Projects Implementation, *Procedia - Social and Behavioral Sciences*. pp. 99–107

[5] Ismail F, Baharuddin H E A, Marhani M A 2013 Factors Towards Site Management Improvement for Industrialised Building System (IBS) Construction, *Procedia - Social and Behavioral Sciences*. pp. 43–50.

[6] Bohn J S, Teizer J 2009 Benefits and Barriers of Construction Project Monitoring using Hi-resolution Automated Cameras, *Construction Research Congress 2009*,

[7] Gan Y, Shen L, Chen J, Tam V W Y, Tan Y, Illankoon I M C S 2017 Critical Factors Affecting the Quality of Industrialized Building System Projects in China, *Sustainability,* p. 1-13

[8] Lou E C W, Kamar K A M 2012 Industrialized Building Systems: Strategic Outlook for Manufactured Construction in Malaysia, *Journal of Architectural Engineering*, Vol 18(2), pp. 69–74.

[9] Hackman J K, Acheampong A, Agyekum K, Ayarkwa J 2015 Factors Affecting Site Supervision in the Ghanaian Construction Industry, *Conference: ICIDA 2015*, Kumasi Ghana.

[10] Callistus T, Felix A L, Ernest K, Stephen B, Andrew A C 2014 Factors Affecting Quality Performance of Construction Firms in Ghana: Evidence from Small–Scale Contractors. *Civil and Environmental Research,* Vol 6(5). Pp. 18-23.

Acknowledgments

The authors would like to thank Universiti Malaysia Pahang for supporting this study through financial grant RDU190341 as well as the industry practitioners that agreed to participate in this work.

IOP Conf. Series: Materials Science and Engineering **712** (2020) 012040 doi:10.1088/1757-899X/712/1/012040

Waste Cooking Oil as Bio Asphalt Binder: A Critical Review

P J Ramadhansyah[1], K A Masri[1], W N A Wan Azahar[2], N Mashros[3], A H Norhidayah[3], M N Mohd Warid[3], M K I M Satar[3] and H Yaacob[3]

[1]Faculty of Civil Engineering Technology, Universiti Malaysia Pahang, 26300 Gambang, Pahang, Malaysia
[2]Kulliyyah of Engineering, Department of Civil Engineering, International Islamic University Malaysia, 50728 Kuala Lumpur, Malaysia
[3]Faculty of Engineering, School of Civil Engineering, Universiti Teknologi Malaysia, 81310 Johor Bahru, Malaysia

Abstract. The modification of binder with waste cooking oil (WCO) for paving materials indicates the implementation of recycling practice with environmental issues concern, thus improving the proper management of this waste product. This paper presents a critical review of the WCO utilisation as a modifier to enhance binder properties. The review was focuses on the evaluation of WCO in asphalt binder modification and asphaltic concrete mixture. Basically, oil-based modification by using WCO in paving material provides a promising waste material potential in improving the engineering value in terms of rheological and mechanical performance for modified binder and asphalt mixture. It is expected that the addition of this modifier in binder gives superior performance and is comparable with the conventional binder.

Content

1. Introduction

Waste materials are widely used as a constituent part or a substitution element in the production of modified asphalt binder to modify conventional binder properties. This waste material is obtained from household, commercial, and industrial by-products. In fact, these waste products have a discreet and specific property, which enable them to be an additive or a modifier element. These additives come in liquid and powder forms that are desirable for modifying and upgrading construction material properties. In order to prevent pavement distress and minimise the total maintenance work cost, the bituminous layer should be strengthened with high quality recycled waste materials to improve its mechanical properties in terms of permanent deformation resistance, fatigue and so on. The application of waste and recycled product has fulfilled the sustainable development requirement so that a green highway can also be achieved.

1.1. Asphalt Binder Modification

An effort for binder performance improvement has induced the evaluation, development and use of various asphalt binder modifiers. Currently, a number of low-cost asphalt modification and extension technique is proposed to minimise the fluctuating cost of binder production, while at the same time enhance the asphalt binder performance. Modified asphalt offers properties enhancement in mitigation of pavement distress and reduces life cycle costs when compared to the conventional binder. Besides, in addressing various arising issues regarding pavement performance, therefore asphalt binder modification is suggested as one effective technical solution to solve the problem faced by asphalt industry [1]. According to Hainin et al. [2], binder properties alteration and improvement can be achieved with modified asphalt binder which in turn ensures a long-term pavement performance. Due to this fact, many researches were carried out to investigate the potential application of waste materials in the construction industry. Basically, a number of noteworthy worldwide researches were conducted in applying a modifier from waste oil source. In recent years, a wide range of oil-based modifications was introduced in the asphalt industries to achieve some desirable properties in material for pavement construction. The oil-based source can be categorised as bio-oil from biomass production industry, waste engine oil from automotive industry and waste cooking oil from agricultural industry. Based on these oil types, numerous researches have explored the potential of modifiers from biomass supply [3]. Meanwhile some were focused on waste oil product, such as waste cooking oil (WCO) [4]. Due to its high availability and low cost feedstock, WCO source has become intensively applied in asphalt binder modification. Most previous research has recorded WCO application for the conventional bitumen replacement, bitumen extender and modifier [5].

2. Waste Cooking Oil

Waste cooking oil is non-edible oil and can be recognised as a by-product of fresh oil produced during food frying [6] as shown in Figure 1. Fresh cooking oil is physically light yellow in color with low viscosity, while used cooking oil has high viscosity and dark brown in color [7]. WCO source is collected from food industries, household disposables, restaurants and recycling centre by authorised companies [8]. The annual WCO collection is approximately three billion gallons [9]. Meanwhile, according to Math et al. [10] about 15 million tonnes of WCO production per year was recorded based on statistics. In Malaysia approximately 0.5 million tonnes of WCO is generated per year and derived from palm oil source. An increase in demand for food consumption (Phan and Phan, 2008), besides rapid and tremendous growth of human population [11] is identified as the main factor that attributed to the huge amount of WCO production.

2.1. Environmental issues

At present, the WCO recycling is at an empirical stage, in which only a small amount of used cooking oil is properly collected and recycled especially from residential area and food restaurants. In fact, enormous WCO quantity, without any appropriate treatment, is illegally dumped into landfills and rivers which have caused serious adverse environmental impact, especially environmental pollution issues. The most common environmental issue that arise is the eutrophication process, whereby this phenomenon is caused by the blockage of sunlight from passing through the river surface because of the presence of a thin layer of oil. Apart from sunlight, oxygen supply to aquatic life is also disrupted by this process [12], thus leading to the excessive growth of micro-organisms, phytoplanktons and algae. River water quality is measured by the existence of these micro-organisms wherein their huge growth rate affects the intrinsic equilibrium of aquatic ecosystem. In residential area community, clogging problem has grabbed other human's attention because of blocked drains. This blockage happens when waste frying oil is poured into the kitchen sink and accumulates in the drains. This issue has affected waste water treatment plants and indirectly maximised the processing costs since routine maintenance work must be done in order to mitigate and monitor this clogging problem. The

IOP Conf. Series: Materials Science and Engineering **712** (2020) 012040 doi:10.1088/1757-899X/712/1/012040

estimation of about 40% of sewerage system blockages are contributed by pouring of waste cooking oil into the kitchen sinks. In addition, the soil has high potential to be contaminated by waste frying oil as the eco-toxic possessed in this oil content can damage plants. Due to dumping problems and possible water and land resources pollution, supervision and management of waste cooking oil has created a significant challenge that needs consideration. Therefore, waste cooking oil recycling is becoming an effective option to mitigate pollution issues due to illegal dumping problem with the greatest concern in high construction cost and natural resource conservation [13]. According to Alcantara et al. [14], the mitigation of major environmental issues, such as waste oil disposal and waste water treatment, can be encountered by the recyclability of WCO as a modifier in asphalt binder modification.

Figure 1: Waste cooking oil sample after frying process

2.2. Chemical characteristic of WCO

The majority of study stated that the predominant compounds found in WCO are Oleic acid, Palmitic acid and Linoleic acid as listed in Table 1. For example, Udomsap et al. [15] and Wen et al. [16] reported that the major components of WCO are Linoleic acid, followed by Oleic acid and Palmitic acid. Meanwhile, Ma and Hanna [17] discovered that the dominant component found in WCO is Palmitic acid, Oleic acid and Linoleic acid. Basically, Lauric acid, Myristic acid, Palmitic acid and Stearic acid are recognised as the saturated fatty acids, while Linoleic acid, γ Linolenic acid and Linolenic acid are identified as polyunsaturated fatty acids [15]. Oleic acid and Cis-11-Eicosenoic are monounsaturated fatty acids. The saturated component can be depicted as the non-existence double-bond in WCO fatty acids. This implies that maximum number of possible hydrogen atoms per atom carbon is contained in these chains [18]. Otherwise, the presence of double-bond in unsaturated component indicates insufficient maximum number of possible hydrogen atoms, thereby will induce the electronegativity to react during chemical reaction in which this unsaturated component tend to accept the hydrogen atoms.

Table 1: Free fatty acids in waste cooking oil

Fatty Acids	Chemical Formula	Waste Cooking Oil (%)		
		Ma and Hanna [17]	Udomsap *et al.* [15]	Wen *et al.* [16]
Palmitic acid	C16:0	42.8	10.2	8.5
Oleic acid	C18:1n9c	40.5	22.8	21.2
Linoleic acid	C18:2n6c	10.1	53.7	55.2

The different fatty acids composition is recorded if different sources of waste cooking oil are used as tabulated in Table 2. These acids are categorised as Free Fatty Acids (FFA) that is naturally contained in fats, oils and greases, in which higher FFA component in WCO leads to higher acid value [19]. Theoretically, the fatty acids chain and number of double bonds affect oil physical and chemical properties [20] since the degree of unsaturated compound is more preferable for chemical reactions. The acid composition in WCO can be categorised as saturated, monounsaturated and polyunsaturated. The comparison of FFA composition between fresh cooking oil and waste cooking oil reported by Maneerung et al. [21] indicated that the reduction of unsaturated fatty acids consists of Oleic acid, Linoleic acid and Linolenic acid in WCO after frying process, as shown in Table 2. The decreasing trends of unsaturated fatty acids are attributed to the chemical degradation reaction during frying in which these acids tend for hydrogen acceptance thereby increases the saturated fatty acids.

Table 2: FFA composition in fresh and waste cooking oil [21]

Free Fatty Acid Composition (%)	Fresh Cooking Oil	Waste Cooking Oil
Oleic ($C_{18}H_{34}O_2$)	43.9	43.2
Linoleic ($C_{18}H_{32}O_2$)	30.4	30.1
Palmitic ($C_{16}H_{32}O_2$)	20.3	19.4
Linolenic ($C_{18}H_{30}O_2$)	4.8	4.7
Stearic ($C_{18}H_{36}O_2$)	2.9	2.6

2.3. Physiochemical Properties

Ullah et al. [22] conducted test properties to compare the properties of unused and used cooking oil as presented in Table 3. The various properties test conducted were acid value, calorific value, peroxide value, density, kinematic viscosity, dynamic viscosity, flash point and moisture content. All these properties were measured to identify the WCO quality. The main parameter for physiochemical properties was evaluated based on the acid value test, density and viscosity. The acid value, density and viscosity (dynamic and kinematic) result obtained indicates an increment in value from unused to used cooking oil. Increasing acid value represents a higher value of free fatty acid content in WCO as these properties are related to each other. According to Chhetri et al. [23], the free fatty acids contained in waste cooking oil are higher as compared to fresh cooking oil. These findings coincide with Choe and Min [6], in which deep fat frying causes parameter alteration, where an increase in free fatty acids content, density and viscosity was observed in WCO. In addition, the density increase is attributed to the decrease in unsaturated fatty acids because the chemical reaction during frying oil has changed into saturated fatty acids. This can be observed by the physical WCO appearance wherein the unused cooking oil is light in condition while used cooking oil is thicker and more viscous. During frying, the oil will undergo oil polymerisation, producing cyclic monomer and polymer. This process depends on the degree of unsaturation, in which the increase of cyclic compound formation is recorded as increase in the linolenic acid, since these types of acid is categorised as unsaturated fatty acid. The polymer compound produced from the polymerisation accelerates oil degradation, thereby will increase the WCO viscosity [24]. Besides, the other factor that attributed to viscosity value is increase in polar formation and oxygen containing molecule. Therefore, it can be noticed that the used cooking oil viscosity is higher in comparison with unused cooking oil.

Table 3: Physiochemical properties of unused and used cooking oil [22]

Properties	Unused cooking oil values	Used cooking oil values

Acid value (mg KOH/gm)	0.3	4.03
Calorific value (J/gm)	-----	39658
Peroxide value (meq/kg)	< 10	10
Density (gm/cm^3)	0.898	0.9013
Kinematic Viscosity (mm^2/s)	39.994	44.956
Dynamic Viscosity (mpa.s)	35.920	40.519
Flash point (^{o}C)	161-164	222-224
Moisture content (wt %)	0.101	0.140

3. Recyclability of WCO as Bio-Binder

A modifier can be defined as a potential material which enhances the service life performance and can withstand the loading factor exposure for improvement of asphalt mixture engineering properties [1]. According to Arabani et al. [25], the substitution of a modifier as part of the modification mechanism in asphaltic concrete mixture has reduced the deterioration rate due to presence of asphalt pavement distress. Based on the listed previous research, WCO was mostly applied as a rejuvenator since the fluidity factor in WCO has the capability for restoration aged binder by reducing its viscosity. Apart from these, the modification of binder by using WCO as a modifier was focused. Basically, the application of a rejuvenator is desirable to determine the cracking resistance at low temperature, while the function of a modifier is preferable for rutting resistance at high temperature. Maharaj et al. [26] performed an assessment for rutting and cracking resistance performance of modified asphalt binder by incorporating WCO as a modifier. The two factors which attributed to the rheological performance, represented as WCO content and test temperature, were measured to assess the potential of these parameters affecting rutting and cracking performance. The rheological performance demonstrated that the decreasing G*.sin δ was observed (see Figure 2) which indicated an improvement in fatigue cracking resistance. The increasing trend of fatigue cracking resistance was recorded as the WCO concentration was increased in the modified binder. In contrast, with the increasing WCO concentration in control binder, the G*/sin δ was decreased which explained the decrement of rutting resistance performance, as shown in Figure 3.

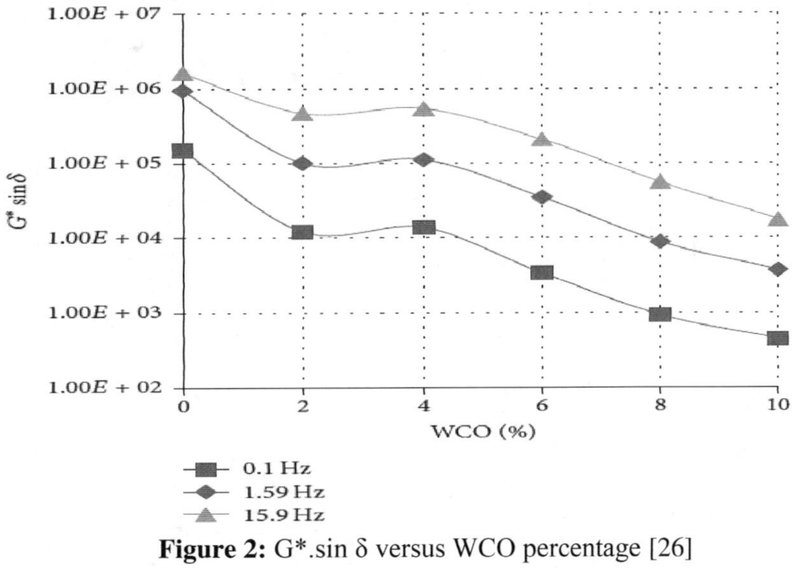

Figure 2: G*.sin δ versus WCO percentage [26]

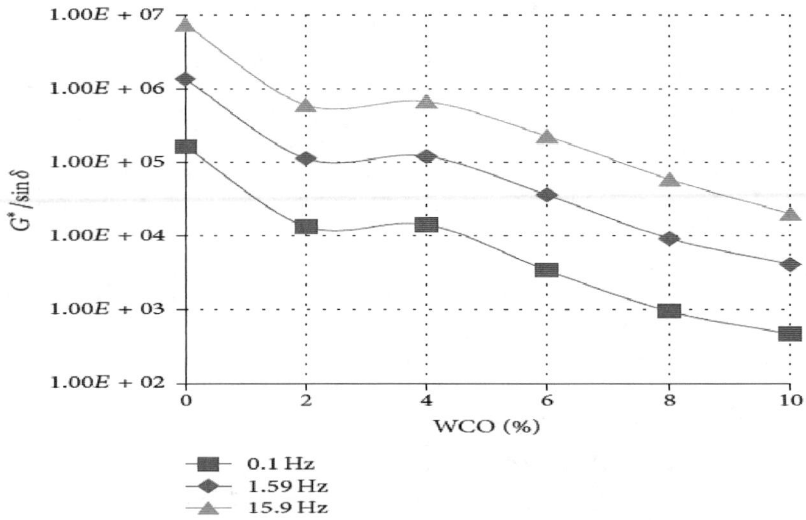

Figure 3: G*/sin δ versus WCO percentage [26]

Previously, Wen et al. [16] conducted a laboratory evaluation for binder containing bio-asphalt from the conversion of WCO through a thermochemical process. There were three different types of base binder used in this study which is PG 58-28, PG 76-22 and PG 82-16. About 30% and 60% of bio-asphalt was blended with PG 58-28 meanwhile the bio-asphalt was mixed at 10% and 30% (by the weight of total binder) for PG 76-22 and PG 82-16 binders. The rutting resistance performance achieved in this research coincided with the previous study, wherein increasing non-recoverable creep compliance (Jnr) value from multiple stress creep and recovery (MSCR) result was recorded with the addition of bio-asphalt as listed in Table 4. The increment of Jnr indicated a high susceptibility of modified binder with bio-asphalt to rutting exposure. Besides, the lower critical strain energy density (CSED) attained in this study resulted in the reduction of fatigue cracking resistance with the addition of bio-asphalt content in the conventional binder. The low value of CSED was attributed by the lower shear strength possessed in the modified binder with bio-asphalt as compared to the control binder. The rheological findings indicated that the modified binder with bio-asphalt reduced the resistance to permanent deformation and fatigue cracking.

Table 4: Non-recoverable creep compliance (J_{nr}) results [16]

PG	Bio-asphalt (%)	Stress level (Pa)	J_{nr} (1/Pa)
58-28	0	100	0.94
58-28	0	3,200	1.14
58-28	30	100	1.24
58-28	30	3,200	1.60
58-28	60	100	0.50
58-28	60	3,200	1.07
82-16	0	100	1.02
82-16	0	3,200	2.17
82-16	10	100	1.30
82-16	10	3,200	3.61
82-16	30	100	1.94
82-16	30	3,200	7.04
76-22	0	100	0.94

76-22	0	3,200	1.36
76-22	10	100	1.45
76-22	10	3,200	2.43
76-22	30	100	2.37
76-22	30	3,200	4.70

3.1. WCO in Pavement

A laboratory evaluation conducted by Wen et al. [16] reported that the addition of waste cooking oil based bio-asphalt to produce alternative binder for hot mix asphalt (HMA) reduced the dynamic modulus, wherein the stiffness was linearly decreased with the addition of bioasphalt concentration. The decrement of flow number with increasing amount of bioasphalt (see in Figure 4) also proved the reduction in rutting resistance performance, which is more prone to deflection than a control mixture. Another interpretation indicated that adding bio-asphalt reduces high-temperature performance as expected.

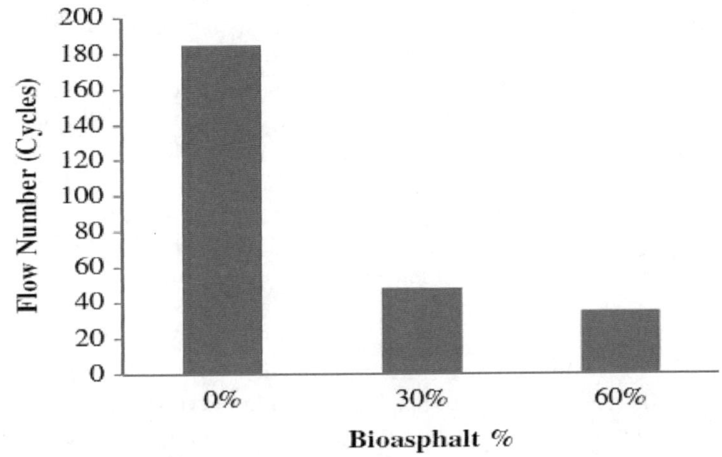

Figure 4: Flow number at different percentage of bio-asphalt from WCO [16]

Other investigation by Bailey and Zoorob [27] proved an opposite findings, where the addition of Used Vegetable Oil (UVO) has increased the Indirect Tensile Stiffness Modulus (ITSM) performance which implied that the modified asphalt mixture with UVO is more durable as compared to the control mixture. The laboratory result was validated with the field evaluation in which the wheel tracking test proved that the lowest rut depth was exhibited by the mixture incorporated with UVO relative to the control mixture. A similar research finding was reported by Some et al. [28], in which a lower rutting depth was observed for modified binder with oil derived from sunflower and rapeseed source (vegetable oil) in comparison with the control mixture. This implied that, an improvement in the resistance to permanent deformation was recorded with the addition of waste vegetable oil. The trend of the compression strength was predicted since a decreasing complex modulus was observed in the binder test [28].

4. Discussions

From the review presented, it can be noticed that WCO can be categorised as one waste material that is mostly utilised for partial substitution as a modifier in binder modification as reported by the previous research work. Oil-based modification by using WCO is expected to improve the engineering value of modified asphalt binder in comparison with the conventional binder. Apart from that, waste disposal problem could be solved and also promotes environmental preservation. A significant and noticeable finding reported by previous researchers indicating that the WCO application is recorded as a superior improvement for cracking resistance performance at low temperature. The superior performance of the modified binder incorporating WCO is attributed to the natural fluidity characteristic of WCO. Theoretically, at low temperature exposure, bitumen undergoes high brittleness rate and is prone to experience cracking failure at this temperature. As a result, the presence of cracking increases the tendency for noticeable stress. The WCO fluidity properties in binder facilitates high flow rate and decreases the viscosity of modified binder within the mix, thereby assisting the reduction in induced stress or cracking existence. However, available information on the effect of modified binder containing WCO on rutting resistance at high temperature is very limited as the adverse effect is widely reported and still unresolved at the moment. Basically, the addition of WCO in binder causes the softening effect on the modified binder and it is expected that the soft binder contributes to poor performance in permanent deformation resistance due to unable to withstand with high temperature exposure. The justification is presently not well documented and there is no conclusion can be drawn in explaining the reason for this mechanism. Borhan et al. [29] justified the main factor that attributed to this issue where it is due to the incompatibility characteristic of the oil source and asphalt binder, thereby inducing instability problem in the chemical reaction between oil in modified binder and asphalt mixture. Despite that, this factor is not a broad concern of other research as WCO is applied directly in the binder without determining the WCO fundamental quality parameter that might affect the modified binder with WCO performance. Due to the arising issue, several valuable efforts were triggered in this study to determine WCO quality and investigate the compatibility properties between WCO and asphalt binder for the modified binder incorporating WCO improvement. Therefore, it is significant to conduct a chemical WCO quality test and perform chemical analysis to discover whether WCO is compatible with the binder for binder modification. Once the WCO quality and compatibility properties are determined, chemical pre-treatment is proposed in producing treated WCO with improved performance than untreated WCO. For this reason, it is vital to evaluate the effect of untreated and treated WCO on the binder and asphalt mixture performance after conducting the chemical treatment.

4.1. Proposed solution

The potential solution is proposed based on the improvement in certain properties, such as oxidative stability, free fatty acids content, low temperature and antioxidant content. One of the major challenges when dealing with oil is low oxidative stability which requires improvement to mitigate the problem. In combating poor oxidative stability performance, three types of solution were proposed, such as genetic modification, chemical modification and various additives with antioxidant materials usage. Genetic modification is explored by the soybean seed alteration and suggested by producing the low linolenic acid which is recognised as one potential alternative solution for hydrogenation (Mounts et al., 1994). The mechanism for chemical modification is conducted by shifting the fatty acids component in oil. The chemical modification through the trans-esterification process is proposed for minimising the high free fatty acids content in oil [30].

5. Conclusions

The significant noticeable findings indicated that WCO application has recorded the greatest improvement in cracking resistance performance at low temperature. Contrast, an adverse effect was observed on high temperature for rutting resistance performance. However, many studies were conducted and are still being developed to discover more about the WCO potential in modified asphalt binder performance, besides finding another alternative solution to enhance rutting resistance performance. Nevertheless, there is still a long journey to go through before the proposed modifier is widely accepted and broadly utilised in pavement industry. More expanding research work is required in future to prove the superiority of WCO application in pavement material.

6. References

[1] Teymourpour P, Sillamäe S and Bahia H U 2015 *Road Mater. Pavement.* **16** pp. 50-74.
[2] Hainin M R, Warid M N M, Izzul R, Ruzaini M K and Yusak M I M 2014 *Adv. Mater. Res.* **911** pp. 449–453.
[3] Chen M Xiao F, Putman B, Leng B and Wu S 2014 *Constr. Build. Mater.* **59** pp. 10–16.

[4] Yang X, You Z, Dai Q and Mills-Beale J *Constr. Build. Mater.* **51** pp. 424–431.
[5] Ramadhansyah P J, Azman M K, Idris A M, Nur Aifa W A W, Ekarizan S, Hainin M R, Norhidayah A H and Haryati Y 2019 *IOP Conf. Series.: Earth. Environ. Sci.* **244**(1) pp. 1-6.
[6] Choe E and Min D B 2007 *J. Food Sci.* **72**(5) pp. 1–10.
[7] Singhabhandhu A and Tezuka T 2010 *Energy.* **35**(6) pp. 2544–2551.
[8] Zhang H, Wang Q and Mortimer S R 2012 *Renew. Sust. Energ. Rev.* **16**(7) pp. 5225–5231.
[9] Putra Jaya R, Masri K A, Awang H, Ali M I, Ramli N I, Wan Azahar W N A, Shaffie E, Yaacob H, Mohd Warid M N and Ramli 2019 *Int. J. Recent Technol. Eng.* **7**(6) pp. 16-19.
[10] Math M C, Kumar S P and Chetty S V 2010 *Energy. Sustain. Dev.* 14(4), 339–345.
[11] Chen Y, Xiao B, Chang J, Fu Y, Lv P and Wang X 2009 *Energy Convers. Manag.* **50**(3) pp. 668-673.
[12] Hamad B S, Rteil A A and El-fadel M 2003 *Constr. Build. Mater.* **17** pp. 311–318.
[13] El-Fadel M and Khoury R 2001 *Resour. Conserv. Recy.* **33** pp. 75–91.
[14] Alcantara R, Amores J, Canoira L, Fidalgo E, Franco M J and Navarro A 2000 *Biomass Bioenergy.* **18** pp. 515–527.
[15] Udomsap P, Chollacoop N, Topaiboul S and Hirotsu T 2009 *I. J. Renew. Energ.* **4**(2) pp. 47-59.
[16] Wen Z, Yu X, Tu S T, Yan J and Dahlquist E 2010 *Bioresour. Technol.* **101** pp. 9570-9576.
[17] Ma F and Hanna M A 1999 *Bioresour. Technol.* **70** pp. 1-15.
[18] Lam M K, Lee K T and Mohamed A R 2010 *Biotechnol. Adv.* **28** pp. 500-518.
[19] Silitonga A S, Masjuki H H, Mahlia T M I, Ong H C, Atabani A E and Chong W T 2013 *Renew. Sust. Energ. Rev.* **24** pp. 514–533.
[20] Zlatanic A, Lava C, Zhang W and Petrovic Z S 2004 *J. Polym. Sci. Pol. Phys.* **42**(5) pp. 809–819.
[21] Maneerung T, Kawi S, Dai Y and Wang C H 2016 *Energy Convers. Manag.* **123** pp. 487–497.
[22] Ullah Z, Bustam M A and Man Z 2014 *Int. J. Chem. Eng.* **5**(2) pp. 134-137.
[23] Chhetri A B, Watts K C and Islam M R 2008 *Energies.* **1**(1) pp. 3-18.
[24] Tseng Y C, Moreira R G and Sun X 1996 *J. Food. Sci. Technology.* **31** pp. 287– 294.
[25] Arabani M, Mirabdolazimi S M and Sasani A R 2010 *Constr. Build. Mater.* **24** pp. 1060-1068.
[26] Maharaj R, Harry V R and Mohamed N 2015 *Sci. World. J.* **15** pp. 1-7.
[27] Bailey H K and Zoorob S E 2012 5th *Eurasphalt - Eurobitume Congress.* Pp. 1-12, Istanbul.
[28] Some S C, Pavoine A and Chailleux E 2016 *Int. J. Pavement Res. and Technol.* **9** pp.368-375.
[29] Borhan M N, Suja F, Ismail A and Rahmat R A O K 2009 *Eur. J. Sci. Res.* **28**(3) pp. 398-411.
[30] Phan A N and Phan T M 2008 *Fuel.* **87** pp. 3490–3496.

Acknowledgments
The support provided by Malaysian Ministry of Higher Education and Universiti Malaysia Pahang in the form of a research grant (RDU/UMP) vote number RDU1803160 for this study are highly appreciated.

IOP Conf. Series: Materials Science and Engineering **712** (2020) 012041 doi:10.1088/1757-899X/712/1/012041

The effects of bottom ash in coastal sand

D Youventharan, O Rokiah and S M Arif

Faculty of Civil Engineering Technology, Universiti Malaysia Pahang, 26300 Gambang, Pahang, Malaysia

Abstract. In the construction industry, breakage of sand particles is very common especially at the very early stage of the pile driving work. The breakage of sand particles will result in change of size spectrum and alter its engineering properties. In this research, the effectiveness of bottom ash in improving the engineering properties of crushed sandy soil was studied. The bottom ash of 5%, 10%, 15% and 20% was added into the sand and crushed with 500 and 1000 blows under the Automatic Compactor machine. The engineering properties of different proportion of bottom ash mixture before crushing and after crushing with 500 and 1000 blows were tested. The engineering properties which were measured include the coefficient of uniformity, breakage index, fouling index, permeability and bearing capacity. Results show that adding of bottom ash into the crushable sand improved the load bearing capacity. By adding 20% of bottom ash, the bearing capacity of the crushed sandy soil can be improved by 55% meanwhile the permeability of the sandy soil can be improved by 200%.

1. Introduction

Sand is a natural granular construction material which is formed by the weathering of rocks. Due to its high accessibility, sand has been widely used in construction for many centuries. During construction, breakage of sand particles and reduction in size of particles may occur under crushing impact. For instance, when the piles are driven into the sand, sand will normally experience stresses which are high enough to break the particles. The breakage of particles will result in the increase of percentage of fine particles and change the grading [1]. The design of foundations is usually done using soil parameters obtained from in-situ or the laboratory test. The calculations are done assuming that the soil parameters will always remain unchanged or at least with some minor error after the post design work. However, in reality the sand will experience crushing impact hence the size of sand may change during the construction work. The changes in size of sand will alter the mechanical behaviour and engineering properties of the granular sand such as strength, volume change, pore pressure developments and permeability [2]. These changes are highly depends on the amount of sand particle crushed due to the applied crushing load. The occurrence of the changes in gradation of sand is uncontrollable and irreversible. This change indicates that the quality of sand which exists at the end of the piling activity will be different from the original sand that exists on site. The changes of mechanical behaviour in soil may result in failure of the construction due to the under estimation of bearing capacity. Therefore it is vital to determine the crushability of granular sand that is used for geotechnical applications and predict the bearing capacity accurately. Over the decades, researchers have carried out many studies on how to improve the impact and the quality of crushable sand and one of the possible solutions would be adding bottom ash as partial replacement to the sand filler. Studies of engineering properties of the physical and chemical characteristic of bottom ash had shown that the engineering behaviour of the bottom ash is almost similar with the natural granular sand. More extensive research has to be carried out to find the specific usage and the benefits of bottom ash as an alternative material in construction industry.

IOP Conf. Series: Materials Science and Engineering **712** (2020) 012041 doi:10.1088/1757-899X/712/1/012041

For example, in 1972 an experimental research [3] was carried out using the standard proctor compaction test and one dimensional compression test on West Virginia bottom ash. The laboratory data showed that at low stress levels, the compressibility of bottom ash was comparable to natural granular soils at similar relative densities. Another study [4] pointed out the possibility of coal bottom ash to replace sand in embankments and the result also showed that the bottom ash had acceptable environmental properties when testing done to its leachate [5]. The overall review from the past reports shows the possibility of bottom ash to be used as to replacement of sand. In this research, the possible effect of the bottom ash to act as the partial sand replacement was studied. A series of tests was conducted to determine the effect of bottom ash in improving the crushability and permeability of the crushable sandy soil as well as improving the bearing capacity of the crushable sand.

2. Literature review

2.1. Sand particle breakage

The particle crushing of sand often affects the stiffness and strength of soil and particle size distribution [6]. The particle crushing is proven as an irrecoverable process, and the sand have become more compacted through inter-particle slip and rotation [7]. Some of the engineering practice in estimating the bearing capacity of pile had to consider the variation of mechanical properties of sand due to particle crushing. In practice, the deep penetration problem such as pile driving and cone penetration testing will create a sufficient stress levels for particle breakage to occur [8]. Figure 1 shows the image of how the sand grains was before and after the event of crushing through laboratory testing.

2.2. Effect of sand particle breakage

Over the past few decades, many theoretical and experimental studies have been conducted to understand the effect of sand particle breakage on engineering properties. It has been proven that the breakage of sand particle had affected the engineering properties of sand such as particle size distribution, shear strength and hydraulic conductivity behaviour [10-12]. Understanding the mechanism and simulation of sand particle breakage will assist in many ways to design a proper structure in places where abundance of granular soil in found.

2.2.1. Effect on particle size distribution

The sand particle could be crushed into smaller size under high stress. It was found that small grains from fractures could lead to the shift of particle size distribution. From the previous study, a uniform grading of sample will shift to a well graded distribution as stress increased and breakage progressed further [9]. As the breakage of sand particles towards the critical state, the soil gradation eventually tends to shift to well-graded [10]. These shifts affect the mechanical properties of granular sand and the crushing strength against further breakage. This explain that a very well-graded and fractal soil will not experience dramatic breakage.

Figure 1. SEM carbonate sands: (a) grains from the original and (b) grains after testing [11]

2.2.2. Effect on shear strength

In the past a research was carried out using monotonic test on variation mixture of crushed non plastic silica fines contents with Ottawa sand [12]. These tests showed that when the fines content in sand increase, the behaviour of Ottawa sand will become more dilative, with less strain-softening reaction occurrence. The steady line in void ratio-effective stress space will be altered as the fines content is changed. The study of the effect of static liquefaction behaviour of four different sands had shown that as the fine content of sand increased, the specimens will behave with lower peak stress and exhibited more contractive behaviour [12]. Many studies have been carried out to understand the effect of particle breakage in sands on stress strain conditions. The strength and bearing capacity were reduced when particle breakage increased [8].

2.2.3. Effect on hydraulic conductivity

The particle size of the crushable sandy soil will affect the permeability of the sand. The particle crushing reduces the mean pore size and the hydraulic conductivity of the soil will decrease and become less permeable when the fines content increase. For instance, the sand particles which located at underlying the layers of dam are subjected to high stress may cause the sand particles to break, and slow down the pore-water pressure dissipation and pore pressure distribution in earth dam [12].

2.3. Quantifying particle breakage

The experimental testing of sand particle crushing is not an easy task to interpret since the breakage of sand particle is not visually clear even after a careful inspection. Therefore a theoretical model of the evolution of crushing is needed to make post-test observation so that the sand crushing strength can be determined.

2.3.1. Breakage index and crushing coefficient

The index of particle breakage, B_g was defined with the assumption that the sand particles which retained in larger sieve size will decrease meanwhile there will be an increase in the amount of sand particles retained in smaller size sieve. In this experiment, the change in weight of soil retained minus the weight of soil retained before the test was plotted for each particle size fraction with respect to its corresponding size [13]. The general shape of the sand is normally determined by the D_{50}, the 50% of the sand by weight. The change in D_{50} can also be used to assess the rate of particle breakage. The breakage index of D_{50} is defined as the ratio of 50% of the size of sand before the test to the 50% of the size of the sand after the test. Since a log scale is used to plot the particle size distribution curve, the ratio of breakage index can also be determined by the horizontal distance between the initial and final sand size curves at 15% finer [14].

2.3.2. Fouling index and percentage of fouling

Fouling refers to the condition of railroad ballast when voids in this unbound aggregate layer are filled with finer materials. Railroad ballast is uniformly-graded coarse aggregate placed between and immediately underneath the crossties. The purpose of ballast is to provide drainage and structural support for the heavy loading applied by trains. As ballast ages, it is progressively fouled with fine-grained materials filling the void spaces [15]. The fouling index and percentage of fouling is often use in determining the suitability of the ballasted track in railways. In order to sustain good performance it is essential to maintain proper drainage in the ballasted track. Railroad ballast usually contains uniformly graded material that creates a sufficiently large pore structure to facilitate rapid drainage. When it is degraded and aged, fine particles accumulate within the voids (ballast fouling) and decrease its drainage capacity. There are numerous fouling indices available in literature to quantify ballast fouling. The fouling Index is defined as a summation of percentage (by weight) passing the 4.75mm (No.4) sieve and 0.075mm (No.200) sieve. This may lead to a misinterpretation of the actual quantity of fouling if the fouled material contains more than one type of material having considerably different specific gravities [16].

IOP Conf. Series: Materials Science and Engineering **712** (2020) 012041 doi:10.1088/1757-899X/712/1/012041

2.4. Crushing of bottom ash and sandy soil

Some experimental work data of mixture of sand and bottom ash have been studied. For example, study on the effect of two types of bottom ash mixed with different types of soil in different proportion to the value of California Bearing Ratio. The results show that the California Bearing Ratio value increased as the proportion of bottom ash increased [17]. Besides that, there were also attempts to investigate the relationship of the percentage of bottom ash and the permeability. From the study, the result show that as the percentage of bottom in the mixture of various incinerator mixtures increased, the permeability of the mixture was increased too [18]. These previous researches showed the possibility of bottom ash to improve the load capacity of crushed sand.

3. Methodology

The collected sand was divided into three portions which were the sand without crushing, sand crushed with 500 blows and sand crushed with 1000 blows using automatic compactor. Each of the portions was tested with experiment of sieve analysis, California bearing ratio and permeability. The tests are repeated for each portion by adding bottom ash in percentage of 5%, 10 %, 15% and 20%.

The sieve analysis was conducted according to ASTM D422. About 500g of sand samples mixed with different proportion of bottom ash were shaken in mechanical sieve shaker for 10 minute and the weight of each sieve set was measured. The particle size distribution curves were drawn. The experiment was repeated with the sand that crushed for 500 and 1000 blows. The permeability test was carried out using ASTM D234 method, the constant head test. The time required for the water in permeameter that filled with sand to flow out and fill up a measuring cylinder of 1000ml was recorded. The test was repeated by adding different proportions of bottom ash and whole process is repeated with sand that crushed with 500 and 1000 blows. For the California Bearing Ratio measurement, the test was done according to ASTM 1883 method. The sand was mixed with different proportions of bottom ash and compacted in the mould by preparing at optimum moisture content accordingly. For unsoaked condition, the compacted soil was tested without soaking in water while for soaked condition, the compacted soil was immersed in water for 3 days. The test sample was tested on the penetration test machine. The load readings at penetration of every 0.25mm was recorded until it reached a 7.5mm penetration. The test was repeated with sand that was crushed with 500 and 1000 blows.

4. Results and Discussions

4.1. Crushability of sand

The data obtained from the particle size distribution curve were used for crushability analysis. The analysis includes the breakage index, crushing coefficient, fouling index and percentage of fouling. Figure 2 and Figure 3 show the breakage index and crushing coefficient for the sand that mixed with different proportion of bottom ash and undergo crushing for 500N and 1000N. The result shows that when the percentage of bottom ash added increased, the breakage index and crushing coefficient of the sand was decreased. This indicates an improvement in reducing the crushability of the sand where the sand will experience less crushing when bottom ash is added into the sand.

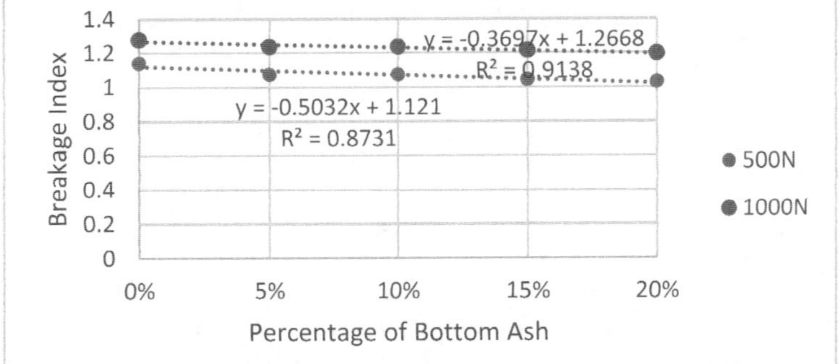

Figure 2. Breakage index for sand underwent crushing with different proportion of bottom ash

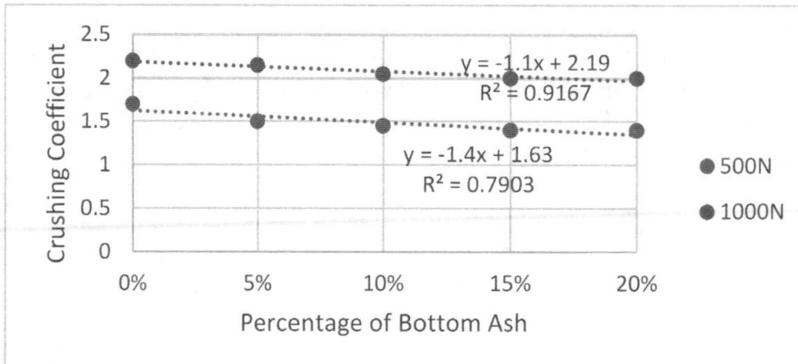

Figure 3. Crushing Coefficient of sand mixed with different proportion of bottom ash underwent crushing of 500N and 1000N

The fouling index and percentage of fouling were analysed from the particle size distribution curve. Figure 4 and Figure 5 showed the fouling index and percentage of fouling of sand mixed with different proportion of bottom ash that subjected to crushing of 500N and 1000N. The result shows that when bottom ash is added into the sand, the fouling index and percentage of fouling were reduced. This indicates that the bottom ash is able to reduce the fouling of the sand. The sand underwent less fouling when bottom ash is added into the sand under crushing.

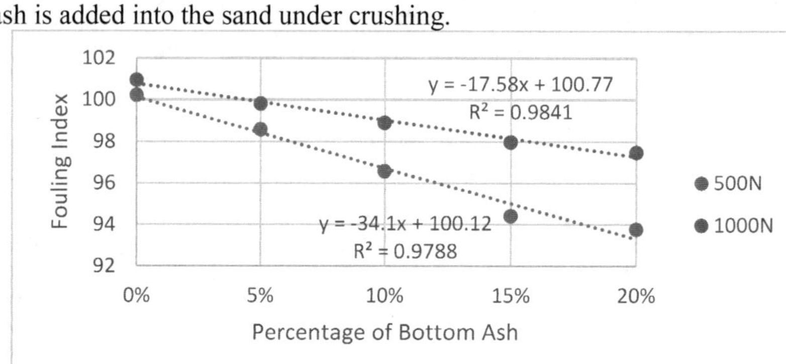

Figure 4. Fouling Index of sand mixed with different proportion of bottom ash underwent crushing of 500N and 1000N

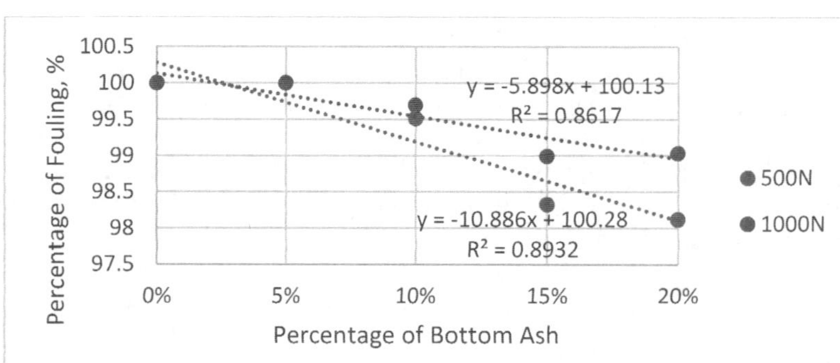

Figure 5. Percentage of fouling of sand mixed with different proportion of bottom ash underwent crushing of 500N and 1000N.

4.2. California bearing ratio

Figure 6 and Figure 7 showed the unsoaked CBR value and soaked CBR value when the sand is mixed with different proportion of bottom ash. By adding the bottom ash, the CBR percentage value of the sand had increased. The higher the percentage of bottom ash added into the sand, the greater the increase in the CBR value, which indicates a better improvement in the bearing capacity of the sand.

The sand which crushed with 500 blows and 1000 blows had the same engineering properties where the bottom ash improved the bearing capacity. The result obtained was similar with the literature review when different percentage of bottom ash mixed with uncrush sand.

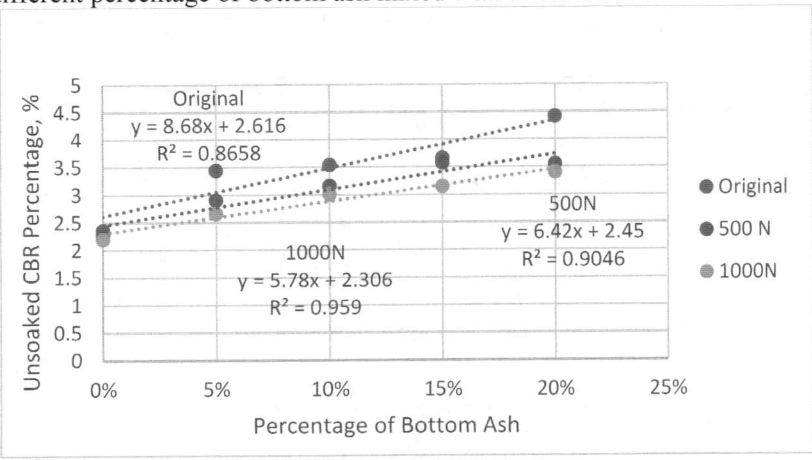

Figure 6. Unsoaked CBR percentage of sand mixed with different proportion of bottom ash sand mixed with different proportion of bottom ash.

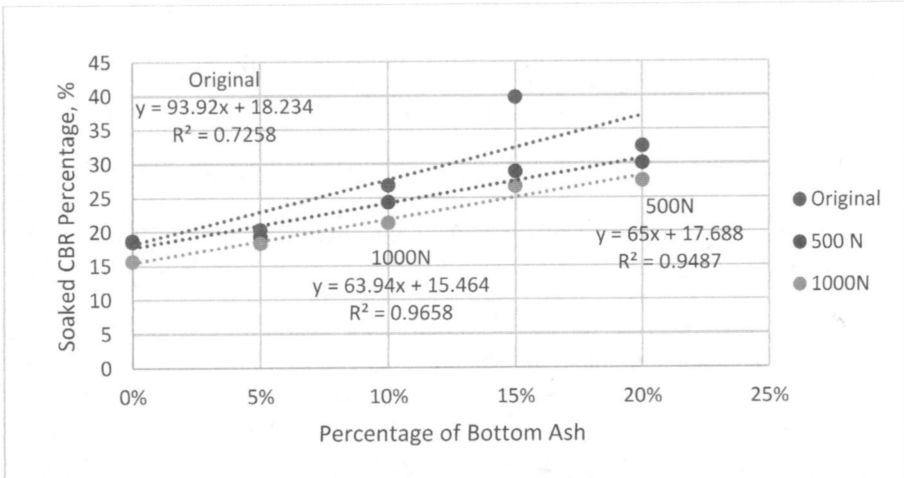

Figure 7. Soaked CBR percentage of sand mixed with different proportion of bottom ash sand mixed with different proportion of bottom ash.

4.3. Permeability

The sand is mixed with different proportion of bottom ash and the permeability of each sample was tested. Figure 8 show the permeability of sand mixed with different proportion of bottom ash that underwent crushing of 500N and 1000N. From Figure 8, when the percentage of bottom ash added into the sand increased, the permeability of the sand increased. This is because the bottom ash has larger particle size compared with the sand. Adding a well graded bottom ash into sand, will create more space between the sand particles and thus the permeability of the sand increases. Result is further supported in the literature review when bottom ash used in percentage of 0%, 20%, 40%, 60%, 80% and 100%. As the percentage of bottom ash increased, the coefficient of permeability also increased, which is similar to findings obtained in this research. This result is similar with result in this research where the permeability of bottom ash mixture increases when the percentage of bottom ash increased.

IOP Conf. Series: Materials Science and Engineering **712** (2020) 012041 doi:10.1088/1757-899X/712/1/012041

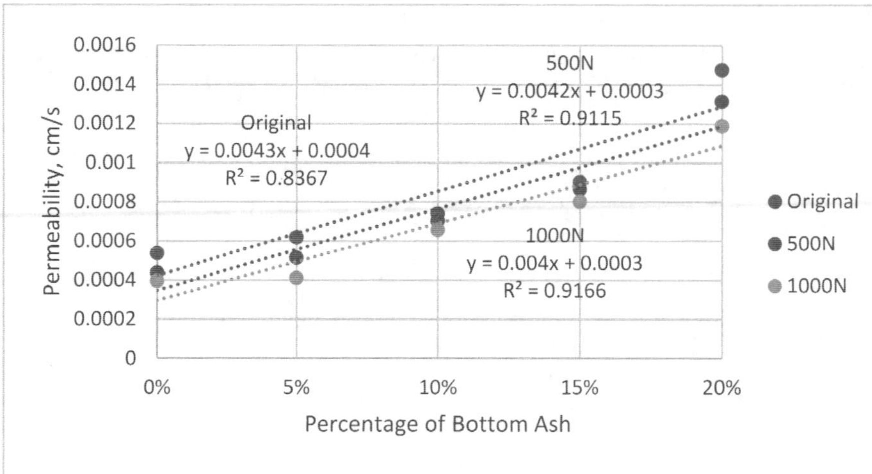

Figure 8. Permeability of sand mixed with different proportion of bottom ash underwent crushing of 500N and 1000N.

5. Conclusion

The ultimate aim of this research is to determine the effect of bottom ash when added to the crushed sand. Based on the preliminary result and analysis done, it can be concluded that:

1. The bottom ash is able to reduce the crushability of the sand by reducing the breakage index, crushing coefficient, fouling index and percentage of fouling. The higher the percentage of bottom ash added into the sand, the greater the reduction of crushability and the better the improvement to the capacity of the sand.
2. The bottom ash is able to increase the bearing capacity of the sand for both the soaked and unsoaked condition. The higher the percentage of bottom ash added into the sand, the greater the increment of the bearing capacity and the better the improvement of the capacity of the sand. This improvement in bearing capacity by adding bottom ash also shown in crushed sandy soil which are subjected to 500 and 1000 blows.
3. The bottom ash is able to increase the permeability of the sand. When the amount of bottom ash added into the sand increases, the permeability of the sand also increased. The greater the amount of bottom ash added, the better the improvement of the permeability of the sand. The improvement in permeability of crushed sandy soil was observed when subjected to 500 and 1000 blows.

Reference

[1] Kikumoto M, Wood DM, Russell A. Particle crushing and deformation behaviour. Soils and foundations. 2010;50(4):547-63.
[2] Di Emidio G, Flores RV, Van Impe WF. Crushability of Granular Materials at High Stress Levels.
[3] Seals RK, Moulton LK, Ruth BE. Bottom ash: An engineering material. Journal of Soil Mechanics & Foundations Div. 1972 Apr;98(sm4).
[4] Consoli NC, Heineck KS, Coop MR, Da Fonseca AV, Ferreira C. Coal bottom ash as a geomaterial: Influence of particle morphology on the behavior of granular materials. Soils and foundations. 2007;47(2):361-73.
[5] da Fonseca AV, Cruz RC, Consoli NC. Strength properties of sandy soil–cement admixtures. Geotechnical and geological engineering. 2009 Dec 1;27(6):681-6.
[6] Russell AR, Wood DM, Kikumoto M. Crushing of particles in idealised granular assemblies. Journal of the Mechanics and Physics of Solids. 2009 Aug 1;57(8):1293-313.
[7] Wu Y, Yamamoto H, Izumi A. Experimental investigation on crushing of granular material in one-dimensional test. Periodica Polytechnica. Civil Engineering. 2016;60(1):27.
[8] Kikumoto M, Wood DM, Russell A. Particle crushing and deformation behaviour. Soils and foundations. 2010;50(4):547-63.

[9] Afshar T, Disfani M, Narsilio G, Arulrajah A. Changes to Grain Properties due to Breakage in a Sand Assembly using Synchrotron Tomography. InEPJ Web of Conferences 2017 (Vol. 140, p. 07004). EDP Sciences.
[10] Vilhar G, Jovičić V, Coop MR. The role of particle breakage in the mechanics of a non-plastic silty sand. Soils and Foundations. 2013 Feb 1;53(1):91-104.
[11] Xiao Y, Liu H, Chen Q, Ma Q, Xiang Y, Zheng Y. Particle breakage and deformation of carbonate sands with wide range of densities during compression loading process. Acta Geotechnica. 2017 Oct 1;12(5):1177-84.
[12] Huang CW. *Study on particle breakage of sands subjected to various confining stress and shear strain levels* (Doctoral dissertation, Ph. D. Thesis, Tokyo Metropolitan University).
[13] Indraratna B, Salim W. Modelling of particle breakage of coarse aggregates incorporating strength and dilatancy.
[14] Sun Y, Zheng C. Breakage and shape analysis of ballast aggregates with different size distributions. Particuology. 2017 Dec 1;35:84-92.
[15] Tutumluer E, Dombrow W, Huang H. Laboratory characterization of coal dust fouled ballast behavior. InAREMA 2008 Annual Conference & Exposition 2008 Sep 21 (pp. 21-24).
[16] Tennakoon N, Indraratna B, Rujikiatkamjorn C, Nimbalkar S. Assessment of ballast fouling and its implications on track drainage.
[17] López EL, Vega-Zamanillo Á, Pérez MA, Hernández-Sanz A. Bearing capacity of bottom ash and its mixture with soils. Soils and Foundations. 2015 Jun 1;55(3):529-35.
[18] Muhunthan B, Taha R, Said J. Geotechnical engineering properties of incinerator ash mixes. Journal of the Air & Waste Management Association. 2004 Aug 1;54(8):985-91.

Acknowledgments
Authors would like to acknowledge the assistance from the technical staff and also the motivation from colleagues. This research work was carried out with financial support from UMP grant RDU1703308.

IOP Conf. Series: Materials Science and Engineering **712** (2020) 012042 doi:10.1088/1757-899X/712/1/012042

Effect of Dried Sewage Sludge on Compressive Strength of Concrete

A L Amminudin[1], P J Ramadhansyah[1], S I Doh[1], S A Mangi[2] and W I Mohd Haziman[3]

[1]Faculty of Civil Engineering Technology, Universiti Malaysia Pahang, 26300 Gambang, Pahang, Malaysia
[2]Department of Civil Engineering, Mehran University of Engineering & Technology, SZAB Campus Khairpur Mir's, 76060 Sindh, Pakistan
[3]Faculty of Civil and Environmental Engineering, Universiti Tun Hussein Onn Malaysia, 86400 Batu Pahat, Johor Bahru, Malaysia

Abstract. Sewage sludge is a waste product generated from the wastewater treatment process at the treatment plant. The amount of sewage sludge produced is increases every year as the population increased. Poor management of sewage sludge will give negative impact to the environment. Replacing cement with sewage sludge ash is more significant to reduce the amount of waste material from sewage treatment plants. In this study, the various percentage of dried sewage sludge (DSS) has been added to replace the cement. Sewage sludge was dried in the oven with 100°C for 24 hours, then sieve through the sieve size 300 μm. Then, DSS was used in the concrete instead of cement with the replacement percentage of 0%, 5%, 10%, and 15% by weight. The compressive strength of concrete cube specimens was investigated after 7 and 28 days of curing. The results showed that the compressive strength of concrete specimens increased with the replacement of cement in concrete with 5 % to 10 % DSS, but the achieved maximum strength still lower compared to the control sample. However, the compressive strength of concrete specimens decreased when the DSS replacement of cement is more than 10 % by weight. The result of XRF test also showed that DSS has good potential to replace cement in concrete.

1. Introduction

It is estimated that the population in Malaysia is increasing to 32.5 million in the year on 2019. The increasing of the total population will automatically increase the amount of wastewater generated by the people. Normally, the wastewater from the residential building and the commercial building will be treated before it is released to the natural waterways that are called effluent. The quality of the effluent must be controlled in order to ensure the quality of the natural water resources in good condition for the population in Malaysia. According to the Department of Environment Malaysia, the standard effluent must be at Standard A and B. Thus, the process to control the quality of effluent is started at the sewage treatment plant. Sewage sludge is a waste product generated from the wastewater treatment process at the treatment plant. Most of the wastewater treatment plants in Malaysia are managed by Indah Water Konsortium (IWK). According to IWK, about 5.3x 10^6 m^3 was produced every year and the amount of sewage sludge is still increasing annually [1]. A large amount of sewage sludge produced every year will affect the environment and also social development[2,3]. Thus, the sewage sludge will be sent to the landfill for disposal process. Besides, the common methods of sewage sludge disposal are ocean dumping, landfilling and also spreading on reclaimed land [1]. However, these methods of disposal of sewage sludge will affect to the environment such as water contamination, air, and land pollution because the sewage sludge contains toxic material and heavy metal. The disposal of sewage sludge at landfill area also can cause problem to the soil due to leachate that contains high potentially toxic and heavy metal produced by the sewage sludge[2,4,5,6,7]. Thus, studies have been done to utilize the sewage sludge ash as a partial replacement in the concrete mix. The incinerated sewage sludge with 600°C to 900°C will produce the sewage sludge ash and minimized the volume of sludge up to 50 %.

Different researchers use different temperatures to produce sewage sludge ash (SSA). A study was done by M. Cyr et al., the temperature of 850°C was used to burn the sewage sludge in order to produce SSA [8]. Monzo et al reported that incinerating the sewage sludge (SS) up to 800°C can produce an amorphous SSA [9]. It was suggested that the sludge should be burnt to a temperature of 550°C for 3 hours to produce SSA as cement replacement [10]. Lin and Weng also incinerate the sludge at 800°C in the combustion chamber to remove the organic substance then was used as a clay substitute in brick manufacturing [11]. Kartini et al in their study, domestic waste sludge was burnt under uncontrolled burning in a ferro cement furnace for 72 hours to produce residue ash that was used in the concrete mix [12]. The temperature used to produce sewage sludge ash does not exhibit any significant effect on chemical properties of the ash but slightly affect the physical shape of the SSA. The study conducted by Siew et al shows that the sludge burnt at 600˚C exhibited needle-shaped particles whereas a smooth structure was found in sludge burnt at 800°C due to the pozzolanic reaction which filled the void and pores in the mortar [13]. Moreover, sludge burnt at 800°C also provides additional strength to the mortar where the compressive strength has increased after 28 and 90 days [13]. An investigation done by Doh et al found that the SSA produce at the temperature of 600˚C for the duration of three hours has the potential to be used as partial cement replacement in concrete since there are high similarities in the major chemical component of SSA compared to cement [14]. Sharif and Attom used burned sludge ash as soil stabilizing agent. They burned the sewage sludge at 550˚C and mix with the different type of clay soil. From the test result, they suggested that the burned sludge ash can be used as soil stabilizer[15] . Carriona et al in his study on using SSA to manufacture pre-cast concrete block burned the SS at 800˚C to produce fine SSA [16]. This fine SSA material then was found out apt to be used in pre-cast concrete block manufacturing. Donatello et al in his investigation, the temperature from 800˚C to 900˚C is used to develop sustainable and low energy constructions product incorporated incinerated sewage sludge ash. Franz in his study conducted in Switzerland used temperature of 830˚C to 850˚C to produce SSA. From the study, he found that SSA can be used as an effective phosphate fertilizer that meets the demands of the agriculture industry. While Jamshidi et al used 650˚C temperature in his study to investigate the mechanical performance of concrete with partial replacement of sand by SSA [17]. However, Kazberuk used two different temperature of the furnace chamber wherein the bottom section the temperature is above 600˚C and the upper section is more than 850˚C to produce sewage sludge ash [18].

Apart from that, the use of SSA as a partial replacement of cement can be seen as a better solution to overcome the problem of disposal of sewage sludge. Using SSA as a partial replacement of cement has the potential to improve the mechanical performance of concrete. The workability and compressive strength of concrete after 28-days were improved with the addition of SSA [10,14] and lower the water absorption [14] compared to the control samples. According to [14,17,19] the optimum amount of SSA is 5 % by weight of ordinary Portland cement to promote the highest performance of compressive strength of concrete after 28 days. However, the high temperature for the combustion process to produce sewage sludge ash contributed to the increase of energy consumption. The manufacturing of ordinary Portland cement also requires a high degree of temperature in the burning process. Thus, by using sewage sludge ash to replace the cement is still increasing energy consumption. Based on the previous study, it found that the researchers burn the sewage sludge to form ash as a cement replacement in the concrete mix. Limited study has been conducted incorporating dried sewage sludge as a partial replacement of cement in the concrete mix. Therefore, this study aims to investigate the compressive strength of concrete containing dried sewage sludge as a partial replacement of cement by weight in concrete.

2. Materials

2.1. Sewage Sludge

Sewage sludge was collected from wastewater treatment plant at student's hostel, UiTM Pahang, Malaysia. The sewage sludge collected from the sludge dry bed at the wastewater treatment plant. Then, sewage sludge is oven dried at a temperature of 100°C for a period of 24 hours to ensure that the samples are dried [14]. The oven drying process is needed in order to remove up to 90% of its moisture content in the sewage sludge[14]. The impurities such as plastics, woods, and trash in the sludge were removed to ensure the best quality of sewage sludge. The dried sewage sludge was sieved passing 300µm.The process to produce dried sewage sludge was shown in Figure 1. In this research, 0%, 5 %, 10% and 15% of the dried sewage sludge were added as a partial replacement of cement in the concrete mix. Then, the compressive strength of the 100 x 100x 100mm concrete cubes were measured and compared to the control samples. The control samples were prepared with 0% dried sewage sludge. Table 1 shows the complete concrete mix design according to the percentage of dried sewage sludge. The weight of aggregates, water, and cement obtained according to 0.008 m^3 of the total concrete mix volume.

2.2. Course and Fine Aggregates

Crushed granite aggregates were used to include in the specimen mixture. The sources of aggregate were obtained from the Sungai Jan Kuari Sdn Bhd, Jerantut, Pahang. The coarse aggregates were graded in a range of 20 mm to 5 mm nominal size. While the fine aggregate or sand were sieve passed 5mm opening sieve size. Meanwhile, the fine aggregates or sand were graded is passing 5mm sieve size.

2.3. Aggregate Crushing Value (ACV) and Aggregate Impact Value (AIV)

The durability of the aggregate samples was tested in order to obtain the ACV and AIV. The ACV and AIV test is based on the BS 812 Part: III. The result obtained that the ACV and AIV is 25% and 17 % respectively.

2.4. Cement

Type 1 ordinary Portland cement was used in this study as a binder material in the concrete preparation process. The cement was produced by YTL Cement Bhd. This particular cement was used for entire experimental work in this study.

3. Experimental Procedure

3.1. Sample Preparation

The concrete mix design consists of coarse aggregate, fine aggregates, ordinary Portland cement added with dried sewage sludge with four different weights as a partial replacement of cement which is 0 %, 5%, 10 %, and 15 % respectively. A control mix was prepared using ordinary Portland cement with 0 % dried sewage sludge as a replacement of cement in the concrete mix. Then, the dried sewage sludge was blended with cement and mixing with aggregates and water. Then the mixture was cast in the 100 mm x 100mm x 100mm steel mould. Six (6) specimens were prepared for each content of dried sewage sludge to measure the compressive strength after 7 days and 28 days. After the specimens cast in the steel mould, the specimens were kept for 24 hours. Then the specimen was removed from the mould and sank in the water for curing process.

Figure 1. Process of Production of Dried Sewage Sludge

Table 1. Mix proportion of concrete containing dried sewage sludge

DSS (%)	w/c	Cement (kg)	DSS (kg)	Water (kg)	Aggregates		
					Fine (kg)	10mm (kg)	20mm (kg)
0 %	0.54	3.112	0	1.68	4.322	3.362	6.724
5 %	0.54	2.956	0.156	1.68	4.322	3.362	6.724
10 %	0.54	2.801	0.311	1.68	4.322	3.362	6.724
15 %	0.54	2.645	0.467	1.68	4.322	3.362	6.724

3.2. Curing Condition
The specimens were placed in the clean water after removing from the mould. The period of the curing process is 7 days until 28 days. After the curing period, the specimens were drained and ready for the compressive strength test.

3.3. Compressive Strength Test
One of the performance characteristics of concrete is compressive strength. 100 x 100 x 100 mm cube specimen was prepared to measure the compressive strength of concrete. The specimens were compressed using a compression strength test machine. The specimens were tested with a loading rate of 3.0kN/s according to British Standard Test method BS EN 12390:2009.

4. Results and Discussion
4.1. X-Ray Flourescene Test (XRF)
Based on the oxide content in SSA, it is suitable to replace the Portland cement content in normal concrete. The XRF result in Table 2 shows that the major oxide of DSS (Fe_2O_2, SiO_2, P_2O_5, CaO, Al_2O_3). According to D. F. Lin et al., the major oxide component in Portland cement is SiO_2, CaO, and Al_2O_3 [11]. The production of DSS is easier and save-energy compared to the production of sewage sludge in the ash form. Thus, it will promote a great potential to reduce the carbon emission to the environment. According to the result of XRF test, it shows that the DSS also consist of the high content of oxide component and potentially to replace the cement content in concrete. Furthermore, the DSS form is more environmentally and considered a feasible alternative solution to the sludge disposal problem [14].

Table 2. Oxide content in DSS and Portland cement

Oxides	DSS	Portland Cement[14]

Iron Oxide (Fe_2O_2)	15%	2.3 %
Silicon Dioxide (SiO_2)	14.9%	21.0%
Phosphorus Pentoxide (P_2O_5)	6.03%	0%
Calcium Oxide (CaO)	1.81%	66%
Aluminium Oxide (Al_2O_3)	7.34%	4.6%

4.2. Concrete Density

The density of concrete specimens were measured for each curing dates and percentages of DSS added as a cement replacement. The result of the concrete density for each percentage of DSS is shown in Figure 2. The density of concrete specimens for 7- and 28-days curing dates is lower compared to the density of the control sample (0% DSS). After 28 days curing period, the maximum density of concrete is 2350 kg/m^3 was recorded compared to the 7 days curing period, the maximum density is 2290 kg/m^3. However, the density of concrete started to drop after 5% of DSS was added for each curing periods as compared to the control sample. Based on the results obtained, it can be noted that the density of concrete decreased when the DSS with 300 μm were used to replacement the cement content in concrete.

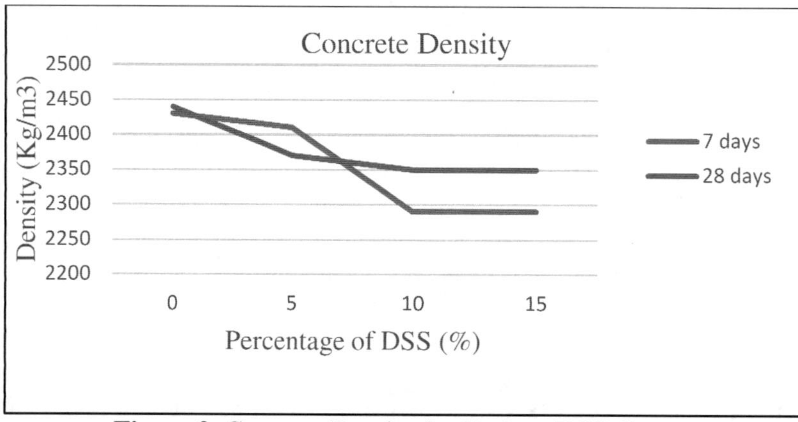

Figure 2. Concrete Density for Various DSS Content

4.3. Compressive Strength

In this study, the test age of concrete is 7 days and 28 days and 3 cubes were tested for each of the same proportion in order to obtain the average data. The results of the compressive strength of each mix are shown in Table 3 and Figure 3. As can be seen, the compressive strength of concrete is higher after 28 days compared to the 7 days curing periods. The result obtained indicates that the compressive strength of concrete is increased when the DSS was added as a cement replacement up to 10 %. Then, the compressive strength of concrete was starting to decrease when the content of DSS is more than 10 %. It also noted that the concrete mixture with DSS have a lower value of compressive strength compared to a control mixture.

Table 3. Compressive strength of various mix design

No.	Sample	Compressive Strength (N/mm^2)	
		7 Days	**28 Days**
1	0%	13.10	16.72
2	5%	8.00	9.33
3	10%	8.30	11.74
4	15%	5.86	9.95

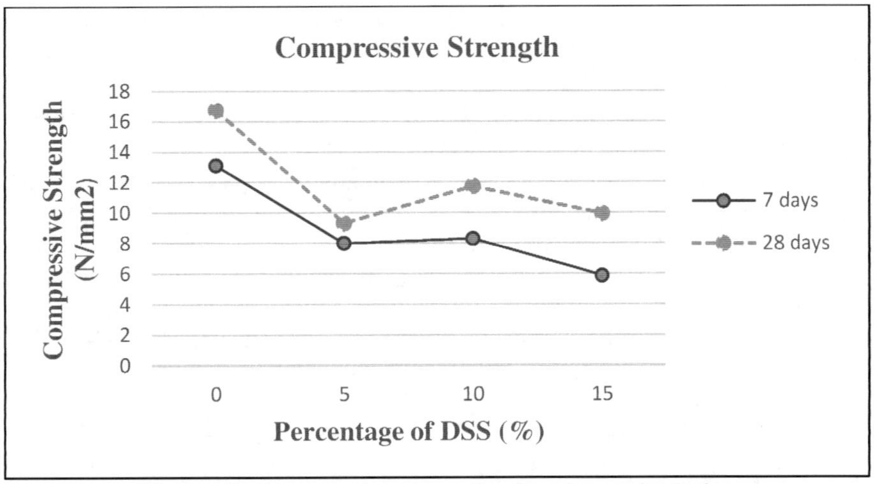

Figure 3. Compressive strength of various mix design

5. Conclusion

This study was done to assess the compressive strength of concrete containing DSS as a partial replacement of cement by weight in the concrete mixture. Results showed that the use of DSS as cement replacement in concrete resulted in good compressive strength compared to the control samples. The compressive strength of concrete increased when the DSS was added as a cement replacement up to 10 % in the concrete mixture. However, compressive strength of concrete was starting to decrease when the content of DSS is more than 10 %. Meanwhile, the density of concrete was decreased when the DSS with 300 μm particle size were used to replacement the cement content in concrete. Based on XRF result, it was found that the major oxide of DSS (Fe_2O_2, SiO_2, P_2O_5, CaO, Al_2O_3) is approximately similar to the Portland cement. Thus, the use of DSS in concrete production has great potential to reduce the amount of sewage sludge at the disposal area. The possibilities to reuse of sewage sludge in other applications also can be explored such as for road and highway construction, ceramic or glass production and brick production. Thus, reuse of sewage sludge also encourages the public or stake holders to save the environment by using the sustainable materials.

6. References

[1] T. Krishta David and S. Karan Nair, "Compressive Strength of Concrete with Sewage Sludge Ash (SSA)," *IOP Conf. Ser. Mater. Sci. Eng.*, vol. 371, no. 1, 2018.

[2] B. M. Cieślik, J. Namieśnik, and P. Konieczka, "Review of sewage sludge management: Standards, regulations and analytical methods," *J. Clean. Prod.*, vol. 90, pp. 1–15, 2015.

[3] J. Werther and T. Ogada, "Sewage sludge combustion," *Prog. Energy Combust. Sci.*, 1999.

[4] C. Wang, X. Hu, M. L. Chen, and Y. H. Wu, "Total concentrations and fractions of Cd, Cr, Pb, Cu, Ni and Zn in sewage sludge from municipal and industrial wastewater treatment plants," *J. Hazard. Mater.*, 2005.

[5] R. P. Singh and M. Agrawal, "Potential benefits and risks of land application of sewage sludge," *Waste Manag.*, 2008.

[6] C. Adam, B. Peplinski, M. Michaelis, G. Kley, and F. G. Simon, "Thermochemical treatment of sewage sludge ashes for phosphorus recovery," *Waste Manag.*, 2009.

[7] R. C. Kistler, P. H. Brunner, and F. Widmer, "Behavior of Chromium, Nickel, Copper, Zinc, Cadmium, Mercury, and Lead during the Pyrolysis of Sewage Sludge," *Environ. Sci. Technol.*, 1987.

[8] M. Cyr, M. Coutand, and P. Clastres, "Technological and environmental behavior of sewage sludge ash (SSA) in cement-based materials," vol. 37, pp. 1278–1289, 2007.

[9] J. Monzó, J. Payá, M. V. Borrachero, and I. Girbés, "Reuse of sewage sludge ashes (SSA) in cement mixtures: The effect of SSA on the workability of cement mortars," *Waste Manag.*, vol. 23, no. 4, pp. 373–381, 2003.

[10] C. M. . Fontes, M. . Barbosa, R. . T. Filho, and J. P. Goncalves, "Potentiality of Sewage Sludge Ash As Mineral Additive in Cement Mortar and High Performance," *Int. RILEM Conf. Use Recycl. Mater. Build. Struct.*, 2004.
[11] D.-F. Lin and C.-H. Weng, "USE OF SEWAGE SLUDGE ASH AS BRICK MATERIAL," *J. Environ. Eng.*, no. October, pp. 922–927, 2001.
[12] K. Kartini, A. M. Dahlia Lema, A. A. Dyg. Siti Quraisyah, A. D. Anthony, T. Nuraini, and R. Siti Rahimah, "Incinerated domestic waste sludge powder as sustainable replacement material for concrete," *Pertanika J. Sci. Technol.*, 2015.
[13] S. C. Chin, D. S. Ing, A. Kusbiantoro, Y. K. Wong, and S. W. Ahmad, "Characterization of sewage sludge ASH (SSA) in cement mortar," *ARPN J. Eng. Appl. Sci.*, 2016.
[14] D. S. Ing, S. C. Chin, T. K. Guan, and A. Suil, "The use of sewage sludge ash (SSA) as partial replacement of cement in concrete," *ARPN J. Eng. Appl. Sci.*, vol. 11, no. 6, pp. 3771–3775, 2016.
[15] M. M. Ai-sharff and F. A. Mousa, "The Use of Burned Sludge as a New Soil Stabilizing Agent," *Environ. Pipeline Eng. 2000*, pp. 378–388, 2000.
[16] M. Pérez-Carrión *et al.*, "Potential use of sewage sludge ash (SSA) as a cement replacement in precast concrete blocks," *Mater. Construcción*, 2014.
[17] A. Jamshidi, M. Jamshidi, N. Mehrdadi, A. Shasavandi, and F. Pacheco-Torgal, "Mechanical Performance of Concrete with Partial Replacement of Sand by Sewage Sludge Ash from Incineration," *Mater. Sci. Forum*, vol. 730–732, pp. 462–467, 2012.
[18] M. Kosior-Kazberuk, "Application of SSA as partial replacement of aggregate in concrete," *Polish J. Environ. Stud.*, 2011.
[19] C. M. A. Fontes, R. D. Toledo Filho, and M. C. Barbosa, "Sewage sludge ash (SSA) in high performance concrete: characterization and application," *Rev. IBRACON Estruturas e Mater.*, 2016.
[20] A. Jamshidi, M. Jamshidi, N. Mehrdadi, A. Shasavandi, and F. Pacheco-Torgal, "Mechanical Performance of Concrete with Partial Replacement of Sand by Sewage Sludge Ash from Incineration," *Mater. Sci. Forum*, 2012.

Acknowledgments
The support provided by Malaysian Ministry of Higher Education, Universiti Teknologi MARA Pahang and Universiti Malaysia Pahang in the form of a research grant (RDU/UMP) vote number RDU190339 for this study is highly appreciated.

IOP Conf. Series: Materials Science and Engineering **712** (2020) 012043 doi:10.1088/1757-899X/712/1/012043

Ultimate bearing capacity of peat treated with cement columns in physical model

D Youventharan, S M Arif and O Rokiah

Faculty of Civil Engineering Technology, Universiti Malaysia Pahang, 26300 Gambang, Pahang, Malaysia

Abstract. Peat is classified as very soft and problematic soil due to its natural behaviour of high compressibility, low in shear strength and with high natural water content. Cement column method is commonly used to stabilise the soil by improving the engineering properties. This research aimed to study the unconfined compressive shear strength and the ultimate vertical bearing capacity of stabilized tropical peat of East Coast of Peninsular Malaysia. A group of cement columns were tested in a series of physical model testing. To study the strength behaviour of peat with or without the cement stabilization, physical models were prepared with different variables such as; the number of cement column, proportion of binders (Ordinary Portland Cement-OPC) and the pozzolanic materials (Palm Oil Fuel Ash-POFA). In this research, a total of 9 physical models testing of peat soil including one without the stabilized cement column were conducted. A group of 4 and 6 cement columns with 25 mm diameter and 200mm in length each was installed. Samples were cured for 28 days as soon as the mixing of cement columns completed. After curing, a series of axial loads were applied uniformly on the pre-fabricated steel plate footing from the top of the cement column in order to study the ultimate bearing capacity of the stabilised soil. The change in strength of the soil was evaluated using Unconfined Compressive Strength (UCS) test. Based on the results, the highest ultimate bearing capacity was recorded in Model 6 with 6 cement columns with area improvement ratio of 18.83 %. The sample with 300kg/m^3 of OPC has the highest UCS value of 106.88kPa. Hence, the cement columns mixed with binder and pozzolanic improved the strength and the ultimate bearing capacity of the peat soil.

1. Introduction

In Malaysia, there is around 3 million hectares or approximately 8% of the land is covered with peats. In state of Sarawak alone there are some 1.66 million of peat covering the land [1,2]. Peat is considered as the extreme form of an organic soil where the organic content is higher than 75%. The extensive formation of peat soil layer is due to the rate of accumulation of plant's remains is higher and faster than the rate of decomposition. The type of peat also represents the accumulation of disintegrated plant remains which have been preserved under conditions of incomplete aeration and high water content. The formation of peats is more favourable when the area is waterlogged, with excess rainfall and low permeability ground, irrespective of latitude or altitude [1-3]. The peat soil is classified as problematic soil due to its natural properties of high compressibility, low shear strength and high natural water content. The high compressibility of peat soil will poses problem of secondary settlement and even tertiary settlement in long term due to continuous decomposition activity. It was found that peat has the potential to decompose and its decaying rate could be accelerated by controlling the influencing factors like oxygen supply, C:N ratio, pH value and temperature for optimum condition [2]. The ever high water content condition of peat makes it unsuitable for construction. The water content of peat normally varies from 500% to 2000% while for peat where the water content are less than 500% are usually indicates that there are high mineral fractions within the peat sample [1]. According to Hwang [3],when dealing with organic and peat soil, few options are available to improve or strengthen the soil, the engineer or owner can choose either (1) to strengthen the foundation, (2) eliminate the problematic soil (cut & replace), (3) apply soil treatment or relocate the project to better ground. Several methods such as surface reinforcement, preloading, vertical drain, deep stabilization, piling and chemical stabilization had been introduced by previous researchers in order to improve organic soil and peat soil. Dry Deep Soil Mixing (DSM) or simply the Cement Column mixing is a common method for soft soil stabilization. This in-situ stabilization method involves the mixing of cementitious compound such as Ordinary Portland Cement (OPC) and lime with peat soil. In this research, the cement column using OPC and palm oil fuel ash (POFA) is used to stabilize the peat soil. Physical models are constructed to simulate the field condition of the peat soil. The bearing capacity of the footings and the strength of the stabilized peat using a group of 4 and 6 cement column are analysed.

2. Materials and Methods

2.1. Sampling and soil classifications

Both the disturbed and the undisturbed soil samples were collected from Pekan, Pahang, Malaysia at a depth of 0.5 to 1m from ground surface. The soil samples were preserved in air tight container to prevent any loss of moisture content. A list of laboratory tests such as moisture content, organic content, fiber content, pH value, specific gravity, liquid limit, plastic limit, and unconfined compressive strength test were conducted to determine the engineering properties of the soil.

Moreover, the field density test was conducted using balloon density method to determine the actual field unit weight of the soil. The peat soil has been classified using the field using the Von Post Scale and based on the results obtained from laboratory tests.

Figure 1. In-situ peat classification technique

2.2. Physical model test chamber

A series of small scale physical models have been constructed with a dimension of 335mm x 525mm in area and 300mm in depth. The model was made of 8mm Perspex plate. In this research, the model was set up to create the undrained field condition. A layer of canvas sheet was placed at the bottom of the chamber to prevent water from draining out. A layer of lubricating oil was smeared on the inner surface of the chamber in order to reduce the skin friction between the wall and the soil. The peat soil was mixed thoroughly to reach homogenous condition. The soil was placed into the chamber in 3 layers and each layer was compacted with 25 blows of free fall steel plate.

Figure 2. Physical model chamber

2.3. Model design and cement column installation

The PVC pipe with diameter of 25mm and 200mm in length was used in this research. A thin layer of silicone grease was applied along the inner and outer surface of the PVC pipe casing before the insertion to decrease the friction between the pipe and the soil. The PVC pipe was pushed into the soil at predetermined location and depth. The soil-filled pipe was removed slowly from the model to create a hole. The removed soil was mixed with the different dosage of OPC and POFA to form cement-soil column as shown in Table 1. Finally, the cement-soil column was placed in the holes to make a composite foundation. There is no compaction was done during the insertion process of the mixture. The procedures were repeated until the column was completed to design length. The stabilized soil specimens in the chamber were cured for 28 days prior to the loading phase.

Table 1. Test conditions and dosage of admixtures

Model	Number of Column (n)	α (%)	Dosage	
			OPC (kg/m^3)	POFA (kg/m^3)
1	-	-	-	-
2	4	12.27	300	0
3	4	12.27	250	50
4	4	12.27	200	100
5	4	12.27	150	150
6	6	18.38	300	0
7	6	18.38	250	50
8	6	18.38	200	100
9	6	18.38	150	150

α is defined as the area improvement ratio based on the cement column area and treatment area

Figure 3. Installation of cement columns using PVC tube

In this phase, a rigid square steel plate with dimension of 200mm x 80mm and thickness of 20mm was designed to represent a rigid body resting on the stabilized soil. A series of loadings as shown in Table

2 were applied on the footing continuously and the settlement of the footing was recorded by 2 dial gauge placed on opposite sites across the centre of the footing. All the cement columns in each physical model were extruded out for the UCS test in order to obtain the UCS value. The reading for stress and deformation were taken every 10 seconds.

Table 2. Applied loading on steel plate footing

Day	Mass of steel plate (kg)	Loading (kN)
1	2	1.23
2	4	2.45
3	8	4.91
4	16	9.81
5	32	19.62
6	64	39.24
7	100	61.31

3. Results and Discussions

3.1. Soil and admixtures properties

Von Post Scale was used for the classification of peat soil. This scale is based on the degree of humification, appearance of soil and colour of water dripping out when the soil is being squeezed. The peat was classified as H3 according to the system as it is slightly decomposed peat soil, which, when squeeze, release muddy brown water, but from which no peat passes between fingers. Plant remains still identifiable and no amorphous material present. The engineering properties of the soil are shown in Table 3. It can be seen that most of the values obtained were not in the range of the published data. According to researchers [4,5], the soil can only be classified as peat when the organic content is greater than 75% while for organic content between the range of 25-75% should be classified as organic soil. The chemical composition and properties of both admixtures have been analysed using X-Ray Fluorescence (XRF). The Table 4 showed the chemical composition for the OPC and POFA.

Table 3. Engineering properties of tropical peat soil

Properties	Results	Published Data	Reference
Water Content (%)	362.12	198-417	[5-9]
Organic Content (%)	33.24	41-99	[5-11]
Fiber Content (%)	30.54	31-77	[6-11]
Liquid Limit (%)	69.85	202.3-220.7	[7-11]
Specific Gravity	1.91	0.95-1.34	[7-11]
pH value	3.3	3-5	[5-11]
Dry Unit Weight (kN/m^3)	10.64	10-13	[5-11]

Table 4. Chemical composition of admixtures

Properties	OPC	POFA
Silicon Dioxide (SiO_2)	21.45%	42.21%
Calcium Oxide (CaO)	60.98%	9.60%
Potassium Oxide (K_2O)	0.51%	7.59%

Iron Oxide (Fe_2O_3)	3.50%	4.88%
Phosphorus Pentoxide (F_2O_5)	-	4.35%
Aluminium Oxide (Al_2O_3)	3.62%	3.74%
Magnesium Oxide (MgO)	0.59%	3.67%
Sulphur Trioxide (SO_3)	-	1.69%
Loss of Ignition (LOI)	1.37%	2.55%

It can be seen that most of the chemical values of POFA is close to the value of OPC. According ASTM standard [12], POFA used for this research can be classified as Class C in pozzolanic material which can be used to partially replace the OPC in soil improvement. The Silica or the Silicon Dioxide (SiO_2) in POFA will react with the residue from the hydration process of cement to produce secondary C-S-H gel. The secondary C-S-H gel will contribute to the strength of the cement column and increase the strength of the soil.

3.2. Physical model test results

The ultimate bearing capacity (UBC) of the improved soil is contributed by many factors. In this research, the factors of area improvement ratio and shear strength of the soil and column were studied. Figure 4 and figure 5 showing the relationship between vertical stress and settlement for model with 4 cement columns (α=12.27%) and 6 cement columns (α=18.38%). The footings with higher ultimate bearing capacity, Q_{ult} for both 4 and 6 cement columns models were compared with unstabilised peat. Model 8 showed the greatest improvement towards the ultimate bearing capacity of the footing as it has the stiffest line when compared to other models. Area improvement ratio is correlated to the number of cement columns used for the treatment zone. When the number of cement columns used increases, the area improvement ratio increased too. In this research, there are only 2 groups of cement columns being considered; 4 cement columns group and the 6 cement columns group. The area improvement ratio corresponding to models with 4 and 6 cement columns group were 12.27% and 18.38% respectively. Compared to the untreated soil, the percentage of increment in terms of ultimate bearing capacity using 4 cement columns group varied from 200-280%, meanwhile for 6 cement columns group, the percentage of increment varied from 300-780%. From the results obtained, it clearly shows that the area improvement ratio effect the stabilised peat soil in terms of the bearing capacity. The major benefit for this performance is when the area improvement ratio is high, the capacity to distribute load through the cement column are more efficient and extensive.

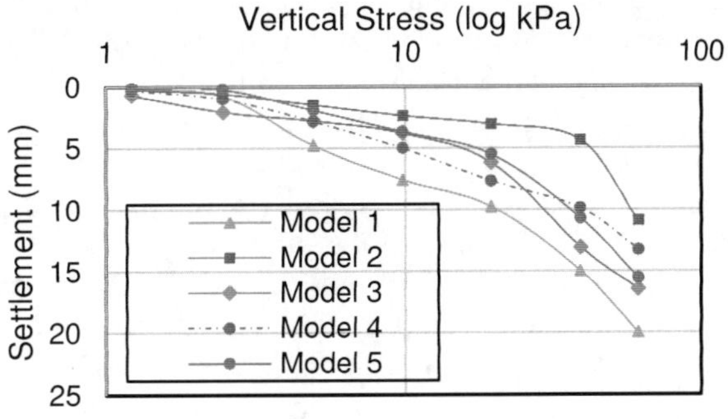

Figure 4. Relationship between vertical stress and settlement for α=12.27%

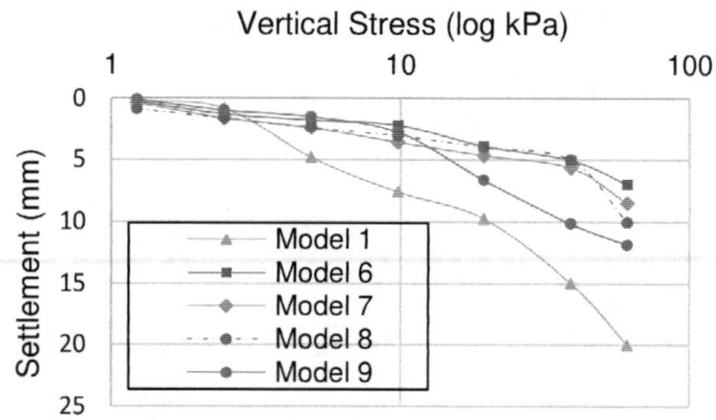

Figure 5. Relationship between vertical stress and settlement for α=18.38

Figure 6 illustrates the stress and strain curve for the UCS test. From the results obtained, it shows that Model 4 (Peat + 200kg/m^3 OPC + 100kg/m^3 POFA) has the highest UCS value of 92.34kPa. Moreover, the UCS value obtained in Model 4 is greater than Model 2 (Peat + 300kg/m^3 OPC). This clearly shows that the strength achieved in Model 4 is greater than Model 2. From the same figure, it was observed that the UCS value decreased when sample was mixed with the same amount of different type of binders such as in Model 5 (Peat + 150kg/m^3 OPC + 150kg/m^3 POFA). In all cases the best performance was achieved when the OPC used was two times the amount of POFA.

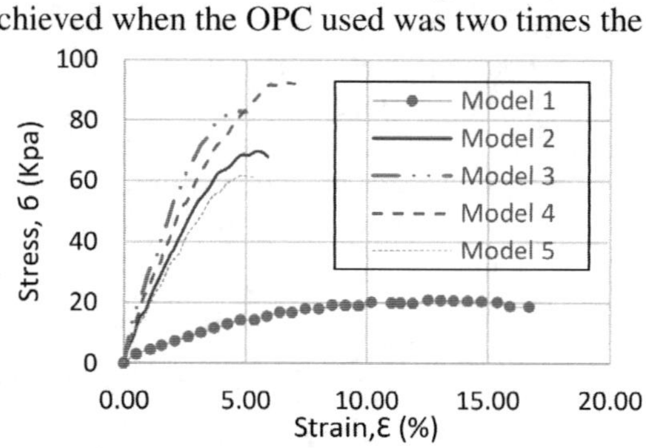

Figure 6. Stress and strain response during UCS test

4. Conclusion

The main objective of this research was to study the effectiveness of cement columns in increasing the shear strength and the ultimate vertical bearing capacity of footing on peat soil. This research proved that the location of where the peat soil is plays an important factor when it comes to choosing the right technique. Findings from this research confirmed that the combination OPC and POFA could serve as the most cost effective binder for the cement column in improving the ultimate bearing capacity and shear strength of the peat soil. The ultimate bearing capacity value for the stabilised peat soil treated with the 6 cement columns of 300 kg/m^3 OPC was 39 kPa as compared to untreated soil which only had 5 kPa for the ultimate bearing capacity. This research proved that the ultimate bearing capacity of the soil effected by the area of improvement and the number of cement columns used for peat soil stabilisation. The ultimate bearing capacity of peat soil treated with 6 cement columns was higher as compared to the ultimate bearing capacity of peat soil treated with only 4 cement columns. This research proved that peat soil can be improved in terms of shear strength and bearing capacity using OPC with two times the amount of POFA as binders when mixing cement column.

Reference

[1] Huat BB, Prasad A, Asadi A, Kazemian S. Geotechnics of organic soils and peat. CRC press; 2014 Feb 18.

[2] Pichan S, O'Kelly BC. Effect of decomposition on the compressibility of fibrous peat. InGeoCongress 2012: State of the Art and Practice in Geotechnical Engineering 2012 (pp. 4329-4338).

[3] Hwang J, Humphrey A, Bobet A, Santagata MC. Stabilization and improvement of organic soils.
[4] ASTM, "ASTM D4427-13 Standard Classification of Peat Samples by Laboratory Testing," *Am. Soc. Testings Methods, ASTM*, vol. 92, no. Reapproved, pp. 5–7, 2013.
[5] Huat BK. Organic and peat soils engineering. Penerbit Universiti Putra Malaysia; 2004.
[6] Duraisamy Y. *Compressibility behavior of tropical peat reinforced with cement column* (Master Sc. dissertation, Universiti Putra Malaysia).
[7] Huat BB, Maail S, Mohamed TA. Effect of chemical admixtures on the engineering properties of tropical peat soils. American Journal of Applied Sciences. 2005;2(7):1113-20.
[8] Kazemian S, Huat BB, Prasad A, Barghchi M. Effect of peat media on stabilization of peat by traditional binders. International Journal of Physical Sciences. 2011 Feb 4;6(3):476-81.
[9] Duraisamy Y, Huat BB, Muniandy R. Compressibility behavior of fibrous peat reinforced with cement columns. Geotechnical and Geological Engineering. 2009 Oct 1;27(5):619-29.
[10] Kolay PK, Sii HY, Taib SN. Tropical peat soil stabilization using class F pond ash from coal fired power plant. International Journal of Civil and Environmental Engineering. 2011;3(2):79-83.
[11] Adon R, Bakar I, Wijeyesekera DC, Zainorabidin A. Overview of the sustainable uses of peat soil in Malaysia with some relevant geotechnical assessments. International Journal of Integrated Engineering. 2012;4(4).
[12] ASTM, "C618-05 Standard specification for coal fly ash and raw or calcined natural pozzolan for use in concrete.," pp. 1–3, 2005.

Acknowledgments
Authors would like to acknowledge the assistance from the technical staff and also the motivation from colleagues. This research work was carried out with financial support from UMP grant RDU1703308.

Carbonation Resistance of Concrete with Fly Ash as Fine Aggregate

D Zhang[1], M Mao[1], Q Yang[1] and K S Lim [2]

[1]School of Civil and Hydraulic Engineering, Ningxia University, Yinchuan 750021, China
[2]Faculty of Civil Engineering Technology, Universiti Malaysia Pahang, 26300 Gambang, Pahang, Malaysia

Abstract. In order to investigate the carbonation resistance of concrete with fly ash as fine aggregate (CFA), the rapid carbonation test was conducted to explore the influence of the curing age, water-to-cement (w/c) ratio, fly ash replacement rate and the activator on the carbonation depth. The experimental results show that prolonging the curing time is beneficial to improving the carbonation resistance of the CFA. The carbonation depth of the CFA correlates positively with the w/c ratio. With an increase in the fly ash replacement rate and w/c ratio, the carbonation resistance of the CFA increases to a remarkable extent. After 28 days of standard maintenance of the CFA, the compactness of the CFA is boosted due to the pozzolanic effect of the fly ash, and its carbonation resistance is enhanced compared to that of ordinary concrete. In addition, the incorporation of the activator marginally contributes to the improvement of the carbonation resistance of the CFA.

1. Introduction

China is the world's largest coal consumer and fly ash is the main bulk industrial waste of coal [1]. As energy demands have continued to increase, by the end of 2016, the stock of fly ash has reached three billion tons in China. This substantial accumulation of fly ash not only occupies a large amount of farmland, but also pollutes the environment, destroys the ecological balance, and causes serious environmental and social problems. At the same time, river sand is an important ingredient in producing concrete, and excessive river sand has caused some geological disasters, such as the collapse of river banks, and river bridges being scrapped. Under this scenario, researchers have proposed to blend fly ash into concrete as a replacement for fine aggregate, thus turning the waste into something valuable[2-8]. Besides, the compressive strength, flexural strength, splitting resistance, and wear resistance of the concrete can also be improved over those of ordinary concrete [7-8]. At present, the research on CFA mainly focuses on its strength, blending, and shrinkage properties, and lacks systematic research on the carbonation performance of CFA. Carbonation is an important prerequisite for steel corrosion; corrosion causes a protective layer to separate, causing cracks along the steel bars[9-10]. This will further result in a reduction in the bond strength and effective area of the steel bar, seriously threatening the bearing capacity and durability of the reinforced concrete structure. In order to investigate the durability of the CFA, the rapid carbonation test is conducted to explore the influence of the curing age, water-to-cement (w/c) ratio, fly ash replacement rate and the activator.

The research results provide a reference for further research on the carbonation of CFA in complex environments.

2. Materrials and experiments

2.1. Materrials and experiments

The cement was P·O 42.5R cement produced by Ningxia Saima Cement Plant. The fly ash was Glass III fly ash (equivalent to Class C fly ash) from Xixia Thermal Power Plant, with density of 2.058 g/cm^3 and ignition loss of 5.74%, and the chemical composition was shown in Table 1. The fine aggregate was natural river sand. The fineness modulus of the river sand was tested to be 2.71, which was medium sand. The coarse aggregate adopts the continuous graded gravel with particle size of 5~31.5 mm, mud content of 0.6% and crushing index of 10.2%. A powdery naphthalene-type superplasticizer that was suitable for CFA was used to ensure a desirable slump value, and the mixing amount was 2%-3.5% of the mass of the cement in the mixture. The water was tap water.

Table 1. Chemical composition of fly ash (%)

SiO_2	Al_2O_3	CaO	Fe_2O_3	MgO	SO_3	P_2O_5	Na_2O	K_2O
50.35	29.65	5.85	6.61	1.83	1.72	1.13	0.335	2.11

2.2. Mixture Proportion

According to the experimental design, the w/c ratio of concrete was determined as 0.60, 0.55, and 0.50, The equivalent volume replacement rate of fly ash were 0% and 15%. The mix design is shown in Table 2. Formula (1) represents the replacement of 15% fine aggregate with an equal volume of fly ash:

$$M_{FA} = \frac{M_S \cdot 15 \cdot \rho_{FA}}{\rho_S} \tag{1}$$

where ρ_S and ρ_{FA} are the densities of fine aggregate and fly ash, respectively; M_S and M_{FA} are unit consumptions of fine aggregate and fly ash, respectively;

Table 2. Mixture proportions for a cubic-meter of concrete

Water to cement ratio	Fly ash replacement ratio:%	Water: kg/m^3	Cement: kg/m^3	Fly ash: kg/m^3	Sand: kg/m^3	Gravel: kg/m^3	Superplasticizer : kg/m^3
0.60	0	185.3	308.8	0	840.3	1027.0	6.2
0.60	15	185.3	308.8	96.4	743.9	1027.0	6.2
0.55	15	185.3	336.9	96.0	590.0	1030.5	6.7
0.50	15	185.3	370.6	96.1	563.2	1030.1	9.3

2.3. Experimental method

The prism test block of 100 mm×100 mm×400 mm was used, with three blocks in one group. The experimental procedures comply with the rapid carbonation method in accordance with the Chinese standard GB/T 50082-2009. The specific procedures were as follows. The test specimens were placed in a standard maintenance room with curing age of 3 d and 28 d, respectively. Two days before the test, the test specimens were taken out and placed in a 60°C oven for 48 h, and then placed in a carbonation box, as shown in Figure 1. When the test specimens were carbonized to 3 , 7 , 14 and 28 d, the prism specimens were taken out and the carbonation depth was measured. A rock cutter was used to break the shape and brush the powder on the section, and the phenolphthalein solution was sprayed. The region with purple color was considered as the carbonated zone and if it did not change color, it was carbonized zone, as shown in Figure 2. The carbonation depth of each point was measured every 10 mm using a digital vernier caliper with accuracy of 0.1 mm, and the average value was taken as the carbonation depth of the test specimen at the carbonation time. Table 3 shows the results of the rapid carbonation test of CFA.

Figure 1. Carbonation testing machine.

Figure 2. Carbonation depth measurement

Table 3. Carbonation depth of CFA

Ingredients	Fly ash replacement ratio:%	Water-to-cement ratio	Carbonation depth (mm)			
			3d	7d	14d	28d

A1	0	0.60	3.3	4.7	5.8	7.4
A2	15	0.60	4.1	5.8	7.2	9.1
A3	15	0.55	3.7	5.1	6.2	8.3
A4	15	0.50	3.1	4.2	5.1	6.5
C1	0	0.60	0.9	1.7	2.2	3.7
C2	15	0.60	1.0	1.6	2.0	3.1
C3	15	0.55	0.9	1.4	1.7	2.5
C4	15	0.50	0.7	1.2	1.6	2.1
C4-1	15	0.50	0.6	1.0	1.2	1.5

Note： The standard curing age number is 3d for class A and 28d for class C. The C4-1 is doped with 1% activator (anhydrous sodium sulfate)

3. Results and discussion

3.1. Influence of curing age on carbonation depth

The rapid carbonation experiment of CFA was carried out according to the mixture in Table 2. The relationship between the curing age and the carbonation depth is shown in Figure 3. It can be seen from Figure 3 that when the concrete is not sufficiently cured, that is, the standard curing age of 3 d, the carbonation of CFA became more serious as the carbonation time increased. Under the same w/c ratio (0.6), the carbonation of CFA was rapid while its carbonation resistance was weak. The main reason was that when CFA was not sufficiently cured, the cement hydration would not be complete; the initial holes were massive and large. As a result, the internal structure of CFA was not compact, and CO_2 could easily entered CFA, increasing the carbonation speed[11-12]. Although the products such as C-S-H gel formed by the pozzolanic reaction of the fly ash in the later stage were favorable for refining the pore structure, the large pores caused by insufficient early curing were difficult to be filled, and the channels for high concentration of CO_2 entering were still there [13]. With the increase of carbonation age, the lower the $Ca(OH)_2$ content in CFA, the more easily the CFA was carbonized and the carbonation developed rapidly. For the curing time of 28 d, the w/c ratio was 0.6, the fly ash replacement rate was 0% and 15%. When CFA was cured for 28 d (i.e., sufficient curing), the carbonation depth was significantly lower than that of the concrete cured for 3 d (under-maintenance), with the carbonation depth reducing from 7.4 mm and 9.1 mm to 3.7 mm and 3.1 mm, respectively.

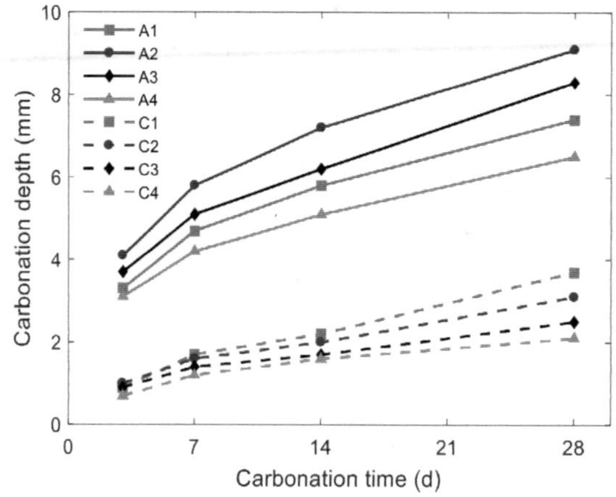

Figure 3. Influence of curing age on carbonation depth

3.2. Influence of w/c ratio on carbonation depth

It is well known that the w/c ratio is one of the important factors affecting the carbonation resistance of concrete [14-15]. The value of w/c ratio determines the internal microstructure and macroscopic properties of concrete. In order to investigate the effect of w/c ratio on the carbonation of CFA, the water-to-cement ratio are the A2~A4, C2~C4 series shown in Table 3. The relationship between w/c ratio and carbonation depth is shown in Figure 4. It can be seen from Figure 4 that the carbonation depth of CFA is positively correlated with the w/c ratio. When the carbonation time did not exceed 7 d, the effect of w/c ratio was not obvious. But when carbonized to 28 d, the carbonation depth increased obviously with the increase of w/c ratio. The main reason was that the larger the w/c ratio, the more pores in the concrete, the more easily the CO_2 would diffuse [16]. In addition, as the Ca^{2+} concentration in the concrete decreased, in order to maintain the balance, $Ca(OH)_2$ would continue to dissolve, resulting in a reduction in the liquid alkalinity and alkali reserve. When the pH value or $Ca(OH)_2$ decreased to a certain level, the other calcium-containing hydration products would decompose and carbonize, resulting in an increase in the carbonation value of the concrete and a decrease in the carbonation resistance of CFA.

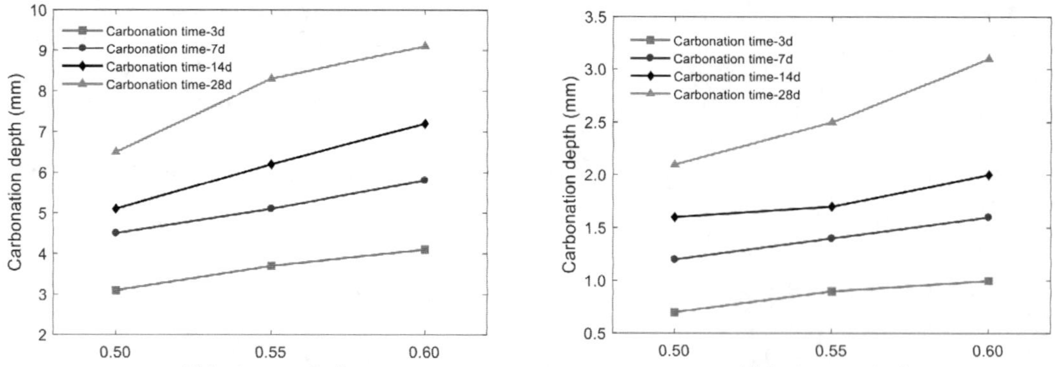

(a) 3 days of curing age (b) 28 days of curing age

Figure 4. Influence of w/c ratio on carbonation depth. (a) 3d of curing age and (b) 28d of curing age

3.3. Influence of fly ash replacement rate on carbonation depth

Figure 5 shows the effect of the fly ash replacement rate on the carbonation depth of the concrete. As can be seen from Figure 5a, when the concrete was cured for 3 d, the carbonation depth of CFA under different carbonation times was larger than that of ordinary concrete, and the carbonation was accelerated in the later stage. This was because when the curing was insufficient; the hydration of the fly ash was insufficient, resulting in a significant increase in the porosity, number of coarser pores and air permeability of the researched concrete than ordinary concrete. Therefore, the penetration and diffusion rate of CO_2 in the air was accelerated, and the carbonation depth at each curing age was also increased. It can be found from Figure 5b that when the concrete was cured for 28 d, the carbonation depth of CFA was slightly larger than that of ordinary concrete except for the carbonation time of 3 d, which might be due to experimental error. By contrast, the carbonation depth of CFA under other carbon age was smaller than that of ordinary concrete, and the difference became more obvious with the increase of carbonation time. This was because in the concrete where fly ash replaced fine aggregate, the amount of cement in the concrete was not reduced compared with the traditional fly ash instead of cement. In addition, with the progress of cement hydration, the hydration product $Ca(OH)_2$ increased, which in turn stimulated the participation of SiO_2 and Al_2O_3 in the fly ash, and the gel products such as CSH and CAH were filled into the pores of the interfacial transition zone[17-18]. This had improved the compactness of the concrete and thus improved the carbonation resistance of the concrete to some extent.

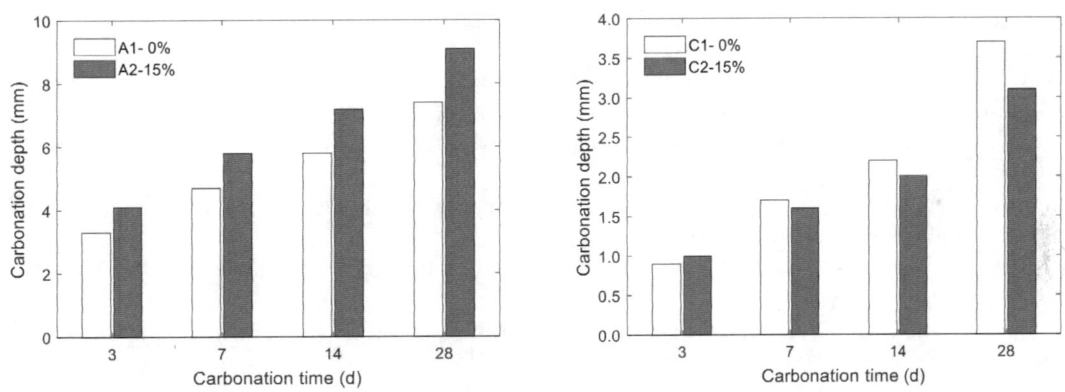

(a) 3 days of curing age (b) 28 days of curing age

Figure 5. Influence of fly ash replacement rate on carbonation depth. (a) 3d of curing age and (b) 28d of curing age

3.4. Influence of activator on carbonation depth

Since the low w/c ratio is beneficial to the carbonation resistance of CFA, the C4 sample was separately doped with 1% anhydrous sodium sulfate activator (labeled as C4-1), followed by the rapid carbonation test with standard curing time of 28 d. The results are shown in Table 4 and Figure 6. For the same carbonation age, the carbonation depth of CFA doped with the activator was slightly reduced; the carbonation depth of the C4 group for rapid carbonation of 28 d was 2.1 mm; the carbonation depth of the C4-1 group for rapid carbonation of 28 d was 1.5 mm, indicating that the activator could slightly improve the carbonation resistance of CFA. In actual engineering, due to the low concentration of CO_2 in the atmosphere, the carbonation process is very slow[19]. When the curing is sufficient, the pozzolanic reaction of the fly ash in the concrete continues, the internal compactness of the concrete is continuously increased, the porosity is reduced, and the carbonation speed is also reduced. Therefore, the carbonation resistance of CFA is continuously enhanced.

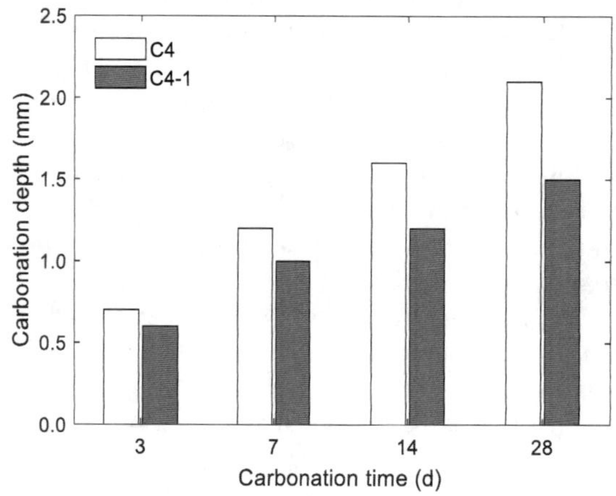

Figure 6. Influence of activator on carbonation depth

4. Conclusion

a) The curing system has a great influence on the carbonation resistance of concrete, and prolonging the curing time is beneficial to improving the carbonation resistance of concrete.
b) The carbonation depth of CFA is positively correlated with the w/c ratio. The carbonation depth increases as the w/c ratio increases.
c) When the curing is insufficient, the carbonation is accelerated in the later stage. Compared with ordinary concrete, the carbonation of the CFA is greater. When the curing is sufficient the pozzolanic effect of fly ash improves the compactness of concrete, thus improving the carbonation resistance of the CFA to some extent.
d) The incorporation of the activator marginally contributes to the improvement of the carbonation resistance of the CFA.

References

[1] Xun G, Hong Y, Dan Z, et al. 2010. Leaching characteristics of heavy metals in fly ash from a Chinese coal- fired power plant. Asia-Pacific Journal of Chemical Engineering. Volume 5(2), pp. 330-336
[2] MingJie, M, DongSheng Z. et al. 2019. Study of Durability of Concrete with Fly Ash as Fine Aggregate under Alternative Interactions of Freeze-Thaw and Carbonation. Advances in Civil Engineering. Volume 2019
[3] Nair, Deepa. 2015. Fly ash as a Fine Aggregate Replacement in Concrete Building Blocks. International Journal of Engineering and Advanced Research Technology. Volume 2(1), pp. 47-51
[4] Seo, T.; Lee, M.; Choi, C.; Ohno, Y., 2010. Properties of drying shrinkage cracking of concrete containing fly ash as partial replacement of fine aggregate. Magazine of Concrete Research. Volume 62(6), pp. 427-433

[5] Mihashi, H. 2007. Characteristic Points of AIJ Recommendations for Practice of Crack Control in Reinforced Concrete Buildings (Design and Construction). Journal of Concrete Journal. Volume 45(2), pp. 9-15

[6] DongSheng Z, MingJie M. et al. 2019. Experimental Investigation of the Neutralisation of Concrete with Fly Ash as Fine Aggregate in a Freeze-Thaw Environment. Advances in Civil Engineering. Volume 2019

[7] Rafieizonooz M, Mirza J, Salim M R, et al. 2016. Investigation of coal bottom ash and fly ash in concrete as replacement for sand and cement. Construction and Building Materials, Volume 116, pp.15-24

[8] Siddique R. 2003. Effect of fine aggregate replacement with Class F fly ash on the mechanical properties of concrete. Cement and Concrete research, Volume 33(4), pp. 539-547

[9] Otieno M, Ikotun J, Ballim Y. 2019. Experimental investigations on the influence of cover depth and concrete quality on time to cover cracking due to carbonation-induced corrosion of steel in RC structures in an urban, inland environment. Construction and Building Materials, Volume 198, pp. 172-181

[10] Carsana M, Canonico F, Bertolini L. 2018. Corrosion resistance of steel embedded in sulfoaluminate-based binders. Cement and Concrete Composites, Volume 88, pp. 211-219

[11] Zhang D, Cai X, Shao Y. 2016. Carbonation curing of precast fly ash concrete. Journal of Materials in Civil Engineering,Volume 28(11)

[12] Ekolu S O. 2016. A review on effects of curing, sheltering, and CO2 concentration upon natural carbonation of concrete. Construction and Building Materials, Volume 127, pp. 306-320

[13] Dias W P S. 2000. Reduction of concrete sorptivity with age through carbonation. Cement and Concrete Research, Volume 30(8), pp. 1255-1261

[14] Mohammed A, Sanjayan J G, Nazari A, et al. 2018. The role of graphene oxide in limited long-term carbonation of cement-based matrix. Construction and Building Materials, Volume 168, pp. 858-866

[15] Li D, Chen B, Sun H, et al. 2018. Evaluating the effect of external and internal factors on carbonation of existing concrete building structures. Construction and Building Materials, Volume 167, pp. 73-81

[16] Liu J, Qiu Q, Chen X, et al. 2017. Understanding the interacted mechanism between carbonation and chloride aerosol attack in ordinary Portland cement concrete. Cement and Concrete Research, Volume 95, pp. 217-225

[17] Shaikh F U A, Supit S W M. 2015. Compressive strength and durability properties of high volume fly ash (HVFA) concretes containing ultrafine fly ash (UFFA). Construction and building materials, Volume 82, pp. 192-205

[18] Malvar L J, Lenke L R. 2006. Efficiency of fly ash in mitigating alkali-silica reaction based on chemical composition. ACI materials journal, Volume 103(5)

[19] Bertos M F, Simons S J R, Hills C D, et al. 2004. A review of accelerated carbonation technology in the treatment of cement-based materias and sequestration of CO2. Journal of hazardous materials, Volume 112(3), pp. 193-205

Acknowledgement

This study was supported by the National Natural Science Foundation of China (51568055), the Special Projects for Scientific and Technological Cooperation with Foreign Countries of Ningxia Province of China (2018BFH03002), and the Graduate Innovation Project of Ningxia University (GIP2019022).

IOP Conf. Series: Materials Science and Engineering **712** (2020) 012045 doi:10.1088/1757-899X/712/1/012045

Success Factors of Design-Build Public Sector Projects in Malaysia

Z P Lee[1], R A Rahman[1,2] and S I Doh[1]

[1]Construction Management Research Group, Faculty of Civil Engineering Technology, Universiti Malaysia Pahang, 26300 Gambang, Pahang, Malaysia
[2]Earth Resources & Sustainability Centre, Universiti Malaysia Pahang, Lebuhraya Tun Razak, Gambang, Pahang, 26300, Malaysia

Abstract. Various procurement methods are employed in the construction industry, with each having varying success rates. The popularity of design and build (D-B) method has been increasing worldwide because its distinctive features of integrating design and construction have successfully addressed the problems inherent in the traditional design-bid-build (D-B-B) method. Previous research efforts in Malaysia have been focused on establishing a D-B strategy, including its challenges and risks and defining a broader scope for implementing D-B in public projects. However, not many empirical studies have identified the D-B success factors in Malaysia's construction industry. This paper aims to determine the main factors attributing to the successful application of the D-B method for public projects in Malaysia's construction industry. Specifically, this paper answers the main research question of "What are the success factors of the D-B project delivery method in Malaysian public-sector projects?". To achieve that objective, this study involves interviewing industry practitioners that have direct hands-on experience in managing D-B public sector projects in Malaysia. The interview data are analyzed to identify the D-B success factors in construction projects. The findings suggest that the underlying factors for successful D-B public project include client capabilities, competency of the construction leader, competency of the consultant team, organizational behavior, and effective project management system. The success factors in each underlying categories are discussed in the paper. This paper contributes a better understanding of the D-B project success factors in Malaysia's construction industry.

1. Introduction

The construction industry is a significant contributor to Malaysia's economy. Numerous studies have suggested that the construction industry play a significant role towards a nation's Gross Domestic Product (GDP), economic activity, government revenue, private investment, and employment [1,2,3,4]. In Malaysia, the construction industry contributes about 5.4% of the country's annual Gross Domestic Product (GDP) in 2018. Based on the data generated by the Central Bank of Malaysia in 2018, the highest National GDP was recorded in 2014 (6.0%), while the GDP attributed to the construction industry was 18.1% in 2012. The construction industry in Malaysia has grown consistently from 2011 until Q1 2018, inclusive of its rapid growth noted in 2012. These GDP values are shown in Table 1. In other words, the construction industry is vital to a nation's economy, and Malaysia is no exception. While the construction industry plays a vital role in Malaysia's economy, construction projects usually face severe cost and time overruns [5,6]. Projects fall behind schedules because of various reasons, but the most exceptional causes are poor site management and supervision, the project team's slow rate in making decisions, changes in job scope, financial conditions, lack of planning skills, and changes in the design made by clients [7,8,9,10]. Therefore, identifying approaches to improvise the construction industry is crucial to support Malaysia's economic growth.

Table 1: National GDP and Construction GDP in Malaysia from 2011-2018

Year	2011	2012	2013	2014	2015	2016	2017	Q1 2018
National GDP (%)	5.3	5.5	4.7	6.0	5.0	4.2	5.9	5.4
Construction GDP (%)	4.6	18.1	10.6	11.7	8.2	7.4	6.7	5.5

The Design and Build (D-B) method is one of the alternative approaches for delivering construction projects. It is a procurement method in which one entity is contractually responsible for both the design and construction of the project [11,12,13,14,15]. This established procurement approach was introduced and developed to cope with the complexity of construction projects [16]. In developed countries such as the United Kingdom, 42% of its total projects are procured using D-B [16]. However, the D-B implementation in Malaysia has been inefficient, mainly due to the unfamiliarity and inexperience of the project team in dealing with such an innovative approach [17]. Also, project clients in Malaysia are still clueless about the numerous procurement methods that are available and must be carefully selected at the start of the project [18]. Besides that, the lack of project management knowledge, the project size, and the diversity of the construction industry, together with the lack of skills and expertise to conduct a thorough assessment of available procurement systems are other issues that have been associated with D-B in Malaysia [19]. Hence, there is a need to identify elements that can improve the success rate for implementing D-B in Malaysia. The above discussion highlights the significant differences in construction industries based on different economic, management, and people attributes. Following that, the objective of this study is to identify the key success factors for implementing public sector D-B projects in Malaysia. To achieve this objective, this paper addresses the main research question, "What are the success factors for implementing the "Design-Build" project delivery method in Malaysian public-sector projects?" This study categorized the D-B success factors based on responses from individual interviews with industry practitioners. The similarities and differences from the feedback obtained are segregated into main themes. This paper contributes a better understanding of the success factors for implementing the D-B project delivery method in the Malaysian construction industry.

2. Literature Review

A contractor's expertise and performance significantly impact the successful delivery of a project [20]. While clients strive to make the best decisions in selecting the right contractor for the right job, a clear understanding of the underlying attributes associated with contractor selection, in the context of achieving successful project outcomes, is critical. The increasing globalization of projects and project management add to a diverse mix, creating intercultural challenges for project managers [17]. There is growing recognition that different types of projects require different approaches to their management and management procedures that are tailored to the needs of the project [20], as well as the selection of project managers with the right competencies [17,21]. Therefore, it is crucial to identify project success factors so that project owners and managers are provided with a specific and measurable framework for tracking key project outcomes [22,23]. The crucial success factors for project delivery include front-end planning and project definition, on-time performance, on-cost performance, and projects that fit the work program [24]. For example, a study in Vietnam showed that the construction parties' capabilities played an essential role in determining the success or failure of D-B projects [24]. In addition, the results of causality from the study highlighted six success factors, which significantly

affected D-B project performance, including resolving conflicts quickly; effective overall managerial actions in planning, organizing, leading and controlling; project participants' satisfaction with the financial return from the project; competent multidisciplinary project team; project team members' good and active attitudes toward the job; and adequate funding throughout the project. They also found no significant difference between the project performances of public and private sector D-B projects [17]. Therefore, identifying success factors are crucial for construction projects.
Every construction project is unique and comprises unique complexities and risks across the various levels of the construction process. The increasing complexity in design and the involvement of a multitude of stakeholders in modern construction projects further add pressure for both clients and contractors to match required skills and capabilities so that the project is successfully delivered [21]. Incorrect selection of a contractor for a given project may potentially lead to a bad legacy of failure coupled with consequences such as poor business practices or bankruptcy [21]. Studies suggest that the stakeholders, including clients, contractors, and consultants, have a major impact on the success factors of D-B projects in the public sector. Additionally, the people and operational are also the main criteria determining the success of D-B public projects. Finally, the implementation of D-B method in public projects has been identified in developed countries such as the United States of America, the United Kingdom, China, Japan, and Singapore. For instance, a construction project is considered successful when it is completed within the given time and budget, and is to the client's satisfaction [17]. Additionally, the project success is defined as the ability of the project to meet the objectives of time, cost, and quality, as well as satisfying the project stakeholders [22]. In developed Western countries, project success is defined as an intangible perceptive feeling, which varies with different management expectations, among persons, and with the phases of the project [25]. Meanwhile, in Asia, the measures of project success should also include the satisfaction of project team members [17]. Most studies use success factors as a tool to determine the performance of a project [21,26]. Project success factors refer to a set of circumstances, facts, or influence, which contribute to the project outcomes. The project manager's managerial skills, team members' commitment and their technical background, project attributes, and environmental factors are all critical project success factors [24]. In summary, various studies have shown that a successful project meets all four crucial objectives including cost, time, people or satisfaction, and quality of construction. Hence, these objectives are also the success criteria that must be identified to rectify failure and to improve the construction project's productivity deficiency during liability periods as a result of defects. The previous studies concluded that key success factors identified during the construction stage would strengthen the project management skills of the project leader. To further improve D-B adoption, understanding the success factors that influence the client decision in accepting the D-B method is crucial. To provide this knowledge, the success factors that suit the project characteristics of Malaysia's construction industry such as fragmentation, complexity, dynamism, and lack of standardization should be identified for the construction project success [27]. Thus, this study will address this gap by identifying the key success factors for implementing D-B public sector projects in Malaysia.

3. Methodology
The data collection is based on the qualitative approach of individual interviews. Also, data analysis involves content analysis. Figure 1 outlines this study's methodology. The following subsections discuss the methods used to collect and analyze the data.

3.1. Data Collection
The interview sessions were conducted with relevant industry practitioners such as clients and consultants, and contractors. The main reason for conducting the interview is to understand the actual practice of D-B in the Malaysia construction industry. The data collection involves 20 interviewees that hold at least senior management positions with direct hands-on experiences on D-B public sector projects in Malaysia. The interviewees' average working experience in the construction industry is around 12 years. Thus, their opinions and findings are considered representative, reliable, and valid for this study. Open-ended questions were provided to the participants because it allows the solicitation of

actual information by encouraging the interviewees to express their issues or areas of concern freely. The main question during the interview was, "What are the success factors for implementing the D-B method in Malaysian public-sector construction projects?". Follow-up questions relating to the subject matter were also provided during the discussion. After each interview session, the responses were summarized and returned to the interviewee for verification and approval. The interview process was conducted in the first quarter of 2019.

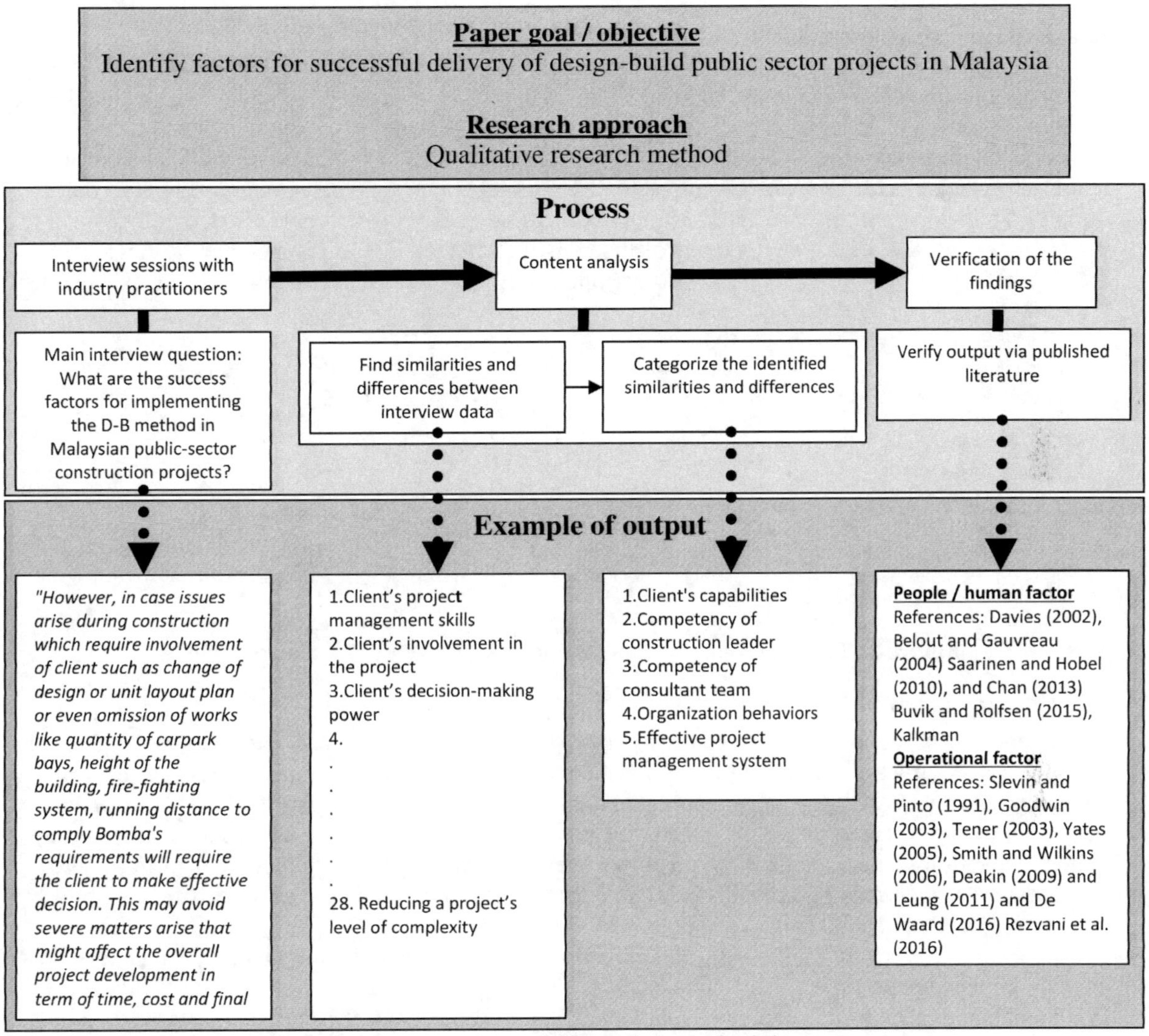

Figure 1: Overview of this study's methodology

3.2. Data Analysis

The data analysis involves content analysis. Six stages of content analysis were applied in this study: (1) formulate the research questions; (2) decide the interview dialogue to be examined; (3) decide the size or type of response to be counted in the analysis; (4) develop a sampling plan; (5) determine the categories from the responses to be divided; and (6) verify the process data based on a reliability check. By elaborating on this process, the data obtained from the in-depth interviews were archived in a Microsoft Excel workbook for subsequent analysis. The interview dialogues were duly analyzed based on the concepts of content analysis in a matrix table format whereby each question was posed against the answers of each interviewee. The interview dialogue could be classified and reduced into more relevant and manageable data as subject themes [28,29]. This method can be applied in a situation in which information and understanding of issues relevant to the general aims and specific research objectives have been obtained [30]. Hence, for this study, the answers were classified into different groups according to the nature of the content and by capturing the similarities and differences as the main theme. Specifically, this study categorized the identified parties into three main groups, "client," "contractor," and "consultant" during the data analysis because these stakeholders play a significant role in delivering a successful project, especially in decision-making and in terms of technical expertise [31].

4. Results and Discussion

Figure 2 illustrates the success factors for implementing D-B public-sector construction projects in Malaysia. In summary, a total of 30 independent variables (i.e., success factors) were identified. Based on the data analysis, the success factors can be categorized into five categories (i.e., underlying factors): 'client capabilities,' 'competency of the construction leader,' 'competency of the consultant team,' "organizational behaviour,' and 'effective project management system.' Finally, the five underlying factors can be associated with either "people" or "operational." Specifically, "people" relates to client capabilities, competency of the construction leader, and competency of the consultant team. Meanwhile, "operational" relates to organizational behavior and effective project management system.

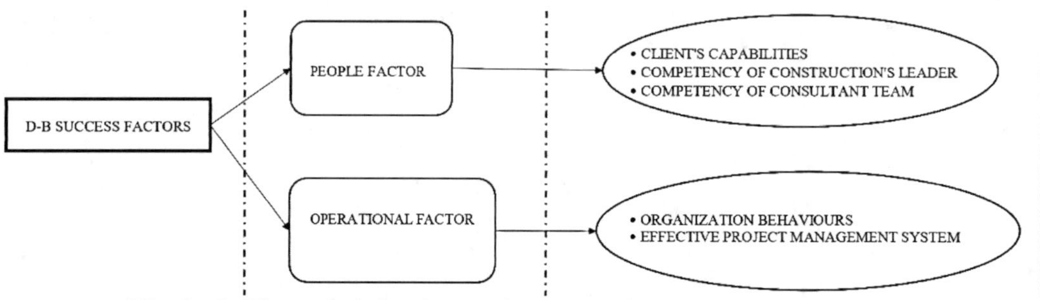

Figure 2: The underlying factors for successful D-B public sectors projects

4.1. Data Analysis

The client can be defined as the person or firm responsible for commissioning and paying for the design and construction of a facility [32]. Clients are also parties that initiate the projects, make decisions, create project objectives, and encourage others to achieve project goals. Therefore, clients play a vital role in determining the successful outcome of a project [40]. The success factors under "client capabilities" are: (1) client's project management skills; (2) client's involvement in the project (3) client's decision-making power; (4) client's experience and technical skills; (5) client's ability to brief the project team; (6) client's ability to provide clear objectives; (7) client's ability to transfer risk to single point; (8) client's ability to understand economic factors such as market demand; and (9) client's knowledge on government policies. The client capabilities are defined as the skills, abilities, and processes needed to develop close relationships with project stakeholders in construction projects [33]. It consists of effective project management skills, the level of involvement in the project, the decision-making power, experience, technical skills, emphasis on the fundamental project criteria such as time, cost and quality, the determination to transfer risk to other party, and the ability to understand the government policy relating to the construction industry. The conception of construction briefs is an essential issue in the construction industry, as the unproductive preparation of briefs significantly increases the total cost of a building. A lack of knowledge of the briefing process has rendered some clients incapable of describing their requirements, resulting in uncertain project objectives [7,34]. Poor decisions during briefings include inadequate involvement of project participants and insufficient time for briefing, which causes delays in the project [35]. Delays are generally regarded as the most common, complex, risky, and most frequently encountered problem in a construction project [35]. Hence, to overcome delay issues, most studies suggested that clients should be able to deliver a good project brief and must be prepared to take an active role in the construction process to prevent gaps between the requirements that they set and that of other project team members [22]. Other necessary skills required by the client are project management skills to lead, plan, and organize, as well as coordinating skills [7,34,36]. However, any poor decisions made by the project client will cause a delay in the construction works and lead to bad relations between project parties as well as increased costs of the construction project. In other words, client involvement in project decision-making is vital for the smooth running of project activities and is crucial to prevent unnecessary changes to the project. These suggest client capabilities play a crucial role in ensuring the success of implementing D-B in public sector construction projects.

4.2 Competency of the Construction Leader

The construction leader refers to the person who leads the design and construction of the D-B project. From the results of this study, the success factors related to the competency of construction leader are: (1) leader's project management skills; (2) leader's experience; (3) construction leader's capabilities; (4) construction leader's commitment to the project; (5) construction leader's support system from the upper management. Apart from working within the constraints of the project itself, project team leaders should have certain 'human' skills in dealing with stress, in establishing good relationships among team members, and in developing a harmonious working atmosphere [25,37,38,39]. For example, project team leaders are increasingly required to have team-building skills that can increase the project team's effectiveness [40]. Moreover, project leaders should be adaptable to the rapid change in the industry from developments in technology, construction, and property markets, regulations, and socio-economic factors. Finally, project managers' success at managing the project is suggested to be dependent on their competence, particularly their leadership style comprising emotional intelligence, management focus, and intellect [15,41,42]. These prior findings illustrate that the competencies of a project leader play a vital role in ensuring successful construction projects.

4.3 Competency of the Consultant Team

One of the main stakeholders in a construction project is the consultant, also known as the design team. Although the consultant team can leave most of the responsibilities to the client and contractor to operate the entire project, the consultant team is still required to have sufficient design personnel committed to the project. This study's analysis suggests that the success factors under "competency of consultant team" for D-B projects have consultant teams with the appropriate level of experience, capabilities in design works, project management skills, and commitment towards the project. In a nutshell, the success factors for the 'competency of the consultant team' are similar to the success factors for D- B in Sri Lanka, Africa, China, and Singapore. These similarities suggest that the consultant team plays a crucial role in the success of implementing D-B in construction projects. At the pre-construction stage, the consultant team is required to have sufficient resources to define the project fully and to prepare accurate specifications for the tendering job [43]. During the construction stage, in addition to minimizing design discrepancies, consultant design teams are required to supporting other project teams if any technical errors [44]. In other words, in addition to project leaders with adequate project management skills, consultant teams with in-depth understandings of technical construction activities are necessary.

4.4 Organizational Behaviour

Organizational behavior provides an understanding of why people behave as they do in organizations. From the results of this study, the success factors under organizational behaviour are: (1) project team members working together as a group (2) a harmonious working environment; (3) project team members working with confidence; and (4) organizational structure that allows transferring of the decision-making authority to the top management. The core of all successful organizations is the effectiveness with which people work together. Organizations should be able to manage individuals with differing characteristics and personalities to meeting organizational objectives. Construction projects are usually complex and highly differentiated, with a wide range of specialists with diverse professional skills working in a highly integrated way to deliver projects successfully. Therefore, understanding how the people involved in construction behave and work together is necessary for projects to achieve successful outcomes [31,32]. In other words, having the right organizational behavior is essential in improving the success rate of delivering projects, including D-B projects.

4.5 Effective Project Management System

The D-B approach encourages the adoption of innovative building techniques, which enable the faster delivery of projects [17]. There is a wide range of variables in a project management system of the project's packages offered by the client. The procurement methods have turned out to be an important subject in the construction industry because construction involves a series of processes that are interrelated and sequential, especially the procurement of construction projects [45,46]. In other words, the success or failure of projects depends on the effectiveness and efficiency by which the project management system executes these processes. The six success factors under effective project management system include having processes for: (1) a systematic up-front planning; (2) effective communication; (3) effective control management system; (4) effective team organization; (5) attractive work packages; and (6) reducing a project's level of complexity. These factors are similar to the findings on the critical success factors for implementing D-B in China and Singapore [44]. These similarities suggest that organizational behavior is a critical underlying factor to implement D-B successfully.

5. Conclusion

This study illustrates the success factors for D-B public sector projects in Malaysia by analyzing individual interviews with industry practitioners that have hands-on experience in managing D-B projects. In conclusion, the success factors for D-B public projects in Malaysia involve three main project parties, which are the client, the contractor, and the consultant. Clients must have a clear understanding of the project scope, develop a clear brief, assess the contractor's proposal thoroughly, and develop a clear understanding between project costs, as well as limit any changes during construction. Meanwhile, the contractor must be capable and experienced in managing D-B projects. For consultants, they must have the experience and the capability for design works. Finally, the satisfaction of the people from the project's stakeholder is vital to enhance the work productivity and growth sustainably by the mode of systematic D-B approach. In a more practical sense, the findings from this study highlighted the importance of the success factors required for fulfilling the user requirements, functionality, and workability of the end-products, the reputation of the project, and meeting the allocated time frame and construction costs. This paper contributes a better understanding of the success factors of the D-B project delivery method in Malaysia.

6. References

[1] Dlamini S 2012 *Relationship of construction sector to economic growth* Proceedings of International Congress on Construction Management Research 26-29 June 2012 Montreal Canada

[2] Hassanabadi A, Chang A., Wiewiora A, Ashkanasy N, M Jordan, P J and Zolin R 2011 *Manager emotional intelligence and project success:* The mediating role of job satisfaction and trust International Journal of Project Management 34(7) 1112-1122

[3] Khann R A 2008 *Role of construction sector in economic growth: Empirical evidence from Pakistan economy* Proceedings of 1st International Conference on Construction in Developing Countries 4-5th August 2008 Karachi Pakistan 279-290

[4] Wibowo A 2009 *The contribution of the construction industry to the economy of Indonesia: A systemic approach* Monograph, Diponegoro University Indonesia

[5] Ding R., Chen P H, Chew D A and Teo C C 2011 *Critical project management factors of AEC firms for delivering green building projects in Singapore* Journal of construction engineering and management 137(12) 1153-1163

[6] Khoo M C and Yong Y C 2014 *Factors contributing to project time and hence cost overrun in the Malaysian construction industry* Journal of Financial Management of Property and Construction 19(1) 55-75

[7] Abdul-Rahman, H Rahim, F A M and Mohd M S 2007 *A study on quality management during*

the pre-construction stage of design-and-build projects In CME 25 Conference Construction Management and Economics 753

[8] Al-Momani 2002 *Causes of construction delay: traditional contracts* International journal of project management 20(1) 67-73

[9] Chan A P and Kumaraswamy N 2012 *Analysis of factors critical to construction project success in Malaysia* Engineering, Construction and Architectural Management 19(5) 543-556

[10] Olawale Y A and Sun M 2010 *Cost and time control of construction projects: inhibiting factors and mitigating measures in practice* Construction management and economics 28(5) 509-526

[11] Chen H L, Buvik M P and Rolfsen M 2015 *Prior ties and trust development in project teams A case study from the construction industry* International Journal of Project Management 33(7) 1484-1494

[12] Design-Build Institute of America 2016 *Professional's Guide To Managing The Design Phase* (No.14) The Charles Pankow Foundation

[13] Fredrickson K 1998 *Design guidelines for design-build projects:* Journal of Management in Engineering 14(2) 77-80

[14] Kumaraswamy M and Dulaimi M 2001 *Empowering innovative improvements through creative construction procurement*: Engineering, Construction and Architectural Management 8(5) 325-334

[15] Merchant A and Bajaj D 2002 *An analysis of the use of the "Design-Build" project delivery system by interior design firm and its effectiveness in delivering fit-out projects*: Journal of Construction Procurement 8(1) 17-34

[16] Chan E H and Yu A T 2010 *Design management in design and build projects: The new role of the contractor* In Construction Research Congress 2005: Broadening Perspectives 1-11

[17] Chan A P and Tam C M 2012 *Factors affecting the quality of building projects in Hong Kong* International Journal of Quality & Reliability Management 17(4/5) 423-442

[18] Carvalho M M, Patah L A and De Souza Bido D 2015 *Project management and its effects on project success: Cross-country and cross-industry comparisons* International Journal of Project Management 33(7) 1509-1522

[19] Khairuddin A 2012 *Design and Build and its significance for Malaysia* Commonwealth-Governance and Growth 54-57

[20] Adnan H, Azmi, W N, S W M Rashid, Z Z A and Yusuwan N M 2018 *Housing Developer's Perspective on Social Sustainability* Malaysian Construction Research Journal (MCRJ) 63

[21] Doloi, H. 2010 *Cost overruns and failure in project management: Understanding the roles of key stakeholders in construction projects* Journal of construction engineering and management 139(3) 267-279

[22] Yu A T, Chan E H, Chan D W, Lam P T and Tang P W 2010 *Management of client requirements for design and build projects in the construction industry of Hong Kong* Facilities 28 (13/14) 657-672

[23] Khang D B and Moe T L 2008 *Success criteria and factors for international development projects: A life-cycle-based framework* Project Management Journal 39(1) 72-84

[24] Taherdost J P 2015 *Inter-organisational disaster management projects: Finding the middle way between trust and control* International Journal of Project Management 35(5) 889-899

[25] Clarke N 2010 *The impact of a training programme designed to target the emotional intelligence abilities of project managers* International Journal of Project Management 28(5) 461-468

[26] Zawawia E, Kamaruzzaman S, Ithnina Z and Zulkarnaina S A 2012 *"Conceptual framework for describing CSF of building maintenance management* The 2nd International Building Control Conference (1) 18-22

[27] Hwang B G and Lim E S J 2012 *Critical success factors for key project players and objectives: Case study of Singapore* Journal of Construction Engineering and Management 139(2) 204-215

[28] Riff D, Lacy S, Fico F, Riffe D and Fico F G 2008 *Analyzing media messages: Using quantitative content analysis in research* Routledge 82-92

[29] Weber R P 1990 *Basic content analysis* (No. 49) Sage 101-125

[30] Gillham F 2001 *The use of qualitative content analysis in case study research* In Forum: Qualitative Social Research Vol. 7(1) 1-30

[31] Yong Y C and Mustaffa N E 2010 *Critical success factors for Malaysian construction projects: an empirical assessment* Construction Management and Economics 31(9) 959-978

[32] Kalkman J P and De Waard E J 2016 *Inter-organisational disaster management projects: Finding the middle way between trust and control* International Journal of Project Management 35(5) 889-899

[33] Morgan B A, Leonoudakis R and Mcgraw 2009 *Project Time Management: The Foundation for EffectiveResourceManagement*Availablefrom:http://www.rbryanpeterson.com/files/Project_Time_Management_v2_2 Feb 2009-1.pdf [Assessed on 07/04/2019]

[34] Slevin J K and Pinto M B 1991 *Determinant of cross-functional cooperation in the project implementation process*: Project Management Journal 22(2) 13-20

[35] Khairulzan Y and Syahira Nabilah A H 2015 *The significant factors causing delay of building construction projects in Malaysia* Malaysia Construction Journal 20 593-599

[36] Williams T 2016 *Identifying success factors in construction projects: A case study* Project Management Journal 47(1) 97-112

[37] Dainty A R, Cheng M I and Moore D R 2005 *Competency-based model for predicting construction project managers' performance* Journal of Management in Engineering 21(1) 2-9

[38] Edum-Fotwe F T and McCaffer R 2000 *Developing project management competency: perspectives from the construction industry* International Journal of Project Management 18(2) 111-124

[39] Radzi A R, Bokhari H R, Rahman R A and Ayer, S. K. 2019 *Key Attributes of Change Agents for Successful Technology Adoptions in Construction Companies: A Thematic Analysis* Computing in Civil Engineering 2019: Data, Sensing and Analytics pp. 430-437 Reston VA: American Society of Civil Engineers

[40] Xia B, Chan A P and Zuo J 2012 *Comparison of Key Competences of Clients and Design-build Contractors in the Construction Market of the People's Republic of China (PRC)* Doctoral dissertation Ceylon Institute of Builders 114-155

[41] Rahman R A, Alsafouri S, Tang P and Ayer S K 2016 *Building Information Modelling skills for career success* Proceedings of the Academic Interoperability Coalition: 10th BIM Academic Symposium Orlando Florida pp. 35-42

[42] Rahman R A, Alsafouri S, Tang P and Ayer S K 2016 *Comparing building information modelling skills of project managers and BIM managers based on social media analysis* Procedia Engineering 145 812-819

[43] Songer A D and Molenaar K R 2010 *Model for public sector design-build project selection* Journal of Construction Engineering and Management 124(6) 467-479

[44] Palaneeswamy E and Kumaraswamy M 2010 *Knowledge mining of information sources for research in construction management* Journal of Construction Engineering and Management 129(2) 182-191

[45] Oladinrin T O, Adeniyi O and Eboreime I V 2013 *Analysis of non-excusable delay factors influencing contractors' performance in Lagos state* Nigeria Journal of Construction in Developing Countries 18(1) 53

[46] Olawale Y A and Sun M 2010 *Cost and time control of construction projects: inhibiting factors and mitigating measures in practice* Construction management and economics 28(5) 509-526

Acknowledgments

Authors wishing to acknowledge assistance or encouragement from colleagues, special work by technical staff or financial support from organizations should do so in an unnumbered Acknowledgments section immediately following the last numbered section of the paper.

Experimental Study on Flexural Behaviour of Reinforced Foamed Concrete Square Hollow Beam

R Othman, K Muthusamy, P J Ramadhansyah, D Youventharan, M A Sulaiman, A A G Nadiatul Adilah and B W Chong

Faculty of Civil Engineering Technology, Universiti Malaysia Pahang, 26300 Gambang, Kuantan, Pahang, Malaysia

Abstract. Nowadays construction industry continuously looking for new, better and efficient construction material and method due to environmental impact. This research reported the effect of reinforced foamed concrete square hollow beam in order to reduce the effect of the application of concrete materials. Foamed concrete with density of 1600kg/m^3 is now widely used as a structural component like slab and wall due to it lighter than the normal concrete. Meantime, it helps in saving cost and reduce the use of formwork. The hollow section also can reduce the dead loads which contribute to seismic effect in high rise buildings. In addition, the Processed Spent Bleaching Earth (PSBE) is an industry waste which possess a good pozzolanic reaction is used as partial cement replacement in order to make it more environmentally friendly. The experimental program consists of casting and testing of foamed concrete beams of size 150 x 200 x 1500 mm with and without square hollow section. To study the flexural behaviour, all beams are tested after 28 days of curing subjected to four-points bending test in order to analyse the effect of the reinforcement of hollow section on the strength, deflection and mode of failure of foamed concrete beam. The performance of reinforced foamed concrete hollow square beam (RHB) under flexural shows better when compared with conventional solid beam (CSB) and hollow square beam without reinforcement (HB). The RHB exhibited greater strength with lesser deflection and cracking.

1. Introduction

In a double story house construction, there is lots of formwork being used to support the structural element on the upper floors where formwork constituent 60% of the total cost [1] of the construction. It will be the largest waste of the construction where the formwork like wooden formwork can only be used for one to two times. Even nowadays, there is other type of formwork being used in construction, but mostly construction still prefer the use of wooden as a formwork in construction. Generally, formwork is used to support the wet concrete to dry out and form to the shape. The heavier the concrete, more material will be used in formwork to support the wet concrete as a mould. Thus, a lightweight structural element can be a good way to solve the problem by having a hollow section in the middle and using a light weight foamed concrete. A hollow foamed concrete is a duality way to reduce the weight of a concrete elements. As a result, lightweight concrete needs less formwork to support the concrete. However, unreinforced hollow foamed concrete will largely reduce the strength of the concrete. Thus, reinforcement is introduced to boost the strength of hollow foamed concrete to the original strength or even stronger. Foamed concrete is a lightweight concrete used construction industry for light loaded structure like slab and wall. Foamed concrete is type of concrete with large amount of foam or air void inside the concrete in order to reduce the volume of concrete and make the foamed concrete light. Foamed concrete is constituent from cement, foaming agent, sand and water.

There is no aggregate used in foamed concrete to enhance its lightweight properties. Furthermore, foaming agent is the key to produce foam in the concrete and make concrete light. Only small amount of the foaming agent needed to mix with concrete to produce foam. The density of foamed concrete is around 400 kg/m³ to 1600 kg/m³ whereas foamed concrete with density of 400 kg/m³ is used for non-loading structure and foamed concrete with density of 1600 kg/m³ [2] can be used for light loading structure. In this study, since the design is based on the beam, thus, density of foamed concrete is chosen for 1600 kg/m³. Besides, concrete is indispensable in construction especially for the high-rise building. However, concrete need cement as an ingredient to mix up with other materials like aggregate and sand. Yet, the production of cement contributes lots of Carbon Dioxide (CO_2) [3] which will cause the greenhouse effect and global warming[4]. To make the concrete more environmentally friendly, the use of cement can be partial replace by waste materials. In this study, the waste material will be used is Processed Spent Bleaching Earth (PSBE). It is a refinery waste from palm oil industry. PSBE is chosen as a replacement of cement due to its pozzolanic properties.[5] Besides, it will be a good way to solve the landfill disposal problems cause by PSBE. By partially replacing cement with PSBE will surely make the concrete doubly friendlier to the environmental. In short, this study is conducted to determine the effect of reinforcement to the hollow section on the strength of foamed concrete beam with the flexural strength, deflection and the mode of failure.

2 Method of Research

2.1 Materials

Among the materials used to prepare specimens in this research are cement, water, silica sand, foaming agent and PSBE. Ordinary Portland cement (OPC) from a single source was used throughout the experimental work. Tap water was used for mixing and curing purposes. Foaming agent was manufactured by LCM Technology Sdn. Bhd. Kuantan, Silica sand was supplied by Johor Silica Industries Technology Sdn. Bhd and PSBE was provided by Eco Innovation Sdn. Bhd.

2.2 Sample Preparations and Testing

Table 1 shows the summary details of specimens. In this research, three types of beam specimens were prepared with three number for each specimen. Specimen 1, also known conventional solid beam (CSB) as control beam with no hollow section. The reinforcement for CSB was 2T12 to both top and bottom of the beam; and shear reinforcement R6 with range of 250 mm. Specimen 2 is the hollow beam (HB) with hollow section (50 mm x 50 mm) to the centre of the beam in the view of x-axis. Reinforcement of HB was used similar to CSB which is 2T12 to both tension and compression reinforcement and R6-250mm for shear reinforcement. Specimen 3 also known as reinforced hollow beam (RHB) with additional reinforcement at the hollow section (50 mm x 50 mm). The reinforcement for RHB was 2T12 to both tension and compression reinforcement and R6-250mm for shear reinforcement also. In this research, the reinforcement 1T12 was added to the top and bottom of the hollow section. All the beams were casted with 30% of PSBE as partially cement replacement. Flexural strength of the beam was calculated using the formula for four-point bending test which involved the maxing loading of the beam. Equation (1) shows below was the formula used to calculate the flexural strength.

$$\sigma = \frac{FL}{bd^2} \quad \text{Eq. 1}$$

The beams were air cured for 28 days and painted to ease the observation of cracking pattern. All the specimens were tested using 4 points bending test to collect the data of maximum loading, deflection and cracking pattern. The load was applied constantly until the beams were failed to resist loading. Three LVDT were used to record the deflection of the beams. two of the LVDT were placed under the loading point while one LVDT was placed to the centre of the beams. The supports were placed to 150 mm from both end of the beam. Meanwhile, the loading points were set to 200 mm from the centre of the beam to the both sides. The data was record and analysed to obtain the result of testing. The testing

IOP Conf. Series: Materials Science and Engineering **712** (2020) 012046 doi:10.1088/1757-899X/712/1/012046

was according to the ASTM D6272 where the span of loading was one third of the support span. Figure 1 below shows the testing set up of 4 points bending test. Figure 2 shows the setup of LVDT under the beam. Figure 3 shows the specimen setup of 4 points bending test.

Table 1: Summary detail of specimens

Specimen	Hollow Section	Shear Link	Details
1 Conventional Solid Beam (CSB)	-	R6-250mm	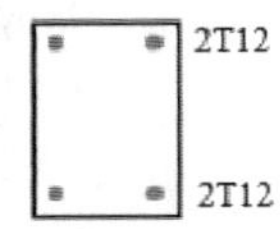
2 Hollow Beam (HB)	50 x 50	R6-250mm	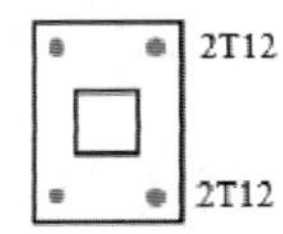
3 Reinforced Hollow Beam (RHB)	50 x 50	R6-250mm	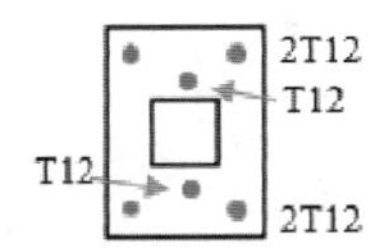

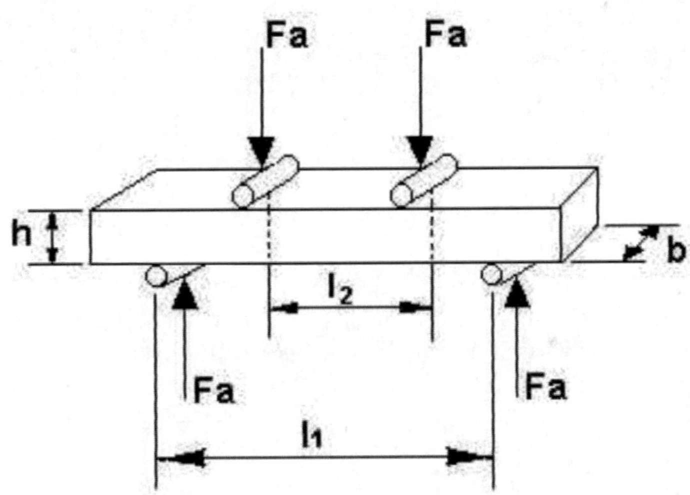

Figure 1: Four points bending test

Figure 2: Setup of LVDT

Figure 3: Specimen Setup of four points bending test

3 Result and Discussion

3.1 Flexural Strength

Flexural strength was determined from the maximum loading obtained from the testing and the parameter of the beam. Table 2 shows the summary of maximum loading obtained from testing and the flexural strength calculated based on equation 1. Specimen 1 (CSB) can resist highest loading of 31kN and continuously by specimen 3 (RHB) and specimen 2 (HB) with 20kN and 18kN respectively. While, the values of flexural of each specimen of CSB, HB and RHB were 6.20 N/mm², 3.70N/mm² and 4.00 N/mm² respectively. Hence, the flexural strength presents the strength capacity [7] of the beam.

Table 2: Summary of maximum loading and flexural strength of specimens.

Specimen	Max Loading (kN)	Flexural Strength (N/mm²)
1 (CSB)	31.00	6.20
2 (HB)	18.00	3.70
3 (RHB)	20.00	4.00

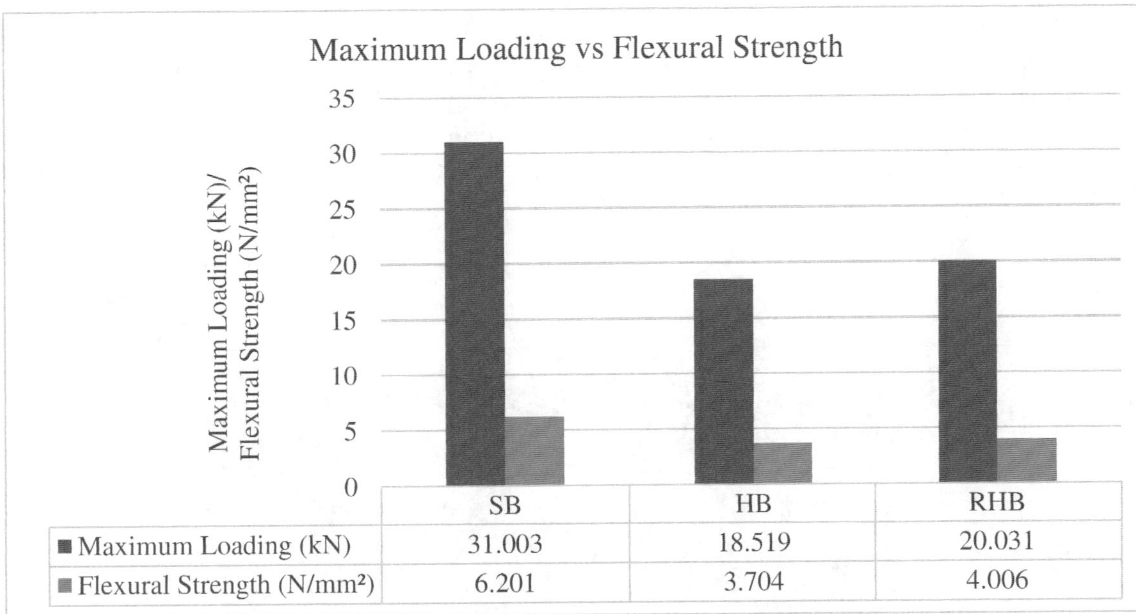

Figure 4: Maximum loading and Flexural strength

Figure 4 shows the comparison between the maximum loading and flexural strength for each specimen. Based on the analysis, the maximum loading HB and RHB specimen were lower 40% and 35% respectively compared to CSB. It shows that the hollow section was affecting the strength of the beam. Even though there was reinforcement added to both top and bottom of the hollow section, but solid beam still resists more loading compared to the beam with hollow section. However, RHB specimen with reinforcement around the hollow section having 7.5% higher flexural strength than the HB specimen. Thus. the reinforcement increased the strength of the beam. Nevertheless, the flexural strength of hollow beam was less compared with solid beam [8].The solid beam can resist more loading compared to the hollow beam.

3.2 Deflection profile

In this study, deflection profile was analysed through linear variable displacement transducer (LVDT) while beam undergoes deformation. Table 3 shows the deflection data of the specimens. From the analysis, the CSB specimen was the highest maximum deflection compared to HB and RHB specimen with 8.60 mm, 5.10mm, and 4.0 mm respectively. By comparison of the specimens, the deflection of beam decreased with beam with hollow section. It should be noted that the beam with hollow section exhibits in lesser deflection compared to solid beam[9,10]. Through comparison of HB and RHB, deflection decreased with beam with higher reinforcement.

Table 3: Summary of deflection data.

Specimen	**Deflection (mm)**
1 (SB)	8.60
2 (HB)	5.10
3 (RHB)	4.0

3.3 Crack pattern

Figure 5 shows the crack pattern in the all the tested beams. The beams were crack while it reaches the limit to resist the loading. The crack pattern of the beam was observed after the loading test. The crack pattern for all the specimens were slanting crack as shown in Figure 5. The slanting crack was also known as a shear crack which the beams were failed in shear due to insufficient shear reinforcement designed to the beam.

Specimen	Cracking Pattern	Failure Mode
1 (SB)	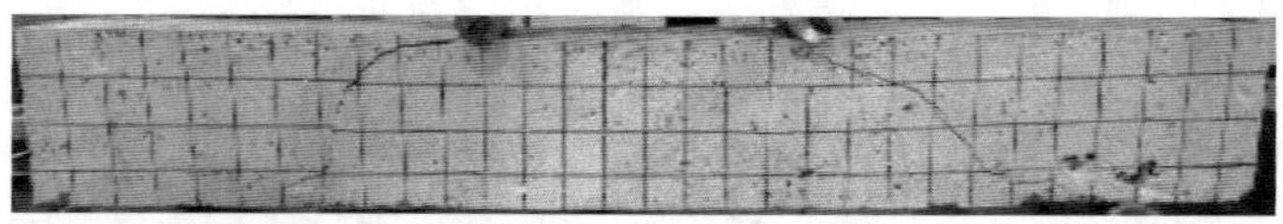	Shear crack
2 (HB)	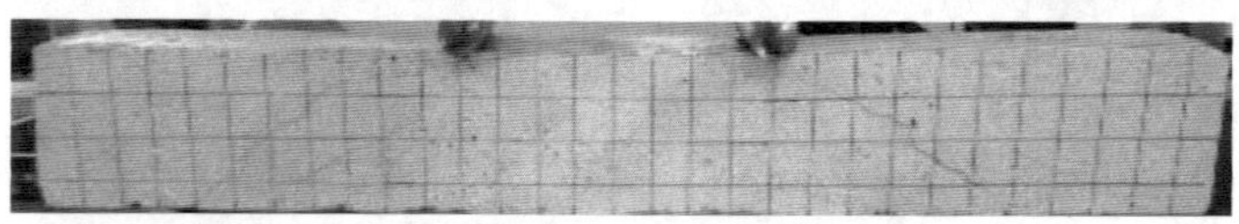	Shear crack
3 (RHB)	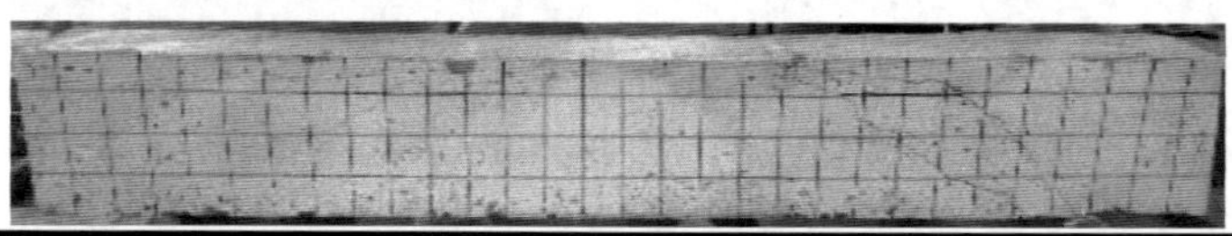	Shear crack

Figure 5: Crack pattern of beam

4 Conclusions

Based on the experimental results, the hollow section affects the behaviour of the beam even more reinforcement added to reinforce the hollow section. Conventional solid beam (CSB) exhibits the highest maximum flexural strength and deflection as compared to hollow beam HB and RHB. However, the reinforced hollow beam (RHB) gain 5% strength compared to hollow beam (HB) due to slightly effect of the reinforcement around the hollow section. In addition, the RHB exhibits lower deflection compared to CSB and HB. While, all the beams experienced shear cracking as slanting crack due to insufficient design shear reinforcement. Thus, the RHB can be used as a light structure beam of a building especially for the roof beam which receive less loading from roof.

References

[1] Nemati KM 2007 Formwork for concrete University of Washington
[2] Harith IK 2018 Study on polyurethane foamed concrete for use in structural applications *Case studies in construction materials* **8** 79-86
[3] Andrew RM 2018 Global CO 2 emissions from cement production. *Earth System Science Data* **10(1)**:195.
[4] Ali N, Jaffar A, Anwer M, Khan S, Anjum M, Hussain A, Raja M, Ming X 2015 The greenhouse gas emissions produced by cement production and its impact on environment: A review of global cement Processing *International Journal of Research* **2(2)** 488-500
[5] Arpornpong N, Charoensaeng A, Khaodhiar S, Sabatini DA 2018 Formulation of microemulsion-based washing agent for oil recovery from spent bleaching earth-hydrophilic lipophilic deviation concept. Colloids and Surfaces A *Physicochemical and Engineering Aspects* **541** 87-96.

IOP Conf. Series: Materials Science and Engineering **712** (2020) 012046 doi:10.1088/1757-899X/712/1/012046

[6] Loh SK, James S, Ngatiman M, Cheong KY, Choo YM, Lim WS 2013 Enhancement of palm oil refinery waste–Spent bleaching earth (SBE) into bio organic fertilizer and their effects on crop biomass growth *Industrial crops and products* **49** 775-81.

[7] Nakamura H, Nanri T, Miura T, Roy S 2018 Experimental investigation of compressive strength and compressive fracture energy of longitudinally cracked concrete *Cement and Concrete Composites* **93** 1-8

[8] Alnuaimi AS, Al-Jabri KS, Hago A 2008 Comparison between solid and hollow reinforced concrete beams. *Materials and Structures* **41(2)** 269-86.

[9] Al-Azzawi AA, Mahdy AH, Farhan OS 2010Finite element analysis of deep beams on nonlinear elastic foundations *Journal of the Serbian Society for Computational Mechanics* **4(2)** 13-42.

[10] Abidin AF 2014 The Behaviours of Lighweight Foamed Concrete Beam with Different Size of Square Hollow Section (Doctoral dissertation, UMP).

Acknowledgement

The authors would like to acknowledge the help and co-operation received from the lecturers and technical staff of Universiti Malaysia Pahang in conducting the experimental work. The financial support received from Universiti Malaysia Pahang through grant RDU150383 is highly appreciated.

IOP Conf. Series: Materials Science and Engineering **712** (2020) 012047 doi:10.1088/1757-899X/712/1/012047

Effect of temperatures and loading rates on direct shear strength of asphaltic concrete using layer-parallel direct shear test

C K Lung[1], M R Mohd Hasan[1], M O Hamzah[1], A Sani[1], S Poovaneshvaran[1] and P J Ramadhansyah[2]

[1]School of Civil Engineering, Engineering Campus, Universiti Sains Malaysia, 14300 Nibong Tebal, Penang, Malaysia
[2]Faculty of Civil Engineering Technology, Universiti Malaysia Pahang, 23600 Gambang, Pahang, Malaysia

Abstract. This paper evaluates the influence of temperatures and loading rates on the shear strength of dense asphalt (AC14) prepared based on the Malaysian Public Works Department specification. In this study, different specimen conditioning temperatures, 10°C, 20°C, 30°C, and loading rates of 20.88mm/min, 40.88mm/min, 50.88mm/min, and 60.8mm/min were used to assess the performance of specimens by using Layer-Parallel Direct Shear (LPDS) test. Based on the results, both parameters have significant effects on the shear strength of the asphalt specimens. The results indicate that the loading rate is directly proportional to the shear strength of asphalt specimens. It is because the stiffness of the asphalt mix increases as the loading rate increases. Hence, it enhanced the shear strength of asphalt specimens. On the other hand, the temperature is inversely proportional to the shear strength of asphalt mixes. As the temperature increases, the stiffness of the asphalt mixture reduces due to the lower resilient modulus; therefore, it reduces the shear strength of the specimen.

1. Introduction

The durability and long-term satisfactory pavement performance are greatly affected by asphalt paving materials and the materials' inherent properties. The uppermost layer of a pavement structure is important to support traffic load or stress alleviation and protect the structure. It is conceived as layers of superior quality materials in the structure and constructed accordingly. In bituminous pavements, bituminous binders and aggregates are the key ingredients. Hence it is desiring that the materials are in good quality, making the selection of materials an important task, which is often given less attention due to certain limitations. Asphalt pavement cannot be constructed in a single lift if the thickness of the pavement is higher than 2.5 to 3 inches. It is due to the compaction difficulties offered by thicker lifts. Hence, asphalt pavements are built in layers instead, making the construction of interfaces between layers inevitable. Being a layered structure, the life span of asphalt pavement is dependent on its bond strength and stiffness between individual layers. Similarly, the bond quality and strength between the layers depend on the construction technique, the texture, and placement procedure [1]. Poor bonding between pavement layers can lead to various types of distresses. One of the most common distress is slippage failure. Slippage failure usually occurs at the intersections or sharp curves where heavy vehicles are often accelerating, decelerating, or turning. Vehicle load are applied in both horizontal and vertical direction, which creates dynamic normal and tangential stresses. The interface between the asphalt pavement layers is vital to the pavement's integrity when vehicle loads are being transferred to each underlying layer of asphalt. Slippage failure develops when pavement layers begin to slide on one another, whereas the top layer separated from the lower layer. It is caused by the lack of bonding and due to high horizontal forces that cause two layers to separate [2]. These problems does not only reduce the pavement serviceability, but also adversely affect the short-term and long-term performance of pavements. Therefore, high shear resistance and adequate interlayer bonding are vital in maintaining the pavement's structural and functional integrity.

In the past, many studies have been carried out to evaluate pavement interlayer bonding strength. Raab and Partl [3] evaluated the influence of tack coats on interlayer adhesion of gyratory specimens by using nearly 20 different types of tack coat. The tests were conducted by using the Layer-Parallel Direct Shear (LPDS) test device, and the behavior of specimens with and without tack coat on a rough and a smooth surface were compared. The influences of compaction, surface condition, moisture, heat, and water on interlayer shear behavior, were also evaluated. The study revealed that the application of tack coats did not necessarily result in better interlayer shear strength, and this supported the outcome of a study performed by Molenaar et al. [4] at Delft University of Technology (TUDelft) using a simple shear test. Raab and Partl [3] further mentioned that by using a certain type of tack coat, it was possible to improve the shear adhesion potential for top layer compaction of 204 gyrations up to 5 kN or 10%, while such improvement was not observed for specimens compacted at 50 gyrations. Apart from that, the test also showed that all specimens with smooth surfaces sustained higher shear forces than specimens with rough surfaces. Hu et al. [5] evaluated the effect of tack coat dosage and temperature on the interface shear properties bonded with emulsified asphalt binders. The tests were conducted using a self-designed device with the aid of the Universal Testing Machine (UTM). The results showed that the shear strength at the asphalt interface decreases gradually with an increase in temperature. Additionally, the results designated that the brittleness at low temperatures, the flowability at intermediate temperatures, and the viscosity at high temperatures of tack binder have the most substantial effects on the shear property of the asphalt pavement interface. Collop et al. [6] assessed the pavement bond conditions using the Leutner Shear test. The study investigated the bond between two interfaces in a typical flexible pavement structure. The result showed that there was an approximately linear relationship between interface shear strength and the corresponding shear displacement for a particular combination of materials, which was independent of the bonding condition. Furthermore, the highest average shear strengths were obtained when a normal amount of tack coat was applied at the interface. While, it occurred to have the lowest average shear strength when the interface is dirty and additional tack coats were added. Canestrari et al. [7] used Ancona Shear Testing Research and Analysis (ASTRA) equipment to evaluate the interlayer pavement shear resistance when subjected to various influence parameters such as tack coat type, temperature and applied normal load. The results revealed that the shear strength of asphalt interface increased with an increment of normal stress at a given temperature. Meanwhile, the strength of the sample also increased with a decrease in temperature. In addition, Yao et al. [8] used the direct shear method to investigate the shear characteristics of steel asphalt interface under the influence of temperature, normal stress level and tack coat. The findings showed that the shear strength and shear reaction modulus increase with decreasing temperature and increasing normal stress levels.

Based on research conducted by Hachiya and Sato [2], the result indicated that the loading rate was a significant influence factor on pavement interlayer bond strength. The results demonstrated that the loading rate of 100 mm/min yielded much higher bond strengths than 1 mm/min. Sholar et al. [9] have evaluated the shear strength of field cores at 25°C with 50.8 mm/min by using Simple Direct Shear test equipment. The findings indicated that the presence of moisture at the material interface had significantly reduced the bond strength of the specimens. Furthermore, the shear strength of specimens was found to be directly proportional to curing time. As for surface texture, it considerably affected the shear strength of the pavement interfaces. Additionally, the HMA coarse graded mixture gives a higher shear strength compared to fine graded mixes. While Mohammad et al. [10] mentioned that the shear strength of the pavement interface yielded at 25°C was approximately five times the shear strength at 55°C. Chen and Huang [11], in their study, declared that the usage of a tack coat was more effective for dense-graded mixtures to increase peak shear strength compared to gap-graded and open-graded mixtures. The increase of mean texture depth has resulted in a higher peak shear stress and greater interlayer residual stress. However, thicker asphalt film thickness had caused a reduction in bonding properties. The shear strength reduces when temperature increases due to the reduction of viscosity of asphalt cement that causes greater deformation and lowered the interlayer resistance at elevated temperatures [12, 13]. Raab et al. [14] have postulated that shear strength reduced due to deterioration of pavement as a results of traffic volume exceeds the allowable design limit over a span period of time. The proper bonding within the pavement is not always achieved, and several pavement failures were linked to poor bond conditions. Due to a lack of pavement bonding and shear strength, pavement structural integrity had compromised and hastens the pavement failure. Hence, it results in frequent maintenance and high rehabilitation costs. Based on the previous studies, this study was initiated to investigate the effects of temperatures and loading rate on the direct shear strength of dense asphalt mixture prepared according to Malaysian Public Works Department (PWD) specifications. In this study, different temperatures and loading rates were adopted to assess the performance of AC14 specimens using the LPDS test. The ANOVA and Duncan's Post Hoc test analyses were adopted to analyze the data.

2. Materials and Methods

2.1 Materials

The aggregates used in this study were crushed granite aggregate supplied by Kuad Quarry Sdn. Bhd. The aggregates were sieved into the selected size range according to PWD specifications on designated sieve sizes, as shown in Table 1. The virgin asphalt binder used was a conventional binder 80/100 penetration grade with 5% optimum binder content (OBC), and Ordinary Portland Cement (OPC) was used as a filler material in this study. The target air void of asphalt mixes is 4 ± 1 %.

Table 1. Gradation limit for asphaltic concrete AC14.

Sieve Size (mm)	Percentage Passing (%)
20	100
14	90-100
10	76-86
5	50-62
3.35	40-54
1.18	18-34
0.425	12-24
0.150	6-14
0.075	4-8

2.2 Experimental works

During the specimen's preparation, the aggregate and asphalt binder were mixed and compacted at 160°C and 150°C, respectively. Before the sample compaction, the loose asphalt mixture was conditioned at the compaction temperature of 150°C for 2 hours to provide a realistic simulation of short-term aging and to allow binder absorption which occurs during construction. The mixes were subjected to compaction in a 100 mm diameter mold using a Servopac gyratory compactor. The specimens were then allowed to cool overnight after compaction had ceased. Before performing the test, the dimensions of the specimen were measured to the nearest mm. Based on the volumetric properties evaluation, the theoretical maximum specific gravity and the bulk specific gravity of the asphalt mixes obtained are 2.421 and 2.323, respectively. Based on those values, the calculated average air void is equal to 4.16 %, which has fulfilled the required air voids limit specified by

Malaysian PWD (4 ± 1%). Then, the specimens were conditioned at the designated temperature; 10°C, 20°C, and 30°C for a minimum of five hours. After setting up the apparatus, the specimen was placed on the U-bearing, and the interface are aligned between the lower and upper shear rings. The sample support are tightened, and the jig are placed into the loading frame. The upper shear ring was adjusted until it nearly touched the specimen. The loading rate was adjusted to the required rates, which were 20.8mm/min, 40.8mm/min, 50.8mm/min and 60.8mm/min. The shear strength of the mix are recorded once it reached its maximum shear strength.

2.3 Testing device

The LPDS test device (Figure 1) is a Swiss Federal Laboratories for Materials Science and Technology (EMPA) modified version of similar equipment developed in Germany by Leutner [15]. The LPDS is incorporated into the Swiss Standard SN 671961 (Schweizer Norm SN 671961, 2000) since the year 2000. This test device fits into an ordinary servo-hydraulic Marshall testing machine and allows the testing of cores with a diameter of about 150 mm [3]. One part of the core is laid on a circular u-bearing and held with a well-defined pressure by a semi-circular pneumatic clamp. The other part, the core head, remain unsuspended. The shear load was induced to the core head by a semi-circular sheer yoke with a specified deformation rate, thus producing fracture within the pre-defined shear plane. The specimens were conditioned in a climate chamber for a minimum of five hours before testing, and the tests was performed at the room temperatures. Figure 2 shows the Servopac gyratory compactor and the LPDS equipment that was used in this study.

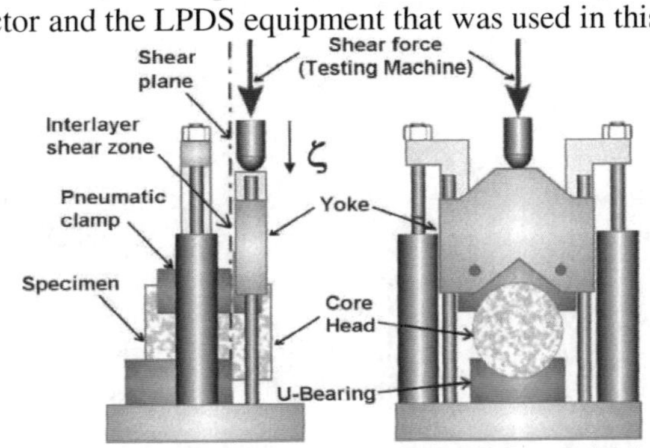

Figure 1. Schematic diagram of the LPDS test device [3]

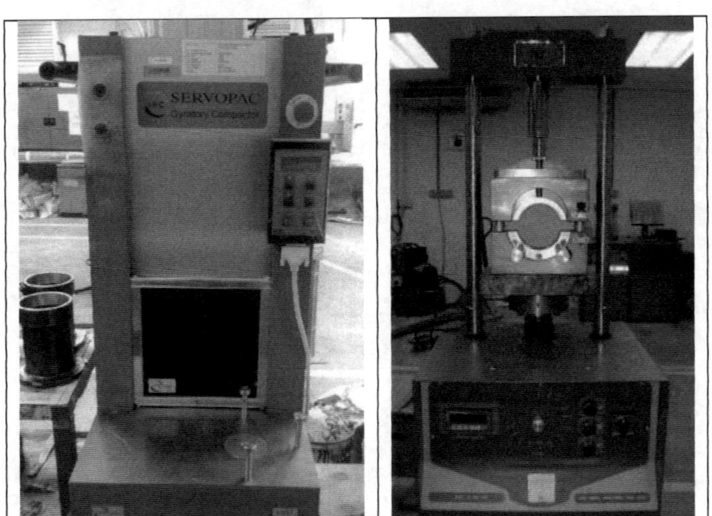

(a) (b)

Figure 2. (a) Servopac gyratory compactor, and (b) LPDS equipment used in this study

2.4 Statistical analysis

One-way analysis of variance (ANOVA) is used to compare the means of two or more samples by using the F distribution. The ANOVA tests the null hypothesis that samples in two or more groups are drawn from the same population at a 95 % confidence level ($\alpha = 0.05$). The parameters used in the ANOVA analysis are presented in Table 2.

Table 2. Parameters in one-way ANOVA analysis.

Variable	Parameters				Hypothesis
	μ_1	μ_2	μ_3	μ_4	
Temperature (°C)	10	20	30	-	$Ho{:}\mu_1=\mu_2=\mu_3$ $Ho{:}\mu_1\neq\mu_2\neq\mu_3$
Loading rate (mm/min)	20.8	40.8	50.8	60.8	$Ho{:}\mu_1=\mu_2=\mu_3=\mu_4$ $Ho{:}\mu_1\neq\mu_2\neq\mu_3\neq\mu_4$

Duncan's Post Hoc test was carried out to provide specific information on which parameters it significantly different from each other. The mean shear strengths which do not show a significant difference from each other will be grouped together. The two-way ANOVA is also used for further statistical analysis. This analysis measures the effect of two factors simultaneously. From this analysis, it does not only assess the effect of temperature and loading rate on the shear strength separately but also evaluates the interaction between these two parameters.

3. Result and discussion

3.1 Effects of temperature on shear strength of asphalt mixes

The interface shear strength test of asphalt mixture was carried out at a designated temperature of 10°C, 20°C, and 30°C. The resistance of the asphalt mixture towards shear is among the factors that are affected by changes in temperature. Figure 3 shows the effects of temperature on the shear strength of dense asphalt specimens. The figure indicates that temperature has a significant effect on the shear strength of the asphalt mix, where the specimen shear strength is inversely proportional to the temperature. It is due to the rheological behavior of the bitumen. The bitumen becomes less stiff as the temperature rises, and this phenomenon causes the binder to lose its ability to hold the aggregates together at high temperatures and hence results in lower resilient modulus. As the temperature increases, the stiffness of the asphalt mix reduces due to the lower resilient modulus; therefore, it reduces the shear strength of the mixes. There is only a small fluctuation of shear strength at the temperature of 20°C and 30°C. Rapid changes of shear strength was recorded at the testing temperature of 10°C. Variation of shear strength at different temperatures validates that the bond strength of asphalt mixture is a temperature-dependent component.

The One-way ANOVA result is as shown in Table 3. The analysis shows that the temperature has a significant effect on the shear strength of asphalt mixtures, which is indicated by a p-value less than 0.05. Duncan's Post Hoc test results is presented in Table 4. Generally, there is no substantial difference in the results of the specimens tested at 20°C and 30°C, which is designated as Group 1 with the p-value equal to 0.194. Noticeable improvement of the asphalt mixture shear strength is recorded in the results when the specimen was tested at 10°C (Group 2) with the shear strength value of 4824 N.

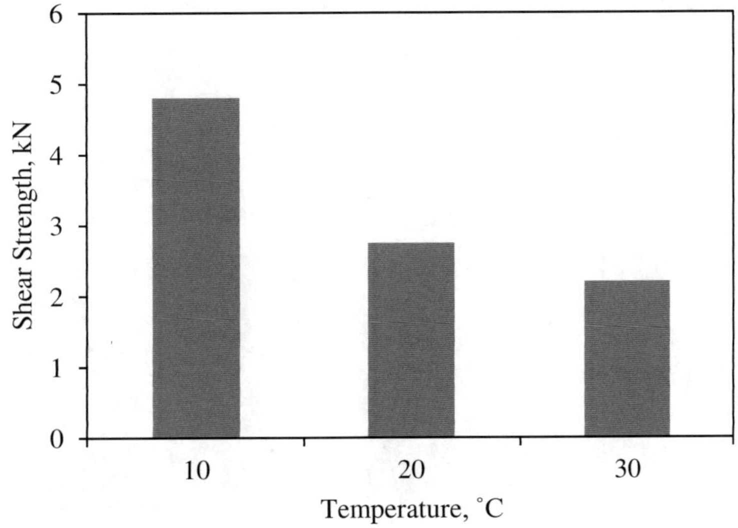

Figure 3. Relationship between shear strength and temperature.

Table 3. One-way ANOVA of temperature and shear strength.

Variable		Sum of Squares	Df	Mean Square	F	p-value
Temperature	Between Groups	4.614E7	2	2.307E7	21.324	<0.001
	Within Groups	3.571E7	33	1081990.455		
	Total	8.158E7	35			

Table 4. Duncan's Post Hoc test analysis for effects of temperature on shear strength.

Temperature	N	Subset for alpha = 0.05	
		1	2
30	12	2191.0000	
20	12	2753.5000	
10	12		4824.0000
p-value		0.194	1.000

3.2 Effects of loading rate on shear strength of asphalt mixes

The effects of strain rate on the shear strength of the asphalt mixture were also studied. The direct shear test by using LPDS seems to be the most commonly used laboratory test method to investigate the interface properties. This test incorporates the application of a normal force in addition to a constant rate of shear loading until the interface failure occurs. Four different strain rates were adapted in this study, namely 20.80 mm/min, 40.80 mm/min, 50.80 mm/min and 60.80 mm/min. Figure 4 shows the effect of strain rates on the shear strength of asphalt mixtures. The result shows that the different strain rate has a significant effect on the shear strength of the asphalt specimens, which is presented by the p-value in Table 5. Figure 4 shows that shear strength is directly proportional to the strain rate. It is due to the fact that the stiffness of asphalt mixes increases as the strain rate increases, hence enhancing the shear strength of asphalt mixes, as presented by Duncan's Post Hoc test result in Table 6. At the strain rate of 20.80 mm/min, the lowest mean bond strength associated to shear strength was recorded which is 1813.33 N. While the highest mean shear strength was recorded at the strain rate of 60.80 mm/min which is 4393.33 N. The difference in shear strength under different strain rate is closely related to the visco-elastic response of asphalt mixture. However, there is no significant difference between groups based on the p-value higher than 0.05.

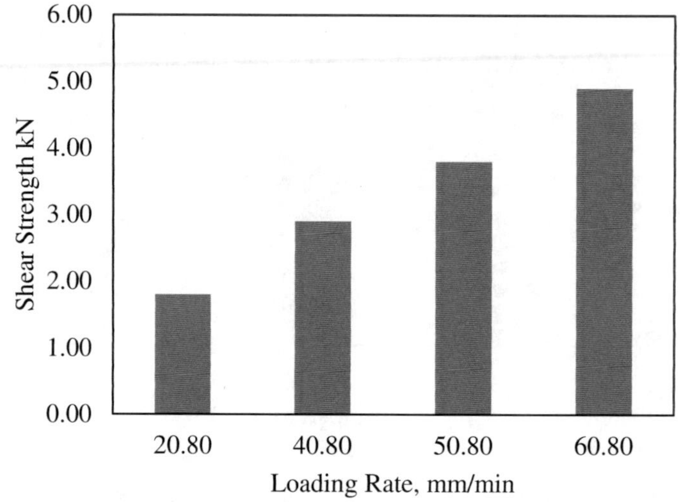

Figure 4. Relationship between shear strength and loading rate.

Table 5. One-way ANOVA of loading rate and shear strength.

Variable		Sum of Squares	Df	Mean Square	F	p-value
Loading Rate	Between Groups	3.430E7	3	1.143E7	7.693	<0.001
	Within Groups	4.755E7	32	1486081.250		
	Total	8.185E7	35			

Table 6. Duncan's Post Hoc test analysis for effects of loading rate on shear strength.

Loading Rate (mm/min)	N	Subset for alpha = 0.05		
		1	2	3
20.80	9	1813.3333		
40.80	9	2968.0000	2968.0000	
50.80	9		3850.0000	3850.0000
60.80	9			4393.3333
p-value		0.053	0.135	0.351

3.3 Interaction factor and correlation between factors

The two-way ANOVA result is as presented in Table 7. It shows that the significance level of temperature and loading rate effects upon the shear strength of the asphalt specimens. The results indicate that the temperature and loading rate have a p-value of less than 0.05. Thus, both parameters should be taken into account as factors that will affect the shear strength of the asphalt specimen. Apart from that, there was also significant effect on the interaction between the temperature and loading rate as the p-value is less than 0.05 (confidence level 95%). Figure 5 presents the relationship between temperature and loading rate to the shear strength of asphalt mixtures. It shows that shear strength is increasing due to the increment of loading rate and decrease of temperature, which complies with Song et al. [16] study.

IOP Conf. Series: Materials Science and Engineering **712** (2020) 012047 doi:10.1088/1757-899X/712/1/012047

Table 7. Two-way ANOVA of temperature and loading rate on shear strength.

Source	SS	Df	MS	F	p-value
Corrected Model	8.185E7	11	7440601.000	53529.504	<0.001
Intercept	3.817E7	1	3.817E8	2746002.655	<0.001
Temperature	4.614E7	2	2.307E7	165986.554	<0.001
Loading Rate	3.430E7	3	1.143E7	82243.038	<0.001
Temperature * Loading Rate	1407002.00	6	234500.333	1687.053	<0.001
Error	3336.000	24	139.000		
Total	4.635E8	36			
Corrected Total	8.185E7	35			

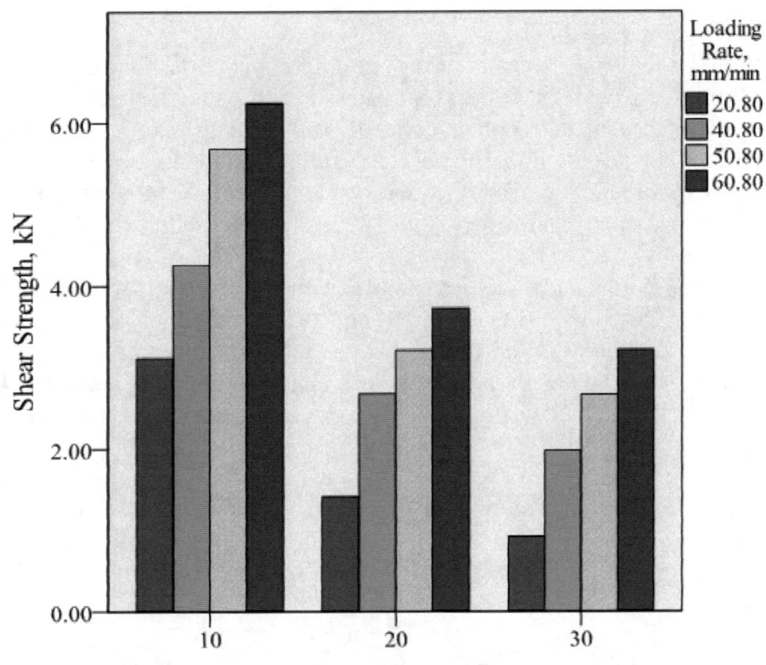

Figure 5. Relationship between shear strength with temperature and loading rate

The relationships between the factors and the pavement shear strength have been analyzed using the Pearson correlation analysis. The Pearson correlation analysis result is typically presented by the correlation coefficient and p-values as shown in Table 8. The correlation coefficient (r) is a numerical summary of a bivariate relationship that ranges from -1.00 to 1.00. A positive value indicates a positive relationship between two variables while a negative value indicates an inverse relationship between variables. The r-value typically divides into various scales depending on the level of correlation ranging from very weak to very strong.

Table 8 presents the Pearson correlation between temperature and loading rate to the shear strength of asphalt specimens. The result shows a strong inverse correlation between temperature and shear strength based on the r-value equal to -0.713. Meanwhile, the shear strength increases as the loading rate increases, as shown by the strong and positive correlation coefficient ($r = 0.645$). Theoretically, the Pearson correlation values obtained from this study are valid to use based on two-tailed p-values less than 0.05 (95% confidence interval).

Table 8. Correlation between shear strength with temperature and loading rate.

Variable		Temperature	Loading Rate	Shear Strength
Temperature	Pearson Correlation	1	0.000	-0.713
	p-value (2-tailed)		1.000	<0.001
	N	36	36	36
Loading Rate	Pearson Correlation	0.000	1	0.645
	p-value (2-tailed)	1.000		<0.001
	N	36	36	36
Shear Strength	Pearson Correlation	-0.713	0.645	
	p-value (2-tailed)	<0.001	<0.001	
	N	36	36	

4. Conclusion

High shear strength and strong interlayer bonding are important in designing asphalt pavement. It is also vital in maintaining the pavement's structural and functional integrity. Thus, this study is carried out to assess the performance of the asphalt concrete specimen by using the Layer-Parallel Direct Shear test. The study investigates the influence of temperature and loading rate on the shear strength of the specimen. In the experimental works, the dense mixes type AC14 was prepared according to Malaysian Public Works Department specifications and tested under different temperatures and loading rates.

Based on the experimental works and statistical analyses, the results indicate that the temperature and loading rate significantly influences the shear strength of conventional dense asphalt due to the rheological behavior of the binder. Where the shear strength is inversely proportional to the test temperature, in contrast, shear strength positively correlated to the loading rate. The increases in loading rate enhance the stiffness of asphalt specimens, hence increasing the shear strength of the asphalt specimens. The finding from this study can be related to the slippage failure distress that commonly occurs at the intersections or sharp curves where heavy vehicles are often accelerating, decelerating, or turning. Plastic deformation of asphalt pavement typically occurs under loading at high ambient temperatures and sudden action of heavy vehicles. However, a test method that incorporates cyclic loading in both the vertical and shear directions should be considered in order to have a better result (e.g., Superpave Shear Tester).

References

[1] Kim Y K and Lee S W 2013 Performance evaluation of bonded concrete overlay, Construction and Building Materials, **49** pp. 464-470.
[2] Hachiya Y and Sato K 1997 Effect of tack coat on binding characteristics at interface between asphalt concrete layers, in, University of Washington, Seattle, pp. 349-362.
[3] Raab C and Partl M N 2004 Effect of tack coats on interlayer shear bond of pavements, in: Proceedings of the 8th conference on Asphalt Pavements for Southern Africa (CAPSA'04), 2004, pp. 16.
[4] Molenaar A A A, Heerkens J C P and Verhoeven J H M 1986 Effects of stress absorbing membrane interlayers, in: Proceedings of the Association of Asphalt Paving Technologies, pp. 453-481.
[5] Hu X, Lei Y, Wang H, Jiang P, Yang X and You Z 2017 Effect of tack coat dosage and temperature on the interface shear properties of asphalt layers bonded with emulsified asphalt binders, Construction and Building Materials, **141** pp. 86-93.
[6] Collop A C, Thom N H and Sangiorgi C 2003 Assessment of bond condition using the Leutner shear test, Proceedings of the Institution of Civil Engineers - Transport **156** pp. 211-217.

[7] Canestrari F, Ferrotti G, Partl M N and Santagata E 2005 Advanced testing and characterization of interlayer shear resistance, Transportation Research Record, **1929** pp. 69-78.
[8] Yao B, Li F, Wang X and Cheng G 2016 Evaluation of the shear characteristics of steel–asphalt interface by a direct shear test method, International Journal of Adhesion and Adhesives, **68** pp. 70-79.
[9] Sholar G A, Page G C, Musselman J A, Upshaw P B and Moseley H L 2004 Preliminary investigation of a test method to evaluate bond strength of bituminious tack coats, in: Journal of the Association of Asphalt Paving, Technologists. Association of Asphalt Paving Technologists (AAPT), pp. 771-806.
[10] Mohammad L N, Raqib M and Huang B 2002 Influence of asphalt tack coat materials on interface shear strength, Transportation Research Record, **1789** pp. 56-65.
[11] Chen J-S and Huang C-C 2010 Effect of surface characteristics on bonding properties of bituminous tack coat, Transportation Research Record, **2180** pp. 142-149.
[12] Hou D, Han M, Muhammad Y, Liu Y, Zhang F, Yin Y, Duan S and Li J 2018 Performance evaluation of modified asphalt based trackless tack coat materials, Construction and Building Materials, **165** pp. 385-394.
[13] Park P, El-Tawil S and Naaman A E 2017 Pull-out behavior of straight steel fibers from asphalt binder, Construction and Building Materials, **144** pp. 125-137.
[14] Raab C, Grenfell J, Abd El Halim A O and Partl M N 2016 Comparison of Interlayer Bond Behavior Due to Ageing, in, Springer Netherlands, Dordrecht, pp. 323-334.
[15] Leutner R 1979 Untersuchung des schichtenverbundes beim bituminoesen oberbau, Bitumen, **41** pp. 84-91.
[16] Song W, Shu X, Huang B and Woods M 2015 Factors affecting shear strength between open-graded friction course and underlying layer, Construction and Building Materials, **101** pp. 527-535.

Acknowledgments
The authors would like to acknowledge Universiti Sains Malaysia (USM) that has funded this research through the Short-Term Research Grant Scheme (Grant No.304/PAWAM/60313048). The authors would also like to acknowledge the technicians of the Highway Engineering Laboratory at the School of Civil Engineering, USM, for their kind assistance. The authors would like to thank the material suppliers for providing the raw ingredients (aggregate and bitumen) and ordinary Portland cement that enables this research to be carried out.

IOP Conf. Series: Materials Science and Engineering **712** (2020) 012048 doi:10.1088/1757-899X/712/1/012048

Finite Element Analysis on the Effect of Hollow Section on the Strength of Foamed Concrete Beam

R Othman, K Muthusamy, P J Ramadhansyah, D Youventharan, M A Sulaiman, A A G Nadiatul Adilah and B W Chong

Faculty of Civil Engineering Technology, Universiti Malaysia Pahang, Lebuhraya Tun Razak, 26300 Gambang, Kuantan, Pahang, Malaysia

Abstract. The environmental impact of cement production has prompted innovative measures to reduce the usage of cement in accordance with the principle of sustainable construction. At the same time, the palm oil industry generates about 2 million tons of spent bleaching earth (SBE) are generated annually with most of it ends up in the landfill, causing environmental degradation. This research focuses on the effect of hollow section on the strength of foamed concrete beam by using processed spent bleaching earth (PSBE) as partial cement replacement. Hollow section of two shapes were implemented at three different locations along the beam specimen. The foamed concrete beams with 30% cement replaced with PSBE were subjected to tensile test at 28 days to observe the effect of hollow section on the maximum load and tensile strength. The deflection of the specimen was compared with theoretical value and the mode of failure was studied. Hollow section of circular and square shape was studied while the locations were on neutral axis, above neutral axis, and below neutral axis. Verification of experiment was conducted using Finite Element Analysis (FEA) software, ANSYS. Test result shows that beams with circular hollow section has higher strength than those with circular hollow sections. The best position of hollow was above the neutral axis as to not impact the weak tensile strength of foamed concrete. FEA analysis shows great accuracy in predicting the performance of the specimens.

1. Introduction

Concrete is the most-used material in the construction industry. Cement is the basic material for producing concrete. However, the production of cement involves greenhouse gas emission and is very dusty. Dust particle from about 1 to 100 μm in diameter is a hazard to human health and also causes smog [1]. In the modern day, the Global Goals, or Sustainable Development Goals (SDGs) had called for action to reduce the environment impact of urbanization. Foamed concrete has many advantages such as high strength-to-weight ratio, low dead weight, and economical [2]. Waste material can be used in foamed concrete mix to promote green construction. Researchers had attempted to recycle palm oil fuel ash, fly ash, glass powder, and more in concrete mix [3]. Spent Bleaching Earth (SBE) is a hazardous waste that causes environmental degradation. About 2,000,000 tons of SBE are generated annually [4]. As one of the biggest producers of oil, Malaysia generates huge quantity of the waste. However, most SBE in Malaysia ends up in the landfill. That is a huge waste because energy may be recovered from SBE as it contained about 20-40% residue oil [5]. The usage of cement can be reduced by using lightweight concrete as load-bearing members. With aggregate taking up about 65-80% of the weight of normal concrete, foam concrete which does not use aggregate is significantly lighter [6]. Hollow concrete beam optimizes the strength and weakness of concrete material. Hollow section can serve as a pathing for mechanical and electrical works [7]. Concrete is strong in compression but weak in tension. According to the loading condition of a concrete member, the concrete at the tension zone is often disregarded in design [8]. This has prompts various research reducing the volume of the member with hollow section at the area in which the strength loss is minimal [7-9]. Hollow lightweight concrete beam is applicable to the upper storey of buildings or the roof level, whereby the imposed load is the lowest. The lightweight nature of foamed concrete eases the usage of Industrialize Building System (IBS) in the project. The concept of IBS started in the 1960s, but Malaysia is now encouraging the usage of it due to various benefits. IBS saves time and labor cost while ensuring better quality and workmanship of the work [11]. IBS also reduces the usage of formwork, which may contribute up to one-third of the cost in a project [12].

2 Method of Research

2.1 Materials

The materials used to prepare concrete specimens in this research are cement, water, fine aggregate, foaming agent and PSBE. Ordinary Portland Cement (OPC) of grade 52.5N from YTL Cement was used. Silica sand was used as fine aggregate in this study. The size of silica sand was from 0.001mm to 0.6mm. Casting and curing of foamed concrete was done using Tap water from the laboratory. Protein forming agent was used. The forming agent was prepared with raw material in the presence of $Ca(OH)_2$ and a small portion of $NaHSO_3$. Foaming agent has no reaction with concrete, but it served as a layer to trap air for producing formed concrete. It also forms no fume and is non-toxic. The forming agent was diluted with a ratio of 1L forming agent with 25L water. PSBE was used as the pozzolanic material of foamed concrete. It is in fine powder form and mixes well with concrete to enhance secondary pozzolanic reaction which improves the performance of concrete. Figure 1 and 2 shows the forming agent and PSBE used in the experiment.

Figure 1: Forming agent before dilution

Figure 2: Processed spent bleaching earth

2.2 Sample Preparations and Testing

Seven concrete beams with different hollow shape and position were prepared in this experiment. The size of the specimen was 150mm × 200mm ×1500mm. The mix proportion of the specimens were the same, expect the control, which was shown in Table 1. The various condition of hollow was presented as in Figure 3. The size of square hollow section was 50mm × 50mm while the size of circular hollow section was 50mm diameter. The design density of formed concrete was 1600 kg/m^3 and the cement sand ration used was 1:1.5. Water cement ratio was set to be 0.50. The control sample was casted without PSBE replacement and hollow section, while 30% PSBE was used to replace cement in the six studied specimens. Casting was done with mixing drum and the specimen was demoulded after 24 hours. All specimen was immerged in water for 28 days. Flexural test was conducted in accordance to BSEN 12390-5 [16]. At the same time, the specimens were modelled using ANSYS to conduct FEA. The result of flexural strength and deflection was compared for accuracy.

Table 1: Mix proportion of specimens

Mixes	**Cement (kg/m^3)**	**Silica sand (kg/m^3)**	**Foam**	**PSBE**	**Water (kg/m^3)**
Control	288	432	12.96	-	144
Specimens	201.6	432	12.96	86.4	144

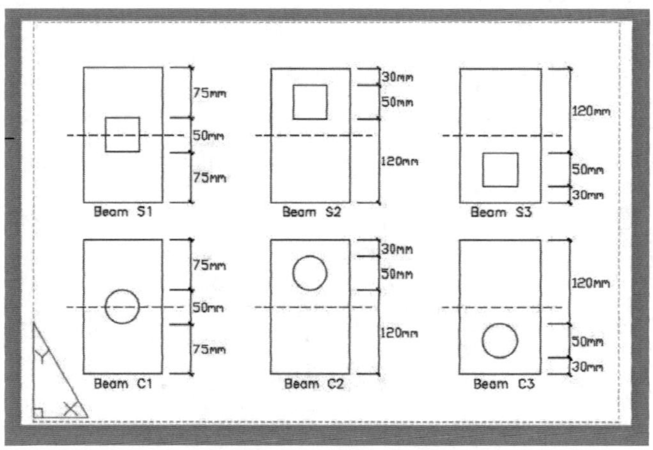

Figure 3: Hollow shape and position of specimens

3 Result and Discussion

3.1 Experimental Result

Figure 4 and 5 shows the result of flexural test. It was clear that the control without any hollow section sustained the highest amount of load before failure. For specimens were hollow section, two treads could be observed. First, specimens with circular hollow section performed slightly better than those with square hollow section. Second, specimen number 2 had the highest maximum load and flexural strength within the hollow shape group. The trend can be attributed to the properties of concrete. Concrete has strong compressive strength but weak tensile strength. The tensile strength of concrete material is only about 20% of its compressive strength. This applies even to form concrete. When an axial load was applied on the beam, the top half of the beam's cross section experienced compression while the bottom half of the beam experienced tension. In this study, the foam concrete beams were unreinforced, making them vulnerable at the tension zone, which is the section below the neutral axis [13]. The maximum load of the specimen decreased because the volume of concrete at the tension zone reduced. The effect followed the expectation of the study. When the hollow section was located above neutral axis however, the ultimate load became higher. This can be explained with the compressive strength of the specimen was adequate to resist the ultimate loading. All specimen failed due to tensile force instead of compression force. Hence, shifting the hollow section to above neutral axis had no effect of weakening the beam [9]. Instead, the foam concrete beam resisted maximum loading as there was no hollow section on the tensile zone. As a result, the ultimate loading increased slightly.

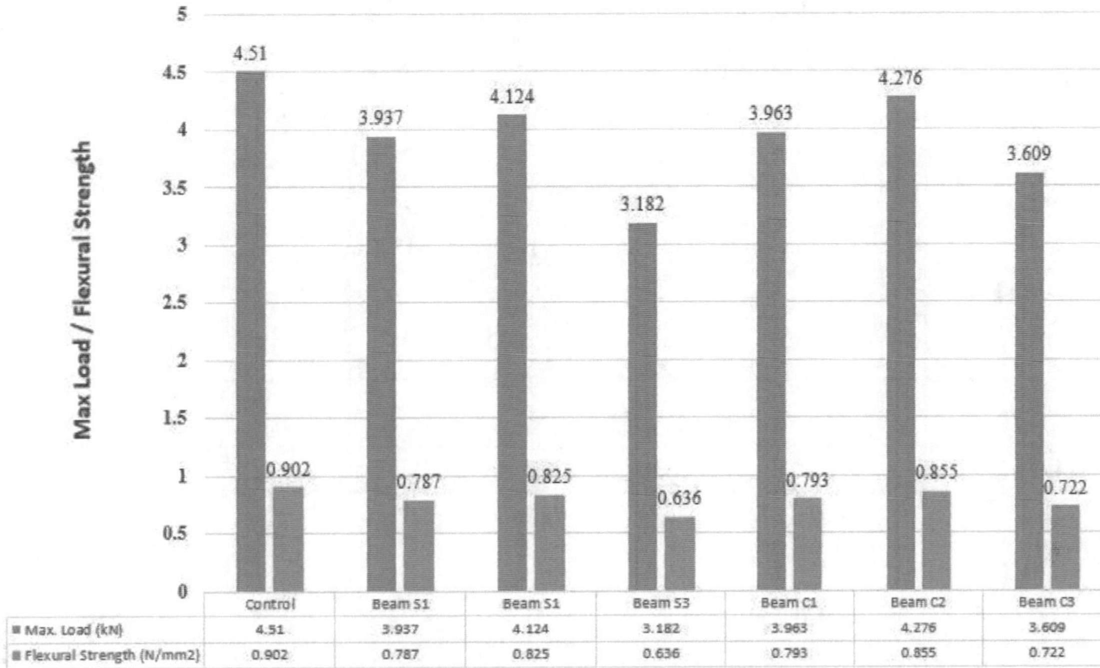

Figure 4: Maximum load and flexural strength

Figure 5: Flexural test of specimen

3.2 Finite Element Analysis

Table 2 and 3 shows the comparison between experimental, FEA, and theoretical data. For flexural strength, the greatest differences between laboratory experiment and ANSYS modelled value was only 6.31%. This indicates that FEA method is a very accurate approach to conduct structural analysis of foamed concrete with hollow sections. The result also agreed with similar study of finite element analysis on flexural behaviour on concrete beam [14]. For deflection, approximately 30% error was recorded when comparing the result between ANSYS software and theoretical value. The error was

IOP Conf. Series: Materials Science and Engineering **712** (2020) 012048 doi:10.1088/1757-899X/712/1/012048

however minor, and more importantly the result pattern agreed with theoretical value. The consistent deviation happened because hollow foam concrete beam did not fully obey the theory of an ideal concrete beam.

Table 2: Flexural strength differences

	Experimental		FEA		Differences
Sample	**Maximum Load (kN)**	**Flexural Strength (N/mm^2)**	**Maximum Load (kN)**	**Flexural Strength (N/mm^2)**	**(%)**
Control	4.510	0.902	4.571	0.914	1.35
Beam S1	3.937	0.787	3.940	0.788	0.08
Beam S2	4.124	0.825	3.980	0.796	3.49
Beam S3	3.182	0.636	3.338	0.668	4.90
Beam C1	3.963	0.793	3.989	0.798	0.66
Beam C2	4.276	0.855	4.015	0.803	6.10
Beam C3	3.609	0.722	3.837	0.767	6.31

Table 3: Deflection differences

Sample	Maximum Deflection (mm)		Differences
	Modelled Value	**Theoretical Value**	**(%)**
Control	0.0865	0.125	30.8
Beam S1	0.0882	0.126	30.0
Beam S2	0.0758	0.106	28.5
Beam S3	0.0858	0.126	31.9
Beam C1	0.0870	0.127	31.5
Beam C2	0.0849	0.122	30.4
Beam C3	0.0865	0.125	30.8

3.3 Mode of Failure

Finite Element Analysis did not show the failure of specimen directly, but it was possible to predict the development of crack through the deformation profile. Since this study simulated four-point bending test, the critical region where stress was the highest was the area between the two loading points. From Figure 6, it was shown that the 'red zone' of the deformation chart fell in between two points where load was being applied. The maximum deformation occurred at the mid-span of the beam, so the first crack of the specimen was located at the mid-spam and propagated through the beam in a line. This agreed with the laboratory study. All beam failed under the same condition of ultimate crack. Ultimate crack refers to a crack that occurs abruptly due to the lack of reinforcement on a concrete member. The result is similar with the similar study of crack propagation using finite element method [15].

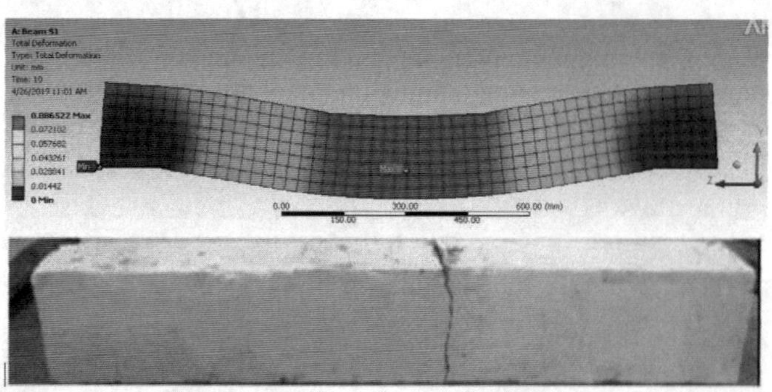

Figure 6: Deflection profile of modelled and actual specimen

4 Conclusions

Based on the results from both experimental and modelling, it is concluded that beam with circular hollow section has better performance compared to square hollow section. The best position for unreinforced foamed concrete beam is above the neutral axis. Verification with ANSYS shows great accurate for flexural strength and deflection of beam with minor deviation.

References

[1] Ahmad S and Ismail S 2014 Management of Air Pollution Control in *J. Middle East Appl. Sci. Technol.* **1** 30–6

[2] Amran Y H M, Farzadnia N and Ali A A A 2015 Properties and applications of foamed concrete; A review *Constr. Build. Mater.* **101** 990–1005

[3] Chandni T J and Anand K B 2018 Utilization of recycled waste as filler in foam concrete *J. Build. Eng.* **19** 154–60

[4] Beshara A and Cheeseman C R 2014 Reuse of spent bleaching earth by polymerisation of residual organics *Waste Manag.* **34** 1770–4

[5] Eliche-quesada D and Corpas-iglesias F A 2014 Utilisation of spent fi ltration earth or spent bleaching earth from the oil re fi nery industry in clay products *Ceram. Int.* **40** 16677–87

[6] Kamaruddin M A, Abdullah M M A, Zawawi M H and Zainol M R R A 2017 Potential use of plastic waste as construction materials: Recent progress and future prospect *IOP Conf. Ser. Mater. Sci. Eng.* **267**

[7] Jabbar S, Hejazi F and Mahmod H M 2016 Effect of an opening on reinforced concrete hollow beam web under torsional, flexural, and cyclic loadings *Lat. Am. J. Solids Struct.* **13** 1576–95

[8] Dhinesh N P and Satheesh V S 2017 Flexural Behaviour of Hollow Square Beam 1–7

[9] Bhattarai B P and Bhattarai N 2017 Experimental Study on Flexural Behavior of Reinforced Solid and Hollow Concrete Beams **3** 1–8

[10] Kumbhar U N and Jadhav H S 2018 Flexural Behaviour Of Reinforced Concrete Hollow Beam With Polypropylene Plastic Sheet Infill 1517–21

[11] Kamar K and Hamid Z 2011 Industrialized Building System (IBS): revisiting issues of definition and classification *Int. J. ...* **1** 120–32

[12] Ko C H, Wang W C and Kuo J D 2011 Improving Formwork Engineering Using the Toyota Way *J. Eng. Proj. Prod. Manag.* **1** 13–27

[13] N parthiban M N 2017 Flexural Behavior of Reinforced Concrete Beam With *Int. Res. J. Eng. Technol.* **4** 2263–74

[14] Wahyuni A S, Vimonsatit V and Nikraz H 2012 FEM modelling and analysis of reinforced concrete section with lightweight blocks infill **4** 375–9

[15] Min F, Yao Z and Jiang T 2014 Experimental and Numerical Study on Tensile Strength of Concrete under Different Strain Rates **2014**

[16] B.S. EN 12390-5:2019 Testing hardened concrete–Part 5: Flexural strength Test

Acknowledgement

The authors would like to acknowledge the help and co-operation received from the lecturers and technical staff of Universiti Malaysia Pahang in conducting the experimental work. The financial support received from Universiti Malaysia Pahang through grant RDU150383 highly appreciated.

AUTHOR INDEX